Panic Disorder

Antonio Egidio Nardi
Rafael Christophe R. Freire
Editors

Panic Disorder

Neurobiological and Treatment Aspects

 Springer

Editors
Antonio Egidio Nardi
Laboratory of Panic and Respiration,
 Institute of Psychiatry
Federal University of Rio de Janeiro
Rio de Janeiro, Brazil

Rafael Christophe R. Freire
Laboratory of Panic and Respiration,
 Institute of Psychiatry
Federal University of Rio de Janeiro
Rio de Janeiro, Brazil

ISBN 978-3-319-79180-7   ISBN 978-3-319-12538-1   (eBook)
DOI 10.1007/978-3-319-12538-1

Printed on acid-free paper

This Springer imprint is published by Springer Nature
The registered company is Springer International Publishing AG Switzerland

To Andrea
A.E.N.

To Ana Catarina
R.C.R.F.

Foreword

Needless to say, I am immensely gratified to be asked to write the foreword to this volume, which exemplifies the multiple, burgeoning approaches to the understanding of panic.

Your editors' historical approach refers to ancient descriptions of the panic attack and the various attempts to develop an understanding of anxiety related conditions. They state that, "In 1954, Mayer-Gross associated anxiety disorders to hereditary, organic and psychological factors, dividing them in simple anxious states and phobic anxious states. In 1959, Donald Klein observed that patients with these disorders responded favorably to imipramine, a tricyclic antidepressant."

Naturally, the protagonist of this event has a more detailed and complicated recollection. Our concern in the late 1950s was to understand the effects of this new agent, imipramine. European studies stated it to be an antidepressant, so our expectation was that it would be a super cocaine and blow the patients out of their rut. At Hillside Hospital, a psychoanalytic hospital, the premium treatment was psychoanalytic psychotherapy without medication. The average length of stay was 10 months. After the failure of therapy, the patient was voluntarily referred to the Department of Experimental Psychiatry where the senior scientist was Max Fink MD and I was a very junior associate. Max and I were confirmed anti-diagnosticians since all the excellent studies of the 1950s indicated gross diagnostic unreliability. Psychoses could be barely discriminated from neuroses and that was about it. Therefore, it seemed foolish to attempt to fit medications to diagnoses.

So our first effort was a pilot trial trying to treat the entire range of patients that had not responded to treatment with psychoanalytic psychotherapy, chlorpromazine or imipramine. Our results led to a tentative classification based upon response to medication [1, 2]. We noted that several patients, plagued by anxiety attacks, had not responded to chlorpromazine, considered the most potent anti-anxiety agent, but had responded during treatment with imipramine. We were not sure if these patients might have been depressed. We looked therefore for nondepressed patients with manifest anxiety to see if imipramine had specific anti-anxiety effects, independent of their effects on depression.

The first patient that met our criteria was incessantly complaining of his fears of being alone, traveling, and dying. He demanded that he be accompanied at all times. His hospitalization was initiated by his family. They were no longer able to put up with his demands for constant reassurance and having a companion at all times. Yet, he was not depressed. He was radically pessimistic about his fate but also had a lively interest in his circumstances, took pleasure in gossip, spoke well, ate well, slept well, and laughed at jokes.

He was placed on imipramine 75 mg daily to be weekly increased by 75 mg. He remained in psychotherapy. I examined him before the clinical open trial. The patient stated that by being placed on medication the doctors had given up on him. I also weekly interviewed the therapist, the supervisor, and the ward staff. The first 2 weeks were negative. The patient bitterly stated the medication had done him no good, and his therapist, supervisor, and ward staff agreed. His therapist indicated that he saw signs of loose associations and was sure the patient was actually psychotic. Perhaps pseudo-neurotic schizophrenia, a recently fashionable diagnosis applied to seemingly neurotic patients who did not respond to therapy. The supervisor felt that the therapist had not gone deep enough but did not elaborate.

After the third week of treatment, the patient, therapist, and supervisor remained in pessimistic agreement. Nothing was happening.

Most of the ward staff persisted in negative evaluation, but one nurse claimed that he was better, pointing out that for the past 10 months the patient had run to nursing station 3–4 times a day, saying that he was dying and had not done that for the past week. Previously, they had held his hand, told him for the thousandth time that his heart was fine and that it was just his terrible anxiety. After 20 min, he left quite dissatisfied to reappear 6 h later and go through the same routine. Other nurses said this was unimportant since whenever you talked to the patient, it was the same litany of complaints.

Discussing this with the patient, I mentioned that he seemed to be better.

"Who told you that?," he snarled. "A nurse" I stuttered. "What do they know?," he dismissed. When I asked him about not running to the nursing station, he seemed puzzled, and quickly affirmed that he was anxious as ever. Pressed about the change in his behavior, he finally noted, "I guess I learned that they can't do anything for me." I questioned, "You mean after 10 months you learned just this week?"

"Well, you have to learn sometime," he replied, thus anticipating the development of cognitive behavioral therapy.

It became clear that the run to the nursing station was precipitated by an attack. What confused me was that the overwhelming crescendo of the attack was considered the worst anxiety but imipramine took the top off, while leaving marked anxiety behind. It seemed reasonable that a medication would be most effective at the moderate level of illness rather than the worst form. It was at that time I realized that anxiety was heterogeneous and that the anxiety attack should be called something else, to distinguish it from ordinary chronic anxiety, itself considered an inappropriate manifestation of fear. Notably, the chronic anxiety persisted. This led to my renaming the attack as a panic attack. It also seemed, by reviewing detailed histories, that the phobic manifestations only occurred after the onset of repeated panic attacks. The phobias of these hospitalized patients were limited to situations where, if they had a panic attack, they couldn't get to help. This was not due to conditioning since these patients often refused to fly, although they never had a panic attack on an airplane. It was the possibility that paralyzed them. They were not afraid of dogs, or thunder, or heights. Also, it became apparent that the time between repeated panic attacks and phobic manifestations was very variable. Some patients withstood the attacks for many months while others folded up immediately and became housebound.

In general with repeated attacks, chronic anticipatory anxiety developed about when the next panic attack would happen. This led to severe travel limitations by ensuring that help was always easily available. It could be radically diminished by a companion. The propensity for chronic high levels of anticipatory anxiety, unleashed by the panic, seemed an independent component of the phobic development. A final confusing note was the frequent history of separation anxiety disorder that became manifest when school was required, leading to the misnamed school phobia. These children were not afraid of school. School was a forced separation from their mother and they were overwhelmed, which they explained by concern for the mother's welfare." Maybe she got sick," they would whimper. However they usually got over this and did not have panic attacks during this period. At sometime later, frequently when they had to change schools or move the separation anxiety would recur. Panic attacks occurred rarely before puberty and raised the suspicion of cerebral dysrhythmia. Later we found that Panic Disorder waxed and waned.

These early observations were followed by a series of controlled trials regarding panic disorder and separation anxiety disorder [3–8].

However, an independent placebo controlled trial confirming the anti-panic effect of imipramine awaited Sheehan and Ballenger [9], two nervy, smart residents at Massachusetts General Hospital, part of the Harvard empire. Hillside Hospital had no academic links and nowhere near Harvard's prestige.

This necessary independent study took place about 17 years after our initial presentations to a largely dismissive world. At Yale a senior psychoanalyst explained that these phobic patients did not travel because that requires walking the streets and since they had an unconscious desire to be streetwalkers (prostitutes) they couldn't do it. He thought imipramine must be a chemical straight jacket. Others were not as articulate but just as disbelieving.

The publication of DSM-III was also in 1980. By including Panic Disorder in a manual for clinical diagnosis, it legitimatized research in Panic Disorder, which took off after 1980. That research in psychiatry requires the prior development of a clinical category is illogical. However, that this should lead to rejection of clinical categories, as in the NIMH RDoC program, is glib and misleading [10].

Emeritus Professor Psychiatry   Donald F. Klein
Columbia University Medical Center
New York, NY, USA

References

1. Klein DF, Fink M. Psychiatric reaction patterns to imipramine. Am J Psychiatry. 1962; 119:432–8.
2. Klein DF, Fink M. Behavioral reaction patterns with phenothiazines. Arch Gen Psychiatry. 1962;7:449–59.
3. Klein DF. Delineation of two drug-responsive anxiety syndromes. Psychopharmacologia. 1964; 5:397–408. [Japanese version appeared in Archives of Psychiatric Diagnostic and Clinical Evaluation 1993; 4:505–8.]
4. Klein DF. Importance of psychiatric diagnosis in prediction of clinical drug effects. Arch Gen Psychiatry. 1967;16:118–26.
5. Gittelman-Klein R, Klein DF. Controlled imipramine treatment of school phobia. Arch Gen Psychiatry. 1971;25:204–7.
6. Zitrin CM, Klein DF, Woerner MG. Behavior therapy, supportive psychotherapy, imipramine and phobias. Arch Gen Psychiatry. 1978;35:307–16.
7. Zitrin CM, Klein DF, Woerner MG, Ross DC. Treatment of phobias (I): comparison of imipramine hydrochloride and placebo. Arch Gen Psychiatry. 1983;40(2):125–38.
8. Klein DF, Zitrin CM, Woerner MG, Ross DC. Treatment of phobias (II): behavior therapy and supportive psychotherapy: are there any specific ingredients? Arch Gen Psychiatry. 1983;40:139–45.
9. Sheehan DV, Ballenger J, Jacobsen G. Treatment of endogenous anxiety with phobic, hysterical, and hypochondriacal symptoms. Arch Gen Psychiatry. 1980;37(1):51–9.
10. Weinberge DR, Glick ID, Klein DF. Whither Research Domain Criteria (RDoC)?: the good, the bad, and the ugly. JAMA Psychiatry. 2015;72(12):1161–2.

Preface

This book was carefully edited to condensate the accumulated knowledge on panic disorder produced by a network of very productive researchers at this subject in recent years. This network includes distinguished professors and renowned researchers who enthusiastically participated from Brazil, United States, Italy, Spain, United Kingdom, Mexico, and Switzerland.

We present several aspects of panic disorder including historical aspects, neurobiology findings, connections with the respiratory and cardiovascular systems, pharmacological and nonpharmacological treatments, psychopathology and genetics, among others.

We recommend it to anyone who has an interest in anxiety and panic disorder, specially psychiatrists, clinical psychologists, postgraduate students, and researchers in this area.

Rio de Janeiro, Brazil

Rafael Christophe R. Freire
Antonio Egidio Nardi

Contents

Contributors

Roman Amrein Private Practice, Basel, Switzerland

Laboratory of Panic and Respiration, Institute of Psychiatry, Federal University of Rio de Janeiro, Rio de Janeiro, Brazil

Claudio Gil Soares de Araújo Heart Institute Edson Saad, Federal University of Rio de Janeiro, Rio de Janeiro, Brazil

Exercise Medicine Clinic (CLINIMEX), Rio de Janeiro, Brazil

Oscar Arias-Carrión Unidad de Trastornos del Movimiento y Sueño (TMS), Hospital General Dr. Manuel Gea González, México, DF, Mexico

Unidad de Trastornos del Movimiento y Sueño (TMS), Hospital General Ajusco Medio, México, DF, Mexico

Tathiana Pires Baczynski Laboratory of Panic and Respiration, Institute of Psychiatry, Federal University of Rio de Janeiro, Rio de Janeiro, Brazil

Daniela Caldirola Department of Clinical Neuroscience, Villa San Benedetto Menni, Hermanas Hospitalarias, FoRiPsi, Albese con Cassano, Italy

Patricia Cirillo Laboratory of Panic and Respiration, Institute of Psychiatry, Federal University of Rio de Janeiro, Rio de Janeiro, Brazil

Fiammetta Cosci Department of Health Sciences, University of Florence, Florence, Italy

Gisele Pereira Dias Laboratory of Panic and Respiration, Institute of Psychiatry, Federal University of Rio de Janeiro, Rio de Janeiro, Brazil

Elfi Egmond Department of Psychiatry and Psychology, Hospital Clinic, Barcelona, Spain

Department of Clinical and Health Psychology, Faculty of Psychology, Universidad Autònoma de Barcelona, Cerdanyola del Vallés, Barcelona, Spain

Rafael Christophe R. Freire Laboratory of Panic and Respiration, Institute of Psychiatry, Federal University of Rio de Janeiro, Rio de Janeiro, Brazil

Thalita Gabínio Dom Bosco Catholic University (UCDB), Campo Grande, Brazil

Silvia Hoirisch-Clapauch Department of Hematology, Hospital Federal dos Servidores do Estado, Ministry of Health, Rio de Janeiro, Brazil

Giuseppe Iannone Department of Clinical Neuroscience, Villa San Benedetto Menni, Hermanas Hospitalarias, FoRiPsi, Albese con Cassano, Italy

Jeffrey P. Kahn Department of Psychiatry, Weill-Cornell Medical College, Cornell University, New York, NY, USA

Eduardo Lattari Laboratory of Panic and Respiration, Institute of Psychiatry, Federal University of Rio de Janeiro, Rio de Janeiro, Brazil

Michelle Levitan Laboratory of Panic and Respiration, Institute of Psychiatry, Federal University of Rio de Janeiro, Rio de Janeiro, Brazil

Fabiana Leão Lopes Laboratory of Panic and Respiration, Institute of Psychiatry, Federal University of Rio de Janeiro, Rio de Janeiro, Brazil

Sergio Machado Laboratory of Panic and Respiration, Institute of Psychiatry, Federal University of Rio de Janeiro, Rio de Janeiro, Brazil

Physical Activity Neuroscience, Physical Activity Sciences Postgraduate Program, Salgado de Oliveira University, Niterói, Brazil

Gisele Gus Manfro Anxiety Disorders Program, Hospital de Clínicas de Porto Alegre (HCPA), Porto Alegre, Brazil

Department of Psychiatry, Federal University of Rio Grande do Sul (UFRGS), Porto Alegre, Brazil

Rocío Martín-Santos Department of Psychiatry and Psychology, Hospital Clinic, Institut d'Investigació Biomèdica August Pi I Sunyer (IDIBAPS), Centro de Investigación Biomédica en Red en Salud Mental (CIBERSAM), G25, Universidad de Barcelona, Barcelona, Spain

Marina Dyskant Mochcovitch Laboratory of Panic and Respiration, Institute of Psychiatry, Federal University of Rio de Janeiro, Rio de Janeiro, Brazil

Antonio Egidio Nardi Laboratory of Panic and Respiration, Institute of Psychiatry, Federal University of Rio de Janeiro, Rio de Janeiro, Brazil

Ricard Navinés Department of Psychiatry and Psychology, Hospital Clinic Institut d'Investigació Biomèdica August Pi I Sunyer (IDIBAPS), Centro de Investigación Biomédica en Red en Salud Mental (CIBERSAM), G25, and Universidad de Barcelona, Barcelona, Spain

Flávia Paes Laboratory of Panic and Respiration, Institute of Psychiatry, Federal University of Rio de Janeiro, Rio de Janeiro, Brazil

Marcelo Papelbaum State Institute of Diabetes and Endocrinology of Rio de Janeiro, Rio de Janeiro, Brazil

Giampaolo Perna Department of Clinical Neuroscience, Villa San Benedetto Menni, Hermanas Hospitalarias, FoRiPsi, Albese con Cassano, Italy

Department of Psychiatry and Neuropsychology, Maastricht University, Maastricht, The Netherlands

Department of Psychiatry and Behavioral Sciences, Leonard Miller School of Medicine, University of Miami, Miami, FL, USA

AIAMC (Italian Association for Behavioural Analysis, Modification and Behavioural and Cognitive Therapies), Milan, Italy

Maurice Preter Department of Psychiatry, College of Physicians and Surgeons, Columbia University, New York, NY, USA

Department of Neurology, Mount Sinai School of Medicine, New York, NY, USA

Aline Sardinha Laboratory of Panic and Respiration, Institute of Psychiatry, Federal University of Rio de Janeiro, Rio de Janeiro, Brazil

Cognitive Therapy Association of Rio de Janeiro (ATC-Rio), Rio de Janeiro, Brazil

Luiz Carlos Schenberg Department of Physiological Sciences, Laboratory of Neurobiology of Mood and Anxiety Disorders, Federal University of Espírito Santo, Vitória, ES, Brazil

Cristiano Tschiedel Belem da Silva Anxiety Disorders Program, Hospital de Clínicas de Porto Alegre (HCPA), Porto Alegre, Brazil

Department of Psychiatry, Federal University of Rio Grande do Sul (UFRGS), Porto Alegre, Brazil

Sandrine Thuret Department of Basic and Clinical Neuroscience, Laboratory of Adult Neurogenesis and Mental Health, Institute of Psychiatry, Psychology and Neuroscience, King's College London, London, UK

Tatiana Torti Department of Clinical Neuroscience, Villa San Benedetto Menni, Hermanas Hospitalarias, FoRiPsi, Albese con Cassano, Italy

AIAMC (Italian Association for Behavioural Analysis, Modification and Behavioural and Cognitive Therapies), Milan, Italy

André B. Veras Dom Bosco Catholic University (UCDB), Campo Grande, Brazil

Morena Mourao Zugliani Laboratory of Panic and Respiration, Institute of Psychiatry, Federal University of Rio de Janeiro, Rio de Janeiro, Brazil

Abbreviations

5-HT	Serotonin
A1	A1 noradrenergic group
A5	A5 noradrenergic group
ACTH	Corticotropin
AIC	Anterior insular cortex
Amb	Nucleus ambiguous
AP	Area postrema
APA	American Psychiatry Association
BLA	Basolateral amygdala
BNST	Bed nucleus of stria terminalis
BZD	Benzodiazepine
C1	C1 adrenergic group
CCHS	Congenital central hypoventilation syndrome
CCK	Cholecystokinin
CeA	Central amygdala
CePAG	Juxtaqueductal and medial sectors of periaqueductal grey
CnF	Cuneiform nucleus
CO	Carbon monoxide
CO_2	Carbon dioxide
COR	Cortisol
CORT	Corticosterone
CRH	Corticotropin releasing hormone
CSA	Childhood separation anxiety
ACC	Anterior cingulate cortex
dACC	Dorsal anterior cingulate cortex
DGH	Deakin/Graeff hypothesis
DLPAG	Dorsolateral periaqueductal grey
DLSC	Deep layers of superior colliculus
DMH	Dorsomedial hypothalamus
DMHd	Dorsomedial hypothalamus *pars diffusa*
DMPAG	Dorsomedial periaqueductal grey matter
DMX	Dorsal motor nucleus of vagus
DPAG	Dorsal periaqueductal grey matter
DS	Dorsal striatum
DSM	Diagnostic and statistical manual of mental disorders
DSM-I	Diagnostic and statistical manual of mental disorders – first edition

DSM-II	Diagnostic and statistical manual of mental disorders – second edition
DSM-III	Diagnostic and statistical manual of mental disorders – third edition
DSM-III-R	Diagnostic and statistical manual of mental disorders – third edition, revised
DSM-IV	Diagnostic and statistical manual of mental disorders – fourth edition
DSM-IV-TR	Diagnostic and statistical manual of mental Disorders – fourth edition, text revision
DSM-5	Diagnostic and statistical manual of mental disorders – fifth edition
eLPB	External division of the lateral parabrachial area
EPM	Elevated plus-maze
ES	Escapable shock
ETM	Elevated T-maze
FLI	Fos-like immunoreactivity
FS	Fictive shock
FST	Forced swimming test
GABA	Gamma-aminobutyric acid
GAD	Generalized anxiety disorder
GiA	Gigantocellular reticular area
HPA	Hypothalamus-pituitary-adrenal axis
IS	Inescapable shock
KCN	Potassium cyanide
KF	Kölliker-Fuse nucleus
LAC	Sodium lactate
L-AG	*l*-allylglycine
LC	Locus coeruleus
LDTg	Laterodorsal tegmental nucleus
LLPDD	Late luteal phase dysphoric disorder
LPAG	Lateral periaqueductal grey
LPAGcv	Caudoventral lateral periaqueductal grey
LPBA	Lateral parabrachial area
LPGi	Lateral paragigantocellular nucleus
MAOI	Monoamine oxidase inhibitor
MDD	Major depression disorder
MRI	Magnetic resonance imaging
NE	Norepinephrine
NMDA	*n*-methyl-D-aspartic acid
NRD	Nucleus raphe dorsalis
NRDd	Nucleus raphe dorsalis *pars dorsalis*
NRDlw	Nucleus raphe dorsalis "lateral wing"
NRDv	Nucleus raphe dorsalis *pars ventralis*
NRM	Nucleus raphe magnus
NRMn	Nucleus raphe medianus
NRO	Nucleus raphe obscurus
NRPa	Nucleus raphe pallidus

NRPo	Nucleus raphe pontis
NSI	Neonatal social isolation
NTS	Nucleus tractus solitarii
OVLT	Organum vasculosum lamina terminalis
$PACO_2$	Carbon dioxide alveolar partial pressure
$PaCO_2$	Carbon dioxide arterial partial pressure
PAG	Periaqueductal grey matter of the midbrain
PaO_2	Oxygen arterial partial pressure
PBA	Parabrachial area
pBöC	Pre-Bötzinger complex
PCC	Posterior cingulate cortex
PD	Panic disorder
PeF	Perifornical hypothalamus
PET	Positron emission tomography
$P_{ET}CO_2$	Carbon dioxide end-tidal partial pressure
PFC	Prefrontal cortex
PH	Posterior hypothalamus
PHA-L	Leucoagglutinin of *Phaseolus vulgaris*
PIC	Posterior insular cortex
PMD	Premammillary dorsal nucleus of the hypothalamus
PPy	Parapyramidal area
PRL	Prolactin
PRV	Pseudorabies virus
PVN	Paraventricular nucleus of hypothalamus
PVT	Paraventricular nucleus of the thalamus
rCBF	Regional cerebral blood flow
RVLM	Rostroventrolateral medulla
SAH	Separation anxiety hypothesis
SFA	Suffocation false alarm hypothesis
SFO	Subfornical organ
sgACC	Subgenual anterior cingulate cortex
SI	Social interaction test
SON	Supraoptic nucleus
SSRI	Selective serotonin reuptake inhibitor
SNRI	Serotonin and noradrenaline reuptake inhibitor
TASK	Tandem acid-sensitive potassium channels
TCA	Tricyclic antidepressants
TH	Tyrosine hydroxylase
VHZ	Ventral hypothalamic zone
VLM	Ventrolateral medulla
VLPAG	Ventrolateral periaqueductal grey matter
VMH	Ventromedial hypothalamus
VMHdm	Dorsomedial ventromedial hypothalamus
VMS	Ventral medullary surface
WHO	World Health Organization

The Panic Disorder Concept: A Historical Perspective

1

Antonio Egidio Nardi
and Rafael Christophe R. Freire

Contents

Abstract

Panic disorder was described in several literary reports and folklore. Perhaps one of the oldest examples lies in Greek mythology, in which the Pan god was responsible for the term *panic*. Before 1850, the symptoms for anxiety were still usually associated with signs and symptoms of depression. In the second half of the nineteenth century, a progressive change began to take place in the field of anxiety symptoms. Henry Maudsley, in 1879, described a *melancholic panic*, and this was the first time the term panic was technically used in psychiatry. Jacob Mendes DaCosta, during the American Civil War described *the irritable heart*. At 1894, Sigmund Freud described the *angstneurose* (anxiety neurosis) and was impressed with the symptoms and associated phobias. In 1954, Mayer-Gross, associated anxiety disorders to hereditary, organic and psychological factors, dividing them in simple anxious states and phobic anxious states. In 1964, Donald Klein, published that patients with these disorders responded favorably to imipramine, a tricyclic antidepressant. His observations and descriptions influenced the third edition of the Diagnostic and Statistical Manual of Mental Disorders (DSM-III, 1980), in which the term *panic disorder* appears for the first time in an official classification. In 1993, Donald Klein described the "False Suffocation Alarm Theory". This

A.E. Nardi • R.C.R. Freire (✉)
Laboratory of Panic and Respiration, Institute
of Psychiatry, Federal University of Rio de Janeiro,
Rio de Janeiro, Brazil
e-mail: antonioenardi@gmail.com;
rafaelcrfreire@gmail.com

© Springer International Publishing Switzerland 2016
A.E. Nardi, R.C.R. Freire (eds.), *Panic Disorder*, DOI 10.1007/978-3-319-12538-1_1

theory has been widely accepted, based on laboratory studies of respiratory, cognitive and biochemical tests. In the last 50 years, the mysteries of panic disorder have been revealed through basic and clinical research.

Keywords

Panic disorder • Agoraphobia • Anxiety disorders • Anxiety • Anti-anxiety agents

1.1 Introduction

Although a common mental disorder [1], panic disorder has received little historical attention. Berrios [2] pointed that this may be due to its relative newness; or to the fact that the historical model used to account for the traditional mental disorder is inappropriate for the new disorders. Our aim is to describe some important points in the history of the concept of panic disorder and highlight the importance of the presence of this diagnosis in the official classifications for the clinical and research developments.

The word *anxiety* derives from the Indo-Germanic root a*ngh* for narrowness or constriction. This rootword was the source for the Greek term *anshein*, meaning to strangulate, suffocate, oppress and correlated Latin terms, such as *angustus*, to express discomfort, *angor*, meaning oppression or lack of air, and *angere*, signifying constriction, suffering, panic [2].

1.2 Historical Overview

1.2.1 Classical Era

Several literary reports and folklore demonstrate that anxiety symptoms observed in the past are what we call panic disorder nowadays. Perhaps one of the oldest examples of panic symptoms lies in Greek mythology – the legend of the Pan god. He was responsible for anxiety attacks and originated the term *panic* [3]. Although he was born in Arcadia, Pan roamed the Greek mountains and roads. According to the legend, he was the god of flocks and shepherds, and since he was half man and half goat, with horns and goat legs, his appearance was frightening. He was very active and full of energy, but very irritable. He loved music and played a small reed pipe, Syrinx. In several stories, Pan is reported to cause fright, screams, fears, terror and suffering. Like some other gods, Pan harassed nymphs who ran from him, maybe because of his mien or due to his always unexpected and sudden apparition [3]. Like other woodland gods, he was feared by those who had to go through the forest. Meeting one of these deities could provoke an overwhelming and irrational fear, for no reason at all, or what was known as *panic terrors* or *panic attacks*. Fear of meeting Pan again and fear of being startled once more made travelers stop journeying through roads and avoiding going to the market (in Greek, *ágora*), thus developing agoraphobia (fear of large open or public places) [2].

In ancient Greece, Plato presents in *Timaeus* a case of anxiety associated to the *wandering* [4]. Although this description is often associated to hysteria, the original text describes a woman with acute anxiety that is very similar to panic disorder. "The uterus is an animal desirous of procreating children. When it remains unfruitful long periods beyond puberty, gets discontented and angry, begins to wander throughout the body, closing the air passages, impeding breathing, bringing about painful distress, and causing a variety of associated diseases". Plato not only associated panic disorder to women of reproductive age, but also stated that respiratory symptoms (dyspnea, difficulty in breathing) are common symptoms and pregnancy improves them (in some cases) [4].

From the time of Hippocrates up to the seventeenth century, the description and interpretation of the signs and symptoms of anxiety were determined by the principle of body fluids described by Hippocrates in *Corpus Hippocraticum* or by reports in layman and religious literature [2]. *Corpus Hippocraticum* consisted of a series of seventy medical treatises dating to the fifth century AD, attributed to Hippocrates and his disciples. The traditional Egyptian idea of blood, yellow bile, black bile and phlegm would be the four cardinal humors or fluids, and would be associated to distinctive features of each individual, including diseases. Hippocrates associated the excess of black bile to depression [2].

1.2.2 Medieval Period and Renaissance

During the medieval period and the Renaissance, what we today consider to be severe anxiety syndrome was associated to the signs and symptoms of depression. In the seventeenth century, the English doctor Robert Burton [5] described in his book *"The Essential Anatomy of Melancholy"* an acute anxiety episode, which he considered to be a type of fear: *"This fear cause in man, as to be red, pale, tremble, sweat; it makes sudden cold and heat to come over all the body, palpitations of the heart, syncope, etc. It amazed many men that are to seek or show themselves in public assemblies...Many men are so amazed and astonished with fear, they know not where they are, what they do, and which is worst, it tortures them many days before with continual affrights and suspicious ..."*

Burton [5] described several types of pathological anxiety in a style quite different from today's scientific texts; associating philosophy and beliefs of the time. He listed the fear of death, fear of losing a loved one and *paranoid anxiety*. He described anxiety based on delirium, associated to depersonalization, to hyperventilation, to hypochondria and even to agoraphobia, as well as anticipatory anxiety and several types of phobias – public speaking, acrophobia and claustrophobia.

The renaissance was not only a time of man's rediscovery of man through art, but it was also a time marked by science's rise in prestige, and, at this time, alchemy received scientific criteria. Paracelsus (1493–1541) was one of the leaders in the new art to question the hegemony of Hippocrates' humors [6].

1.2.3 Nineteenth Century

Modern medical description of panic disorder (PD) began prior to the nineteenth century, during a period when psychiatry was established

itself as an independent branch of study. Berrios [2] states that those who took care of patients since immemorial time knew anxiety symptoms and syndromes. However, each symptom was treated as a separate medical complaint, as if it were an isolated physical problem. Symptoms were considered for its *face value* and treated as symptoms associated to the disorder of an organ; for example, palpitation was associated as a disease of the heart. No mental disorder was considered when these complaints were presented. Because of its physical characteristics – tachycardia, precordial discomfort, nausea, sweating, paresthesia, etc. – these anxiety manifestations were identified and described by clinical physicians and not by psychiatrists [2].

In French psychiatry, panic disorder has been studied since the nineteenth century under the name of *acute episodes of distress*, and were part of the description for various nosological entities such as Benedict Morel's *emotional delirium* (*delire émotif*), Henri Le Grand Du Saulle's *fear of spaces* (*peur des espaces*) based on Karl Westphal's texts on agoraphobia; Doyen's *morbid terrors* (*terreurs morbides*), Henry Ey's *severe anxiety* (*grande anxiété*); from Brissaud's *paroxystic anxiety* (*anxiété paroxystique*) to the *fear in the army* (*peur dans les armées*) described by Brousseau [7]. In the beginning of the nineteenth century, the French doctor, Landré-Beauvais defined *distress* as "a certain discomfort, restlessness, excessive physical activity", and that the symptoms could be associated to severe or chronic diseases [7].

Maurice Krishaber (1836–1883), an otorhinolaryngologist who practiced in Paris, described "the cerebral-cardiac neurosis" associating some symptoms – dizziness, tachycardia, restlessness, among others – to one unique neurocirculatory disease. The term neurosis referred to nerves, an organic, somatic disturbance, with no association to a psychiatric disorder [2, 7].

Benjamin Rush (1745–1813), an American physician from Philadelphia, and considered as the "father of American psychiatry", described in his psychiatry book (1812) the association between somatic causes and phobias, relating depression (*tristimania*) to hypochondriasis [2, 7]. Before 1850, the symptoms for anxiety were still associated with signs and symptoms of depression. In 1858, Littré and Robin defined *distress* as "a feeling of oppression or weight in the epigastrium, associated to a great deal of difficulty in breathing or excessive sadness, this being the most advanced degree of anxiety". They also described anxiety as a "problematic and agitated state, with difficulty in breathing and precordial pressure: restlessness, anxiety and distress are three stages of the same phenomenon, in order of seriousness" [2, 7].

In the second half of the nineteenth century, a progressive change began to take place in the field of anxiety symptoms [6, 7]. Somatic causes, that up to that time were fully accepted, began do divide the attention with possible psychological causes. In 1872, Karl Friedrich Otto Westphal (1833–1890) described agoraphobia, the fear of wide and open places. He cited three male patients who demonstrated fear in wide streets and open spaces and who, at times, were compelled to ask passersby for help. In 1878, Le Grand Du Saulle published a paper on the *fear of spaces* (*peur des espaces*), broadening Westphal's original concept. Du Saulle stated that, "although we can observe now and previously that these patients fear open places, they can feel fear of theaters, churches, high balconies in buildings or whenever they are found near wide windows, or buses, boats or bridges" [2, 6, 7].

In the second half of the nineteenth century, definitions of states of anxiety and agoraphobia became more specified and detailed [6, 7]. Benedict Morel's (1809–1873) studies were very important during this period. According to Morel in 1866 [8], the category of *emotional delirium* combined the organism's physical and moral sensibility symptoms, which have no relation whatsoever with the signs and symptoms that are today considered to be psychotic. Morel listed as physical symptoms: hyperesthesia, paresthesia, hot and cold flashes, sweating, pain, etc. Phobias were among the moral symptoms. The explanation given by Morel for this *emotional delirium* was recorded in his general theory of degeneration. Among Morel's cases, we can observe what would today be described as PD and generalized anxiety disorder. The physical and moral causes would be combined with hereditary factors, in order to appear as a disease, from the moment that determining conditions – moral and physical – were present. Base alteration would be the functional fragility of the visceral, ganglionary, nervous system [8]. Henry Maudsley (1835–1918), in 1879, described the *melancholic panic*, and this was the first time the term *panic* was technically used in psychiatry [2]. It is important to highlight the notion that all the symptoms could be manifestations of a unitary construct – anxiety – and this new concept had limited acceptance in nineteenth century psychiatry.

At the end of the nineteenth century, two American doctors played a fundamental role in the future make up of Freud's proposed *anxiety neurosis*: George M. Beard and Jacob Mendes DaCosta [2, 6, 7]. The first studied neurasthenia, and was cited by Freud in his articles on the anxiety neurosis. Beard's 1869 article entitled "Neurasthenia" was the starting point for a series of papers with the objective of establishing the specificity of neurasthenia, both in the clinical description as well as in the explanatory hypothesis. According to Beard, neurasthenia would be centered on physical fatigue of nervous origin (related to nerves, a neurological concept) – functional weakness of the brain due to sexual energy drain from abnormal sexual activity, such as excessive masturbation, and would be accompanied by other symptoms such as pain, digestive problems, paresthesia, depression, reduced libido, apathy and indifference. However, what was most important, from the anxiety point of view was that Beard's ample description of neurasthenia considered acute anxiety a very important part. The *morbid fears of particular type*, especially agoraphobia, anthropophobia (fear of society) and the phobia of traveling alone. Freud's criticism of Beard's work lead Freud to introduce his own description for anxiety neurosis.

Jacob DaCosta (1833–1900) [6], a military doctor, during the American Civil War described what he called "the irritable heart, a functional cardiac disorder". Men affected by this disorder demonstrated acute symptoms with palpitations of variable intensity, lasting from minutes to hours, accompanied by thoracic pain and general discomfort. Since he could not find an organic cardiac lesion or subjective conditions due to war, Da Costa concluded that this was a functional disorder of the sympathetic nervous system. Da Costa described a detailed clinical evaluation of almost 300 patients. When the heart is submitted to extremely intense effort and tension, it becomes physiologically "irritable", and as a result, palpitations arise.

Sigmund Freud (1856–1939) was also impressed with the symptoms and phobias associated with what we call today panic disorder. At around 1894, he [9] described the *anxiety neurosis* (*Angstneurose*). The origin of the term goes back to the studies by E. Hecker, who in 1893, had already demonstrated the presence of anxious states in neurasthenia. Freud's merit was to separate anxiety neurosis from neurasthenia and describe it with a specific clinical presentation. On several occasions, Freud states that certain acute symptoms, such as dizziness, cardiac activity disorders, sweating, tremors, shock, diarrhea and *pavor nocturnus*, are special forms of anxiety attacks, which he now denominates "anxiety equivalent". Freud [9] stated "...these patients symptoms are not mentally determined or removable by analysis, but they must be regarded as direct toxic consequences of disturbed sexual chemical processes".

Since Freud's first papers [9] on *anxiety neurosis*, the anxiety attack was considered as one of two fundamental forms of clinical manifestation of distress, the other being a *chronic state*. Freud also associated agoraphobia to anxiety neurosis. To him, agoraphobia referred to an effort made by the patient with the intention of not being stricken by *anxiety attacks* in unfamiliar circumstances, when he was not be sure he could get help. It is interesting to note that Freud also discussed the relation of agoraphobia and panic attacks, widely studied nowadays, in his texts on the neurosis of anxiety.

1.2.4   Twentieth Century – First Half

In the twentieth century, although some concepts on psychological factors in anxiety having already been discussed, the symptoms of anxiety were still largely associated to hereditary and biological factors. In 1903, Pierre Janet [10] described *psychasthenia*, a case of anxiety with somatic and obsessive symptoms, associating these signs and symptoms to a breakdown of feelings and liberation of primitive behavior. Janet believed that the psychological performance of an individual could be divided into five levels. At the superior level would be the harmonious reality function, followed by habitual and automatic actions, then imagination functions, emotional visceral reactions and muscular movements. The term and its definition would encompass a series of mental disturbances, including anxiety. Anxiety and distress were manifestations of this breakdown, but not its main component. In 1926, in his book, *De l'Angoisse a l'Exstase*, Janet [11] cites the case of Madelaine, a 40 year-old patient with severe signs and symptoms of anxiety, with possible panic attacks associated with constitutional factors.

In 1907, Emil Kraepelin [12] (1856–1926), describes the *neurosis of terror* (*Schreckneurose*), in which panic attacks are etiologically associated to the affective state. In the sixth edition of his classic *Psychiatrie. Ein Lehrbuch für Studirende und Aerzte,* 1899, in the chapter on compulsive insanity, Kraepelin associated agoraphobia to anxiety attacks with several somatic symptoms. He related that the improvement of symptoms does not mean a concurrent improvement of the agoraphobia: this state could persist indefinitely.

Ernest Kretschmer introduced in his "Medical Psychology" the panic attack as an "an outburst of attempted impulsive movements", trying to take the individual away from the source of danger or excitement as quickly as possible [6]. Panic attacks appear in the Legrand Du Salle's fear of spaces, which determined the *fear of feeling fear*.

To Adolf Meyer [13] (1866–1950), a psychiatrist who influenced the first and second editions of the Diagnostic and Statistical Manual of Mental Disorders (DSM-I and DSM-II), the individual is a psychobiological being and any pathological manifestation would be a form of reaction to particular characteristics of the environment. Psychosis and neurosis would be different parts of the same psychiatric spectrum of continuous gradation, going from one extreme to the other.

Henrique Roxo [14] a Brazilian Professor of Psychiatry (1946) divided neurasthenia in two groups: psychasthenis and nervousness. Psychasthenis included obsessions, phobias and impulses. Nervousness represented an extraordinary state of anxiety, (the patient) revealing a feeling of indescribable discomfort, in which kinesthesia disturbances play a very important role. Although the anxiety attack is described within the symptoms of nervousness, Roxo [14] associated to the two neurasthenia groups' symptoms that nowadays are considered those of depressive syndromes and chronic and acute anxiety syndromes. Etiology was an association of psychological, environmental and constitutional factors, and treatment consisted of varied experimental medication, which supposedly would act specifically on each subtype.

1.2.5   The Last 65 Years

The second half of the twentieth century marked a revolution in the practice of psychiatry. Not only was psychiatric diagnosis revised and modified, seeking reliability, but treatment also received marked assistance from psychopharmacological agents. In 1954, Mayer-Gross [15] (1889–1961), associated anxiety reaction to hereditary, organic and psychological factors, dividing it in simple anxious states and phobic anxious states. The latter included agoraphobia with associated somatic symptoms. The 50's brought about the discovery of the monoaminoxidase inhibitors and tricyclic antidepressants. This was followed by the advent of the benzodiazepines. Thus, the road was paved for a more efficient panic treatment. In 1959, Donald Klein [16], a New York psychiatrist, observed that patients with depressive-anxiety symptoms responded favorably to imipramine, a tricyclic antidepressant. It is interesting to note that after a few weeks, Klein [16] and his patients were disappointed by the effects of imipramine and were ready to quit the experiment when the nurses pointed out that the patients were less anxious. The nurses noticed that the patients reduced their

trips to the nursing station to complain about *being sick* or *dying*; that they were more independent, walking around the hospital by themselves. Klein [16] concluded that imipramine was efficient in panic attacks, but not on chronic anxiety. He also considered agoraphobia a consequence of panic attacks, where patients who did not fear bridges or closed environments, but feared the possibility of having a panic attack and an immediate way out or help would be difficult or impossible.

Klein [16] more recently also distinguished three types of panic attacks: spontaneous, situational (associated with an agoraphobic situation) and those provoked by a constant phobic stimulus (animals, height, darkness, etc.). His posterior observations and descriptions influenced the third edition of the Diagnostic and Statistical Manual of Mental Disorders (DSM-III) (1980) [17], in which the term "panic disorder" appears for the first time in an official medical classification. In no time, panic disorder became the most studied psychiatric disorder from a diagnostic and therapeutic point of view.

The DSM-III [17] divided anxiety neurosis, also named in this classification of anxiety disorders, as panic disorder (acute anxiety) and generalized anxiety disorder (chronic anxiety), creating operational criteria for each category diagnosed. It also divided phobic neurosis into simple phobia, social phobia and agoraphobia (with or without panic attack). The extensive diagnostic reorganization of DSM-III continued for 7 years, resulting in the revised edition – DSM-III-R (1987) [18]. In this edition, agoraphobia no longer appears as an isolated category, but as a consequence of panic disorder, and now listed under the term: panic disorder with and without agoraphobia. The criteria for panic disorder were simplified and closer to clinical practice. One example was that only one panic attack with phobic repercussion during the previous month was required to establish a positive diagnosis. In other words, greater importance was given to the phobic consequences of the panic attack and not just the physical symptoms of the attack. The DSM-IV (1992) [19] maintained practically the same definitions, but defined panic attacks, demonstrating they could occur associated to other diagnosis and without fulfilling all the criteria for panic disorder; it also distinguished spontaneous panic attacks, situational panic attacks (linked to agoraphobia) and those provoked by a phobic stimulus (more closely linked to specific phobias). In the year 2000, a revised edition of DSM-IV was published, entitled DSM-IV-TR [20], in which some of the concepts were refined but the criteria for panic disorder remained the same. In 2013, the DSM-5 [21] separated again panic disorder from agoraphobia, although the emphasis on the compromise of a panic attack to the diagnosis of panic disorder was kept.

We can divide the evolution of psychopharmacology in relation to panic disorder into three main moments: first, Donald Klein's (1964) observation of the efficiency of the tricyclic antidepressants [16]. The second moment was when the efficacy of the benzodiazepines was perceived [22]. Finally, by the noted efficacy of the selective serotonin reuptake inhibitors [23] in 1990. Today, psychopharmacology has been leading psychiatry in the direction of biology. This biological perspective entails in putting anxiety in the frame of the evolutionist paradigm. Charles Darwin (1872) [24], in *The Expression of Emotion in Man and in Animals*, pointed the way to search for the adaptive value of the behavioral and psychological processes. Anxiety and fear have their roots in the defensive reactions of animals, observed in response to the danger normally found in the environment. The interpretation of a stimulus or a situation as dangerous depends on the nature of cognitive operations. In humans, cognitive factors acquire importance due to the intervention of the system of symbols socially codified, verbal or non-verbal. The behavioral responses to fear are accompanied by intense physiological alterations – physical symptoms – and alterations in the emotional state. The physiological alterations consist of objective measures of anxiety. Thus, cardiac frequency, arterial pressure, respiratory frequency and increased electrical conductivity of the skin brought on by sweating, are the measures mostly used to determine the degree of anxiety.

In the 80's, in their independent research projects, Jack Gorman, in the USA, and Van Den Hout, in the Netherlands, observed that 35 % carbon dioxide mixture induced panic symptoms [25]. This observation afforded the opportunity for greater insight concerning the factors that brought about panic attacks.

In 1993, Donald Klein [26] described the "False Suffocation Alarm Theory". This theory describes panic attack as a disorder of the physiological suffocation alarm. The regulating monitor would inform the central nervous system of an imminent suffocation situation when this was not actually occurring. This theory has been widely accepted, based on laboratory studies of respiratory, cognitive and biochemical tests.

Recent literature indicates that panic attacks may originate from a network of fear with altered sensibility, including in this network; the prefrontal cortex, the insula, the thalamus, the amygdala and amygdala projections toward the brain stem and hypothalamus [25]. When we administer a panicogenic agent, we would not affect a specific autonomic area of the brain stem, but we would be activating the entire neural fear net. Patients with panic disorder often complain of uncomfortable somatic sensations. The administration of a panicogenic agent would correspond to a non-specific activation. Since all these agents produce uncomfortably sharp physical sensations, the hypothesis would be that they stimulate a sensitive brain system conditioned to responding to noxious stimuli. As time passes, the projections of the central nucleus of the amygdala such as the *locus ceruleus*, periaqueductal gray area, and hypothalamus can become more or less sensitive [25]. There can also be an interindividual difference in the strength of these afferent projections. In this way, the standard neuroendocrine and autonomic responses presented during a panic attack may vary from one patient to another, and on the same patient throughout time.

1.3   Conclusion

Panic disorder has been having its mysteries revealed through basic and clinical research, and patients who suffer from this frightening disease can be sure that correct diagnosis and adequate treatment are already part of everyday clinical practices. We have, however, to perfect these practices even more, so we can continue to improve patient prognosis.

References

1. Weissman MM, Bland RC, Canino GJ, Faravelli C, Greenwald S, Hwu HG, et al. The cross-national epidemiology of panic disorder. Arch Gen Psychiatry. 1997;54(4):305–9.
2. Berrios GE. Anxiety and cognate disorders. In: Berrios GE, editor. The history of mental symptoms: descriptive psychopathology since the nineteenth century. Cambridge: Cambridge University Press; 1996.
3. Merivale P. Pan, the goat-god: his myth in modern times. Cambridge: Harvard University Press; 1969.
4. Plato. Plato: complete works. Indianapolis: Hackett Publishing Company; 1997.
5. Burton R. The essential anatomy of melancholy. New York: Dover Publications; 2002.
6. Stone MH. History of anxiety disorders. In: Stein DJ, Hollander E, editors. Textbook of anxiety disorders. Washington, DC: American Psychiatric Press; 2002. p. 3–11.
7. Costa-Pereira ME. Pânico: contribuição à psicopatologia dos ataques de pânico. Sao Paulo: Lemos Editorial; 1997.
8. Morel BA. Traité des dégénérescences physiques, intellectuelles et morales de l' espèce humaine. New York: Arno Press; 1976.
9. Freud S. Autoprésentation, Inhibition, symptôme et angoisse, Autres textess. Oeuvres complètes—Psychanalyse. Paris: Presses Universitaires de France; 1992.
10. Janet P. Les obsessions et la psychasthénie. New York: Arno Press; 1976.
11. Janet P. De l'angoisse à l'extase. Paris: Flammarion; 1923.
12. Kraepelin E. Psychiatrie: ein lehrbuch für studirende und aerzte. New Delhi: Science History Publications; 1990.
13. Brodsky A. Benjamin rush: patriot and physician. New York: Saint Martin's Press; 2004.
14. Roxo HB. Manual de Psiquiatria. Rio de Janeiro: Editora Guanabara; 1946.
15. Mayer-Gross W. Clinical psychiatry. London: Baillière, Tindall & Cassell; 1954.
16. Klein DF. Delineation of 2 drug-responsive anxiety syndromes. Psychopharmacologia. 1964;5(6):397–408. doi:10.1007/Bf02193476.
17. American Psychiatric Association. Diagnostic and statistical manual of mental disorders: DSM-III. 3rd ed. Washington, DC: American Psychiatric Association; 1980.
18. American Psychiatric Association. Diagnostic and statistical manual of mental disorders: DSM-III-R. 3rd ed. Washington, DC: American Psychiatric Association; 1987.
19. American Psychiatric Association. Diagnostic and statistical manual of mental disorders: DSM-IV. 4th ed. Washington, DC: American Psychiatric Press; 1994.
20. American Psychiatric Association. Diagnostic and statistical manual of mental disorders: DSM-IV-TR. 4th ed. Washington, DC: American Psychiatric Association; 2000.
21. American Psychiatric Association. Diagnostic and statistical manual of mental disorders: DSM-5. 5th ed. Washington, DC: American Psychiatric Association; 2013.
22. Sheehan DV. Current perspectives in the treatment of panic and phobic disorders. Drug Ther. 1982;12:179–93.
23. Boyer W. Serotonin uptake inhibitors are superior to imipramine and alprazolam in alleviating panic attacks—a metaanalysis. Int Clin Psychopharmacol. 1995;10(1):45–9. doi:10.1097/00004850-199503000-00006.
24. Darwin C. The expression of emotion in man and in animals. London: Fontana Press; 1999.
25. Gorman JM, Kent JM, Sullivan GM, Coplan JD. Neuroanatomical hypothesis of panic disorder, revised. Am J Psychiatry. 2000;157(4):493–505.
26. Klein DF. False suffocation alarms, spontaneous panics, and related conditions. An integrative hypothesis. Arch Gen Psychiatry. 1993;50(4):306–17.

A Neural Systems Approach to the Study of the Respiratory-Type Panic Disorder

2

Luiz Carlos Schenberg

Contents

L.C. Schenberg (✉)
Department of Physiological Sciences, Laboratory of Neurobiology of Mood and Anxiety Disorders, Federal University of Espírito Santo, Vitória, ES, Brazil
e-mail: luiz.schenberg@gmail.com

© Springer International Publishing Switzerland 2016
A.E. Nardi, R.C.R. Freire (eds.), *Panic Disorder*, DOI 10.1007/978-3-319-12538-1_2

Abstract

Panic disorder (PD) patients are exquisitely and specifically sensitive to inhalations of 5–7% carbon dioxide and infusions of 0.5 M sodium lactate. Another startling feature of clinical panic is the lack of increments of the 'stress hormones' corticotropin, cortisol and prolactin. PD is also more frequent in women and shows high comorbidity with childhood separation anxiety, late luteal period dysphoric disorder and depression. The hypothalamus-pituitary-adrenal axis is nevertheless activated in fear-like panics marked by palpitations, tremor and sweating, that are devoid of suffocation symptoms. These and other data suggest the existence of both respiratory and non-respiratory types of panic attacks. Increasing evidence suggests, on the other hand, that panics are mediated at midbrain's dorsal periaqueductal grey matter (DPAG). Therefore, here we summarized data showing that: (1) the DPAG harbors a suffocation alarm system which is activated by low intravenous doses of potassium cyanide (KCN); (2) KCN evokes defensive behaviors that are facilitated by hypercapnia, blocked by lesions of DPAG and attenuated by clinically effective treatments with panicolytics; (3) DPAG stimulations do not change the stress hormones when escape is prevented by stimulating the rats in a small compartment; (4) DPAG-evoked panics responses are facilitated in neonatally-isolated adult rats, a model of childhood separation anxiety; (5) DPAG-evoked panic-like behaviors are facilitated in diestrus phase of rat ovulatory cycle. It is proposed a neural model of panic attacks in which the PAG is the fulcrum of threatening signals from both forebrain and hindbrain. This model emphasizes the role of PAG as a suffocation alarm system.

Keywords

Panic disorder • Separation anxiety • Periaqueductal gray • Hypothalamo-hypophyseal system • Adrenal glands • Adrenocorticotropic hormone • Hydrocortisone • Prolactin

2.1 A Brief Story of Panic and Panic Theories

The underpinnings of the current classification of anxiety disorders were laid down by Sigmund Freud in the course of his efforts to separate 'anxiety neuroses' (*Angstneuroses*) from both 'neurasthenia' and 'melancholia' [1, 2]. Freud reported that anxious patients had two major syndromes of anxiety, the 'anxious expectation' (*ängstliche Erwartung*), which he considered the most essential syndrome of anxiety, and the less frequent 'attack of anxiety' (*Angstanfall*). According to Freud, in anxious expectation there is '*a quantum of freely floating anxiety which controls the choice of ideas by expectation*'. In contrast, in anxiety attacks '*anxiety breaks suddenly into consciousness without being aroused by the issue of any idea*'. Freud emphasized that anxiety attacks manifest either as '*the anxious feeling alone*' or the combination of this feeling with '*the nearest interpretation of the termination of life, such as the idea of sudden death or threatening insanity*'. Remarkably, Freud noted that patients suffering from anxiety attacks '*put the feeling of anxiety to the background or [described it] rather vaguely … as feeling badly, uncomfortably, etc.*', an observation corroborated by present-day psychiatrists. Freud was also aware of the high comorbidity of panic attacks with agoraphobia, which he linked to a '*locomotion disorder*' associated with the presence of dizziness during panic attacks [1, 2]. Moreover, he stressed that in agoraphobia '*we often find the recollection of a panic attack; and what patients actually fears is the occurrence of such attack under the special condition in which he believes he cannot escape it*' [3].

Freud's descriptions of 'anxious expectation' and 'anxiety attack' were very similar to the present diagnoses of generalized anxiety disorder (GAD) and panic disorder (PD), respectively. Yet, it would take almost a century before PD be accepted as a psychiatric disorder on its own [4]. Consequently, while the anxiety disorders continued to be vaguely diagnosed as 'neurasthenia' until the middle of the last century, panic attacks (or similar reactions) received a bewildering wealth of labels, including anxiety neurosis, neurocirculatory asthenia, vasomotor neurosis, nervous tachycardia, effort syndrome, Da Costa's syndrome, soldier heart, irritable heart and hyperventilation syndrome [5–8], and were even treated as schizophrenics (D.F. Klein, personal communication).

This scenario began to undergo a dramatic change following the publication of Donald Klein's influential studies showing that anxious expectation and panic were treated by different classes of drugs [9, 10]. Serendipitously, Klein observed that, although the panic attacks of hospitalized agoraphobics remitted during the treatment of comorbid depression with the tricyclic antidepressant imipramine, the expectant (anticipatory) anxiety of being overwhelmed by a panic attack was refractory to this medication. Notably, Klein also noted that the onset of PD was very often precipitated by separation and bereavement and that half of agoraphobics of his studies had suffered from severe childhood separation anxiety (CSA) that frequently prevented school attendance [10]. Besides suggesting that panic differs from anxiety, these observations led Klein to propose that CSA predisposes individuals to the later development of PD even before Bowlby's publication of the classical trilogy on attachment, separation anxiety and loss [11].

At approximately the same time, Pitts and McClure [7] showed that panic attacks could be precipitated by a 20-min intravenous infusion of 0.5 M sodium lactate (LAC) in patients prone to panic but not in normal volunteers. Further studies showed that a fraction of LAC-sensitive patients was also sensitive to the inhalation of 5–7% carbon dioxide (CO_2) [12]. The demonstration that PD had both physiological markers (CO_2, LAC) and drug-specific treatments (imipramine) suggested that clinical panic was the outcome of a specific brain circuit discharging maladaptively.

The next breakthrough occurred a few years after the acknowledgment of PD as a separate anxiety syndrome [4]. Indeed, the counterintuitive lack of increments in 'stress hormones' corticotropin (ACTH), cortisol (COR) and prolactin (PRL) in both natural and provoked panic attacks

was a remarkable finding as it discriminated panic from both fear and stress [13–20]. These studies also questioned the cognitive hypotheses that equated clinical panic to the fear brought about by the catastrophic evaluation of bodily symptoms [21–23].

As a result, in the early 1990s Klein proposed that clinical panic is bound to the fear of suffocation of which dyspnea (but not hyperpnea) is the leading symptom [24, 25]. In particular, Klein suggested that the apparently spontaneous clinical panic is the outcome of the misfiring of an as-yet-unknown 'suffocation alarm system', thereby producing sudden respiratory distress (dyspnea), panic, hyperventilation, and the urge to escape from immediate situation. Klein [24] argued that the 'suffocation false alarm' (SFA) theory is an explanation both sufficient and consistent of the hypersensitivity of panic patients to respiratory metabolites (CO_2, LAC) and of the occurrence of panic attacks during hypercapnic conditions of relaxation and sleep. Klein [24] also stressed that the SFA theory is consistent with the reduced incidence of panic attacks during pregnancy, delivery, and lactation and, conversely, of the increased frequency of panic in late luteal phase dysphoric disorder (LLPDD), in which respiration is increased or decreased by parallel changes of respiratory stimulant progesterone, respectively [24]. The SFA hypothesis is also consistent with the high comorbidity of panic with present and antecedent respiratory diseases [26, 27]. Lastly, Klein [24] argued that panics induced by 7% CO_2 and 0.5 M LAC remained the best models of clinical panic because they are not precipitated in healthy subjects [7, 24], obsessive patients (as cited by Griez and Schruers [28]) or patients with social phobia [29]. Moreover, whereas panics produced by CO_2 and LAC were blocked by chronic treatment with imipramine [30–33], those induced by β-carboline and yohimbine were not [24].

Klein's original hypothesis [10] that CSA predisposes to panic was corroborated by evidence both clinical and epidemiological [34–37]. The so-called 'separation anxiety hypothesis' (SAH) [36] was also supported by the demonstration of the clinical effectiveness of imipramine in CSA [34, 38]. Additionally, it was shown that separation-anxious children of parents with PD had respiratory responses to CO_2 similar to those of panickers [35, 39]. Most importantly, twin-based genetically-informative recent studies suggested that CSA shares a common genetic diathesis with both PD and CO_2 hypersensitivity [40, 41]. The necessity of an integrative explanation resulted in the expanded theory of SFA [27, 42] according to which panic attacks are the outcome of the episodic dysfunction of opioidergic systems tonically inhibiting both suffocation and separation alarm systems. The opioidergic hypothesis of panic attacks relied on clinical and preclinical evidence of the crucial role of opioids in both respiration [43–45] and parental bonding [46–49]. The existence of suffocation and separation alarm systems is still vividly debated in present time.

Competing theories propose that panic attacks are either the catastrofization of bodily symptoms by cortically-mediated cognitive processes [21–23] or the mistaken activation of fear-like responses to proximal threats [50–52]. Although all these theories propose that panics are false alarms, factors triggering panics remain largely obscure.

2.2 Idiosyncratic Features of Clinical Panics

The core symptoms of panic attacks were largely conserved ever since they were first described by Sigmund Freud in late eighteens (Table 2.1). Goetz et al. [53] found in addition that panickers rate their experience of LAC-induced panic as very much like the spontaneous panic attacks with respect to both quality (76%) and severity (84%) of symptoms. Although the patients reported that LAC-induced panics were at most moderate, the severity of panic attacks correlated mostly with desire to flee (0.70), fear of losing control (0.57), afraid in general (0.49) and dyspnea (0.48). Curiously, however, the 'desire to flee' is not included in clinical symptomatology (Table 2.1).

The relationships of panic attack symptoms, including anticipatory anxiety, agoraphobia, and 13 clinical symptoms of DSM-III-R, was investi-

gated by Shioiri et al. [55]. Cluster analysis revealed that panic symptoms are clustered in three groups: cluster A (dyspnea, choking, sweating, nausea, flushes/chills); cluster B (dizziness, palpitations, trembling or shaking, depersonalization, agoraphobia, and anticipatory anxiety); and cluster C (fear of dying, fear of going crazy, paresthesias, and chest pain or discomfort). Accordingly, whereas the cluster A includes mostly respiratory symptoms, clusters B and C include physiological and cognitive symptoms alike those of fear.

In particular, Perna et al. [56] examined the different types of dyspnea induced by 35% CO_2 challenges in patients with PD. Factor analysis identified 3 main factors: breathing effort, sense of suffocation, and rapid breath. Factor scores for 'breathing effort' and 'sense of suffocation' significantly discriminated between patients who did and those who did not report CO_2-induced panic attacks, respectively. Factor scores for breathing effort loaded significantly for patients whose reaction resembled 'unexpected' panic attacks. Authors suggested that although the sense of suffocation was linked to an increased sensitivity to CO_2, it may not be the main factor of unexpected panic attacks.

These studies are in line with evidence amassed in the last decades suggesting that panics may be either respiratory or non-respiratory, depending on the prominence of respiratory symptoms [57, 58]. The existence of two types of panic was corroborated by latent class analysis of the temporal stability, psychiatric comorbidity and treatment outcome in a large-scale epidemiological survey [59]. Data showed that while the temporal stability of panic subtypes was mainly observed in females, respiratory panics were associated with both the more severe forms of PD and the increased comorbidity with depression and other anxiety disorders. Yet, treatment outcome did not suggest that panic subtypes respond differentially to imipramine and alprazolam. The latter result disagrees from Klein's [24] proposal that panic sensitivity to tricyclics and benzodiazepines parallel laboratory panics reminiscent of either asphyxia (as provoked by LAC, CO_2, bicarbonate and isoproterenol) or fear, respectively (as provoked by yohimbine, flumazenil, benzodiazepine inverse agonists, caffeine and

Table 2.1 Symptoms of spontaneous and lactate-induced panic attacks

Panic attack symptoms	Freud[a]	RDC[b]	Diagnostic and statistical manual[c]				ICD-10[d]	API[e]
			III	III-R	VI	V		
Palpitations, pounding or tachycardia	+	+	+	+	+	+	+	+
Chest pain or discomfort	+	+	+	+	+	+	+	+
Dyspnea, shortness of breath	+	+	+	+	+	+	+	+
Sensations of smothering or choking	+	+	+	+	+	+	+	+
Trembling or shaking	+	+	+	+	+	+	+	+
Dizzy, unsteady, lightheaded, or faint	+	+	+	+	+	+	+	+
Fear of losing control or going crazy	+	+	+	+	+	+	+	+
Fear of dying	+	+	+	+	+	+	+	+
Paresthesias (numbness, tingling)	+	+	+	+	+	+	+	
Sweating	+	+	+	+	+	+	+	
Chills or hot flushes		+	+	+	+	+	+	
Derealization, depersonalization		+	+	+	+	+	+	
Nausea or abdominal distress			+	+	+	+		
Difficulty in concentrating							+	+
Difficulty of speaking							+	+
Desire to flee								+
Feeling confused								+

[a]Freud [1]
[b]Research diagnostic criteria
[c]American Psychiatric Association (APA)
[d]World Health Organization (WHO)
[e]Lactate-induced symptoms in panicking patients relative to non-panicking controls [53] assessed by a 21-item modified version of the Acute Panic Inventory of Dillon et al. [54]

methyl-phenylpiperazine). As well, Klein [24] suggested that more traumatic spontaneous panics stem primarily from mistaken signals of suffocation and are particularly benefited by selective serotonin (5-HT) reuptake inhibitors (SSRIs). Lastly, Klein [24] proposed that patients who panic during CO_2 inhalations have higher incidence of panic and that respiratory-type panics are both specific to PD and distinct from fear.

Evidence of a two-type panic attack may explain why some neurobiological studies on panic focus on midbrain's periaqueductal gray matter (PAG) [50–52, 60–62] while others emphasize the amygdala [63, 64], the hypothalamus [65, 66], or the locus coeruleus (LC) [67]. Moreover, whereas clinical studies emphasize respiratory-type panics, most preclinical models equate panic

to fear or, at the very best, to a proximal threat-induced fear [50, 52, 68].

Remarkably, however, Beitman and co-workers presented evidence of panic attacks devoid of fear [69–71]. In particular, Beitman et al. [69] showed that 12 of 38 cardiology patients with chest pain and current PD neither experienced intense fear, nor fear of dying, or of losing control, or of going crazy in their last major panic attacks [69]. Notably, as well, Fleet et al. [71] reported that 43 of 104 cardiology patients with pseudoangina were diagnosed as having PD (38 current, 5 past) as part of a prevalence study of panic in cardiology patients. Fleet et al. [70] also compared 48 'non-fearful panic disorder' (NFPD) patients with 60 PD patients and 333 controls at the time of admittance at emergency service and after a follow-up of approximately 2 years. Of note, a significantly greater proportion of PD patients had comorbid GAD and agoraphobia relative to NFPD patients. At follow-up, NFPD patients, like PD patients, were still symptomatic and had not improved or had even worsened according to scores on self-report measures [70]. Authors suggested that NFPD should be recognized as a variant of PD, both because of its high prevalence in medical settings and its poor prognosis.

In turn, the lack of increase in the secretion of 'stress hormones' differentiates clinical panic not only from fear [72] but also from stress [73] and simple and social phobias [74, 75]. Even more remarkably, recent studies reported that fear-unresponsive Urbach-Wiethe disease patients with extensive bilateral calcifications of the amygdala develop panic both spontaneously [76] and in response to a tidal volume inhalation of 35% CO_2 [77]. These data add compelling evidence that panic is not fear nor does it require the participation of the amygdala. By contrast, recent human studies gave strong support to the participation of ventromedial hypothalamus (VMH) in panic attacks [78].

Despite the above evidence, preclinical models very often overlooked the idiosyncratic features of clinical panic, being validated mostly pharmacologically [50, 51, 60, 68].

Because the participation of PAG in non-respiratory type of panic attacks was extensively reviewed elsewhere [51, 52], the following sections emphasize the role of the PAG and dorsomedial hypothalamus (DMH) in the mediation of respiratory-type and LAC-evoked panic attacks. This approach does not exclude the eventual participation of PAG-projecting neurons of the VMH in non-respiratory panic [78]. Neither does it exclude the participation of the amygdala, hippocampus and prefrontal cortex in comorbid anticipatory anxiety and agoraphobia.

2.3 Early Neuroanatomical Models of Panic Disorder

The first neuroanatomical model of panic was proposed by Gorman et al. [67]. This model states that panic attacks are the outcome of the increased activity of noradrenergic neurons of LC. Gorman's model was based on both the panicogenic properties of yohimbine (an α_2-adrenoceptor antagonist that increases the firing of LC) and the elicitation of panic-like behaviors by electrical stimulation of the LC, in humans and monkeys, respectively [79–81]. Moreover, whereas the exposure to CO_2 increased the LC activity [82], treatments with the α_2-adrenoceptor agonist clonidine and tricyclic panicolytics (imipramine, desipramine) reduced panic attacks [83, 84]. Further studies corroborated the CO_2 activation of LC neurons both *in vitro* [85] and *in vivo* [86]. The noradrenergic hypothesis of PD was also supported by studies suggesting that the panicolytic effect of imipramine is due to the desensitization of α-2 adrenoceptors [87, 88]. Moreover, Gorman et al. [67] proposed that the anticipatory anxiety of having a panic attack might be due to the 'kindling' of parahippocampal gyrus brought about by an increased activity of LC. The latter argument was supported by evidence showing that treatments with traditional benzodiazepines block anticipatory anxiety while are ineffective in panic attacks [89], that LAC-induced panics are heralded by the activation of parahippocampal gyrus [90, 91] and that anxiety appears to be mediated by the septo-hippocampal circuit (for review, see Gray and McNaughton [92]). Gorman et al. [67] further speculated that phobic avoidance (agoraphobia) is the outcome of the noradrenergic sensitization of learning mechanisms of prefrontal cortex. The latter argument was supported by data showing that PD patients that present avoidance-oriented coping strategies and fear of anxiety-related symptoms are more sensitive to anxiogenic effects of yohimbine [93]. The panic circuits of prefrontal cortex and hippocampus would in turn be the basis of the cognitive therapy of PD. Gorman's model is illustrated in Fig. 2.1a and b.

The noradrenergic theory of PD was severely criticized by researchers arguing that the LC mediates arousal rather than panic (*pros* and *cons* reviewed in [63, 67]). Most notably, electrical stimulations of the LC of humans that produced four- to ninefold increases in plasma levels of norepinephrine (NE) metabolites failed in eliciting any feeling of fear, anxiety, or discomfort [94, 95].

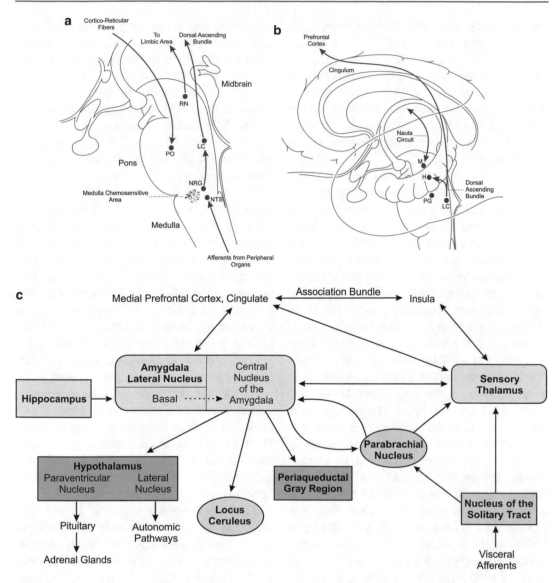

Fig. 2.1 Gorman's early (**a**, **b**) and revised (**c**) models of panic disorder. (**a**) Hypothesized mechanisms of panic attack onset. Gorman and collaborators emphasize the crucial role of both the chemosensitive areas of the medulla and the noradrenergic ascending projections of locus coeruleus (LC); (**b**) Hypothesized pathway of panic disorder anticipatory anxiety. Anticipatory anxiety would result from the noradrenergic sensitization ('kindling') of parahippocampal formation. Reciprocal connections of these circuits with prefrontal cortex would in turn be the basis of phobic avoidance; (**c**) In the revised version, the core structure is the amygdala (redrawn from Gorman et al. [63, 67])

At approximately the same time, the amygdala emerged as a crucial structure in brain processing of conditioned fear [96–98]. As a result, Gorman revised his theory shifting the focus from LC to the amygdala and its efferent connections to brainstem nuclei, including the LC itself [63] (Fig. 2.1). Accordingly, Gorman's revised model considered the brainstem activations and associated autonomic responses as simply an 'epiphenomenon' of the activation of the amygdala. Gorman's panic amygdala model was but profoundly discredited by recent studies showing that fear-unresponsive Urbach-Wiethe disease patients devoid of the amygdala develop panic attacks both spontaneously

[76] and in response to 35 % CO_2 [77]. Remarkably, as well, besides being observed in the three rare patients with Urbach-Wiethe disease, the panic attacks of patients were more intense than the only panic attack recorded in 16 healthy controls. Accordingly, the Feinstein et al. [77] suggested that panic is mediated 'at the brainstem' in spite of the established role of the amygdala in fear and anxiety of both humans and animals.

In the meanwhile, researchers become gradually aware that electrical and chemical stimulations of periaqueductal grey matter (PAG) produce aversive emotions in humans [99–101] and defensive responses in animals [102–104] that bear a striking resemblance to panic attacks. Moreover, the SSRIs had begun to be widely prescribed not only for the treatment of depression, but also of anxiety and panic. Accordingly, in the late 1980s Graeff first proposed that the PAG mediates panic attacks [105]. In a following study, Graeff further suggested that 5-HT had opposite actions in PAG (panicolytic) and amygdala (anxiogenic) [106]. As a consequence, in the early 1990s Deakin and Graeff [50] put forward a daring hypothesis according to which the brain ascending serotonergic systems evolved as a mechanism primarily concerned with adaptive responses to aversive events. Briefly, the Deakin/Graeff Hypothesis (DGH) suggested that whereas the GAD is produced by the overactivity of 5-HT excitatory projections from nucleus raphe dorsalis (NRD) to regions of the hypothalamus, amygdala and prefrontal cortex (PFC) that process distal/potential threats, the PD is the outcome of the malfunctioning of 5-HT inhibitory projections from NRD to dorsal regions of PAG (DPAG) that process defensive responses to innate fear, proximal threat and asphyxia (Fig. 2.2). As well, they proposed that whereas the NRD efferents to the amygdala facilitate avoidance from threat through the activation of 5-HT_2 receptors, NRD efferents to the DPAG restrain flight/fight responses through the activation of both 5-HT_{1A} and 5-HT_2 receptors (presumably, in output neurons and GABA interneurons, respectively). As a consequence, Deakin and Graeff [50] postulated that anxiety inhibits panic. As well, the DGH stated that the stress-induced increase in glucocorticoid secretion lead to depres-

sion by the impairment of 5-HT transmission at 'behavioral resilience system'. Specifically, the DGH states that depression is the outcome of a glucocorticoid-induced downregulation of 5HT_{1A} hippocampal receptors whereby the stressful events are uncoupled from daily life routines.

The DGH was largely supported by animal data on both the pharmacology of DPAG-evoked panic-like responses [51, 60] and the differential functions of 5-HT ascending systems and receptors [107, 108]. The DGH was also supported by positron emission tomography (PET) studies of the brain activations in volunteers exposed to a virtual predator which was otherwise able to inflict real shocks to the subject's finger [62]. In particular, the latter study showed that as the predator approaches the prey, brain activations shift from prefrontal cortex/basolateral amygdala circuit to central amygdala/periaqueductal grey circuit that are the presumptive mediators of anxiety and panic, respectively. The participation of 5-HT_2 receptors in anxiety was in turn endorsed by the rather specific localization of these receptors in ventral areas of the hippocampus [109] that receive dense inputs from basolateral amygdala (BLA) [110]. Despite the several mechanistic propositions that remain to be confirmed, the DGH provided a unified framework for the SSRIs' efficacy in multiple anxiety and affective disorders.

Plenty of evidence suggests, on the other hand, that panic is modulated by multiple transmitters, including not only 5-HT, but also gamma-aminobutyric acid (GABA), norepinephrine (NE), cholecystokinin (CCK), and opioids, among other candidates. Moreover, the putative panicolytic action of high-potency benzodiazepines (alprazolam and clonazepam) is compelling evidence of the prominent role of GABA-A/benzodiazepine (GABA/BZD) transmission. The involvement of the GABA-A/benzodiazepine receptor was also supported by data showing that the intravenous injection of the benzodiazepine antagonist flumazenil precipitates panic attacks in PD patients but not in healthy subjects [111]. Rather than the antagonism of a long-sought endogenous BZD, Nutt et al. [111] suggested that PD patients express a GABA/BZD receptor in which the flumazenil acts as an inverse agonist.

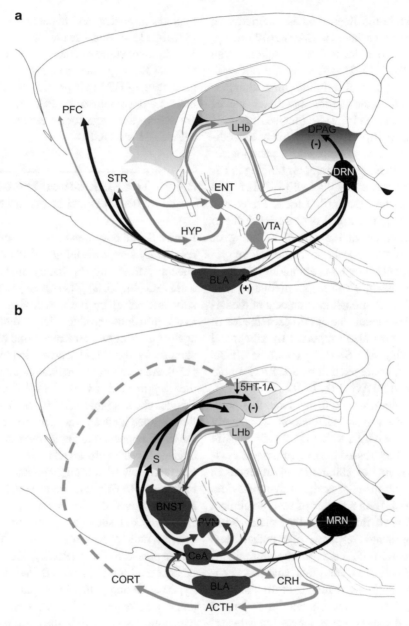

Fig. 2.2 The Deakin-Graeff Hypothesis. *Top panel*: In the early 1990s, Deakin and Graeff [50] suggested that whereas the generalized anxiety disorder is produced by the overactivity of 5-HT (*black arrows*) excitatory projections of dorsal raphe nucleus (DRN) to areas of prefrontal cortex (PFC) and basolateral amygdala (BLA) that process distal threat, panic attacks would be the result of a dysfunction of 5-HT inhibitory projections to dorsal regions of periaqueductal grey matter (DPAG) that process responses to proximal threat, innate fear, or hypoxia. They also proposed that conflict anxiety is the outcome of the simultaneous activation of 5-HT and dopamine projections (*yellow arrows*) from DRN and ventral tegmental area (VTA) that modulate striatum (STR) mechanisms that mediate avoidance and approach behaviors (STR), respectively. *Bottom panel*: Deakin and Graeff [50] also proposed that 5-HT efferents from the median raphe nucleus (MRN) to the hippocampus are the substrate of a behavior resilience system that uncouples stressful events from daily life routine. Stress-increased glucocorticoid plasma level downregulation of 5-HT$_{1A}$ receptors of the hippocampus would result in learned helplessness and depression. Other abbreviations: *BLA* basolateral amygdala, *BNST* bed nucleus of stria terminalis, *CeA* central amygdala, *CRH* corticotrophin releasing hormone, *ENT* nucleus entopeduncularis (internal pallidum); *HYP* hypothalamus, *LHb* lateral habenula; *PVN* paraventricular nucleus of hypothalamus, *S* septum (from Schenberg [259])

However, while the flumazenil is reportedly an agonist in some variants of GABA/BZD receptors, there is no evidence of its action as an inverse agonist [112, 113]. Conversely, however, other studies reported that flumazenil neither precipitates panic attacks, nor facilitates panics induced by LAC [114, 115] (but see reply of Potokar et al. [116]). The high-potency benzodiazepines could also compensate a pathological reduction in GABA/BZD receptor binding in PD patients as suggested by clinical [117–119] and preclinical studies [65, 66]. Lastly, Preter and Klein's [27] hypothesis that panics are the due to the malfunctioning of the opioid buffering of both the suffocation and the separation alarm systems was supported by LAC elicitation of hyperventilation in healthy volunteers pretreated with naloxone [120]. Although the subjects of the latter study did not panic, the opioid participation in panic attacks was also supported by preclinical studies showing that SSRIs' panicolytic effects may be due to an interaction of 5-HT and opioid transmission at DPAG [121, 122].

Whatever the transmitter involved, it is noteworthy that fear from suffocation was barely mentioned in DGH. As well, the role of early-life stress was not addressed in spite of the acknowledged importance of childhood environment in the late development of maladaptive behaviors [3, 11, 123–125]. As a matter of fact, preclinical evidence showed that neonatally-isolated adult rats show augmented responses of hypothalamus-pituitary-adrenal (HPA) axis to stressful stimuli [126–132]. In contrast, adult rats handled as neonates present augmented resilience to stress, as shown by the reduced neuroendocrine responses and increased expression glucocorticoid receptors of the hippocampus that mediate the feedback inhibition of HPA axis [133]. Notably, as well, it appears that decreases [134–136] or increases [133] in behavioral resilience are accompanied by parallel changes in both 5-HT$_{1A}$ receptor expression in the hippocampus and 5-HT turnover in frontal cortex and hippocampus, respectively. Recent data also showed that neonatally-isolated adult rats present facilitations of both panic responses to electrical stimulation of the PAG [137] and respiratory responses to

hypoxia (in males) and hypercapnia (in females) [37, 132, 138]. Because the HPA axis is hyperactive in neonatally-isolated adult rats, Dumont et al. [138] suggested that early-life stress programming of HPA axis predisposes the subject to the later development of PD. The latter studies are setting the stage for an eventual bridging of SFA, SAH and DGH.

2.4 The Suffocation False Alarm Theory: Facts and Questions

Klein's SFA theory was heavily based on both the respiratory symptomatology of PD and the panicogenic effects of respiratory metabolites CO_2 and LAC [24, 25, 27]. The theory was also markedly influenced by the paradoxical finding that LAC infusions produce hyperventilation [13] against a background of metabolic alkalosis presumptively due to LAC conversion to bicarbonate [139]. As a result, the metabolic alkalosis is further aggravated by the respiratory alkalosis. Besides showing that LAC-induced hyperventilation is mediated by mechanisms distinct from those of eupnea, the latter data suggested that LAC-induced panic attacks are secondary to a central hypercarbia brought about by bicarbonate breakdown to CO_2 and water [140]. Bicarbonate infusion proved this incorrect since only patients who panicked showed significant hyperventilation, as indicated by a precursor fall in arterial partial pressure of CO_2 ($PaCO_2$). Conversely, non-panickers showed increased $PaCO_2$ in spite of hyperventilating [140]. These data suggested that panic attacks cannot be ascribed to a central hypercarbia secondary to the peripheral increased level of bicarbonate. The eventual role of central hypercapnia in LAC-induced panics was nevertheless discredited by the unanticipated finding that d-LAC produces panic, hyperventilation and respiratory alkalosis in spite of not being metabolized [141] (note, however, that some studies showed that mammals are able to metabolize up to 75 % d-LAC 5 times slower than they do for l-LAC [142]). Taken together, the above findings suggested a crucial role of alkalosis, the common effect of d- and l-LAC infusions, bicarbonate infusion and hypoxia. To rule out the likely effect of alkalosis, Gorman et al. [12] carried out a study in which the panic patients were asked to hyperventilate a 5 % CO_2 mixture. Unexpectedly, however, patients panicked more during hypercarbia than in room air in spite of the lack of alkalosis in the former condition. Accordingly, in a preliminary presentation of SFA theory, Klein [25] suggested that the infusion of racemic LAC produces panic by two synergistic mechanisms, d-LAC producing a pseudo-hypoxia and l-LAC, both pseudo-hypoxia and central hypercapnia. Because the blood-brain barrier of humans is permeable to both LAC isomers [143, 144], Klein added that the exogenously administered LAC might also mimic the fast increase of brain LAC during asphyxia, thereby triggering a panic attack [24, 25]. These and other evidence led Klein [24, 25] to propose that clinical panic is due to the misfiring of an '*integrated suffocation monitor*' that detects both the increases in $PaCO_2$ and decreases in arterial partial pressure of O_2 (PaO_2), and changes in LAC and pH blood level as well. Klein [24] also suggested that the suffocation alarm system assess the level of muscular exertion while evaluating the LAC blood levels. Indeed, running not only fails in precipitating panic attacks [145] but also aborts panic [146], presumably, by providing countervailing information to the suffocation alarm system [24]. The existence of an integrated suffocation monitor was further supported by studies showing that PD patients are also hypersensitive to hypoxia [147].

However, while suffocation is most often understood as the prevention of air from entering the lungs, asphyxia is generally defined as the local or systemic deficiency of O_2 and excess of CO_2 in living tissues. Therefore, it remains unclear whether panic attacks are triggered by hypoxia or hypercapnia alone, or require both conditions of asphyxia ('integrated suffocation monitor'). Indeed, even though the high-altitude illness lacks panic symptoms [148], high-altitude travelers eventually report limited-symptom panic attacks during sleeping when hypoxia is further aggravated by hypercapnia [149]. Additionally, it is unclear whether suffocation and asphyxia necessarily elicits 'hunger for air' and 'desire to flee'.

For instance, it is a well known fact that tissue hypoxia by inhalation of carbon monoxide (CO) (the 'silent killer') is devoid of both feelings of asphyxia and desire to escape.

Epidemiological studies suggest, on the other hand, that panic is influenced by both genes and environment. Although there is a fairly established consensus that early-life stress [150, 151] and CSA [10, 24, 27, 36, 37, 42, 152] predisposes the subject to PD, existing evidence is nevertheless mixed, either corroborating [153] or disproving [41] the SAH.

Epidemiological studies also showed that panics are highly comorbid with depressive disorders and twice more frequent in women [154]. That the higher incidence of panic in women is due to sex hormones is shown by the reduced frequency of panic attacks either before puberty or after menopause. Panic attacks are also more frequent and severe during the late luteal (premenstrual) phase (the sex-dependent features of panic were reviewed by Lovick [155]). Yet, there are few animal studies addressing the comorbidity of panic with other psychiatric disorders.

2.5   Face Validity of Dorsal Periaqueductal Grey Matter Mediation of Panic Attacks in Humans

The pioneering studies of Nashold et al. [99, 156] provided the best description of feelings and sensations produced by electrical stimulations of human PAG. These studies were carried out in 12 subjects without a history of psychiatric disorders who were chronically implanted with 34 electrodes into the dorsal midbrain. Data showed that electrical stimulations of the PAG produce anxiety, panic, terror and feelings of imminent death, along with marked visceral responses that are a faithful reproduction of the core symptoms of panic attacks. In particular, Nashold et al. [156] reported that electrical stimulations "*in or near the lateral edge of the central grey resulted in strong reactions in most patients. Feelings of fear and death were often expressed. Autonomic activation such as the contralateral piloerection and sweating, increase in the pulse and respiratory rate, blushing of the entire face and neck,...were also noted'*. At higher intensities, a subjected reported that fear "*was so unpleasant that she was not willing to tolerate repeated stimulations*" [99]. Even the most peculiar neurological symptoms of panic, such as tremor and chest pain, were produced by the PAG stimulations. These observations were in line with prior studies reporting that patients stimulated in the medial sites of the mesencephalic tectum complained of a "*choking feeling referred into the chest*" (in Spiegel and Wycis, 1962, as cited by Nashold et al. [99]) which is a cardinal symptom of panic attacks [53]. As confirmed by X-ray analyses, sites effective in producing these symptoms were found into the PAG but not in the nearby tegmentum 5 mm beyond the aqueduct [99]. Ever since the panic-like effects of PAG stimulations were confirmed by several groups [100, 101, 157]. In particular, Amano et al. [100] reported that during a surgery for pain relief, a patient stimulated in the PAG uttered '*somebody is now chasing me, I am trying to escape from him*'. In agreement with the latter observation, the severity of LAC-induced panic is robustly correlated with the 'desire to flee' [53]. The isomorphism of the PAG-evoked responses and human panic attacks, either spontaneous or induced by LAC, is shown in Table 2.2.

2.6 The Periaqueductal Grey Matter of the Midbrain: A Likely Substrate of Respiratory-Type Panic Attacks

In humans, the sites which stimulation provoked panic reactions were localized within 5 mm from the aqueduct [99, 101] into the dorsal half of the PAG [156]. However, the PAG is far from homogeneous. Traditionally, this structure is regarded as four functionally specialised columns extending for varying distances along the aqueduct, namely, the dorsomedial (DMPAG), dorsolateral (DLPAG), lateral (LPAG) and ventrolateral (VLPAG) columns [158–163]. The abundant

intrinsic connections suggest, on the other hand, that columns are not independent [164]. Indeed, existing evidence suggests that rat defensive behaviors are controlled by a concerted though differentiated activity of DMPAG, DLPAG and LPAG generally conflated as the 'dorsal PAG' (DPAG) [162, 165]. On the other hand, it is uncertain whether the PAG columns show further specializations along their length. For instance, the caudal regions of DPAG appear to present increased sensitiveness to agonists of N-methyl-D-aspartic acid receptor [166]. Moreover, whereas the rostral LPAG (LPAGr) does most connections with forebrain [167], the caudal LPAG does it with the medulla [162, 168]. Additionally, the VLPAG harbor multiple transmitter nuclei in spite of its shorter length. It is worth noting that the PAG is also parcelled out according to a cell density-increasing gradient in juxtaqueductal, medial and peripheral regions [162].

The traditional subdivision of the PAG was revised based on the immunocytochemistry for acetylcholinesterase (AChE), NADPH diaphorase (NADPHd), tyrosine hydroxylase (TH) and 5-HT, among other markers [169]. Besides corroborating the NADPHd specific staining of DLPAG [170], Ruiz-Torner et al. [169] showed that the LPAG differs from the VLPAG in that the former is markedly stained for AChE. Additionally, they showed that the medial sectors of LPAG and VLPAG makes up a single neurochemical entity herein termed the central PAG (CePAG). As well, Ruiz-Torner et al. [169] proposed that the medial VLPAG displaces the LPAG at caudalmost levels of PAG (−8 mm from bregma). As a result, whereas the caudal regions of medial VLPAG occupy the entire expanse of former LPAG, the caudalmost region of the LPAG is now in a dorsolateral region neighbour to the remnants of DLPAG. Evidence below discussed suggests that the CePAG and, perhaps, LPAGr, are crucial mediators of respiratory-type panic attacks. As a matter of fact, the CePAG is the only PAG region reportedly targeted by afferents from nucleus tractus solitarius (NTS) [171, 172].

It is also noteworthy that although the caudalmost sectors of VLPAG are believed to play

Table 2.2 Isomorphism of panic attacks in humans and stimulation of dorsal periaqueductal gray matter of both humans and rats

	Spontaneous or lactate-induced panic attacks in humans	Stimulation of dorsal periaqueductal gray matter in humans	Stimulation of dorsal periaqueductal gray matter in rats
Feelings/behaviour	Severe anxiety, panic, terror	Fearful, frightful, terrible	Strong aversion
	Intense discomfort	Intense discomfort	–
	Desire to flee	Beg to switch-off stimulus	Flight behaviour
	"Block while walking"	n.a.	Freezing
	Difficulty in doing job	n.a.	Freezing
	Choking	Choking	–
	Fear of dying	"Scare to death"	–
	Difficulty in speaking	Speech arrest	–
	Fear of going crazy	n.a.	–
	Fear of losing control	n.a.	–
	Feeling confused	n.a.	–
	Dizziness/lightheadness	n.a.	–
	Difficulty in concentration	n.a.	–
	Faintness	n.a.	–
Autonomic responses	Dyspnoea	Dyspnoea, apnoea	Dyspnoea
	Tachypnoea	Tachypnoea	Tachypnoea
	Hyperventilation	Hyperventilation	Hyperventilation
	Difficulty in deep breathing	Sighs, deep breaths	Deep breaths
	Palpitation	Palpitation	Tachycardia
	Hypertension	n.a.	Hypertension
	Sweating	Sweating contralateral	n.a.
	–	Wide-open eyes	Exophthalmus
	–	Bladder voiding urge	Micturition, defecation
	–	Blush in face	–
	–	Piloerection contralateral	n.o.
Endocrine responses	No prolactin	n.a.	No prolactin
	No corticotrophin	n.a.	No corticotrophin
	No cortisol	n.a.	No corticosterone
Neurological responses/paresthesias	Sensation of tremor	Sensation of vibration	–
	Chest pain	Chest and heart pain	–
	Numbness	Numbness	–
	Chills or hot flushes	"Burn/cold" sensations	–
	n.a.	Face pain	–
Brain areas activated	Dorsal periaqueductal gray, deep layers of superior colliculus, amygdala (PET)	Dorsal periaqueductal gray and adjacent tectum (0–5 mm lateral to aqueduct) (X-rays)	Dorsal periaqueductal gray, deep layers of superior colliculus

Abbreviations: *n.a.* not available, *n.o.* not observed (see text for references)

a crucial role in conditioned freezing to a stimulus previously paired to a shock [96, 97, 173], the low-magnitude stimulation of VLPAG and/or NRD appear to inhibit the DPAG [174–179], being even rewarding in both rats [180] and humans [101]. As discussed in following sections, plenty of evidence suggests that panic attacks may be the result from the VLPAG deficient inhibition of PAG output neurons.

2.7 Neuroanatomical Basis of Respiratory-Type Panic Attacks

Breathing is controlled by a widespread network involving structures ranging from caudal medulla to anterior hypothalamus, limbic system and cortex [181]. In caudal medulla, the nucleus tractus solitarius (NTS) is the recipient of glossopharyngeal afferents conveying blood gas signals from peripheral chemoreceptors [182]. The respiratory drive is also critically dependent on the chemosensitive neurons of the ventral surface of the medulla which monitor pH and CO_2 tissue levels [183–186]. The NTS second-order neurons project both to ventral respiratory nuclei of the medulla that control phrenic nerve activity [181] and to suprapontine nuclei that control behavioral and endocrine responses, including the central amygdala (CeA), the paraventricular nucleus of the hypothalamus (PVN) [187] and the CePAG [171, 172]. Projections to the hypothalamus, limbic structures and cortex also appear to be involved in thermoregulatory, emotional, and volitional aspects of respiration, respectively [181]. Peripheral and central chemoreceptors act in concert to maintain the blood PaO_2 and $PaCO_2$ at appropriate levels (i.e., 95 mmHg and 40 mmHg, respectively).

2.7.1 Psychophysics of Asphyxia

Although the blood/brain gases are continuously monitored by both peripheral and central chemoreceptors, panic attacks occur in room-air conditions, being very rare in healthy people. Accordingly, the SFA theory presumes that the suffocation alarm system and chemoreceptors may operate separately under specific circumstances. The eventual uncoupling of these systems might give way to panic attacks either spontaneous or induced by chemicals such as LAC, CO_2, CCK, β-carbolines and others. These considerations raise the question about the conditions whereby the suffocation alarm system is uncoupled from chemoreceptors.

Although the physiology teaches us that 'man is not endowed with any sensory perception of hypoxia that might alert him to impending danger' [188], Moosavi et al. [189] showed that hypercapnia and hypoxia are equally potent in producing hunger for air in healthy subjects matched by ventilatory drive (i.e., at fixed respiratory rate, tidal volume and CO_2 blood level). Similarly, Beck et al. [147] showed that PD patients have increased sensitivity to both hypercapnia and hypoxia relative to healthy controls. These data are consonant with an integrated suffocation alarm system endowed with the capability to translate major respiratory inputs (H^+/CO_2, PaO_2) and, possibly, LAC blood levels, into panic attacks [24, 27, 42].

Because dyspnea (uncomfortable and/or irregular breathing) is the leading symptom of clinical panic [53], the SFA theory predicts that panic attack effectors might also be physiologically activated by degrees of hypoxia and/or hypercapnia that give rise to manifest feelings of breathlessness or hunger for air in both patients and volunteers. In particular, mechanically ventilated humans report hunger for air at CO_2 alveolar partial pressure ($PACO_2$) of 43 mmHg, i.e., only 4 mmHg above the normal CO_2 end-tidal partial pressure ($P_{ET}CO_2$). In turn, the $P_{ET}CO_2$ of 50 mmHg produces unbearable hunger for air [190].

Although the human low tolerance to increases in $PACO_2$ endorses the key role of CO_2 in the elicitation of feelings of suffocation and, probably, clinical panic, Banzett et al. [190] showed that the subjective symptoms of dyspnea do not correlate with ventilatory responses to CO_2. Moreover, whereas the hypoxia ($P_{ET}O_2$ of 60–75 mmHg) reduced CO_2 sensitivity and tolerance by only 2 mmHg, hyperventilation produced robust reductions in hunger for air at any $P_{ET}CO_2$, increasing CO_2 tolerance threshold by 5 mmHg while decreasing CO_2 sensitivity by 50 % [190]. In particular, mechanically-driven hyperventilation reduced hunger for air even when patients hyperventilate in isocapnic conditions. Studies carried out with subjects either curarised or quadriplegic showed, on the other hand, that pulmonary mechanoreceptors play a minor role in respiration-induced decrease of hunger for air [191, 192]. These latter data support the central mediation of the breathing-induced reduction in hunger for air.

This mechanism could be the basis of the traditional practice of treating panic attacks by having the patients rebreathing into a brown paper bag. Although this technique was formerly explained as a compensation of panic-evoked respiratory alkalosis, other studies showed that the eventual benefits are secondary to patient's distraction [13] and/or expectation [193].

Although the dissociation of hunger for air and ventilatory responses to CO_2 [190] suggests the relative independence of underlying mechanisms, the cognitive (cortical) hypotheses state that feelings of hunger for air arise from the forebrain processing of 'corollary discharges' (or 'efferent copies') of respiratory motor neuron activity [191, 194]. If so, the intensity of hunger for air should parallel the respiratory responses to both hypoxia and hypercapnia. Because sustained hypoxia produces a sharp ventilatory increase followed by a slow decay, Moosavi et al. [194] tested the 'corollary discharge' hypothesis by comparing the time course of hunger for air during isocapnic hypoxia in either freely-breathing or mechanically-ventilated healthy subjects. As predicted, during sustained hypoxia, there was a sharp increase in hunger for air followed by a progressive decline, mirroring the biphasic pattern of respiratory response. Otherwise, the parallel time course of respiration and dyspnea could be explained by the parallel processing of chemoreceptor inputs at respiratory centers of the medulla and suffocation alarm circuits elsewhere in the brain, respectively.

As well, the 'corollary discharge' hypothesis was tested in patients with congenital central hypoventilation syndrome (CCHS) [195]. Spengler et al. [195] took advantage that CCHS patients hyperventilate during exercise while lacking CO_2-evoked respiratory (hyperventilation) and subjective (hunger for air) responses. The 'corollary discharge' hypothesis predicts that CCHS patients would experience normal levels of 'shortness of breath' during constant-load exhausting cycling. Conversely, however, CCHS patients showed reduced breathlessness relative to controls. Thus, whereas the hyperventilation of CCHS patients was reduced in only 37 % relative to controls, the hunger for air was reduced in 75 % as shown by the area under the curve of hyperventilation and hunger for air [195]. However, CCHS patients who exercised heavily and increased respiratory activity as much as controls reported '*some sensation akin to shortness of breath*'. Therefore, Spengler et al. [195] concluded that this study '*failed to disprove the corollary discharge hypothesis*'. An alternative interpretation suggests, however, that both 'hunger for air' to hypercapnia and 'shortness of breath' to exercise depend on the activation of a CO_2-sensitive suffocation alarm system. In the same vein, CCHS patients reported no respiratory discomfort during either the CO_2 inhalation or the maximum breath hold which was significantly longer than that of controls [196]. Taken together, these studies make unlikely the 'corollary discharge' hypothesis.

Although Banzett et al. [191, 197] proposed that feelings of hunger for air are processed at primary interoceptive posterior insular cortex (PIC) [198, 199], recent studies in rats showed that neither the escape nor the cardiovascular responses to severe hypoxia (8 % O_2) were attenuated following chemical inactivations of PIC (i.e., the dorsal bank of perirhinal cortex) [200]. Casanova et al. [200] showed in addition that whereas the 8 % hypoxia did not change c-fos immunoreactivity in PIC, it produced robust labeling in both NTS and DPAG.

Although Klein [24] suggested that the suffocation alarm system might be activated by toxic gases as well, exposures to CO neither activate the peripheral chemoreceptors nor produce the panic responses. Peripheral chemoreceptors are actually inhibited by low concentrations of CO [201]. Although the lack of defensive responses to CO (the 'silent killer') supports the crucial role of peripheral chemoreceptors in behavioral responses to both hypoxia and hypercapnia, existing evidence suggests that PD patients present normal thresholds and sensitivities of central and peripheral chemoreceptors [202]. Accordingly, the presumptive abnormality of the suffocation alarm system appears to reside in supramedullary structures that do not participate in the respiratory reflexes of eupnea. The brain defence structures deployed along the brainstem axis are thus in a strategic position to mediate a panic attack.

2.7.2 Neuroimaging 'Hunger for Air' in Humans

Despite the low spatial resolution of PET studies of hypercapnia, these studies provide valuable information regarding the major structures of the suffocation alarm system of humans. To increase spatial resolution, PET studies are often performed with co-registration of magnetic resonance imaging (MRI) of suprapontine structures of healthy volunteers that presented definite feelings of suffocation, breathlessness or hunger for air during the inhalation of CO_2 [203–205]. To study the central activations to hypercapnia, Brannan et al. [204] compared the PET scans of hypercapnia (8 % CO_2/92 % O_2) to those of hyperoxia (9 % N_2/91 % O_2) in which the carotid chemoreceptor peripheral inputs are suppressed. Compared to hyperoxia, CO_2 produced the specific activation of a vast array of cortical or subcortical structures in one or both sides of the brain. In particular, CO_2 produced significant increases in the regional cerebral blood flow (rCBF) in the pons, PAG, midbrain tegmentum, hypothalamus, amygdala, sublenticular region, hippocampus (BA28) and parahippocampal areas (BA20/36), subgenual anterior cingulate cortex (sgACC, BA24), fusiform gyrus (BA19/37), middle temporal gyrus (BA9), anterior insular cortex (AIC), pulvinar, putamen/caudatum, and several sites in the midline and lateral cerebellum. In addition, the activated areas possibly involved the LC, the parabrachial area (PBA), and NRD. Conversely, there were significant deactivations in the dorsal anterior cingulated cortex (dACC, BA24), posterior cingulate cortex (PCC, BA23/31) and lateral prefrontal cortex (PFC, BA9/46).

Structures activated during both LAC- and CCK-induced panic attacks were but a small subset of the network associated with breathlessness and hunger for air in hypercapnia [205, 206]. Most notably, the only regions activated in both hypercapnia and LAC-induced panic attacks were the sgACC, the insula, the hypothalamus, the midline cerebellum, the deep layers of superior colliculus (DLSC) and, possibly, DPAG and LC. Remarkably, the temporopolar cortex was activated by both the LAC

infusion [206] and the CO_2 inhalation [205]. The employment of high-resolution MRI coupled with PET in the latter study makes unlikely the mistaken activation of temporopolar cortex due to changes in extracranial blood flow reported in CCK-induced panic attacks [207–209]. However, whereas the insula/claustrum were regularly activated in LAC- and CCK-induced panic attacks, the temporopolar cortex and parahippocampal gyrus were also activated during anticipatory anxiety in both volunteers and panickers [210]. Accordingly, the activation of the temporopolar cortex is most probably due to anticipatory anxiety and not panic itself.

Recently, Goossens et al. [211] examined the CO_2-evoked brain activations in panic patients, healthy volunteers and experienced divers showing decreasing sensitivity to CO_2. Patients exposed 2 min to hyperoxic hypercapnia (7 % CO_2/ 93 % O_2) showed increased brainstem activations relative to both controls and divers in spite of showing the same $P_{ET}CO_2$ and respiratory rate (changes in tidal volume were not reported). Data also showed that although the PIC activations did not differ significantly, AIC activations correlated with subjective feelings of dyspnea (breathing discomfort), being more intense in patients. Authors also suggested that the lack of differences in PIC was due to the exposure of all groups to the same gas mixture. The differential activation of AIC in PD patients relative to the other groups adds further evidence of the key role of this structure in conscious awareness of the physiological state our body [198, 199]. Of note, the employment of a hyperoxic mixture makes unlikely the contribution of peripheral chemoreceptors to CO_2 activations of above areas.

Although Goossens et al. [211] did not find significant activations in either the PAG or the LC of subjects exposed to 2-min 7 % hypercapnia, Teppema et al. [86] reported marked activations in VLPAG of rats exposed to 2-h 15 % hypercapnia. Because VLPAG activations were not attenuated by hyperoxia (60 % O_2/15 % CO_2), the peripheral chemoreceptors seems not involved as well. The lack of activation of VLPAG of humans exposed to hypercapnia [211] should be ascribed to either the short exposure to CO_2 (2 min) or the inherent difficulty of brainstem scans.

Remarkably, however, a pontine region compatible with nucleus raphe magnus (NRM) was markedly activated in both rats and humans [86, 211]. These findings are consonant with a recent study that employed engineered chemogenetic silencing allele to target the expression of a DREADD (*designer receptor exclusively activated by designer drug*) inhibitory receptor in mouse embryo rhombomeres that give rise to 5-HT specific nuclei of the brainstem [212]. Results showed that the full development of the ventilatory response to hypercapnia in conscious mice requires a NRM specific cell lineage that is sensitive to small changes in $PaCO_2/pH$. Brust et al. [212] showed in addition that although these cells contribute to CO_2 chemosensitivity, they do not participate in respiratory motor responses. It is nevertheless noteworthy that stimulations of NRM of rats failed in producing panic-like behavioral and cardiorespiratory responses [213, 214]. Much the opposite, chemical stimulations of both NRM and nucleus raphe obscurus (NRO) produced long-lasting inhibitions of cardiovascular (NRM, NRO) and respiratory (NRO) responses elicited by electrical stimulation of DLPAG of the anesthetized rat [214]. Interestingly, as well, whereas the VLPAG receives most 5-HT afferents from NRM, the DPAG does it from NRD, NRO, nucleus raphe pontis (NRPo), and nucleus raphe medianus (NRMn) [215]. The CO_2 activations of NRM could in turn modulate the DPAG-projecting neurons of VLPAG as discussed in following sessions.

2.7.3   c-Fos Immunohistochemistry Studies Reveal a Limited Number of Structures Activated by Hypoxia or Hypercapnia

Because the PET signal is a function of rCBF, the vasodilator effect of both LAC and CO_2 [63, 203] is a major drawback of these studies. In fact, the only PET study of a spontaneous panic attack (a case report) revealed only deactivations in right orbitofrontal (BA11), anterior temporal (BA15), anterior cingulate (BA32) and prelimbic (BA25) cortices [216]. Conversely, the c-Fos protein expression is independent from rCBF. Accordingly, c-Fos immunohistochemistry studies in animals exposed either to hypoxia or to hypercapnia might be a reliable control of PET studies in humans. Unfortunately, however, whereas the PET human studies focused mostly on midbrain and prosencephalon, c-Fos animal studies focused on pontine and medullary regions [86, 217–220]. Yet, some studies examined c-Fos expression throughout the midbrain and diencephalon of both rats and cats [221–223]. A second difficulty concerns the different protocols of PET and c-Fos studies in humans and animals, respectively. In particular, while the PET imaging studies employed short exposures (20 min) to moderate degrees of hypercapnia (8 %), c-Fos immunochemical studies employed long exposures (hours) to moderate hypercapnia (15 %) or varying degrees of hypoxia (5–11 %) [86, 221, 222]. However, Johnson et al. [223] examined the c-fos expression throughout the brain of rats exposed to the fast raise of CO_2 (up to 20 % CO_2 in 5 min). Furthermore, Teppema et al. [86] evaluated the contribution of peripheral chemoreceptors in c-fos expression to hypercapnia (15 % CO_2) by adding surplus oxygen (60 % O_2) in gas mixture (hyperoxic hypercapnia).

In any event, data showed that prolonged exposures to severe degrees of hypoxia (8 % O_2) produced significant increases in fos-like immunoreactivity (FLI) of medullary nuclei that play a crucial role in eupnea, including both the commissural and caudal NTS, ventral medullary surface (VMS) and parapyramidal area (PPy) [86, 222] (Table 2.3, Fig. 2.3). Teppema et al. [86] reported additional labeling in both A1 and C1 cathecolaminergic groups. Double labeling for c-fos and TH was mostly found in the rostral and caudal ventrolateral reticular nuclei, but not in the lateral paragigantocellular nucleus (LPGi). The NTS was also activated by 5 % and 15 % CO_2 [86, 222]. Following the exposure to 15 % CO_2, the number of c-fos labelled cells in the rostroventrolateral medulla (RVLM) was nearly 2.5 times the number of labeled cells following the exposure to 9 % O_2, corroborating the higher sensitivity of RVLM neurons to hypercapnia [224]. Most importantly, addition of 60 % O_2 produced robust

Table 2.3 Metanalysis of brain c-fos protein expression of rats exposed to hypoxia, hypercapnia, hyperoxic hypercapnia or tissue hypoxia

Brain level	Exposure	Hypercapnia					Hypoxia		Hyperoxia effects on c-fos expression to hypercapnia	Tissue Hypoxia
		Berquin et al. [222]	Teppema et al. [86]			Johnson et al. [223]	Berquin et al. [222]	Teppema et al. [86]	Teppema et al. [86]	Bodineau et al. [235]
		3 h	2 h			5 min	3 h	2 h	2 h	5 min
	[gás]	5% CO_2	8% CO_2	10% CO_2	15% CO_2	20% CO_2	11% O_2	9% O_2	15% CO_2 + 60% O_2	1% CO
Medulla	cNTS[a]	110	(240)	(120)	1050	–	560	620	–44	389
	iNTS	–	(389)	(344)	1482	–	–	1278	–37	415
	rNTS	–	–	–	–	–	300	–	–	–
	DMX	–	–	–	–	–	–	–	–	275
	AP	–	–	–	–	–	–	–	–	74
	GiA	–	–	1067	633	–	–	(433)	–	–
	LPGi	–	550	550	1556	142	–	–	–31	212
	cVLM	–	–	378	–	–	–	480	(–60)	585
	iVLM	–	–	520	1120	–	–	620	–61	–
	rVLM	–	–	620	2200	650	–	828	–60	292
	RPa	–	192	167	–	–	–	–	–	218
	PPy	300	–	–	–	–	400	–	–	197
Pons	CnF	–	–	87	–	–	–	–	–	–
	A5	–	300	800	550	–	–	550	–	–
	LC	114	767	1233	1433	–	(41)	–	–	137
	LPBA	148	2600	3013	3180	(208)	191	1400	–	149

Midbrain	LDTg	–	–	–	–	–	–	107
	NRD	457	–	457	206	–	–	66
	DMPAG	–	–	–	–	460	–	–
	DLPAG	(97)	–	–	–	300	116	–
	VLPAG	169	–	142	115	279	150	111
Hypothalamus	SUM	–	–	–	–	–	–	157
	PH	179	–	–	–	150	150	129
	PMD	–	–	–	100	–	–	–
	DMH	180	–	–	47	47	120	85
	PeF	–	–	–	31	31	–	152
	PVN	207	–	–	65	65	–	–
	SON	(460)	–	–	–	–	(300)	–
	VHZ	335	–	–	–	250	250	186

Original data were transformed to percent changes relative to air breathing rats of each study. In hyperoxic hypercapnia numbers represent changes relative to CO_2 alone. Values between parentheses are changes that although noticeable did not reach statistical significance. Empty cells represent data either non-significant or not available. Note that the 2 h-exposure to 8 % and 10 % CO_2 produced marked c-fos increases in LPBA in spite of the non-significant changes in NTS

[a]*Abbreviations:* *c* caudal, *i* intermediate, *r* rostral, *A5* caudolateral pons noradrenergic group, *AP* area postrema, *CnF* cuneiform nucleus, *NTS* nucleus tractus solitarius, *VLM* ventrolateral medulla, *DLPAG DMPAG VLPAG* dorsolateral, dorsomedial and ventrolateral periaqueductal grey matter, *DMH* dorsomedial hypothalamus, *DMX* dorsal motor nucleus of vagus, *NRD* nucleus raphe dorsalis, *GiA* gigantocellular reticular area, *LC* locus coeruleus, *LDTg* laterodorsal tegmental nucleus, *LPBA* lateral parabrachial area, *LPGi* lateral paragigantocellular nucleus, *PeF* perifornical hypothalamus, *PH* posterior hypothalamus, *PMD* premammillary dorsal nucleus, *PPy* parapyramidal area, *PVN* paraventricular nucleus of the hypothalamus, *RPa* raphe pallidus, *SON* supraoptic nucleus, *SUM* supramammillar nucleus, *VHZ* ventral hypothalamic zone

reductions in the number of CO_2-activated cells in NTS, RVLM and LPGi, but not in VLPAG, LC and PBA [86] (Table 2.3). The latter data suggest that CO_2 activations of VLPAG, LC and, most surprisingly, PBA do not depend on peripheral chemoreceptors. Similarly, the carotid denervation that suppressed the hypoxia activation of DPAG had no effects in hypoxia activation of VLPAG [225]. Prolonged exposures to hypercapnia also labeled a superficial column of cells of retrotrapezoid nucleus in the VMS (0–100 μm depth) [86]. These neurons may correspond either to the paraolivary cell group described by Berquin et al. [222] or to the VMS sensors of H^+/CO_2 described by Loeschcke and Koepchen [183]. Indeed, Loeschcke [184] noted that pH changes in the VMS of cats produce '*maximal drive of ventilation ... in two areas, one medial to vagal root and the other medial to the hypoglossal root but lateral to the pyramid*'. Surprisingly, however, neither the pre-Bötzinger complex (pBöC) nor the ventrolateral medulla (VLM), the areas currently proposed to play a critical role in the neurogenesis of respiratory rhythm [226], were activated by 15 % hypercapnia [86]. In the dorsolateral pons, exposures to either hypercapnia or hypoxia activated both the external division of the lateral parabrachial area (eLPBA) and the Kölliker-Fuse nucleus (KF) [86, 222]. In contrast, hypoxia did not affect c-Fos expression of medullary areas that do not participate in respiratory and autonomic functions.

Remarkably, midbrain structures not belonging to the established respiratory network also showed significant increases in FLI (Fig. 2.3). In particular, the DLPAG, VLPAG, LC and nucleus subcoeruleus were intensely labelled by moderate degrees of either hypoxia or hypercapnia [86, 222]. In contrast, the cuneiform nucleus (CnF) was inconsistently labeled in spite of being the main target of DLPAG (the CnF showed moderate increases in FLI to both 10 % CO_2 and 15 % $CO_2/60 \% O_2$, but no FLI increase to 15 % CO_2) [86] (Table 2.3, Fig. 2.3).

Teppema et al. [86] showed that VLPAG and PBA were markedly activated by the prolonged exposure to both 10 % and 15 % CO_2, being even more marked than the caudal NTS and RVLM (Fig. 2.3). In fact, the PBA showed the highest increase in c-fos expression as compared to air-breathing rats. Because these activations were not attenuated by hyperoxia, they do not appear to depend on the peripheral input from carotid chemoreceptors. Otherwise, the PBA activations could be due to the hyperventilation increase in the afferent input from pulmonary stretch receptors [181]. Of note, LC fos-labeled neurons of rats exposed to hypercapnia were also double-stained for TH [86]. These findings agree with *in vitro* experiments showing that LC neurons are highly sensitive to CO_2 [85]. Conversely, the evidence of LC activation by hypoxia is contradictory, the activations being either manifest [82, 218] or hardly observed [86, 222]. Moreover, Elam et al. [82] reported that whereas the LC responses to hypoxia were suppressed by peripheral denervation, those to hypercapnia were unchanged. The latter data support the central mediation of the hypercapnia activation of LC.

Although the above studies suggest that the suffocation alarm system lies somewhere between the NTS and LC or PAG, Berquin et al. [222] found significant FLI increases rostrally in DMH, posterior hypothalamus (PH), and along a ventral hypothalamic zone (VHZ) extending from the mammillary nuclei to the retrochiasmatic area. Even though not significant, hypoxia and hypercapnia also produced 184 % and 460 % increases in FLI of PVN and supraoptic nucleus (SON), respectively. The PVN activation is most likely involved in HPA axis response to hypercapnia in both rats [227] and humans [228–230]. Labeling of DMH is particularly important since this nucleus has been proposed to mediate the LAC vulnerability of panic patients [66, 231] (see Sect. 2.9).

In turn, the PH was activated by both hypoxia and hypercapnia [222]. Remarkably, as well, the PH harbors PAG-projecting neurons that are intrinsically sensitive to both hypoxia and hypercapnia [232]. Accordingly, Ryan and Waldrop [232] suggested that the PAG and PBA are the key relay structures connecting the PH with medullary nuclei involved in the ventilatory response to both hypoxia and hypercapnia.

Because burrowing species (Israeli mole rats) can tolerate extreme degrees of hypoxia (7.2 % O_2) and hypercapnia (6.1 % CO_2) in their natural

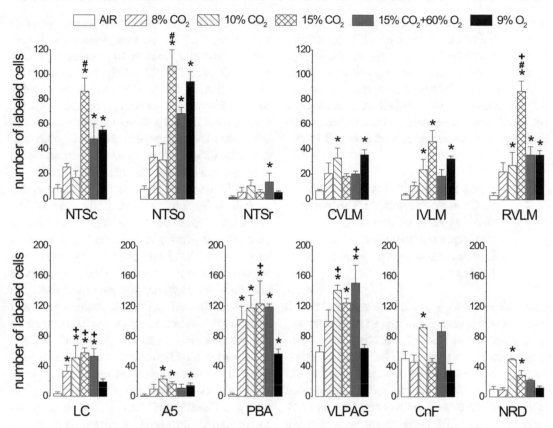

Fig. 2.3 Brainstem areas activated in hypercapnia, hypoxia and hypercapnic hypoxia. Columns represent the mean number of c-fos labeled cells (±SD) per 60-μm hemisections of the braintem of rats subjected to 2-h exposures to room air; 8 %, 10 % and 15 % CO_2; 15 % CO_2 plus 60 % O_2; and 9 % O_2. Although these data are not represented as density of labeled cells, it is noteworthy that the number of c-fos labeled cells of hypercapnic rats in ventrolateral periaqueductal grey (VLPAG) and parabrachial area (PBA) was higher than that of nucleus of tractus solitarius (NTS) and rostroventral lateral medulla (RVLM). The central origin of CO_2 activations of VLPAG and PBA is suggested by the lack of difference between hypercapnia (15 % CO_2) and hyperoxic hypercapnia (15 % CO_2 plus 60 % O_2). Other abbreviations: A5 caudolateral pons noradrenergic cell group, CVLM caudal ventralateral medulla, CUN cuneiform nucleus, NRD nucleus raphe dorsalis, IVLM intermediate ventrolateral medulla, LC locus coeruleus Symbols indicate significant differences (*) from room air; (#) from hyperoxic hypercapnia or (+) from 9 % hypoxia at Bonferroni's 5 % descriptive level (redrawn from Figs. 2, 4 and 8 of Teppema et al. [86])

habitat [233], it is noteworthy that brain activations were quite similar in both rats [222] and cats [221, 234]. In fact, the only difference between these species was the lack of labelling in the retrotrapezoid nucleus of hypoxic rats. The latter observation is nevertheless consonant with the lack of response of VMS neurons to hypoxia [184].

More recently, Johnson et al. [223] examined the c-fos protein expression of selected brain regions of forebrain and brainstem of rats exposed to a sharp increase in CO_2 (up to 20 % CO_2 in 5 min). This protocol is similar to the two-tidal volume inhalation of 35 % CO_2 that elicits panic attacks in both patients and controls. Interestingly enough, the RVLM and LPGi (but not the NTS or caudal VLM) were the only medullary nuclei showing FLI significant increases (Table 2.3). There were also significant increases in c-fos expression of most hypothalamic nuclei associated to stress and/or defensive behaviors, including the PVN, the DMH and the premammillary dorsal nucleus (PMD), but not the VMH or the PH. Surprisingly, as well, although the FLI was markedly increased in DMPAG, DLPAG and VLPAG, FLI was unchanged in LPAG and PBA regions that appear to play a prominent role in suffocation alarm. Most importantly, however, exposures to 20 % CO_2 did not produce escape in spite of the marked activation of both DMPAG and DLPAG. The lack of escape responses suggests that PAG activations were due to the depolarization of inhibitory interneurons rather than of output neurons. Alternatively, CO_2 activations of DPAG output neurons could have been counteracted by concomitant activations of VLPAG.

2.7.4 c-Fos Studies in Rats Exposed to Carbon Monoxide Supports the Key Role of Dorsal Periaqueductal Gray in Respiratory-Type Panic Attacks

Rats subjected to a 5-min exposure to 1 % CO showed widespread activations of respiratory and autonomic nuclei [235] (Table 2.3). These activations occurred in spite of the peripheral chemoreceptor inhibition by low concentrations of CO [201]. In particular, exposures to 1 % CO produced significant increases in c-fos expression in NTS, area postrema (AP), dorsal motor nucleus of the vagus (DMX), VLM, PPy, LPGi, LPBA and NRPa. More rostrally, the CO activated the supramamillary nucleus, the DMH and PH. On increasing the severity of tissue hypoxia, FLI was also observed in the retrotrapezoid nucleus, ventral tegmental area, arcuate nucleus and PVN. Most notably, however, whereas the CO produced robust increases in c-fos expression of NRD, LC and VLPAG, it did not activate the DPAG. The latter result is consonant with the lack of defensive responses during intoxication with the "silent killer".

2.8 Evidence of a Suffocation Alarm System Within the Periaqueductal Grey Matter of Midbrain

2.8.1 Brain Detectors of Hypoxia and Hypercapnia

Data from both PET and c-Fos studies showed that the NTS, the PH, the PAG, the LPBA, and, less clearly, the LC, were the only brainstem nuclei activated by hypercapnia. With the possible exception of LPBA, all these areas possess neurons intrinsically sensitive to changes in H^+/CO_2 and/or O_2 [85, 232, 236–238]. Accordingly, the high sensitivity of PD patients to CO_2 may be due to the abnormal functioning of these nuclei. Conversely, the activations of cerebellum midline cortical areas regularly observed during both hypercapnia and panic attacks of humans were not observed in c-Fos animal studies.

Additionally, Coates et al. [237] showed that phrenic nerve activity is markedly increased by microinjections of acetazolamide (an inhibitor of carbonic anhydrase) in multiple brainstem sites of servo-ventilated rats and cats that were both vagotomised and glomectomised. Responsive sites were found within 800 μm of the VMS and close to NTS and LC. These authors proposed that central chemoreceptors are distributed in several locations within 1.5 mm from outer and inner surfaces of brainstem, which includes the PAG. It is noteworthy, however, that the pH of microinjected sites was compatible with a $P_{ET}CO_2$ of approximately 36 mmHg, which is much lower than the threshold for elicitation of hunger for air in humans.

In turn, extracellular and whole-cell patch-clamp recordings of brain slice preparations showed that neurons of PH are sensitive to both hypoxia and, less markedly, hypercapnia [236]. In particular, extracellular recordings showed that hypoxia stimulated over 80 % of tested neurons in a dose-dependent manner. Furthermore, over 80 % of the cells excited by hypoxia retained the response during synaptic blockade. In contrast, hypercapnia increased the firing frequency of only 22 % of the neurons. Similarly, whole-cell patch-clamp recordings showed that whereas the hypoxia depolarised 76 % of neurons of PH, hypercapnia depolarised and/or increased the discharge of only 35 % of tested neurons. These results suggest that the caudal hypothalamus has separate populations of neurons sensitive to hypoxia and hypercapnia. Importantly, as well, Ryan and Waldrop [232] showed that the PH harbors PAG-projecting neurons which are sensitive to both hypoxia (53 %) and hypercapnia (27 %).

Remarkably, as well, Kramer et al. [238] showed that 74 % (39 of 53) of *in vitro* neurons of

PAG responded to hypoxia, 92% of which with an increase in firing rate. Moreover, hypoxia-sensitive neurons of DLPAG/LPAG regions showed firing rate increases greater than those of VLPAG. By contrast, only 21% (7/33) of PAG neurons responded to hypercapnia. PAG neuron responses to hypoxia were retained in spite of the blockade of synaptic transmission in low-Ca^{2+}/high-Mg^{2+} medium. Taken together, these data show that the PAG harbors neurons intrinsically sensitive to hypoxia.

On the other hand, existing evidence suggests that carotid body chemoreceptor type-I cells sense both hypoxia and hypercapnia through 'tandem acid-sensitive potassium channels' (TASK-1 and TASK-3 channels) that are inhibited by acidic pH and hypoxia and activated by alkalosis [239–241]. Prabhakar [201] showed in addition that these channels might be inhibited by the intracellular gaseous messengers CO and hydrogen sulfide (H_2S). TASK genes are also expressed in brain chemosensitive areas of hindbrain, including the RVLM, raphe nuclei, and LC [242–245]. It remains to be elucidated whether these channels sense pH/brain gas levels in panic-related areas sensitive to both hypoxia and hypercapnia, including the PH and the PAG.

2.8.2   Carbon Dioxide Inhibits DPA-Evoked Panic-Like Responses

Recently, Schimitel et al. [246] examined the behavioral effects of both the hypercapnia (8% and 13% CO_2) and the selective stimulation of carotid chemoreceptors with potassium cyanide (KCN, 10–160 µg, i.v.), administered either alone or combined, in rats either stimulated or lesioned in the DPAG.

Behavioral data showed that although the exposure to 13% CO_2 alone produced manifest increases in respiration, rats only explored the open-field slowly or remained quiet with eyes opened and muscles relaxed, as shown by the lowering of trunk, head and tail. Exposures to 13% CO_2 also reduced grooming while increasing exophthalmus (fully-opened bulged eyes),

thereby suggesting an arousal to environmental cues. The latter responses may be due to CO_2 activations of in vitro [85] and in vivo neurons of LC in rats either intact [86, 222] or denervated [82]. Data of Schimitel et al. [246] are also in agreement with previous observations showing that CO_2 concentrations as high as 13% failed to evoke defensive responses in rats [86]. Despite the significant activation of DPAG, not even the sudden exposure to 20% CO_2 provoked manifest defensive behaviors [223].

Unexpectedly, as well, exposures to both 8% and 13% CO_2 produced significant attenuations of DPAG-evoked freezing (immobility plus exophthalmus) and flight (trotting and galloping) behaviors (Fig. 2.4) [246]. Evidence from c-fos immunohistochemistry studies showed, on the other hand, that 2-h exposures to low concentrations of CO_2 activate the NRD and, markedly, a discrete region of caudal VLPAG atop the laterodorsal tegmental nucleus (LDTg) [86, 222]. The lack of effects of hyperoxia (60% O_2) on CO_2 activations of VLPAG makes unlikely the involvement of peripheral chemoreceptors [86]. On the other hand, chemical stimulations of the same region (i.e., VLPAG atop the LDTg) of freely-moving rats produced only 'hyperreactive immobility' or 'quiescent behavior' [293]. Taken together, data suggest that CO_2 inhibition of DPAG-evoked panic-like responses is mediated by the VLPAG.

Vianna et al. [247] showed, on the other hand, that the defensive behaviors elicited by electrical stimulation of DPAG are unchanged by electrolytic lesions of VLPAG. Although these results make unlikely the VLPAG tonic inhibition of DPAG, plenty of evidence supports the VLPAG phasic inhibition of the latter structure [174–179]. Notably, as well, chemical stimulations of medullary raphe produced prolonged attenuations of cardiovascular responses to electrical stimulations of DPAG [214]. Because the latter effect is presumptively mediated by NRM/NRO efferents to DPAG-projecting neurons of VLPAG [215], the CO_2 inhibition of DPAG-evoked behaviors [246] could be mediated by 5-HT neurons of NRD properly and/or NRD 'lateral wing' (NRDlw) within the VLPAG. Although the

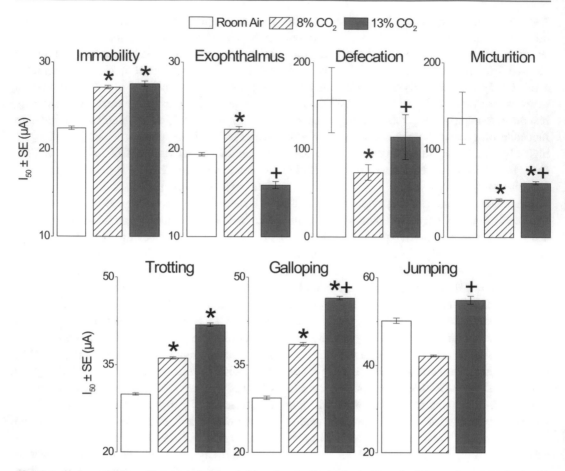

Fig. 2.4 Carbon dioxide effects on the thresholds of defensive responses induced by electrical stimulation of DPAG. Columns represent the population unbiased estimates of median threshold intensities ($I_{50} \pm SE$). *$P < 0.05$, significantly different from DPAG stimulations in room air, +$P < 0.05$, significantly different from DPAG stimulations in 8 % CO_2 enriched-air sessions (Bonferroni's 5 % criterion of likelihood ratio chi-square tests for location of parallel-fitted threshold curves) (from Schimitel et al. [246])

DPAG is targeted by 5-HT afferents from NRD, NRO, NRPo and NRMn [215] (but see Jansen et al. [164]), prolonged exposures to CO_2 increased c-fos expression of only NRPa and, more markedly, NRD [86] (Table 2.3). In contrast, whereas the sudden exposure to 20 % CO_2 did not activate the raphe nuclei, it produced significant activations of ventral medullary areas, PVN, perifornical hypothalamus (PeF), DMH, PMD and caudal PAG, including DMPAG, DLPAG and CePAG ('*medial VLPAG*') [223]. Because the exposures to 20 % CO_2 did not produce escape, the PAG activations were most probably due to the depolarization of inhibitory interneurons. If so, the CO_2 activation of GABA

interneurons could inhibit the DPAG output neurons presumptively involved in respiratory-type panic attacks.

2.8.3 Peripheral Injections of Potassium Cyanide Elicits Respiratory-Type Panic Responses

Behavioral effects of injections of KCN were much the opposite of those of CO_2 [246]. Indeed, intravenous injections of KCN alone induced dose-dependent defensive behaviors (Fig. 2.5) similar to those observed with stepwise-increasing

Fig. 2.5 Dose–response curves of KCN-evoked behaviors. Curves are the best-fitted logistic functions of the accumulated frequencies of threshold responses in function of the logarithm of the KCN dose (i.v.). Abscissa in logarithmic scale.
Abbreviations: *r* responders; *n* total number of rats (modified from Schimitel et al. [246])

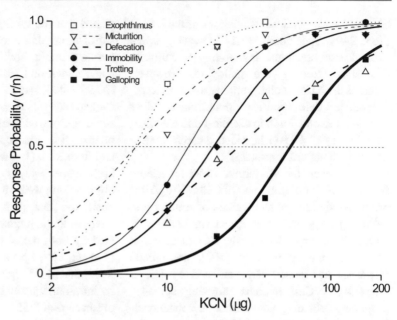

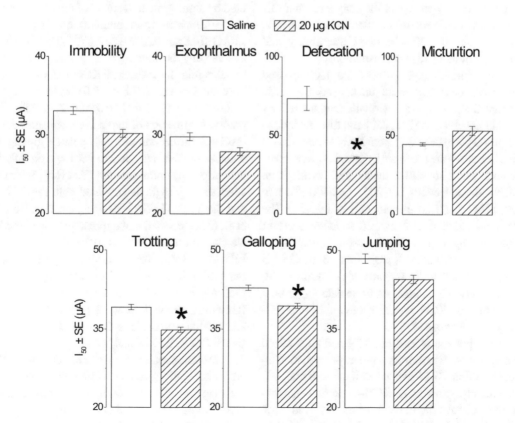

Fig. 2.6 Effects of intravenous injections of 20 μg KCN or vehicle (0.9 % NaCl) on the median thresholds ($I_{50} \pm SE$) of defensive behaviors induced by electrical stimulation of PAG. *$P < 0.05$, significantly different from saline sessions. Details as in Fig. 2.4 (from Schimitel et al. [246])

electrical and chemical stimulations of the DPAG [165, 249]. Specifically, KCN increasing doses evoked exophthalmus, immobility, micturition, defecation, trotting and galloping. However, whereas the KCN elicited jumping in one rat only, it produced defecation and micturition at rather low doses as compared to chemical and electrical stimulations of DPAG in which these responses showed the highest thresholds.

Schimitel et al. [246] further showed that slow infusions of a subliminal dose of KCN ($20\,\mu g/30\,s$) produced significant facilitations of trotting and galloping responses to electrical stimulations of DPAG (Fig. 2.6). Because the DPAG does not project to the spinal cord [162], the escape responses are most likely mediated by DPAG efferents to CnF neurons targeting spinally-projecting cells of gigantocellular and magnocellular reticular nuclei [250]. Conversely, the inconsistent labeling of CnF nucleus by both hypoxia and hypercapnia [86] suggests that this nucleus is not involved in detection of asphyxia. Facilitations of DPAG-evoked immobility and exophthalmus were less conspicuous.

Most importantly, whereas the KCN-evoked defensive behaviors were unchanged by aortic denervation, they were virtually suppressed by carotid denervation [251, 252] and discrete lesions of DPAG [246] (Fig. 2.7). That KCN-evoked defensive behaviors are mediated by peripheral chemoreceptors was also shown by the lack of effect of KCN microinjections (12–200 nmol/100 nL) into the DPAG (C.J.T. Müller, unpublished results). The PAG mediation of KCN-evoked behaviors is in turn supported by tract-tracing and c-fos immunohistochemical studies of NTS projections to the CePAG [171, 172] and KCN activations of DPAG [253], respectively. Taken together, these data support the mediation of KCN-evoked behavior by CePAG-projecting neurons of NTS.

The opposite effects of KCN and CO_2 raised the question whether the pre-exposure to CO_2 could inhibit KCN-evoked defensive behaviors. Conversely, however, KCN-evoked behaviors were markedly facilitated by both 8% and 13% CO_2 (Fig. 2.8) [246, C.J.T. Müller, unpublished results]. Because CO_2 alone did not produce defensive behaviors, Schimitel et al. [246] sug-

gested that the PAG harbors a hypoxia-sensitive suffocation alarm system which activation in humans might both precipitate a spontaneous panic attack and render the subject sensitive to CO_2. Indeed, recent studies showed that KCN-evoked escape is attenuated by clinically-effective acute and chronic treatments with the putative panicolytics clonazepam ($0.01–0.3$ mg kg^{-1}, i.p.) and fluoxetine ($1–4$ mg kg^{-1} day^{-1} along 21 days, i.p.), respectively [254] (Fig. 2.9). The latter observations were further extended by the demonstration that panic-like responses to severe environmental hypoxia ($7\%\ O_2$) are also attenuated by both the acute injections of alprazolam ($1–4$ mg kg^{-1}, i.p.) and the chronic injections of fluoxetine ($5–15$ mg kg^{-1} day^{-1}, 21 days, i.p.), and by intra-periaqueductal injections of 5-HT antagonists as well [255].

In the same vein, escape responses to severe environmental hypoxia ($8\%\ O_2$) were accompanied by significant activations of intermediate levels (-7.08 mm from bregma) of DLPAG and LPAG (but not of DMPAG and VLPAG) [200]. Conversely, however, prior studies found that hypercapnia, hypoxia and KCN activate mostly the caudal regions (-7.8 to -8.0 mm from bregma) of DLPAG, LPAG and VLPAG [222, 253]. The precise localization of hypoxia-responsive sites of PAG was further complicated by data showing that whereas the 2-h exposures to 9% hypoxia failed in activating any column of the PAG [86], 3-h exposures to 11% hypoxia activated both the DLPAG and the VLPAG [222]. Besides, whereas Berquin et al. [222] reported that hypoxia produced moderate activations of DMH (120%), PH (150%) and LPBA (191%), Teppema et al. [86] reported marked activations of A5 noradrenergic neurons (550%) and LPBA neurons only (1440%). Remaining activations were found in medullary areas unlikely to mediate the defensive behaviours (NTS, DMX, RVLM and VMS).

The demonstration that *in vitro* neurons of PAG are intrinsically sensitive to hypoxia [238] raises the question whether the hypoxia activations of PAG resulted from local changes in tissue partial pressure of O_2 (PO_2). Moreover, prior attempts to localize the KCN-responsive regions within the PAG were compromised by the employment of

Fig. 2.7 Effects of electrolytic lesions of PAG on the defensive behaviors produced by intravenous injections of 80 μg KCN. Columns represent the proportion (±SE) of rats presenting a given response before (*hatched*) or 24 h after (*black*) the application of discrete electrolytic lesions to the PAG (anodal current, 1 mA, 5 s) to the PAG. *P<0.05, **P<0.01, ***P<0.005, proportions significantly different from pre-lesion values (Pearson's χ^2) (from Schimitel et al. [246]). Abbreviations: *EXO* Exophthalmus, *IMO* immobility, *DEF* defecation, *MIC* micturition, *TRT* trotting, *GLP* galloping, *JMP* jumping

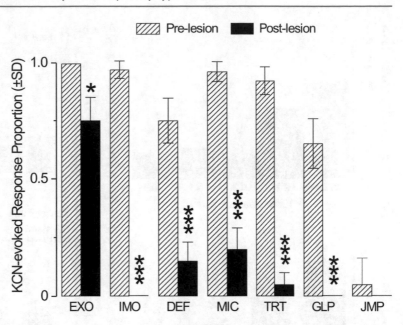

Fig. 2.8 Effects of 8 % and 13 % CO_2 on KCN-evoked flight behavior. The duration of flight behavior is the sum of the duration of trotting, galloping, and jumping responses evoked by intravenous injections of 40 μg KCN. **$P<0.005$, ***$P<0.0001$, significantly different from room air; ++$P<0.005$, significantly different from 8 % CO_2 (Student's one-tailed paired t-tests) (from Schimitel et al. [246])

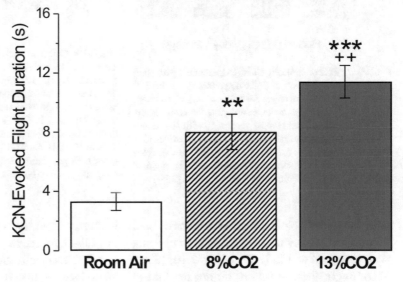

repeated intravenous injections of anesthetized rats with high doses of KCN (60–120 μg, i.v.) [253]. Indeed, whereas the latter study found widespread activations of PAG, a recent study from our laboratory showed that the commissural NTS (−14.4 mm from bregma), the LPAGr (−6.48 mm from bregma) and the VLPAGc (−8.16 mm from bregma) were the only brainstem structures activated following a short-lasting escape response (2–3 s) elicited by a single injection of a low dose of KCN (40 μg, i.v.) (C.J.T. Müller,

unpublished results). Although most afferents of LPAGr come from hypothalamus and limbic cortex [167], the LPAGr is also recipient of afferents from caudal LPAG and LPBA that could mediate the KCN-evoked peripheral signals of hypoxia. Taken together, data suggest that whereas the DLPAG and LPAGr are predominantly activated by hypoxia, the VLPAGc is activated by both hypoxia and hypercapnia.

Although the molecular mechanisms underlying the suffocation alarm system are unsolved, it

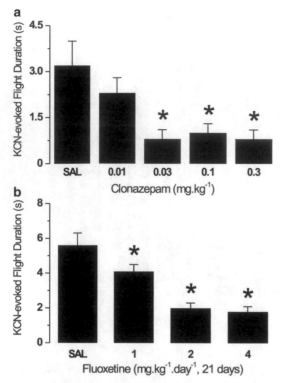

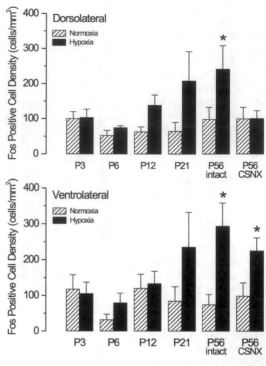

Fig. 2.9 Effects of clinically effective treatments with pani-colytics on the duration of flight responses (mean±SEM) produced by an intravenous injection of KCN (80 μg). *Upper*: Effects of the acute treatment with clonazepam. *Bottom*: Effects of the chronic treatment with fluoxetine. *$P<0.05$, significantly different from saline-treated rats (5% Bonferroni's criterion for Student's 1-tailed t-tests). *KCN* potassium cyanide, *s* seconds, *SAL* saline, *SEM* standard error of the mean (modified from Schimitel et al. [254])

Fig. 2.10 Cell densities of DLPAG and VLPAG neurons expressing Fos protein of normoxic and hypoxic rats during post-natal days 3 to 56 (P3-P56) and at P56 in carotid sinus denervated rats. Carotid sinus denervation (CSNX) unmasked a NTS-mediated inhibitory input that suppressed c-fos activation in DLPAG. In contrast, CSNX produced only a non-significant reduction of 22% of c-fos labeled cells of VLPAG. Data are means±S.E.M. ($n=4$–5 for each group). *Significantly larger than the respective normoxic rat group ($P<0.05$) (redrawn from Horn et al. [225])

is a quite remarkable finding that neurons of both PH and PAG solely begin to respond to hypoxia after post-natal day 12 (PN12) (Fig. 2.10) [225]. Accordingly, these structures do not respond to hypoxia during the 'stress hyporesponsive period' (PN2-PN12) in which the HPA axis is likewise quiescent [256–258]. It remains to be elucidated whether the hypoxia-sensitive neurons are also primed by maternal deprivation, a model of CSA disorder.

In turn, Horn et al. [225] showed that whereas the VLPAG responses to hypoxia were unaltered by sinusal denervation of adult rats, DPAG responses were wholly suppressed (Fig. 2.10). Since PAG neurons are intrinsically sensitive to hypoxia [238], Horn et al. [225] proposed that carotid denervation unmasked an extrinsic

inhibitory input that suppressed c-fos expression in DLPAG. Moreover, Teppema et al. [86] showed that VLPAG activations by hypercapnia were unchanged by hyperoxia (60% O_2) (Table 2.3). Altogether, data suggest that whereas the DLPAG response to hypoxia results from a NTS-mediated disinhibitory mechanism, VLPAGc responses depend predominantly (78%) on central inputs (as below discussed, the 'hypoxia activations' of VLPAGc might actually correspond to activations of ventral half of LPAG). Although the latter data contradict our previous argument that KCN-evoked panic responses are partly due to NTS inhibition of VLPAG [61, 259], they do not rule out the core hypothesis that panics result from VLPAG defective inhibition of DPAG output neurons. As a corollary, tissue hypoxia by 1% CO activated the LC, the NRD and VLPAG, but not the DLPAG or LPAG (Table 2.3) [235]. The latter observations are consonant with the absence of defensive reactions during the intoxication with the 'silent killer'.

Concluding, data suggest that *hypoxia* activates the PAG both directly (via NTS excitatory efferents to LPAG/CePAG) or indirectly (via NTS disinhibition of DLPAG). Conversely, *hypercapnia* aborts DPAG-evoked panic reactions via the central activation of VLPAG inhibitory inputs to DPAG/CePAG (Fig. 2.11, upper). Panic responses are nonetheless facilitated in *hypercapnic hypoxia*, which reproduces real-life asphyxia (Fig. 2.11, middle). Be this as it may, panic patients appear to overreact to CO_2 due to the defective functioning of VLPAG and/or NRD inhibitory projections to DPAG (Fig. 2.11, bottom). Of note, recent studies suggest that LAC infusions could facilitate panic via the inhibition of NRDlw projections to DPAG [231, 260].

2.8.4 On the Probable Substrates of Klein's Three-Layer Cake Model of Panic Attack

In a preliminary presentation of SFA theory, Klein [25] suggested that the march of symptoms of panic attacks "*appears to be a three-layer cake. The first layer is the reaction of the smothering alarm system, as it had received an increment of CO_2, by breathlessness and increased tidal volume. When the control system keeps getting signals interpreted as predictive of asphyxiation, then the panic attack, with its feeling of suffocation and urge to flee is released, followed by the increase in respiratory frequency*". In particular, LAC-induced panic attacks are characterised by early dyspnoea followed by panic, desire for flight and a sustained hyperventilation that continues several minutes after the end of the infusion (Fig. 2.12). Although this model is heavily based on LAC-induced panics, this section examines whether the DPAG circuits might explain Klein's three-layer cake model of panic attacks.

Although the respiratory rhythmogenesis is maintained in pontine preparations, PAG stimulations of anesthetized animals produce prominent respiratory effects. Indeed, in the first study with electrical stimulation of the PAG, Sachs [261] reported that among all regions stimulated in thalamus and brainstem, the PAG yielded the most pronounced respiratory responses, being '*undoubtedly … a respiratory center*'. These observations were corroborated by extensive mapping of respiratory responses induced by electrical stimulation of the brainstem of anesthetized cats. Most notably, Tan [262] showed that low-intensity stimulations of LPAG produced phase-locked stimulus-bound respiratory responses without concomitant cardiovascular responses. Subsequent studies suggested that chemical stimulations of the PAG affect mostly the respiratory frequency, decreasing the duration of both expiration and inspiration [263, 264], an effect thought to be mediated by the PBA inspiratory off-switch of 'pneumotaxic center' [265]. However, periaqueductal microinjections of GABA-A receptor antagonists increase both frequency and amplitude of respiration (Fig. 2.13) [266].

The demonstration that hypoxia facilitates the sham-rage behavior of the decerebrated cat was the first evidence that behavioral responses to hypoxia might be mediated at the brainstem [267]. Reciprocally, Hilton and Joels [268] showed that electrical stimulations of 'defence areas' of both PAG and medial hypothalamus facilitated the hyperventilation produced by intravenous injections of KCN. Much later on, Franchini et al. [251, 252] showed that 'alerting' behaviors to KCN (30 µg, i.v.) were suppressed by both sinusal denervation and glomus artery occlusion. In turn, Hayward and Von Reitzenstein [253] showed that repeated injections of high doses of KCN (60–120 µg, i.v.) activate predominantly the caudal sectors of DPAG. However, less than 10 % of these neurons were retrogradely labelled from A5 region believed to mediate the pressor component of chemoreceptor reflex [269]. Lastly, Schimitel et al. [246] presented compelling evidence that KCN-evoked panic-like behaviors are both facilitated by previous exposures to CO_2 and suppressed by restrict electrolytic lesions of DPAG. Schimitel et al. [246] also showed that DPAG-evoked behaviors were

HEALTHY SUBJECT

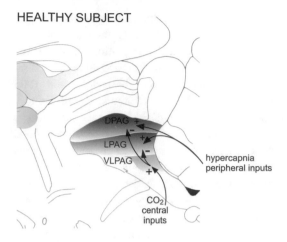

HEALTHY SUBJECT

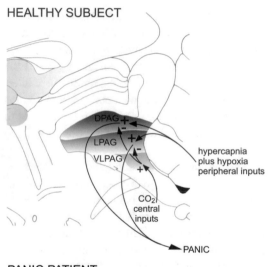

PANIC PATIENT

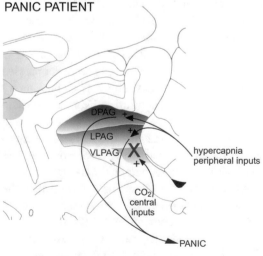

facilitated by the concomitant intravenous infusion of a subliminal dose of KCN (20 µg). Altogether, these data give strong support to the PAG mediation of respiratory-type panic attacks.

Importantly, as well, Hayward and co-workers showed that DPAG-evoked respiratory responses were markedly reduced (−65 % to −90 %) by bilateral microinjections into the LPBA of either a GABA-A receptor agonist or a glutamate receptor antagonist [270, 271]. Moreover, experiments with fluorescent labeling identified the effective site in the inner division of eLPBA (see Fig. 5 of Hayward et al. [270]). In turn, Haibara et al. [272] showed that whereas lidocaine microinjections into the LPBA markedly attenuated the cardiovascular responses to KCN, lidocaine microinjections into DLPAG or LPAG were ineffective. Overall, these data suggest that whereas the DPAG mediates behavioral and respiratory responses to KCN, the LPBA is involved in respiratory and cardiovascular components only. Importantly, as well, these data suggest that respiratory and cardiovascular symptoms of panic attacks are mediated by PAG glutamatergic projections to the LPBA.

Although the DPAG-evoked cardiovascular and respiratory responses have been extensively studied as visceral adjustments of fight and flight behaviors [160, 253, 266, 273–278], the stimulus-response properties of DPAG-evoked respiratory responses [263] suggest that DPAG may be substrate of both the early dyspnea and the late hyperpnea of respiratory-type panic attacks. To study the stimulus-response properties of the respiratory oscillator, Paydarfar and Eldridge [263] examined the changes in phrenic nerve activity brought about by electrical stimulations of DPAG of sinoaortic-denervated anesthetised cats kept at constant $P_{ET}CO_2$ (30–35 mmHg). Data showed that brief threshold stimulations

by the CO_2 activation of inhibitory projections of the ventrolateral periaqueductal grey (VLPAG) to the former structures. *Middle* – During environmental suffocation, peripheral signals of hypoxia and hypercapnia overcomes VLPAG inhibition of DPAG, thereby releasing a panic reaction. *Bottom* – In panic patients, the defective functioning of VLPAG inhibitory projections to DPAG render the subject hyper-responsive to both hypoxia and low doses of CO_2 (modified from Schenberg et al. [51])

Fig. 2.11 Presumptive mechanisms of elicitation of panic attacks by inhalation of low concentrations of carbon dioxide (CO_2) in panic disorder patients. *Upper* – In healthy subjects, CO_2 peripheral inputs to dorsolateral (DLPAG) and lateral (LPAG) periaqueductal grey (PAG) are counteracted

Fig. 2.12 Time-course (min) of respiratory changes in tidal-volume during a panic attack provoked by intravenous infusion with 0.5 M sodium lactate of a patient with panic disorder and agoraphobia. The 'x' indicates the early episode of dyspnea that heralds the panic attack (courtesy of D.F. Klein)

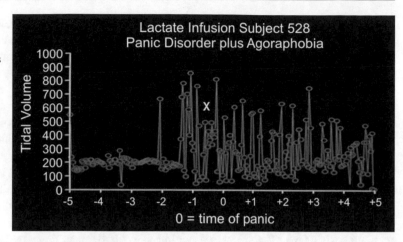

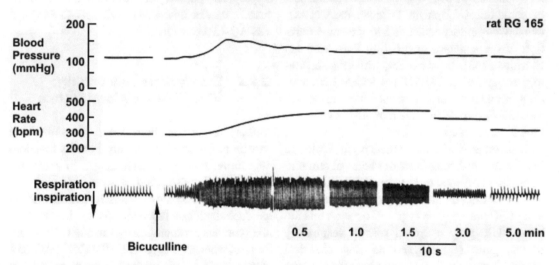

Fig. 2.13 Increases in mean blood pressure, average heart rate and respiration caused by the microinjection of a GABA-A receptor antagonist (bicuculline, 10 nmol, 1 μL/30 s) into the dorsal periaqueductal grey matter of a urethane anesthetized rat. Similar injections in unanesthe- tized rats produce full-blown defensive behaviors. These data show that the structures that mediate panic-like behaviors are tonically inhibited by gabaergic synapses (modified from Schenberg et al. [266])

(1-s trains) of DPAG invariably reset the cycle to inspiratory or expiratory phase. Following cycle resetting, the respiratory rhythm resumed with predictable periodicity (fixed co-phase duration). Yet, if intensity and pulse and train duration were kept at 250 μA, 1 ms and 1 s, respectively, stimuli of intermediate frequency (30 Hz) applied at a definite time of the inspiratory-expiratory transi- tion (approximately at 0.4 cycle) resulted in unpredictable time recovery of the respiratory cycle (random co-phase duration). This response is exactly as expected in dyspnea, the leading symptom of respiratory-type panic attack [24, 27, 42, 53]. Paydarfar and Eldridge [263] also

showed that critically timed stimuli with an appropriate intensity resulted in prolonged inspi- ratory apneusis that could last over 3 respiratory cycles of the decerebrated cat (i.e., approximately 9 s). These data showed that subtle stimulations of DPAG can produce either prolonged episodes of apneusis or irregular breathing which are typi- cal of respiratory-type panic onset (Fig. 2.12).

More recently, Subramanian et al. [278] showed that neuron-selective stimulations of dis- tinct columns of the PAG of precollicularly decerebrated unanesthetised cats produced column-specific patterns of respiration. Indeed, whereas the DMPAG stimulations produced slow and deep breathing along with increased tonus of the diaphragm that resemble the respiratory pattern during the rat confrontation with a conspecific (unpublished observations), DLPAG stimulations produced hyperventilation and tachypnea consistent with the respiratory responses of a full-blown flight behavior. In turn, whereas the chemical stimulations of 'lateral' sectors of LPAG and VLPAG produced respiratory responses associated with emotional vocalization (mews and/or hisses), stimulations of the '*medial part of LPAG*' (i.e., the CePAG) and of VLPAG resulted in irregular breathing or inspiratory apneusis of over 8 s, which are reminiscent of panic-onset dyspnoea. Of note, the CePAG-evoked inspiratory apneusis had the same duration of the apneusis reported by Paydarfar and Eldridge [263]. In any event, while the respiratory responses to CePAG and VLPAG stimulations mimic the panic onset, respiratory responses to DLPAG stimulations are reminiscent of a full-blown panic attack.

In another study, Subramanian and Holstege [248] examined the effects of chemical stimulations (20 nL of D,L-homocysteic acid) of discrete regions of PAG on the activity of both diaphragm and pBöC pre-inspiratory (pre-I) neurons that are believed to play a crucial role in respiratory rhythmogenesis. Data showed that chemical stimulations of DLPAG produce both the increase of firing of pre-I cells and tachypnea. In turn, whereas the stimulation of VLPAG produced expiratory apnea and inhibition of pre-I cells and diaphragm, stimulations of CePAG ('*medial part*' of LPAG) produced inspiratory apnea and converted pre-I neurons into phase-spanning cells that discharged at inspiration/expiration transition. Notably, as well, while the stimulation of the '*lateral part*' of LPAG activated pre-I neurons throughout the inspiration and produced emotional vocalization and diaphragm relaxation, stimulations of '*ventral part*' of caudal LPAG (LPAGcv, −8.0 mm from bregma) produced an early burst of spikes (2 s) followed by a hyperpnea concomitant to a paradoxical sustained inhibition (7 s) of pre-I neuron firing.

The demonstration that LPAGcv and VLPAG mediate opposite responses of hyperpnea and apnea reveals the sheer complexity of PAG.

In particular, the development of hyperpnea in the absence of pBöC activity suggests the relative independence of eupnea and full-blown respiratory panic responses. Moreover, the paradoxical inhibition of pre-I neurons during LPAGcv-evoked tachypnoea suggests that PAG stimulations block respiratory mechanisms of eupnea much as they do with baroreceptor reflex [104, 279–281]. It remains to be elucidated whether the Hering-Breuer reflex is also inhibited during both panics and defensive behaviors, giving way to a full-blown hyperventilation. In any event, these results raise the possibility that the dyspnea/hyperpnea pattern of respiratory-type panic attacks may be mediated by VLPAG-CePAG and CePAG-LPAGcv circuit.

2.8.5 Steady State Functioning of Suffocation Alarm System

During resting conditions, the suffocation alarm system may operate differently from its functioning during full-blown panic attack. In particular, Lopes et al. [282, 283] showed that chemical lesions of either DMPAG/DLPAG or LPAG/VLPAG produce significant reductions (−21 % to −31 %) in the ventilatory response to hypercapnia (7 % CO_2). In turn, whereas lesions of DMPAG/DLPAG produced marked facilitations (67 %) of respiratory responses to hypoxia (7 % O_2), those of LPAG/VLPAG were ineffective.

Therefore, whereas the DMPAG/DLPAG inhibits the respiratory responses to hypoxia, the LPAG/VLPAG facilitates the responses to hypercapnia. As predicted for a suffocation alarm system, these mechanisms appear to reduce oxygen consumption upon the lack of useful air while increasing the responses to hypercapnia, respectively. In particular, the latter mechanism may be responsible for the increased basal respiratory activity of PD patients [24].

Because peripheral signals of hypoxia and hypercapnia are conveyed by the same fibers of glossopharyngeal nerve, the opposite effects of DMPAG/DLPAG and LPAG/VLPAG lesions on respiratory responses to hypoxia and hypercapnia suggest the differential modulation of these areas by peripheral (hypoxia plus hypercapnia) and

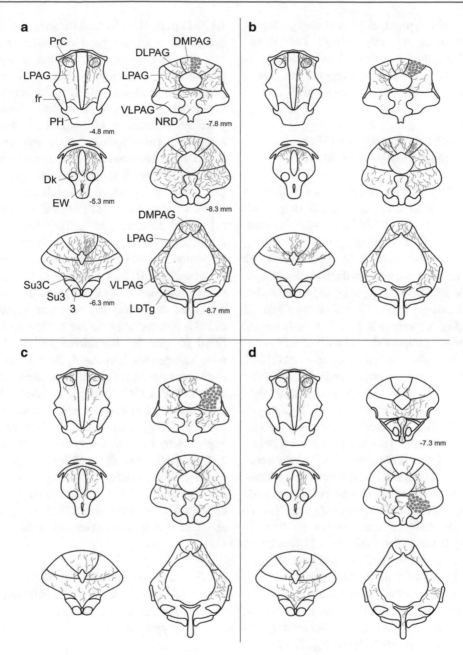

Fig. 2.14 Distribution of PHA-L labeled axons into the PAG of rats whose injections were restricted to dorsomedial (**a**), dorsolateral (**b**), lateral (**c**) or ventrolateral (**d**) columns of PAG. *Solid dots* represent PHA-L labeled cell bodies. (**a**) Following an injection in dmPAG, labeled axons extended throughout the PAG, involving all subnuclei. At rostral levels, the labeling was more concentrated in DPAG and precommissural nucleus (PrC). (**b**) Following an injections into the DLPAG, axons labeling extended mostly throughout the regions caudal to the injection site. (**c**) Following an injection into the LPAG, axonal labeling extended mostly to the caudal PAG, but avoided the DLPAG. (**d**) Following an injection in VLPAG, labeled axons were much more concentrated in the CePAG. Additional abbreviations: *3* oculomotor nucleus, *Dk* nucleus of Darkschewitsch, *EW* Edinger-Westphal nucleus, *fr* fasciculus retroflexus, *Su3* supraoculomotor periaqueductal grey, *Su3C* supraoculomotor cap. Other labels as in the abbreviation list (redrawn from Jansen et al. [164])

central (pH/hypercapnia) respiratory inputs, respectively. Remarkably, as well, PAG lesions affected only the tidal volume. The lack of changes in respiratory frequency suggests that lesion effects are mediated at VMS and/or pBöC.

2.8.6 On respiratory and Non-Respiratory Panic Attacks

Despite the difficulty of interpretation of lesion studies, data of Lopes et al. [282, 283] suggest that DMPAG/DLPAG and LPAG/VLPAG play complementary roles during resting respiratory activity. It should be noted, however, that although these studies are consonant with the lack of activation of VLPAG during hypoxia [86], other studies reported that the VLPAG is activated by hypoxia [225, 253], hypercapnia [222] or both stimuli (C.J.T. Müller, unpublished observations).

Data from chemical stimulations of PAG [248, 278] suggest, on the other hand, that symptom escalation of respiratory-type panic attacks might be due to an early activation of VLPAG projections to CePAG (dyspnea) followed by the sustained activation of DLPAG and LPAGcv output pathways (hyperventilation). The latter mechanisms are supported by tract-tracing of intraperiaqueductal connections with both anterograde (leucoagglutinin of *Phaseolus vulgaris*, PHA-L) and retrograde (pseudorabies virus, PRV) markers [164]. In particular, Jansen et al. [164] showed that whereas the VLPAG projects almost exclusively to CePAG (Fig. 2.14d), the DLPAG and LPAG project to the entire expanse of caudal PAG (Fig. 2.14b and c). By contrast, neither the VLPAG, nor the LPAG, or DMPAG send significant projections to the DLPAG (Fig. 2.14).

Notably, as well, data from PHA-L/PRV double-labeling immunohistochemistry suggested that VLPAG-evoked sympathoinhibition is mediated by a relay neuron of CePAG [164]. The dyspnea-hyperpnea symptoms of respiratory-type panic attack might be mediated by a similar pattern of connections between VMS/VLPAG and CePAG/LPAGcv (Fig. 2.15).

In contrast, fear-like panic attacks would be triggered by PFC, VMH and DLSC inputs to

DLPAG [52]. Therefore, although the respiratory symptoms are more prominent in respiratory-type panics, non-respiratory panics may present similar symptomatology owing to the overlapping output pathways to LPBA, CnF and sympathoexcitatory centers of the medulla. Among the possible differences of respiratory and non-respiratory panic systems, PAG projections to LPBA are presumptively more active in respiratory panic attacks. Indeed, studies with PHAL [284] and c-fos immunohistochemistry [285] showed that whereas the LPAG and VLPAG project to all subnuclei of LPBA, DMPAG and DLPAG project almost exclusively to the superior subnucleus of LPBA. Most notably, however, Krout et al. [284] showed that the CePAG (therein named 'aqueductal PAG' or 'central PAG') sends a unique projection to the inner division of eLPBA. Remarkably, as well, Hayward et al. [270] showed that the microinjection of glutamate antagonists into the eLPBA blocked the respiratory responses to electrical stimulations of DPAG. The eLPBA appears also involved in the entrainment of respiration and afferent muscle activity [286]. The latter mechanism could explain why exercise aborts panic attacks [146, 287]. Indeed, Klein [24] had long proposed that exertion aborts panic by providing countervailing information to the suffocation monitor. It is nevertheless unclear whether the PBA is a monitor or an effector of the suffocation alarm system.

2.8.7 Overview of Suffocation Alarm System of Mammals: An Evidence-Based Hypothesis

The presumptive structure of the suffocation alarm system of mammals is outlined in Fig. 2.16. The diagram depicts the probable pathways involved in respiratory (cyan) and non-respiratory (green) panics, as well as the main inputs (white) and outputs (yellow) of these systems. The 5-HT neurons (pink) are emphasized as well. In turn, the arrows depict nuclei and pathways that are activated by peripheral and/or central signals of asphyxia. In particular, the red arrows

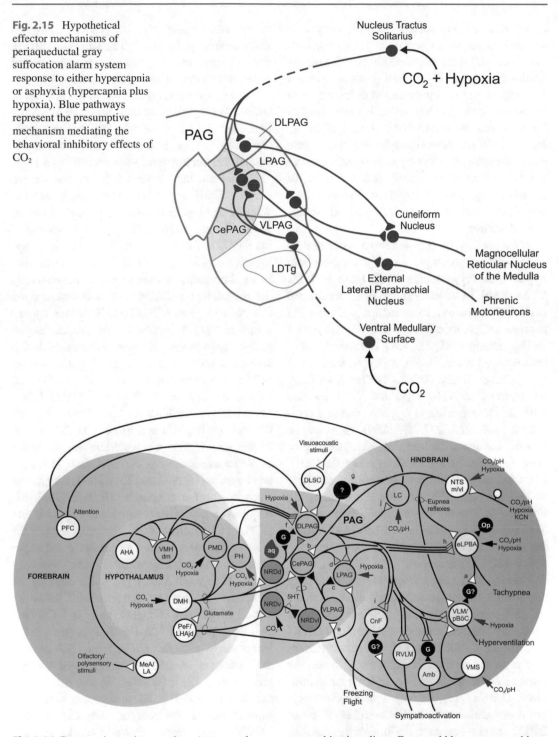

Fig. 2.15 Hypothetical effector mechanisms of periaqueductal gray suffocation alarm system response to either hypercapnia or asphyxia (hypercapnia plus hypoxia). Blue pathways represent the presumptive mechanism mediating the behavioral inhibitory effects of CO_2

Fig. 2.16 Presumptive pathways of respiratory and non-respiratory panic systems. The diagram depicts the probable pathways involved in respiratory (cyan) and non-respiratory (green) panics, as well as the main inputs (white) and outputs (yellow) of these systems. The diagram also emphasizes the 5-HT neurons (pink). Diagram arrows represent structures activated by hypoxia and/or hypercapnia. Red arrows represent structures intrinsically sensitive to hypoxia, hypercapnia or both conditions, as assessed in vitro slices. Green and blue neurons and boutons indicate the core nuclei of non-respiratory (defense-like) and respiratory (suffocation-like) panic attacks, respectively. Black boutons represent inhibitory inputs. White and yellow neurons represent modulatory inputs and effector mechanisms, respectively. See text for explanation. aq - aqueduct, G - gabaergic neuron, Op - opioidergic neuron. Labels as in abbreviation list

depict neurons intrinsically sensitive to hypoxia and/or hypercapnia. Because the diagram activations (arrows) were based mostly on c-fos immunohistochemistry, it remains unclear whether they represent the depolarization of inhibitory or excitatory neurons. Neither is it clear whether these neurons were activated by local (pH, PCO_2, PO_2) or extrinsic inputs (peripheral and/or central chemoreceptors). In any event, boutons were represented either as excitatory (pathway color) or as inhibitory (black) based on immunohistochemical, electrophysiological and pharmacological studies.

Most notably, whereas *in vitro* neurons of neocortex and hippocampus are depressed by both hypoxia and hypercapnia [238], those of NTS, VLM, VMS, LC, PAG, PH and PMD are instead excited (red arrows). In particular, while the PH neurons are intrinsically sensitive to both hypoxia and hypercapnia [85, 232, 238], PAG and LC neurons respond predominantly to hypoxia or hypercapnia, respectively. On the other hand, although the neurons of NTS, VLM and VMS are also intrinsically sensitive to hypoxia and/or hypercapnia [184, 225, 237, 288, 289], there are no reports that the stimulation of these nuclei produce defensive behaviors. Conversely, the PMD, PH, PAG and LC are both intrinsically sensitive to hypoxia and/or hypercapnia and traditionally involved in arousal and/or defensive behaviors. The latter structures are thus the best candidates to translate incoming signals of asphyxia to alerting and/or defensive responses.

Existing evidence also suggests that respiratory responses to hypercapnia are facilitated by both the PH [290] and the PAG [282, 283]. On the other hand, data from lesion studies suggest that DPAG (but not the VLPAG or the PH) inhibits the ventilatory response to hypoxia [283, 290]. While the former mechanism is already expected for a suffocation alarm system, the latter would reduce oxygen consumption upon lack of useful air. Because DPAG lesions attenuate predominantly the tidal volume [283], the DPAG appears to inhibit the pBöC. It is proposed that this effect is mediated by LPBA activation of an inhibitory interneuron (Fig. 2.16a). In particular, Subramanian and Holstege [278] reported that chemical stimulations

of the '*ventral part of LPAG*' produce tachypnea while inhibiting the pBöC. The latter paradoxical effects suggest that the PAG takes over the control of respiration during panic-like reactions.

Although the PAG appears to be the fulcrum of suffocation signals from both forebrain and hindbrain, the CePAG is the only PAG region reportedly targeted by afferents from commissural, medial and ventrolateral subnuclei of NTS [171, 172] that are the terminal fields of sinus nerve afferents [291] (Fig. 2.16b). Interestingly, as well, whereas the repeated injection of high doses of KCN (>60 µg) produced widespread activations of the DPAG of the anesthetized rat [253], the single injection of a low dose of KCN (40 µg) that elicited a short-lasting escape response activated only the rostrolateral (LPAGr) and caudoventrolateral (VLPAGc) regions of PAG (C.J.T. Müller, unpublished results). In contrast, rats that developed escape upon severe ambient hypoxia (8 % O_2) showed significant activations in rostral sectors (−7.08 mm from bregma) of both DLPAG and LPAG, but not in DMPAG or VLPAG [200]. Importantly, as well, the NTS was labeled by both the ambient hypoxia and KCN [200, 253, C.J.T. Müller, unpublished results]. While the differences in PAG activations by KCN (LPAGr and VLPAGc) and hypoxia (DLPAG and LPAGr) may be due to hypoxia direct excitation of DLPAG [238], the only PAG region reliably activated by hypoxia and KCN was the LPAGr. In turn, whereas the LPAGr makes most connections with forebrain and hypothalamus, it is also targeted by afferents from both caudal LPAG and LPBA [167]. Remarkably, as well, the CePAG was markedly labeled following repeated injections of higher doses of KCN (see Fig. 6 of Hayward and Von Reizenstein [253]). Moreover, Sandkühler and Herdegen [292] showed that restricted chemical stimulations of CePAG activate the LPAGr (see Fig. 5b, c in Sandkühler and Herdegen [292]). Taken together, these data support the involvement of both CePAG and LPAGr in respiratory-type panic attacks. In turn, the VLPAGc appear to mediate the behavioral inhibitory effects of normoxic hypercapnia (Fig. 2.16c).

The NTS could also project, either directly or indirectly, to neurons of the ventral part of LPAG

(Fig. 2.16d) mistakenly regarded as part of VLPAG. Subramanian and Holstege [248] showed in addition that whereas the chemical stimulation of the 'ventral part of the LPAG' (−8 mm from bregma) produced tachypnea, that of VLPAG led to expiratory apneusis. Moreover, chemical stimulations of rather close regions of LPAG and VLPAG elicit the opposite responses of quiescence and escape [158, 293]. Taken together, data from chemical stimulations and c-fos immunohistochemistry suggest that the LPAGcv and the VLPAG harbor independent populations of neurons sensitive to peripheral and central signals of hypoxia and hypercapnia, respectively (Fig. 2.16b and e). Data also suggest that VLPAG-CePAG and CePAG-LPAGcv might be crucial in early dyspnoea and late tachypnea of respiratory-type panic attacks, respectively. In this model, the CePAG appears as a comparator of hypercapnia signals from VMS and hypercapnia plus hypoxia signals from NTS. The prevalence of the latter signals would result in a respiratory-type panic attack.

In turn, the hypoxia appears to activate the DLPAG both directly, via local actions (red arrow), and indirectly, via PH excitatory projections and NTS disinhibitory projections [225, 232] (Fig. 2.16f and g). Although highly speculative, afferents from hypothalamus are likely to convey complex signals of suffocation/asphyxia, such as stuffy or stale air, drowning, predators' suffocation bite of nose or throat of preys, or even strangling in humans. The latter possibilities are supported by data of Lopes et al. [283] showing that DMPAG/DLPAG (but not LPAG/VLPAG) inhibits ventilatory responses to hypoxia that could be counterproductive upon hypoxic conditions of preys caught by predators or victims of avalanches, landslides or drowning.

Overall, above data suggest that respiratory-type panics to both environmental hypoxia and KCN [200, 246] are mediated by NTS projections either direct or indirect to CePAG-LPAGcv neurons that project in turn to the LPBA (Fig. 2.16h) and, less markedly, to CnF 'midbrain locomotor region' (Fig. 2.16i). By contrast, non-respiratory type panic attacks appear to be mediated by DLPAG afferents from VMH, PMD and

DLSC that convey pre-processed exteroceptive signals (olfactory, auditory or visual) of an imminent threat [52]. Output neurons of DLPAG would in turn project to CnF and LC neurons that mediate escape and increased in attention to environmental cues, respectively (Fig. 2.16i and j). The apparently spontaneous clinical panic would thus be the result of the activation of either the CePAG-LPAGcv or the VMHdm-PMd-DLPAG efferent systems by as-yet-unknown molecular mechanism.

Krout et al. [284] showed, on the other hand, that the PAG sends profuse projections to most nuclei of LPBA (Fig. 2.16h). Most notably, however, they showed that the CePAG is the only region that projects to the inner division of eLPBA. Interestingly enough, whereas the pharmacological blockade of eLPBA produces robust attenuations of DPAG-evoked respiratory responses [270], chemical stimulations of the CePAG produces inspiratory apneusis [248]. Existing evidence suggests, on the other hand, that the inner division of eLPBA is specifically involved in the entrainment of respiratory and locomotor patterns [286]. It is then proposed that only CePAG activations in the absence of exercise led to the dyspneic symptoms of panic attacks.

Neurons of in vitro slices of PAG are hardly excited by hypercapnia [238]. Conversely, the VLPAG is activated in rats exposed to a wide range of hypercarbic mixtures (5–20 %) [86, 223, 225] (Table 2.3). These data suggest that VLPAG activations by CO_2 are mostly of central origin. Remarkably, as well, hyperoxia (60 % O_2) did not affect VLPAG activations by hypercapnia [86] (Table 2.3). Because glossopharyngeal fibers convey signals of both hypoxia and hypercapnia, the latter data makes unlikely the VLPAG modulation by peripheral chemoreceptor inputs. The VLPAG modulation by pH/CO_2 central inputs is nevertheless supported by the profuse projections of VMS neurons to the caudal sectors of both LPAG and VLPAG [294] (Fig. 2.16e). However, recent results of our group showed that the VLPAGc atop the LDTg nucleus is activated by both 13 % CO_2 and, quite unexpectedly, 40 µg KCN (C.J.T. Müller, unpublished results). The latter author also showed that KCN microinjections

(20–200 pg/100 mL) into the PAG did not affect the rat behavior. The latter results are consonant with the NTS modulation of VLPAGc.

Lastly, although the lesions of caudal VLPAG block conditioned freezing [96, 173], chemical stimulations of VLPAG produce mostly quiescence [293], being pleasurable in both rats [180] and humans [101]. While the latter findings make unlikely the VLPAG mediation of panic attacks, they suggest that the activation of VLPAG upon normoxic hypercapnia might be responsible for the lack of defensive responses of both rats and cats exposed to CO_2 concentrations as high as 20 % [86, 221, 223, 246]. The latter mechanism might also explain the attenuation of DPAG-evoked defensive responses by 8 % and 13 % CO_2 [246]. As a corollary, the hypoxia-induced escape may be explained by the direct excitatory actions of hypoxia on neurons of DPAG. Similarly, the lack of activation of both DLPAG and LPAG in rats exposed to 1 % CO [235] explains why the victims of CO intoxication die quietly in bed.

2.8.8 Insights into a Neurochemical Puzzle

The molecular mechanisms underlying PAG intrinsic sensitivity to hypoxia remain completely unknown. Neurotransmission within the suffocation alarm pathways is likewise unsolved. Difficulties are further aggravated by the apalling complexity of neurotransmission within the PAG, involving aminoacids both excitatory and inhibitory, NE, 5-HT, CCK, nitric oxide (NO), substance P (SP), enkephalins (ENK), endomorphins (EDM), prepro-thyrotropin releasing hormone, orexin, galanin, somatostatin and vasoactive intestinal polypeptide, among other chemicals [162, 170, 172, 295–302].

Notwithstanding, it is a quite remarkable fact that whereas the traditional benzodiazepines were ineffective in DPAG-evoked panic responses [60], low-doses (0.1–1.0 mg kg^{-1}) of clinically-effective benzodiazepine panicolytics (alprazolam, clonazepam) produced significant attenuations of latter responses [303]. DPAG-mediated panic-like responses to chemoreceptor stimulations were even more sensitive than those evoked by electrical stimulation, being markedly attenuated or virtually suppressed by clonazepam doses within the clinical range (0.01–0.3 mg kg^{-1}) [254, 255]. Conversely, microinjections of GABA-A receptor antagonists into the DPAG elicit all behavioral, cardiovascular and respiratory defensive responses of the rat [266, 304–306]. In particular, Behbehani et al. [306] reported that the CePAG (i.e., the 'medial and medioventral parts of the PAG') is the region most sensitive to the microinjection of the GABA-A antagonist bicuculline. These data suggest that DPAG and CePAG are tonically inhibited by gabaergic synapses as would be expected for a system dedicated to cope with emergencies only. Indeed, Lovick and Stezhka [307] showed that although the DPAG harbors two populations of output neurons, only 18 % and 37 % cells of each population fired spontaneously, both at low-firing rate (<4 Hz). Since the CePAG possesses the highest density of GABA-A receptors [306], it is worth examining whether the CO_2 inhibition of DPAG-evoked panic responses is mediated by these receptors.

On the other hand, plenty of evidence suggests that 5-HT exerts opposite roles in DPAG and VLPAG. In particular, whereas the 5-HT is predominantly excitatory in DPAG (72 % of responsive neurons) [308], it is virtually inhibitory in VLPAG (94 % of responsive neurons) [309]. Stezhka and Lovick [177, 310] showed, on the other hand, that the dorsal NRD sends excitatory projections to DLPAG. Brandão et al. [308] also showed that the DPAG harbors both low-firing cells excited by 5-HT$_2$ agonists and high-firing cells inhibited by 5-HT$_{1A}$ agonists. Because the DPAG is tonically inhibited by gabaergic synapses, it is tempting to speculate that high- and low-firing cells correspond to inhibitory interneurons and excitatory output neurons, respectively.

Studies with c-fos immunocytochemistry showed, on the other hand, that whereas the NRD activity is markedly increased (200–450 %) by prolonged exposures (2 h) to moderate hypercapnia (10 % and 15 % CO_2) [86], it is unchanged by both prolonged exposures (3 h) to light hypercapnia (5 % CO_2) [222] and short exposures (5-min) to severe hypercapnia (20 % CO_2) [223] (Table 2.3). Because NRD activations by CO_2 were blocked by hyperoxia (60 % O_2), the NRD seems also sensitive to PO_2 and/or signals from peripheral chemoreceptors [86]. If so, the latter data suggest that the NRD exerts a specific modulation of CePAG-LPAGcv suffocation alarm system. Moreover, although prior studies found that the DPAG is the recipient of 5-HT afferents from NRD, NRO, NRPo and NRMn [215], Jansen et al. [164] reported that the CePAG (i.e., the '*LPAG close to the wall of the aqueduct*') is the main target of NRD projections. Indeed, Ruiz-Torner et al. [169] found that the CePAG is further delimited by 5-HT immunostaining. Not surprisingly, recent studies showed that escape reactions to either hypoxia or KCN are attenuated by chronic treatments with fluoxetine and DPAG microinjections of both $5\text{-}HT_{1A}$ and $5\text{-}HT_{2A}$ receptor agonists [254, 255]. Overall, these data suggest that SSRI's attenuation of panic attacks might be mediated by the facilitation of 5-HT inhibitory actions both direct and indirect on DLPAG and CePAG-LPAGcv circuit, respectively (Fig. 2.16).

Remarkably, as well, the CePAG is the main target of DMH glutamatergic neurons that are presumptively activated by interoceptive cues [311, 312]. In turn, whereas the CePAG is fairly delimited by the expression of EDM [172] and CCK [298], the SP is widely expressed across the full extension of DMPAG and LPAG and also caudal sectors of DLPAG [298]. Similarly, SP receptors (SPr) are particularly dense in cell bodies and dendrites of approximately 6% of neurons of broad areas of DPAG, being observed in postsynaptic and nonsynaptic regions as well. A small proportion of axons (4.2%) and axon terminals (5.3%) also showed SPr immunoreactivity in axo-dendritic synapses predominantly asymmetric (70%) [313]. Whole-cell patch-clamp studies showed, on the other hand, that SP produces dose-dependent depolarizations of 60% of neurons of PAG [314], a greater proportion of which (70%) was inhibited by met-ENK. Taken together, these data suggest that SP may act in a diffuse, nonsynaptic manner, modulating excitatory neurotransmission both presynaptically and postsynaptically.

The balance between CCK, SP and opioid transmission into the PAG appears thus to be of major importance in subject's vulnerability to PD. Most notably, while CCK is a recognized pani-

cogen, Brodin et al. [315] showed that 1- or 7-day isolations of adult rats upregulate SP expression in DPAG. In turn, recent studies of Bassi et al. [316] showed that 1-day isolations significantly increased the number and duration of 22 kHz ultrasonic vocalizations (USVs), which were reversed by resocialization. Conversely, 14-day isolations produced reductions in USVs that could not be reversed by resocialization. The USVs were also reduced by the microinjection of SP into the DPAG (35 pmol/0.2 μL), an effect blocked by pretreatment with the SP antagonist spantide (100 pmol/ 0.2 μL). Bassi et al. [316] concluded that 1- and 14-day isolations recruit distinct brain defensive systems. Most importantly, however, these studies suggest that SP might be implicated in separation anxiety and social loss predisposing effects on the development of PD.

Although the neurotransmission of both DLPAG and CePAG-LPAGcv is largely unsolved, data herein discussed suggests that the DPAG is endowed with the capability to process concomitant signals of hypoxia and hypercapnia that make up real-life asphyxia. It is then proposed that suffocation sensations arise when the VMS-mediated hypercapnia inhibition of CePAG-LPAGcv is surpassed by NTS-mediated excitation of suffocation signals of hypoxic hypercapnia. Indeed, although the high-altitude illness is devoid of panic symptoms [148], Nepalese travelers eventually report nocturnal panic attacks when hypoxia is further aggravated by hypercapnia [149].

From a molecular point of view, spontaneous panic attacks might be triggered by multiple mechanisms, including (1) phasic failure of 5-HT inhibition of CePAG-LPAGcv circuit, (2) CePAG-LPAGcv activations by DMH-mediated interoceptive signals, (3) epigenetic deficiency of EDM/ ENK transmission in CePAG and/or LPAGcv, (4) epigenetic upregulation of CCK transmission in CePAG (5) epigenetic upregulation of CCK in CePAG, (6) epigenetic upregulation of substance P transmission in DPAG. Although the occurrence of panic attacks in room-air conditions makes unlikely the involvement of NTS-CePAG pathway, panics might also be predisposed by a genetically-determined increase in DPAG neurons intrinsic sensitivity to hypoxia.

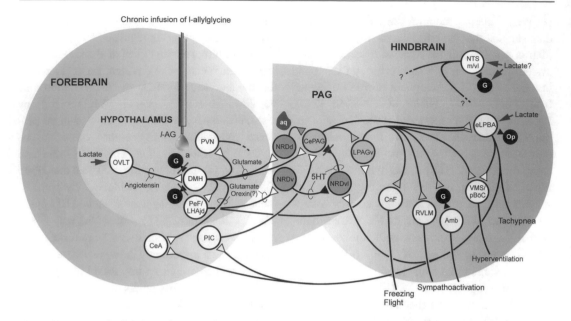

Fig. 2.17 Evidence-based presumptive pathways of lactate induced panic attacks. *Red arrows* represent c-fos labeled structures following either the hypoxia or the LAC infusions in *l*-allylglycine (L-AG) treated rats. *White and black boutons* represent excitatory and inhibitory inputs, respectively. See text for explanation. *G* gabaergic neuron, *Op* opioidergic neuron. Other labels as in abbreviation list

2.9 Modeling Lactate Vulnerability in Rats

Because LAC is a major respiratory metabolite under hypoxic conditions, Pitts and McClure [7] seminal observations that infusions of 0.5 M LAC precipitates panic attacks in predisposed individuals but not in healthy patients had a major impact in panic research and in the development of SFA theory [24, 27, 42]. Surprisingly, however, most pre-clinical research suggests that LAC infusions activate hypothalamic circuits (Fig. 2.17) that show little overlap with pathways presumptively involved in suffocation [65, 66, 231].

The hypothesis of the hypothalamic mediation of panic attacks was based on early observations that inhibitions of gabaergic transmission at PH facilitate freezing in experimental conflict, escape in Sidman's non-signaled avoidance and anxiety in the elevated plus-maze (EPM) [317–322]. Conversely, injections of a GABA-A receptor agonist (muscimol) into the PH released punished behavior [319]. Although the DMH involvement in defensive behaviors was discarded in earlier studies, further research identified

the effective sites in DMH properly [320] and, more recently, in PeF [231]. As well, Shekhar [320] showed that baseline anxiety in EPM is either increased or decreased by blocking or enhancing the gabaergic transmission in DMH, respectively. Shekhar [323] also showed that DMH-evoked defensive behaviors were blocked by subchronic treatments with the clinically effective panicolytics imipramine (5 and 15 mg kg^{-1}, 7 days) and clonazepam (5 mg kg^{-1}, 3 days).

Most notably, however, the DMH depletion of GABA by chronic microinfusions of the inhibitor of glutamate decarboxylase *l*-allylglycine (L-AG) (Fig. 2.17a) rendered rats sensitive to LAC infusions that precipitate panic in patients (i.e., 0.5 M, 10 mL kg^{-1} per 15 min) [64]. In particular, whereas the LAC infusions increased heart rate, blood pressure and anxiety in rats treated with L-AG, they were ineffective in controls treated with the inactive isomer *d*-allylglycine. Shekhar et al. [324] also showed that LAC microinjections into the organum vasculosum lamina terminalis (OVLT) produce anxiety-like behaviors that were blocked by local microinjections of both tetrodotoxin and glutamate antagonists in OVLT and DMH, respectively. In contrast, the microinjections of tetrodotoxin in the subfornical organ (SFO) and medial preoptic area were without effects, making unlikely the involvement of the angiotensinergic pathways that mediate thirst [325].

Further studies showed that LAC-evoked anxiety responses were sensitive to both panicogens and panicolytics. For instance, whereas the LAC-induced anxiety was facilitated by the putative panicogen yohimbine [321, 326], it was blocked by the panicolytic benzodiazepine alprazolam [327]. Moreover, chemical kindling of the amygdala by repeated microinjections of GABA-A and urocortin antagonists rendered rats sensitive to LAC infusions as shown by the increase of anxiety in social interaction test (SI) [328, 329]. Next, Shekhar and collaborators showed that whereas the LAC-induced anxiety was facilitated by the microinjection of angiotensin-II into the DMH, it was attenuated by both blocking of angiotensin-II receptor [322] and silencing of orexin-1 receptor gene [330]. Taken together, these studies provided support to the argument that LAC-evoked anxiety is mediated by OVLT angiotensin and orexin efferents to amygdala-projecting neurons of DMH [322, 330] (Fig. 2.17).

Although the microinjections of LAC in the subfornical organ (SFO) and medial preoptic area were without effects [324], the latter finding does not rule out the question whether the LAC-induced anxiety of rats is a by-product of either thirst [331, 332] or hunger [333] states supposed to be mediated by angiotensin and orexin, respectively. Orexin appears also involved in behavioral arousal responses [334, 335] that could have affected the rat performance in both SI and EPM. As a matter of fact, further studies reported the puzzling finding that L-AG treated 'panic-prone' rats showed increased anxiety to intravenous infusions of 0.5 M sodium chloride [336]. Because isosmolar infusions of D-mannitol were ineffective, the authors concluded that LAC-evoked panic is due to the activation of osmosensitive pathways by the increased concentration of sodium rather than of lactate. Conversely, however, Kellner et al. [337] showed that neither PD patients, nor healthy volunteers developed panic to intravenous infusions of a 2.7 % NaCl solution isosmolar to 0.5 M LAC.

Another drawback of above studies is the evaluation of 'panic' through animal models of GAD (SI, EPM). Accordingly, we cannot rule out the involvement of both angiotensin and orexin in anxieties accompanying thirst and hunger. The hypothesized participation of the amygdala in panic attacks was in turn severely compromised by the demonstration that fear-unresponsive Urbach-Wiethe disease patients lacking the amygdala develop panic attacks both spontaneously [76] and in response to 35 % CO_2 [77]. That amygdala does not mediate panic was also shown by pre-clinical studies in rats showing that kindling of the amygdala facilitates fear-like resistance to capture while inhibiting the panic-like responses to electrical stimulation of DPAG [338].

There are at least two more reasons that make unlikely the DMH mediation of panic attacks properly. First, although the LAC-induced panic attacks of humans are not accompanied by activations of HPA axis, chemical stimulations of DMH in awake rats produced over 650 % increases in ACTH plasma levels [339, 340]. Not surprisingly, the PVN is the main recipient of DMH projections [311]. Second, data from our laboratory showed that high-resolution frequency-varying low-intensity stimulation (<50 µA, a.c.) of DMH *pars diffusa* (DMHd) produced only exophthalmus, whereas that of DMH *pars compacta* (DMHc) produced exophthalmus, immobility, defecation and micturition (A.C. Alves, unpublished results). The limited repertoire of DMHd was confirmed by chemical stimulations with *N*-methyl-D-aspartic acid (NMDA). The latter study showed in addition that while the electrical and chemical stimulations of DMHd elicited robust ingestive behaviors in sated rats, unintentional stimulations of dorsomedial VMH (VMHdm) elicited all defensive responses of the rat, including trotting, galloping and jumping, at thresholds similar to those of DPAG. The latter data suggest that the eventual elicitation of escape during DMH stimulations may be due to the spreading of either current or drug to VMHdm. As a matter of fact, a recent study in humans reported panic attack provocation following electrical stimulation of VMH [78].

Most importantly, Johnson et al. [341] examined c-fos protein expression in L-AG treated rats following the intravenous infusion of either LAC or 0.9% NaCl. Compared to saline treated controls, LAC infusions of panic-prone rats produced significant increases in c-fos expression in OVLT (171%) but not in SFO or AP regions which are also devoid of blood-brain barrier. There were also significant activations of DMH (84%), PVN (150–360%), SON (243%), CeA (241%), PBA (382%) and NTS (700%) (red arrows in Fig. 2.17). Remarkably, the authors stressed that the PBA was '*the only major respiratory center that clearly responded*' to LAC infusions. Moreover, LAC-induced activations of PBA were significantly correlated with increases in respiratory rate, but not with heart rate, blood pressure or anxiety. Although the NTS neurons were also markedly labeled in panic-prone rats, neurons were very often immunoreactive for both c-fos and GAD, thereby suggesting the depolarization of gabaergic interneurons rather than output neurons [341]. In the same vein, LAC infusions failed in activating the ventral medullary areas which are established targets of NTS, including rostral and caudal VLM, pBöC and nucleus ambiguous. Neither did the LAC activated the VMS. The only exception was a light though significant activation of C1 adrenergic neurons involved in both tonic and phasic control of blood pressure.

Lactate brain activations are consonant with respiratory effects of 0.5 M and 2 M LAC infusions in freely-moving Wistar rats [342]. Indeed, 0.5 M LAC increased respiratory rate while decreasing the tidal volume, thereby suggesting that LAC activate predominantly the PBA [265]. Most notably, Johnson et al. [341] showed that LAC infusions activate the same subnuclei of LPBA targeted by the LPAG (i.e., superior lateral, rostral lateral, dorsal lateral, medial and lateral crescent areas) and the CePAG (inner division of eLPBA) (note, however, that the inner division of eLPBA was mistakenly named 'ventrolateral PBN', personal communication of P.L. Johnson). These areas show partial overlap with terminal fields of the NTS efferents to dorsal, central and eLPBA subnuclei of LPBA [343–345]. Conversely, LAC marked activations of

PBA are hardly explained by DMH occasional farthest projections to this nucleus [311].

Recent data suggest, on the other hand, that PBA activations may be independent from NTS afferents. In particular, Kaur et al. [346] showed that 2-h exposures of mice to 10% CO_2 activates VLM glutamatergic projections to both the external lateral and the lateral crescent subnuclei of PBA which project in turn to respiratory motor neurons. These authors also showed that selective delection of the gene of vesicular glutamate transporter-2 blocked respiratory arousal to hypercapnia [347]. Taken together, these studies suggest that VLM glutamatergic projections to LPBA are implicated in respiratory activations to CO_2. However, it is worth remembering that 0.5 M LAC infusions did not activate any medullary respiratory structure.

Importantly, as well, PBA neurons are depressed by opioids [348, 349]. Accordingly, it is tempting to speculate that respiratory activations to naloxone-preceding either the hypoxia or the LAC infusions in healthy volunteers were due to blockade of opioid receptors of PBA [120]. This mechanism could also explain why these infusions produced hyperventilation but not panic [120]. Consonant with the latter finding, LAC infusions of 'panic-prone' rats did not increase c-fos expression in any region of PAG [341]. The latter data are also in agreement with the lack of effects of 0.5 M LAC infusions in KCN-evoked experimental panic presumptively mediated at DPAG (E.A. Moraes, unpublished results). It remains to be examined whether the KCN-evoked panic is facilitated by higher concentrations of LAC. Indeed, Olsson et al. [342] suggested that human infusions with 0.5 M LAC are better modeled by rat infusions with 2 M LAC.

On the other hand, double-labeling of both c-fos protein and tryptophan hydroxylase showed that whereas the LAC infusions activate 5-HT neurons of NRDlw in D-AG treated controls (yellow arrow in Fig. 2.17), they are ineffective in L-AG treated 'panic-prone' rats [231, 260]. Although the DMH does not project significantly to DPAG or NRDlw, it does project to CePAG [310, 312] (Fig. 2.17) and, indirectly, to NRDv via massive projections to PeF/lateral hypothalamus [52]. There is also evidence that nearly 65 % of DMH projections to PAG are glutamatergic [297, 350]. Indeed, the microinjection of glutamate antagonists into the DLPAG/LPAG suppresses the cardiovascular responses to chemical stimulations of DMH [351, 352]. In turn, the NRDv projects mostly to VLPAG and CePAG [164]. Because 5-HT actions are predominantly inhibitory in VLPAG/NRDlw [309], these data suggest that LAC behavioral effects may be due to the NRDv-mediated inhibition of NRDlw inhibitory inputs to CePAG-LPAGcv (Fig. 2.17). It is also proposed that 5-HT excites CePAG via 5-HT$_2$ receptors [309] (Fig. 2.17). The disruption of gabaergic tone in the DMH could thus facilitate the respiratory-type panics both directly, via facilitation of DMH glutamatergic projections to NRDd and CePAG, and indirectly, via NRDv disinhibition of CePAG-LPAGcv panic circuit (Fig. 2.17). These data give support to the argument that subjects vulnerable to LAC infusions present deficiencies in both GABA and 5-HT inhibitory inputs to DMH and CePAG-LPAGcv panic circuit, respectively.

On the other hand, it is a quite remarkable fact that 0.5 M LAC infusions neither activated the PAG [341], nor facilitated the KCN-evoked panic attacks of rats (E.A. Moraes, unpublished results), or produced panics in naloxone-treated healthy volunteers [120]. LAC infusions also did not facilitate KCN-evoked panics of rats that were either pre-treated with naloxone or neonatally-isolated throughout lactation (E.A. Moraes, unpublished results). Nonetheless, LAC infusions produced marked activations of the LPBA and, mostly, of the inner division of eLPBA (see Fig. S3b, d of Johnson et al. [341] which is the specific target of CePAG [284]. Remarkably, as well, the LPBA sends profuse efferents to primary interoceptive areas of PIC [353] that project in turn to the VLPAG [354, 355] (Fig. 2.17). The LPBA-PIC-VLPAG circuit could thus be the basis of panic attack cognitive theories of the 'catastrofization' of bodily symptoms [21–23]. Indeed, the insula is one of few structures activated by both LAC and CO_2 in patients and volunteers, respectively [205, 206].

Concluding, although the evidence is mixed, pre-clinical data suggest that LAC-evoked behavioral effects might be provoked by (1) GABA deficient inhibition of DMH excitatory inputs to CePAG-LPAGv, (2) DMH/NRDv inhibition of NRDlw inhibitory inputs to CePAG-LPAGv, (3) NTS-mediated activation of PBA-CePAG-LPAG, and, (4) LAC direct activations of LPBA-PAG or LPBA-PIC-VLPAG circuits. Admittedly, the mechanisms underlying LAC-evoked panic attacks remain largely obscure almost 50 years from its discovery by Pitts and McClure [7].

2.10 Modeling Neuroendocrine Unresponsiveness of Panic Attacks

The concept of stress was coined by Hans Selye [356, 357] to denote a complex syndrome caused by various noxious agents thereafter named 'stressors'. Selye's 'stress hypothesis' had a widespread influence on medical world and, mostly, immunology and psychiatry. In particular, clinical and experimental studies implicate stress in depression, panic, and post-traumatic stress disorder [358–362]. Selye's hallmark was, however, the doctrine of the non-specificity of the *alarm reaction* according to which "stress" became almost a synonym of the HPA axis activation that follow the exposure of mammals to a perplexing number of stressors. Further studies showed that PRL secretion is also increased in response to a great number of physical and psychological stressors [363–365].

Notably, however, plenty of evidence showed that ACTH, COR and PRL are unaltered during panic attacks. Indeed, neither the situationally provoked panic attacks of agoraphobics [16] nor the experimentally provoked panics to LAC and CO_2 [13, 15, 17–19, 366] increased stress hormones significantly relative to healthy controls. COR plasma levels actually decreased in ten panicking subjects exposed to 7 % CO_2 [366]. Despite a small sample size, Cameron et al. [14] also did not find any change in stress hormones either at peak or 10 and 60 min after nine spontaneous panic attacks in four hospitalized patients.

The neuroendocrine unresponsiveness of clinical panic is both intriguing and an important clue for understanding the neurobiology of PD. In particular, Preter and Klein [27] suggested that upon suffocation HPA axis activation would counter-productively increase catabolic activity beyond the adaptive capacity of the organism.

Cortisol concentrations were nevertheless increased in the saliva of patients having severe panic attacks [367]. The HPA axis is also activated in human, fear-like, panics marked by palpitations, tremor and sweating, but devoid of suffocation symptoms. These fear-like panics can be provoked by drugs that either produce anxiety, such as β-carboline, yohimbine, pentylenetetrazole and fenfluramine [368, 369] or stimulate *in vitro* neurons of the PVN [368]. Therefore, although the CCK-related peptides (CCK-4, CCK-8S, and pentagastrin) produce panic attacks and HPA activations [370–372], neuroendocrine effects of these peptides appear to be due to pharmacological stimulation of PVN neurons by mechanisms unrelated to panic. Indeed, both *in vivo* and *in vitro* studies showed that CCK stimulates DMH and PVN neurons either directly or indirectly via the activation of peripheral receptors [368, 373]. Moreover, the chronic treatment with a clinically effective panicolytic (citalopram) decreased the intensity of the CCK-induced panic attacks; however, it did not change the HPA axis response to this neuropeptide [374]. Lastly, CCK-induced panic attacks were blocked by the 5-HT$_3$ antagonist odasentron [375], vagotomy and CCK-A receptor antagonists [368, 376], thereby suggesting the contribution of a peripherally mediated component.

Because humans are susceptible to both suggestion and procedural anxiety, the demonstration that DPAG-evoked panic responses do not activate the HPA axis would be invaluable evidence of DPAG mediation of panic attacks. Indeed, a preliminary study by our group showed that neither the ACTH nor the PRL secretion increased 5 and 15 min after the panic-like behaviors produced by 1-min electrical stimulation of the DPAG [377]. Nonetheless, further analyses showed that corticosterone (CORT) plasma levels were significantly increased (285.2 ± 8 ng/mL)

one week after the electrode implantation (unpublished results). Accordingly, the lack of ACTH responses might have been due to the CORT increased level inhibition of HPA axis in rats recently implanted with both an electrode (7 days prior to testing) and an indwelling catheter (2 days prior to testing). Additionally, Lim et al. [378] reported a conspicuous increase (160%) in CORT plasma level 30 min after a DPAG-evoked 1-min explosive flight bout ('*running with aimless direction*') in 4-week surgery-recovered rats presenting reduced baseline levels of CORT (70 ng/mL). Such conflicting data might be explained by the different degrees of physical effort during the DPAG-evoked flight behaviors of rats stimulated in arenas of rather different areas (0.3 m^2 versus 1 m^2). That physical effort is relevant in HPA axis responses was already shown by Schenberg et al. [377] report of a moderate (83%) though non-significant increase in ACTH plasma levels following the exhausting effort of DPAG-evoked repetitive flight bouts.

Most importantly, however, a recent study from our group showed that stress hormones remain unchanged when DPAG-evoked escape is prevented by stimulating the rat in a 20-cm diameter roofed cylinder (0.03 m^2) with stimuli that elicited full-blown flight behaviors in a 55-cm diameter open-field (0.3 m^2) [379] (Fig. 2.18). Neither did the ACTH increase when physical exertion was statistically adjusted to the average effort of non-stimulated controls as measured by LAC plasma levels. In contrast, foot-shocks of the same duration as the intracranial stimulus produced marked neuroendocrine responses that were not correlated with muscle activity (Fig. 2.19).

The importance of physical exertion in DPAG-evoked neuroendocrine responses is further evidenced by studies showing that rat escape responses to both DPAG stimulations and 20-kHz artificial alarm calls attain maximum average speed of 1.9 m/s (6.8 km/h) [380, 381]. This speed is over four times the rat treadmill speed (1.5 km/h) required to produce significant increases in both PVN c-fos expression and LAC and ACTH plasma levels [382].

The HPA axis unresponsiveness to DPAG stimulations is further supported by the lack of

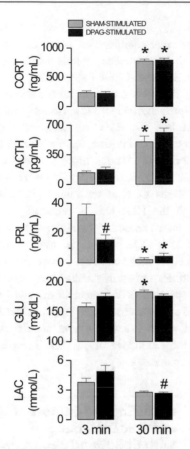

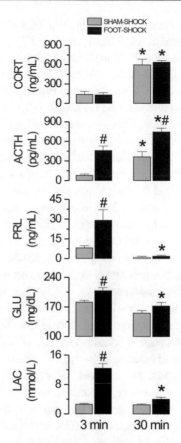

Fig. 2.18 Hormone and metabolite plasma levels following electrical stimulations of DPAG of rats placed in a small compartment (20-cm diameter roofed-cylinder) that prevented flight behavior. Rats were stimulated either fictively (SHAM) or at the flight threshold intensity (DPAG) determined 2 weeks before in a 60-cm diameter open-field. Columns represent means (±SEM) from blood samples collected 3 or 30 min after the end of 1-min intracranial stimulus. Groups further exposed to a 60-cm diameter brightly-lit open-field (30-min groups) showed marked increases in both CORT and ACTH, confirming the HPA responsiveness to a novel environment. Abbreviations: *ACTH* corticotropin, *CORT* corticosterone, *GLU* glucose, *LAC* lactate, *PRL* prolactin. Symbols indicate (*) significant differences between 3- and 30-min plasma levels for the same treatment and (#) significant differences between SHAM and DPAG groups for the same sample interval (from Armini et al. [379])

Fig. 2.19 Hormone and metabolite plasma levels 3 or 30 min after the application of 1-min foot-shocks either effective (1 mA) or fictive. Note the marked increase in ACTH, PRL and LAC 3 min after the end of foot-shock. Symbols indicate (#) significant differences from sham-shocked group for the same blood sampling interval, or (*) significant differences from 3-min plasma level for the same shock condition. Other details as in Fig. 2.18 (from Armini et al. [379])

DPAG excitatory projections to PVN. Indeed, Pittman et al. [383] showed that PAG stimulations activated only 2 of 188 neurons tested in the PVN. In turn, Ziegler et al. [384] recently showed that PAG glutamatergic efferents to the PVN arise only from the VLPAG and the yet undifferentiated commissural PAG. These regions are

either much ventral or much rostral to the stimulated sites of our studies, respectively.

The above data might suggest that HPA axis unresponsiveness in clinical panic is solely due to the lack of DPAG projections to PVN. Remarkably, however, the HPA axis is likewise unresponsive to 20-kHz artificial alarm calls that cause full-blown flight behaviors, tachycardia and widespread activations of DPAG, DMH, BLA, CeA and paraventricular nucleus of the thalamus (PVT) but not of PVN [381, 385]. Because the HPA axis is robustly activated by stimulations of both DMH [340] and CeA [386], the HPA axis is most likely inhibited during both the DPAG stimulation and the 20-kHz artificial alarm call. In particular, Bhatnagar et al. [387] presented evidence of a PVT-mediated inhibition of HPA axis by CCK afferents from VLPAG (note, however, that the CCK neurons of PAG are predominantly localized along the CePAG [298]).

Because the HPA axis is markedly activated in both stressful and non-stressful conditions, a recent multinational authoritative consensus review on stress proposed that the "*use of terms 'stress' and 'stressor' should be restricted to conditions and stimuli where predictability and controllability are at stake*" [388]. Yet, panic attacks are neither controllable nor predictable and do not activate the HPA axis. Koolhaas et al. [388] added that stress should be reconceptualized as "*a stimulus or environmental condition in which the response demands exceed the adaptive capacity of the organism*". As a result, the homeostatic responses should be modified (allostatic changes) to achieve stability "*in anticipation*" of physiological requirements of stress [388]. The latter argument is consonant with Preter and Klein's [27] contention that under suffocation, the acute activation of the HPA axis would counterproductively increase the oxygen demand beyond the adaptive capacity of the organism. Accordingly, Preter and Klein [27] proposed that fear neuroendocrine response should be modified during asphyxia to allow energy conservation until a possible escape.

Nevertheless, whereas the 20-min exposure of anesthetised dogs to low levels of hypercapnia (6% CO_2) did not alter the secretion of both ACTH and CORT, 20-min exposures to moderate hypoxia (10% O_2) produced a 175% increase in the ACTH plasma level that was attenuated (albeit not abolished) by chronic chemodenervation of the carotid body [389]. HPA axis activations of anesthetized animals were even more conspicuous with higher levels of both hypoxia and hypercapnia [390, 391]. Most importantly, however, acute exposures (20 min) of conscious rats to normocapnic hypoxia (7% O_2), hypercapnia (8% CO_2) or hypercapnic hypoxia (7% O_2, 8% CO_2) produced ACTH increases of approximately 100%, 200% and 300%, respectively [227]. While these findings contradict Preter and

Klein's [27] argument, the lack of HPA responses appears nevertheless to hold in prolonged hypoxia. For instance, neither the secretion of ACTH, nor that of CORT showed any change following the 42-h exposure of conscious rats to hypocapnic hypoxia [392]. Consequently, data suggest that the PAG inhibition of HPA axis might be masked during acute asphyxia by neuroendocrine activations brought about by NTS excitatory projections to PVN, DMH, CeA and bed nucleus of stria terminalis [187]. Finally, although the HPA axis is reportedly activated during panics caused by two tidal-volume inhalation of 35% CO_2, NTS projections to both PAG and PVN are believed to be inactive during room-air conditions of spontaneous panic attacks.

Irrespective of the mechanism involved, the lack of stress hormone responses in the DPAG-evoked panic-like behaviours stands out as a compelling evidence of the DPAG mediation of clinical panic.

2.11 Modeling the Comorbidity of Panic Disorder with Childhood Separation Anxiety

Although clinical and epidemiological evidence suggests that CSA predisposes the subject to adult-onset PD (for review, see Aschebrand et al. [371]), other studies reported that neither CSA [393] nor early-life adversity [35, 41] had any effect on later incidence of panic attacks. For instance, Roberson-Nay et al. [41] showed that although PD and CSA share a common genetic diathesis, childhood adversities account for only 1.2% of adult-onset panic attacks. Conversely, Spatola et al. [153] reported that early-life adversities do correlate with CO_2 hypersensitivity provided that the stressful events took place before 18 years of age. Accordingly, the influence of the early-life adversities in the later development of PD remains unclear. Whether genetic or epigenetic, the brain mechanisms whereby CSA predisposes subjects to PD are completely unknown.

Although the comorbidity of psychiatric disorders is rarely studied in animals, a recent study of Quintino-dos-Santos et al. [137] examined whether the neonatal social isolation (NSI), a model of CSA, facilitates panic-like behaviors produced by electrical stimulations of the DPAG in adulthood. To rule out the influences of genetic background and caregiving behaviors, the authors employed a split-litter design in which half of male pups were subjected to 3-h daily isolations throughout lactation (PN2-PN21) while siblings and dam were moved to another box. Because siblings were both handled and kept with a separation-anxious mother, they were considered a sham-isolated group. Experimental groups were further compared with controls that remained undisturbed in home-cages until weaning. At 60 days of age, the rats were implanted with electrodes into the DPAG and, 1 week later, subjected to sessions of DPAG stimulation, EPM and forced-swimming (FST) for the assessment of panic vulnerability, baseline anxiety and depressive mood, respectively.

Results showed that DPAG-evoked panic-like responses of immobility, exophthalmus, trotting, galloping and jumping were significantly facilitated in NSI rats relative to both sham-isolated and control groups (Fig. 2.20). In contrast, NSI rats did not show any change in either anxiety or depression relative to sham-isolated rats. Groups also did not differ with respect to defecation or micturition responses to DPAG stimulations. The latter observations agree with previous studies suggesting that DPAG-evoked defensive behaviors (freezing and flight) and pelvic viscera responses (micturition and defecation) are processed by functionally distinct systems [60, 165, 246, 394, 395]. As it regards, it is pertinent that urges of defecation and micturition are neither experienced by patients during panic attacks [53, 396] nor recognised as symptoms typical of clinical panic [397, 398]. In any event, the NSI facilitation of DPAG-evoked defensive behaviors are the first behavioral evidence in animals that

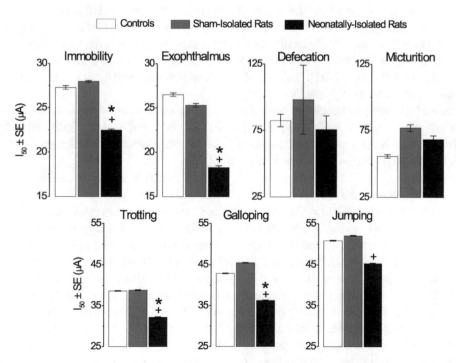

Fig. 2.20 Median threshold intensities ($I_{50} \pm SE$) of DPAG-evoked behaviors of neonatally-isolated rats, sham-isolated rats and non-handled controls. Symbols represent values significantly different from controls (*) and sham-isolated rats (+) for Bonferroni's 5% criterion (likelihood ratio chi-square tests for curve location) (modified from Quintino-dos-Santos et al. [137])

early-life separation stress selectively facilitates panic-like behaviors in adulthood. Importantly, as well, these data implicate the DPAG not only in panic attacks but also in the predispositions of separation-anxious children to the later development of PD.

The study of Quintino-dos-Santos et al. [137] is consonant with previous data showing that neonatally-isolated adult rats present sex-dependent facilitations of panic-like respiratory responses to hypoxia (males) and hypercapnia (females) [37, 138, 399–402]. As a corollary, the PAG is unresponsive to hypoxia during PN2-PN12 [225] 'stress hyporesponsive period' in which the mother shields the pup against early-life adversities [256–258]. Although Dumont et al. [138] suggested that NSI facilitates the central processing of chemoreceptor afferent inputs, Quintino-dos-Santos et al. [137] presented evidence that neonatal separation may sensitise DPAG regions that mediate defensive responses. As a matter of fact, predators are the major threat to the separated pup. Otherwise, NSI facilitations of DPAG-evoked panic could be the result of enduring plastic changes of both ascending and descending projections of DPAG. Lastly, panic facilitations could be the outcome of early-life programming of HPA axis [403]. Indeed, the HPA axis is known to be hyperactive in both panic patients [404, 405] and adult rats subjected to a 24-h single period of mother deprivation [128, 406] or to 3-h daily periods of maternal separations [400, 407] during the 'stress hyporesponsive period'.

The study of Quintino-dos-Santos et al. [137] is reminiscent of Rachel Klein's [36] 15-year double-blind interview-based follow-up study of children presenting manifest symptoms of CSA (school refusal). The latter study showed that although the diagnoses of major depression disorder (MDD) did not differ from controls, separation-anxious subjects showed significant increases in the frequency of both panic attacks and hospitalizations due to depressive episodes.

Unexpectedly, however, recent studies from our laboratory showed that rats subjected to 3-h daily maternal separations along the lactation period (PN2-PN21) were more resilient than both controls

and rats separated during 'stress hyporesponsive period' only (PN2-PN12) (A.C.B. Aguiar, unpublished results). In particular, 21-day separated rats showed longer swimming duration in FST, increased appetite for sugar in sucrose preference test (SPT), higher thresholds of DPAG-evoked panic responses and higher weight gains. Procedures of latter study were nevertheless rather distinct from Quintino-dos-Santos et al. [137]. For instance, whereas the treatments of latter study were applied to siblings from the same litter, those of recent study were applied to pups of independent litters. Moreover, whereas the previous study employed group-reared adult rats, the recent study examined rats reared individually. As well, Quintino-dos-Santos et al. [137] surgery (bone removal for sinus exposure) and recovery period (7-day in glass-walled brightly-lit individual cages) were more stressful than the surgery (burr hole only) and recovery (28-day period in polypropylene individual cages) of the recent study. Yet, because sham-isolated rats of former study did not differ from controls [137], the NSI facilitation of DPAG-evoked behaviors was most probably due to the interaction of early-life and adult-life stress. Indeed, Roberson-Nay et al. [41] presented epidemiological evidence that 60 % of variation of adult-onset panic attacks is due to environmental factors of adulthood.

2.12 Modeling the Comorbidity of Panic and Depression

Panic disorder is highly comorbid with depression [154, 408, 409]. For instance, a 10-year interview-based prospective study showed that 13.6 % and 27.3 % PD patients at age 30 had either MDD or recurrent brief depression (RBD), respectively [154]. Notably, as well, the latter authors found that 10-year prevalence of depression and suicide attempts in PD patients is 6.8 and 4.2 times higher relative to the general population. Clinical data also suggest that panic is facilitated by both acute and post-traumatic stress disorders [410–414]. In contrast to the existing evidence of a common genetic diathesis of PD and CSA [41], mechanisms underlying the comorbidity of panic and depression disorders remain nevertheless completely obscure. It is also unclear whether PD is enhanced by any kind of depressive disorder. In particular, McGrath et al. [415] reported that LAC sensitivity did not differ in controls and depressed outpatients without a history of panic attacks.

Because spontaneous panic attacks are conspicuously uncontrollable stress, a recent study of our laboratory assessed the late effects of uncontrollable shocks, a presumptive model of depression and/or trauma, on DPAG-evoked panic behaviors [416]. Briefly, rats with electrodes in the DPAG were subjected to a 7-day shuttle-box one-way escape yoked training with escapable (ES) or inescapable (IS) foot-shocks. Controls were subjected to fictive shock (FS) sessions. The day after the termination of one-way escape training, the rats were trained in a two-way escape novel task to ascertain the effectiveness of uncontrollable stress. Data showed that

the IS group performed significantly poorer than the ES group in two-way escape task. Unexpectedly, however, IS rats showed a marked attenuation of DPAG-evoked freezing and flight behaviors relative to both ES and FS groups, 2 and 7 days after one-way escape training (Fig. 2.21). Moreover, whereas the threshold of DPAG-evoked freezing and flight behaviors of IS rats remained high or were further increased 7 days after escape training, thresholds of DPAG-evoked defecation and micturition did not change or were even reduced. The latter data support previous studies [246, 395] suggesting the separate processing of defensive behaviors (freezing and flight) and pelvic viscera responses (micturition and defecation) produced by electrical stimulation of DPAG. Most importantly, IS inhibited DPAG-evoked defensive behaviors in spite of the striking differences in the aversive stimulus (foot-shock vs intracranial

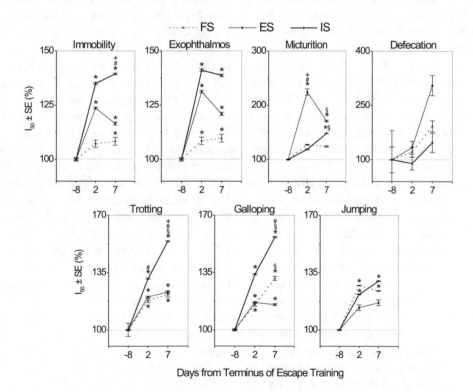

Fig. 2.21 Percent changes of median threshold intensities (I50±SE) of DPAG-evoked defensive behaviors in the 2nd (2) and 7th (7) days after the end of one-way escape training with foot shocks either fictive (FS), escapable (ES) or inescapable (IS). Screening sessions were carried out the day before the onset of one-way escape training (days −7 to 0). Symbols indicate significant differences relative to (*) baseline value, (§) DPAG second stimulation session, or to (+) ES and (#) FS groups (Bonferroni's 5 % criterion of likelihood ratio chi-square tests for location of parallel-fitted threshold curves) (from Quintino-dos-Santos et al. [416])

stimulus) and context (shuttle-box vs open-field) [416]. Accordingly, effects cannot be attributed to either context conditioning or stimulus sensitisation.

In the same vein, experiments carried out in the elevated T-maze (ETM) detected 'panicolytic effects' the day after the rat exposure to IS, FST or restraint stress [417]. Thus, whereas the ETM anxiety-like behavior (avoidance from open-arms) was enhanced, the ETM panic-like behavior (escape from open-arms) was attenuated. The latter effect closely resembles the attenuation of DPAG-evoked escape in inescapably-shocked rats [416].

Plenty of evidence suggests that subjects exposed to uncontrollable stress develop a depression-like syndrome characterised by decreased motivation to respond to the same or other aversive stimuli, by cognitive deficit (learned helplessness), and by emotion and mood effects including an early increase in anxiety and the later development of depression. Data from yoked experiments further showed that these effects result from the subject learning that stress is beyond his control and not from the stressor aversiveness [418–421]. Similarly, the FST is a widespread procedure for the screening of potential antidepressants [422] based on the assumption that floating correlates with depressive mood. Regardless of whether or not uncontrollable stress produces true depressed mood, IS inhibition of escape to both foot-shock and intracranial stimulus implicates the DPAG as the motivational substrate of both escape responses.

Although most researchers associate the outcomes of uncontrollable stress with putative changes in the amygdala [423–425], hippocampus [424, 426–431], or both structures [432, 433], Lino-de-Oliveira et al. [434, 435] showed that whereas the microinjections of glutamate into the DPAG reduce floating behavior, microinjections of lidocaine had the opposite effect. Moreover, they showed that sub-chronic administrations of antidepressants reduce FST-induced increases in FLI in most columns of the PAG [434, 435]. Most notably, however, evidence from positron-emission tomography in rats (microPET) showed that whereas the PAG is markedly activated during the FST training session, it remains inactive during next day test session [436]. Taken together, these data implicate the DPAG in behavioral effects of uncontrollable stress.

Remarkably, as well, freezing was also attenuated one week after the exposure to IS. Accordingly, the IS attenuation of DPAG-evoked escape behaviors cannot be ascribed to an enhancement of freezing at the expense of trotting and galloping. The impairment of both passive (freezing) and active (flight) defensive behaviors is best explained by a deactivation of a DPAG in-built motivational system. If so, the presumptive 'panicolytic effects' of uncontrollable stress on ETM [417] and DPAG-evoked defensive behaviors [416] are better explained by a decrease in resilience (escape failure) to stress. The latter argument is further supported by the higher degree of attenuation of trotting and galloping which are the common responses to both DPAG stimulations and inescapable foot-shocks.

Conversely, however, Strong et al. [437] presented evidence that 5-HT transmission in the dorsal striatum (DS) plays a crucial role in learning deficits of inescapably-shocked rats. Indeed, whereas the microinjections of a $5-HT_{2C}$ antagonist into the DS prevented escape failures of rats previously exposed to IS, microinjections of a $5-HT_{2C}$ agonist impaired learning even in the absence of the prior exposure to IS. Therefore, it is tempting to speculate that DPAG and DS mediate motivational and cognitive effects of helplessness, respectively. On the other hand, Tannure et al. [338] presented evidence of a two-process flight system. In particular, whereas the kindling of the amygdala facilitated the flight responses from the experimenter (capture resistance), it inhibited the DPAG-evoked flight behaviors. Be this as it may, these systems might be bridged by DPAG profuse efferents to intralaminar and midline thalamic nuclei [438] that exert diffuse excitatory actions on both cortex and striatum [439].

In turn, clinical and epidemiological evidence suggest that the first episode of MDD is very often precipitated by uncontrollable stress, including social loss, bond breakdown, disease and unemployment. There is additional evidence that PD predisposes to the later development of both depression [154, 408–410, 440] and trauma [410–412, 414, 441]. Furthermore, patients with post-traumatic stress disorder (PTSD) not only experience the physiological symptoms of panic but also fear these symptoms [414, 442]. Because the consequences of uncontrollable stress have been used as a model of both depression [418–421, 443] and trauma [444], the DPAG-evoked panic-like behaviors were expected to be facilitated in inescapably-shocked rats. Accordingly, the IS unexpected inhibition of DPAG-evoked panic behaviors may be a idiosyncratic feature of the present model of depression. Indeed, whereas PD is most often associated with MDD and RBD [154], the exposure to uncontrollable stress is reminiscent of the 'reactive depression' nowadays termed 'adjustment disorder with depressed mood' [397]. Indeed, DPAG-evoked panic behaviors were markedly facilitated in olfactory bulbectomy model of depression (J.W. Quintino-dos-Santos, unpublished results).

In conclusion, the conjoint inhibition of passive (freezing) and active (flight) defensive behaviors suggests that IS deactivates a DPAG in-built motivational system that may be implicated in depressed patient's difficulties to cope with the stressors of daily life.

2.13 Modeling Female Vulnerability to Panic Disorder

Despite the high variability of epidemiological survey data, the prevalence of panic attacks is consistently higher in women. In particular, at age 30, the Zurich study found panic attacks in 10 % of the population, with a sex ratio of 3:1 in favor of females [154]. Although panic attacks are rarely observed before puberty or after menopause, they are more frequent and severe during the premenstrual phase [445]. Patients with LLPDD also show higher sensitivity to panicogenic challenges in the late luteal phase [446–448].

These and other data prompted Lovick and collaborators to carry out several studies that examined the neural excitability of the DPAG during the estrous cycle [155, 449–451]. These authors showed that the neurons of the DLPAG increase the expression of the alpha4/beta1/delta ($\alpha 4/\beta 1/\delta$) subunits of the GABA-A receptor in the late dies-

trus relative to the other phases of the estrous cycle or to their expression in male rats [452]. Moreover, they showed that most GABA$\alpha 4/\beta 1/\delta$ receptors of the DLPAG are localised in gabaergic neurons (autoreceptors), as revealed by the double labelling of these subunits and glutamic acid decarboxylase [453, 454]. Because the GABA$\alpha 4/\beta 1/\delta$ receptors are highly sensitive to GABA [452], the upregulation of GABA$\alpha 4/\beta 1/\delta$ autoreceptors is expected to reduce the GABA inhibitory tonus within the DLPAG. Accordingly, Lovick and collaborators suggested that the increased expression of GABA$\alpha 4/\beta 1/\delta$ autoreceptors may contribute both to the development of LLPDD and to the high comorbidity of LLPDD and PD.

Lovick and collaborators also showed that the pressor response, tachycardia, and tachypnea produced by the systemic injection of a CCK-related panicogen (pentagastrin) were enhanced in the late diestrus. For rats in estrus, the pressor response and tachycardia, but not the tachypnea, were significantly larger than the response evoked in the early diestrus [455].

Lastly, extracellular recordings from output neurons in the DPAG showed that the increased firing rate produced either by the intravenous administration of pentagastrin or by the iontophoretic injection of the GABA-A receptor antagonist bicuculline, is significantly increased during the estrus and late diestrus relative to proestrus and early diestrus [456].

On the other hand, the bimodal incidence of catamenial epilepsy suggests that brain excitability increases in the middle of the cycle, between the peak of estradiol and the early surge of progesterone, and during the sharp fall of progesterone prior to menses [457]. Although the neuroexcitant properties of estradiol are responsible for the increased excitability around the midcycle peak, the higher excitability in early and late luteal phases are most likely associated with changes in brain levels of the neuroactive metabolite of progesterone, allopregnanolone. Indeed, whereas the acute injections of both progesterone and allopregnanolone have manifest anxiolytic effects in rats, the 72-h exposure to progesterone increases anxiety, akin to that observed in the early surge of progesterone in women. Thereafter, anxiety decreases to baseline levels in spite of the sustained exposure of rats to progesterone. Anxiety is nevertheless reinstated following the sudden withdrawal of progesterone, which mimics the rapid decay of this hormone preceding menses [457, 458].

According to Smith and collaborators, the increases in anxiety in the early and late luteal phases appear to be due to an allopregnanolone-mediated upregulation of the $\alpha4$ subunit of GABA-insensitive receptors [459–461]. Increases in progesterone withdrawal anxiety in rats can be prevented by indomethacin blockage of the breakdown of progesterone to allopregnanolone. Therefore, allopregnanolone rather than progesterone is the active compound [462]. Indeed, the post-treatment allopregnanolone levels in LLPDD patients are significantly lower in improved subjects relative to unimproved ones. Improvement was also significantly associated with lower allopregnanolone levels for premenstrual depression and appetite changes [463].

Overall, the above studies suggest that the increased incidence of anxiety and panic in the late luteal phase appears to be mediated by the distinct effects of allopregnanolone in GABA transmission, i.e., the upregulation of GABA-insensitive receptors in amygdala areas controlling anxiety and the upregulation of GABA-hypersensitive autoreceptors in periaqueductal areas controlling panic. These mechanisms may be complementary in determining the high incidence of panic in women.

2.14   Conclusion

Existing evidence suggests that DLPAG is an 'exteroceptive column' that integrates both the amygdala-processed odor inputs from VMH and PMD and visuo-acoustic inputs from DLSC [249, 464–468]. These circuits appear to mediate the non-respiratory type of panic attacks [52]. By contrast, evidence herein discussed suggests that the CePAG is an 'interoceptive column' that receives afferents from NTS and projects to both the LPAGcv and to the inner division of eLPBA. As shown in the present study, the

CePAG-LPAGcv circuit may be the very core of respiratory-type panic attacks. This circuit appears to be modulated by NRDlw inhibitory projections that gate asphyxia signals. Recent c-fos immunohistochemistry evidence from our laboratory suggests, on the other hand, that the early escape response to a low dose of KCN is mediated by the LPAGr and the VLPAGc atop the LDTg nucleus (C.J.T. Muller, unpublished observations). It remains to be elucidated whether the KCN-sensitive area of the VLPAGc corresponds to the tachypnea-related area of the LPAGcv [248, 278]. In turn, both DLPAG and LPAGcv project to the CnF (the midbrain locomotor region) [162]. Although electrical and chemical stimulations of CnF produce behavioral and cardiorespiratory responses quite similar to those of the stimulation of the DPAG [280, L.C. Schenberg, unpublished results], it is unknown whether these responses are attenuated by clinically-effective panicolytics, as shown for DPAG-mediated panic-like responses to both electrical stimulations and cyanide injections.

The PAG-evoked panic responses satisfy most criteria of a translational model of clinical panic, i.e., face validity (homology of symptoms and physiological responses), predictive validity (drug sensitivity) and construct validity (facilitation by hypoxia, female hormones and early-life stress). Moreover, PAG stimulations in humans produce emotional, neurological and autonomic responses strikingly similar to those of panic attacks either spontaneous or provoked by intravenous infusions of LAC. The PAG model of panic is also endorsed by the attenuation of DPAG-evoked panic-like behaviors by clinically effective panicolytics given at doses and regimens similar to those of panic therapy. In particular, DPAG-evoked panic was markedly attenuated by a 21-day administration of daily doses of fluoxetine as low as 1 mg kg^{-1} and, conversely, facilitated by systemic injection of the putative panicogen pentylenetetrazol. Remarkably, as well, DPAG stimulations are not accompanied by 'stress hormone responses' in the absence of muscular effort. DPAG-evoked responses are also facilitated in both female rats and neonatally-isolated adult rats. Lastly, recent evidence showed that the

Table 2.4 Truth table of translational features of panic attacks evoked by the stimulation of dorsal periaqueductal gray matter (DPAG) of the midbrain of animals and humans

		Clinical panic	
		YES	NO

DPAG-evoked panic in animals or humans	YES	Symptomatology[a]
		Lack of HPA axis responses[b]
		Lack of PRL axis responses[b]
		Attenuated by:
		  Chronic fluoxetine
		  Chronic clomipramine
		  Acute alprazolam
		  Acute clonazepam
		Facilitated by:
		  Hypoxia[c]
		  Hypercapnic hypoxia[d]
		  Cholecystokinin
		  Yohimbine
		  Pentylenetetrazol
		  Late luteal phase
		  Early-life stress[e]
		  Major depression[f]
			Attenuation by acute SSRIs[j]

	NO	Attenuated by chronic imipramine[g]
		Facilitated by hypercapnia
		Facilitated by LAC i.v. infusions[h]
		Facilitated by reactive depression[i]
			Attenuated by:
			  Acute maprotiline
			  Chronic maprotiline
			  Acute diazepam
			  Acute midazolam
			  Acute buspirone
			  Chronic buspirone
			Predisposed by specific phobias[k]

Notes refer to DPAG-evoked panic attacks in animals
[a]See Table 2.2
[b]Stimulation of rats in a 20-cm diameter arena that prevented flight behavior [379]
[c]KCN-evoked selective cytotoxic hypoxia of chemoreceptors
[d]CO_2 potentiation of DPAG-mediated panic attacks evoked by intravenous injections of KCN
[e]Neonatal social isolation
[f]Olfactory bulbectomy model of depression (unpublished results)
[g]Dose above clinical range
[h]0.5 M only (unpublished results)
[i]Exposure to inescapable shocks
[j]Doses above clinical range
[k]See Tannure et al. [338]

DPAG mediates respiratory-type panic attacks to KCN selective stimulation of chemoreceptors and that these responses are potentiated by hypercapnia and attenuated by acute and chronic treatments with clinical doses of the established panicolytics clonazepam and fluoxetine, respectively [254]. Similarly, escape responses to ambient hypoxia were attenuated by acute and chronic administrations of alprazolam and fluoxetine, respectively, and also by intra-periaqueductal injections of 5-HT and its agonists [255].

There are important misses as well. Indeed, although the DPAG appears to be activated by both LAC infusions and CO_2 inhalations in humans, neither the LAC nor the CO_2 produce panic behaviors in rats. Moreover, DPAG-evoked behaviors are not attenuated by chronic administration of the standard panicolytic imipramine. The negative result may be due to the high dose of imipramine (10 mg kg^{-1} day^{-1}, 21 days). Indeed, the 21-day treatment with 5 mg kg^{-1} fluoxetine was less effective in attenuating

DPAG-evoked panic-like behaviors than the 21-day treatments with 1 or 2 mg kg^{-1} fluoxetine [60, C.S. Bernabé, unpublished results]. Although frequently presented as evidence of DPAG mediation of panic attacks, the attenuation of DPAG-evoked shuttle-box escape responses by acute treatments with SSRIs is another inconsistency of this model [303, 469, 470]. These effects may be due to specific features of the shuttle-box model. Indeed, acute injections of low doses of both clomipramine and fluoxetine failed in attenuating the DPAG-evoked panic-like responses of rats stimulated in an open-field [60, 471]. Although the inhibitory effect of hypercapnia on DPAG-evoked panic responses appears incongruent, pre-exposures to hypercapnia produce a reliable facilitation of KCN-evoked panic responses. Accordingly, the former effect may be due to the non-selective features of electrical stimulation.

Most importantly, however, recent results from our laboratory showed that 0.5 M LAC infusions failed in facilitating the KCN-evoked escape responses presumptively mediated at the PAG (E.A. Moraes, unpublished results). Although the latter results may be due to the smaller effects of 0.5 M LAC infusions in rats [342], they also suggest that LAC vulnerability to panic attacks is mediated by mechanisms other than those of KCN-evoked panic attacks. This possibility is supported by both the provocation of panic attacks by d-LAC [141] and the lack of activation of PAG in rats infused with LAC [341].

In conclusion, although the animal models are not expected to reproduce clinical disorders exactly, a translational model of PD should (1) present face validity, (2) be sensitive to clinically-effective panicolytics in doses and regimens similar to those of panic therapy, (3) be sensitive to clinically-specific panicogens (5–7 % CO$_2$ and 0.5 M LAC), (4) lack neuroendocrine responses, (5) be facilitated in female subjects and (6) reproduce the clinical comorbidity of panic disorder. Together, DPAG and KCN models of panic attacks meet most of these criteria (Table 2.4). Preliminary modeling of the comorbidity of PD and CSA further implicates the PAG in the predisposition of separation-anxious children to the later development of PD.

References

1. Freud S. On the right to separate from neurasthenia a definite symptom-complex as anxiety neurosis. In: Brill JAA, editor. Selected papers on hysteria and other psychoneuroses. New York: The Journal of Nervous and Mental Disease Publishing Company; 1912. p. 133–54. Chapter VI.
2. Freud S. The psychotherapy of hysteria. In: Brill JAA, editor. Selected papers on hysteria and other psychoneuroses. New York: The Journal of Nervous and Mental Disease Publishing Company; 1912. p. 75–120. Chapter IV.
3. Freud S. Obsessions and phobias: their psychical mechanism and their aetiology. Complete psychological works, standard edition. London: Hogarth Press; 1962. p. 81.
4. APA. Diagnostic and statistical manual of mental disorders. 3rd ed. Washington, DC: American Psychiatry Association; 1980.
5. Drury A. The percentage of carbon dioxide in the alveolar air, and the tolerance to accumulating carbon dioxide in cases of so-called "irritable heart" of soldiers. Heart. 1920;7:165–73.
6. Cohen ME, White PD. Life situations, emotions and neurocirculatory asthenia (anxiety neurosis, neurasthenia, effort syndrome). Res Publ Assoc Res Nerv Ment Dis. 1949;29:832–69.
7. Pitts FN, McClure JN. Lactate metabolism in anxiety neurosis. N Engl J Med. 1967;277:1329–36.
8. Cowley DS, Roy-Byrne PP. Hyperventilation and panic disorder. Am J Med. 1987;83:929–37.
9. Klein DF, Fink M. Psychiatric reaction patterns to imipramine. Am J Psychiatry. 1962;119:432–8.
10. Klein DF. Delineation of two drug-responsive anxiety syndromes. Psychopharmacologia. 1964;5:397–408.
11. Bowlby J. Attachment, separation anxiety, loss. 2nd ed. NY: Basic Books; 1969.
12. Gorman JM, Askanazi J, Liebowitz MR, Fyer AJ, Stein J, Kinney JM, et al. Response to hyperventilation in a group of patients with panic disorder. Am J Psychiatry. 1984;141:857–61.
13. Liebowitz MR, Gorman JM, Fyer AJ, Levitt M, Dillon D, Levy G, et al. Lactate provocation of panic attacks: II. Biochemical and physiological findings. Arch Gen Psychiatry. 1985;42:709–19.
14. Cameron OG, Lee MA, Curtis GC, McCann DS. Endocrine and physiological changes during "spontaneous" panic attacks. Psychoneuroendocrinology. 1987;12:321–31.
15. Levin AP, Doran AR, Liebowitz MR, Fyer AJ, Gorman JM, Klein DF, et al. Pituitary adrenocortical unresponsiveness in lactate-induced panic. Psychiatry Res. 1987;21:23–32.
16. Woods SW, Charney DS, McPherson CA, Gradman AH, Heninger GR. Situational panic attacks. Behavioral, physiologic, and biochemical characterization. Arch Gen Psychiatry. 1987;44:365–75.
17. Woods SW, Charney DS, Goodman WK, Heninger GR. Carbon dioxide-induced anxiety. Behavioral,

physiologic, and biochemical effects of carbon dioxide in patients with panic disorders and healthy subjects. Arch Gen Psychiatry. 1988;45:43–52.

18. Hollander E, Liebowitz MR, Cohen B, Gorman JM, Fyer AJ, Papp LA, et al. Prolactin and sodium lactate-induced panic. Psychiatry Res. 1989;28: 181–91.

19. Hollander E, Liebowitz MR, Gorman JM, Cohen B, Fyer A, Klein DF. Cortisol and sodium lactate-induced panic. Arch Gen Psychiatry. 1989;46:135–40.

20. Hollander E, Liebowitz MR, DeCaria C, Klein DF. Fenfluramine, cortisol, and anxiety. Psychiatry Res. 1989;31:211–3.

21. Ehlers A, Margraf J, Roth WT, Taylor CB, Maddock RJ, Sheikh J, et al. Lactate infusions and panic attacks: do patients and controls respond differently? Psychiatry Res. 1986;17:295–308.

22. Margraf J, Ehlers A, Roth WT. Sodium lactate infusions and panic attacks: a review and critique. Psychosom Med. 1986;48:23–51.

23. Clark DM, Salkovskis PM, Ost LG, Breitholtz E, Koehler KA, Westling BE, et al. Misinterpretation of body sensations in panic disorder. J Consult Clin Psychol. 1997;65:203–13.

24. Klein DF. False suffocation alarms, spontaneous panics, and related conditions. An integrative hypothesis. Arch Gen Psychiatry. 1993;50:306–18.

25. Klein DF. Panic may be a misfiring suffocation alarm. In: Montgomery SA, editor. Psychopharmacology of panic. 1st ed. New York: Oxford University Press; 1993. p. 67–73.

26. Shavitt RG, Gentil V, Mandetta R. The association of panic/agoraphobia and asthma. Contributing factors and clinical implications. Gen Hosp Psychiatry. 1992;14:420–3.

27. Preter M, Klein DF. Panic, suffocation false alarms, separation anxiety and endogenous opioids. Prog Neuropsychopharmacol Biol Psychiatry. 2008;32: 603–12.

28. Griez E, Schruers K. Experimental pathophysiology of panic. J Psychosom Res. 1998;45:493–503.

29. Liebowitz MR, Fyer AJ, Gorman JM, Dillon D, Davies S, Stein JM, et al. Specificity of lactate infusions in social phobia versus panic disorders. Am J Psychiatry. 1985;142:947–50.

30. Rifkin A, Klein DF, Dillon D, Levitt M. Blockade by imipramine or desipramine of panic induced by sodium lactate. Am J Psychiatry. 1981;138:676–7.

31. Liebowitz MR, Fyer AJ, Gorman JM, Dillon D, Appleby IL, Levy G, et al. Lactate provocation of panic attacks: I. Clinical and behavioral findings. Arch Gen Psychiatry. 1984;41:764–70.

32. Woods SW, Charney DS, Delgado PL, Heninger GR. The effect of long-term imipramine treatment on carbon dioxide-induced anxiety in panic disorder patients. J Clin Psychiatry. 1990;51:505–7.

33. Yeragani VK, Pohl R, Balon R, Rainey JM, Berchou R, Ortiz A. Sodium lactate infusion after treatment with tricyclic antidepressants: behavioral and physiological findings. Biol Psychiatry. 1988;24:767–74.

34. Gittelman-Klein R, Klein DF. School phobia: diagnostic considerations in the light of imipramine effects. J Nerv Ment Dis. 1973;156:199–215.

35. Battaglia M, Bertella S, Politi E, Bernardeschi L, Perna G, Gabriele A, et al. Age at onset of panic disorder: influence of familial liability to the disease and of childhood separation anxiety disorder. Am J Psychiatry. 1995;152:1362–4.

36. Klein RG. Is panic disorder associated with childhood separation anxiety disorder? Clin Neuropharmacol. 1995;18 Suppl 2:S7–14.

37. Battaglia M, Ogliari A, D'Amato F, Kinkead R. Early-life risk factors for panic and separation anxiety disorder: insights and outstanding questions arising from human and animal studies of CO sensitivity. Neurosci Biobehav Rev. 2014;46:455–64.

38. Bernstein GA, Borchardt CM, Perwien AR, Crosby RD, Kushner MG, Thuras PD, et al. Imipramine plus cognitive-behavioral therapy in the treatment of school refusal. J Am Acad Child Adolesc Psychiatry. 2000;39:276–83.

39. Roberson-Nay R, Klein DF, Klein RG, Mannuzza S, Moulton III JL, Guardino M, et al. Carbon dioxide hypersensitivity in separation-anxious offspring of parents with panic disorder. Biol Psychiatry. 2010;67: 1171–7.

40. Battaglia M, Pesenti-Gritti P, Spatola CA, Ogliari A, Tambs K. A twin study of the common vulnerability between heightened sensitivity to hypercapnia and panic disorder. Am J Med Genet B Neuropsychiatr Genet. 2008;147B:586–93.

41. Roberson-Nay R, Eaves LJ, Hettema JM, Kendler KS, Silberg JL. Childhood separation anxiety disorder and adult onset panic attacks share a common genetic diathesis. Depress Anxiety. 2012;29:320–7.

42. Preter M, Klein DF. Lifelong opioidergic vulnerability through early life separation: a recent extension of the false suffocation alarm theory of panic disorder. Neurosci Biobehav Rev. 2014;46:10–351.

43. Lawson EE, Waldrop TG, Eldridge FL. Naloxone enhances respiratory output in cats. J Appl Physiol. 1979;47:1105–11.

44. Bonham AC. Neurotransmitters in the CNS control of breathing. Respir Physiol. 1995;101:219–30.

45. van der Schier R, Roozekrans M, van VM, Dahan A, Niesters M. Opioid-induced respiratory depression: reversal by non-opioid drugs. F1000Prime Rep 2014;6:79.

46. Kalin NH, Shelton SE, Barksdale CM. Opiate modulation of separation-induced distress in non-human primates. Brain Res. 1988;440:285–92.

47. Kalin NH, Shelton SE, Lynn DE. Opiate systems in mother and infant primates coordinate intimate contact during reunion. Psychoneuroendocrinology. 1995;20:735–42.

48. Panksepp J, Herman B, Conner R, Bishop P, Scott JP. The biology of social attachments: opiates alleviate separation distress. Biol Psychiatry. 1978;13:607–18.

49. Panksepp J, Herman BH, Vilberg T, Bishop P, DeEskinazi FG. Endogenous opioids and social behavior. Neurosci Biobehav Rev. 1980;4:473–87.

50. Deakin JFW, Graeff FG. 5-HT and mechanisms of defence. J Psychopharmacol. 1991;5:305–15.

51. Graeff FG. Serotonin, the periaqueductal gray and panic. Neurosci Biobehav Rev. 2004;28:239–59.

52. Canteras NS, Graeff FG. Executive and modulatory neural circuits of defensive reactions: implications for panic disorder. Neurosci Biobehav Rev. 2014;46:352–64.

53. Goetz RR, Klein DF, Gorman JM. Symptoms essential to the experience of sodium lactate-induced panic. Neuropsychopharmacology. 1996;14:355–66.

54. Dillon DJ, Gorman JM, Liebowitz MR, Fyer AJ, Klein DF. Measurement of lactate-induced panic and anxiety. Psychiatry Res. 1987;20:97–105.

55. Shioiri T, Someya T, Murashita J, Takahashi S. The symptom structure of panic disorder: a trial using factor and cluster analysis. Acta Psychiatr Scand. 1996;93:80–6.

56. Perna G, Caldirola D, Namia C, Cucchi M, Vanni G, Bellodi L. Language of dyspnea in panic disorder. Depress Anxiety. 2004;20:32–8.

57. Briggs AC, Stretch DD, Brandon S. Subtyping of panic disorder by symptom profile. Br J Psychiatry. 1993;163:201–9.

58. Nardi AE, Nascimento I, Valenca AM, Lopes FL, Mezzasalma MA, Zin WA, et al. Respiratory panic disorder subtype: acute and long-term response to nortriptyline, a noradrenergic tricyclic antidepressant. Psychiatry Res. 2003;120:283–93.

59. Roberson-Nay R, Latendresse SJ, Kendler KS. A latent class approach to the external validation of respiratory and non-respiratory panic subtypes. Psychol Med. 2012;42:461–74.

60. Schenberg LC, Bittencourt AS, Sudré ECM, Vargas LC. Modeling panic attacks. Neurosci Biobehav Rev. 2001;25:647–59.

61. Schenberg LC, Schimitel FG, Armini RS, Bernabe CS, Rosa CA, Tufik S, et al. Translational approach to studying panic disorder in rats: hits and misses. Neurosci Biobehav Rev. 2014;46:472–96.

62. Mobbs D, Petrovic P, Marchant JL, Hassabis D, Weiskopf N, Seymour B, et al. When fear is near: threat imminence elicits prefrontal-periaqueductal gray shifts in humans. Science. 2007;317:1079–83.

63. Gorman JM, Kent JM, Sullivan GM, Coplan JD. Neuroanatomical hypothesis of panic disorder, revised. Am J Psychiatry. 2000;157:493–505.

64. Smoller JW, Gallagher PJ, Duncan LE, McGrath LM, Haddad SA, Holmes AJ, et al. The human ortholog of acid-sensing ion channel gene ASIC1a is associated with panic disorder and amygdala structure and function. Biol Psychiatry. 2014;76:902–10.

65. Shekhar A, Keim SR. The circum ventricular organs form a potential neural pathway for lactate sensitivity: implications for panic disorder. J Neurosci. 1997;17:9726–35.

66. Johnson PL, Federici LM, Shekhar A. Etiology, triggers and neurochemical circuits associated with unexpected, expected, and laboratory-induced panic attacks. Neurosci Biobehav Rev. 2014;46:429–54.

67. Gorman JM, Liebowitz MR, Fyer AJ, Stein J. A neuroanatomical hypothesis for panic disorder. Am J Psychiatry. 1989;146:148–61.

68. Blanchard DC, Griebel G, Blanchard RJ. The mouse defense test battery: pharmacological and behavioral assays for anxiety and panic. Eur J Pharmacol. 2003;463:97–116.

69. Beitman BD, Basha I, Flaker G, DeRosear L, Mukerji V, Lamberti J. Non-fearful panic disorder: panic attacks without fear. Behav Res Ther. 1987;25:487–92.

70. Fleet RP, Martel JP, Lavoie KL, Dupuis G, Beitman BD. Non-fearful panic disorder: a variant of panic in medical patients? Psychosomatics. 2000;41(4):311–20.

71. Fleet RP, Lavoie KL, Martel JP, Dupuis G, Marchand A, Beitman BD. Two-year follow-up status of emergency department patients with chest pain: was it panic disorder? CJEM. 2003;5:247–54.

72. Figueiredo HF, Bodie BL, Tauchi M, Dolgas CM, Herman JP. Stress integration after acute and chronic predator stress: differential activation of central stress circuitry and sensitization of the hypothalamo-pituitary-adrenocortical axis. Endocrinology. 2003;144(12):5249–58.

73. Hauger RL, Millan MA, Lorang M, Harwood JP, Aguilera G. Corticotropin-releasing factor receptors and pituitary adrenal responses during immobilization stress. Endocrinology. 1988;123:396–405.

74. Fredrikson M, Sundin O, Frankenhaeuser M. Cortisol excretion during the defense reaction in humans. Psychosom Med. 1985;47:313–9.

75. Furlan PM, DeMartinis N, Schweizer E, Rickels K, Lucki I. Abnormal salivary cortisol levels in social phobic patients in response to acute psychological but not physical stress. Biol Psychiatry. 2001;50:254–9.

76. Wiest G, Lehner-Baumgartner E, Baumgartner C. Panic attacks in an individual with bilateral selective lesions of the amygdala. Arch Neurol. 2006;63:1798–801.

77. Feinstein JS, Buzza C, Hurlemann R, Follmer RL, Dahdaleh NS, Coryell WH, et al. Fear and panic in humans with bilateral amygdala damage. Nat Neurosci. 2013;16:270–2.

78. Wilent WB, Oh MY, Buetefisch CM, Bailes JE, Cantella D, Angle C, et al. Induction of panic attack by stimulation of the ventromedial hypothalamus. J Neurosurg. 2010;112:1295–8.

79. Charney DS, Redmond DE. Neurobiological mechanism in human anxiety: evidence supporting central noradrenergic hyperactivity. Neuropharmacology. 1983;22:1531–6.

80. Charney DS, Heninger GR, Redmond Jr DE. Yohimbine induced anxiety and increased noradrenergic function in humans: effects of diazepam and clonidine. Life Sci. 1983;33:19–29.

81. Charney DS, Woods SW, Goodman WK, Heninger GR. Neurobiological mechanisms of panic anxiety: biochemical and behavioral correlates of yohimbine-induced panic attacks. Am J Psychiatry. 1987;144:1030–6.

82. Elam M, Yao T, Thoren P, Svensson TH. Hypercapnia and hypoxia: chemoreceptor-mediated control of locus coeruleus neurons and splanchnic, sympathetic nerves. Brain Res. 1981;222:373–81.

83. Liebowitz MR, Fyer AJ, McGrath PA, Klein DF. Clonidine treatment of panic disorder. Psychopharmacol Bull. 1981;17:122–3.

84. Hoehn-Saric R, Merchant AF, Keyser ML, Smith VK. Effects of clonidine on anxiety disorders. Arch Gen Psychiatry. 1981;38:1278–82.

85. Pineda J, Aghajanian GK. Carbon dioxide regulates the tonic activity of locus coeruleus neurons by modulating a proton- and polyamine-sensitive inward rectifier potassium current. Neuroscience. 1997;77:723–43.

86. Teppema LJ, Veening JG, Kranenburg A, Dahan A, Berkenbosch A, Olievier C. Expression of c-fos in the rat brainstem after exposure to hypoxia and to normoxic and hyperoxic hypercapnia. J Comp Neurol. 1997;388:169–90.

87. Nutt DJ. Altered central alpha 2-adrenoceptor sensitivity in panic disorder. Arch Gen Psychiatry. 1989;46:165–9.

88. Glue P, Nutt DJ. Benzodiazepine receptor sensitivity in panic disorder. Lancet. 1991;337:563.

89. Muskin PR, Fyer AJ. Treatment of panic disorder. J Clin Psychopharmacol. 1981;1:81–90.

90. Reiman EM, Raichle ME, Butler FK, Herscovitch P, Robins E. A focal brain abnormality in panic disorder, a severe form of anxiety. Nature. 1984;310:683–5.

91. Reiman EM, Raichle ME, Robins E, Butler FK, Herscovitch P, Fox P, et al. The application of positron emission tomography to the study of panic disorder. Am J Psychiatry. 1986;143:469–77.

92. Gray JA, McNaughton N. The neuropsychology of anxiety. 2nd ed. Oxford: Oxford Medical Publications; 2000.

93. Kaplan JS, Arnkoff DB, Glass CR, Tinsley R, Geraci M, Hernandez E, et al. Avoidant coping in panic disorder: a yohimbine biological challenge study. Anxiety Stress Coping. 2012;25:425–42.

94. Kaitin KI, Bliwise DL, Gleason C, Nino-Murcia G, Dement WC, Libet B. Sleep disturbance produced by electrical stimulation of the locus coeruleus in a human subject. Biol Psychiatry. 1986;21:710–6.

95. Libet B, Gleason CA. The human locus coeruleus and anxiogenesis. Brain Res. 1994;634:178–80.

96. Ledoux JE, Iwata J, Cicchetti P, Reis DJ. Different projections of the central amygdaloid nucleus mediate autonomic and behavioral correlates of conditioned fear. J Neurosci. 1988;8:2517–29.

97. Ledoux JE, Cicchetti P, Xagoraris A, Romanski LM. The lateral amygdaloid nucleus: sensory interface of the amygdala in fear conditioning. J Neurosci. 1990;10:1062–9.

98. Hitchcock J, Davis M. Lesions of the amygdala, but not of the cerebellum or red nucleus, block conditioned fear as measured with the potentiated startle paradigm. Behav Neurosci. 1986;100:11–22.

99. Nashold Jr BS, Wilson WP, Slaughter DG. Sensations evoked by stimulation in the midbrain of man. J Neurosurg. 1969;30:14–24.

100. Amano K, Tanikawa T, Iseki H, Kawabatake H, Notani M, Kawamura H, et al. Single neuron analysis of the human midbrain tegmentum. Appl Neurophysiol. 1978;41:66–78.

101. Young RF. Brain and spinal stimulation: how and to whom! Clin Neurosurg. 1989;35:429–47.

102. Fernandez de Molina A, Hunsperger RW. Organization of the subcortical system governing defence and flight reactions in the cat. J Physiol. 1962;160:200–13.

103. Adams DB. Brain mechanisms for offense, defense and submission. Behav Brain Sci. 1979;2:201–41.

104. Hilton SM, Redfern WS. A search for brain stem cell groups integrating the defence reaction in the rat. J Physiol. 1986;378:213–28.

105. Graeff FG. The anti-aversive action of drugs. In: Thompson T, Dews PB, Barret JE, editors. Neurobehavioral Pharmacology, Advances in Behavioral Pharmacology, vol.6. Hillsdale: Lawrence Erlbaum Associates Inc.; 1987. p. 129–56.

106. Graeff FG. Animal models of aversion. In: Simon P, Soubrie P, Wildlocher D, editors. Selected models of anxiety, depression and psychosis. 1st ed. Basel: Karger AG; 1988. p. 115–41.

107. Paul ED, Lowry CA. Functional topography of serotonergic systems supports the Deakin/Graeff hypothesis of anxiety and affective disorders. J Psychopharmacol. 2013;27:1090–106.

108. Paul ED, Johnson PL, Shekhar A, Lowry CA. The Deakin/Graeff hypothesis: focus on serotonergic inhibition of panic. Neurosci Biobehav Rev. 2014;46:379–96.

109. Tanaka KF, Samuels BA, Hen R. Serotonin receptor expression along the dorsal-ventral axis of mouse hippocampus. Philos Trans R Soc Lond B Biol Sci. 2012;367:2395–401.

110. Pitkänen A, Pikkarainen M, Nurminen N, Ylinen A. Reciprocal connections between the amygdala and the hippocampal formation, perirhinal cortex, and postrhinal cortex in rat. A review. Ann N Y Acad Sci. 2000;911:369–91.

111. Nutt DJ, Glue P, Lawson C, Wilson S. Flumazenil provocation of panic attacks. Evidence for altered benzodiazepine receptor sensitivity in panic disorder. Arch Gen Psychiatry. 1990;47:917–25.

112. Barnard EA, Skolnick P, Olsen RW, Mohler H, Sieghart W, Biggio G, et al. International union of pharmacology: XV. Subtypes of gamma-aminobutyric acid-A receptors: classification on the basis of subunit structure and receptor function. Pharmacol Rev. 1998;50:291–313.

113. Olsen RW, Sieghart W. International union of pharmacology: LXX. Subtypes of gamma-aminobutyric acid(A) receptors: classification on the basis of subunit composition, pharmacology, and function. Update. Pharmacol Rev. 2008;60:243–60.

114. Ströhle A, Kellner M, Yassouridis A, Holsboer F, Wiedemann K. Effect of flumazenil in lactate-sensitive patients with panic disorder. Am J Psychiatry. 1998;155:610–2.

115. Ströhle A, Kellner M, Holsboer F, Wiedemann K. Behavioral, neuroendocrine, and cardiovascular response to flumazenil: no evidence for an altered

benzodiazepine receptor sensitivity in panic disorder. Biol Psychiatry. 1999;45:321–6.

116. Potokar J, Lawson C, Wilson S, Nutt D. Behavioral, neuroendocrine, and cardiovascular response to flumazenil: no evidence for an altered benzodiazepine receptor sensitivity in panic disorder (Comment on). Biol Psychiatry. 1999;46:1709–11.

117. Kaschka W, Feistel H, Ebert D. Reduced benzodiazepine receptor binding in panic disorders measured by iomazenil SPECT. J Psychiatr Res. 1995;29:427–34.

118. Malizia AL, Cunningham VJ, Bell CJ, Liddle PF, Jones T, Nutt DJ. Decreased brain GABA(A)-benzodiazepine receptor binding in panic disorder: preliminary results from a quantitative PET study. Arch Gen Psychiatry. 1998;55:715–20.

119. Schlegel S, Steinert H, Bockisch A, Hahn K, Schloesser R, Benkert O. Decreased benzodiazepine receptor binding in panic disorder measured by IOMAZENIL-SPECT. A preliminary report. Eur Arch Psychiatry Clin Neurosci. 1994;244:49–51.

120. Preter M, Lee SH, Petkova E, Vannucci M, Kim S, Klein DF. Controlled cross-over study in normal subjects of naloxone-preceding-lactate infusions; respiratory and subjective responses: relationship to endogenous opioid system, suffocation false alarm theory and childhood parental loss. Psychol Med. 2011;41:385–93.

121. Roncon CM, Biesdorf C, Santana RG, Zangrossi Jr H, Graeff FG, Audi EA. The panicolytic-like effect of fluoxetine in the elevated T-maze is mediated by serotonin-induced activation of endogenous opioids in the dorsal periaqueductal grey. J Psychopharmacol. 2012;26:525–31.

122. Roncon CM, Biesdorf C, Coimbra NC, Audi EA, Zangrossi Jr H, Graeff FG. Cooperative regulation of anxiety and panic-related defensive behaviors in the rat periaqueductal grey matter by 5-HT1A and mu-receptors. J Psychopharmacol. 2013;27: 1141–8.

123. Breuer J, Freud S. Studien über Hysterie. Vienna: Franz Deudicke; 1895.

124. Bowlby J. A secure base: parent-child attachment and healthy human development. New York: Basic Books; 1988.

125. Heim C, Shugart M, Craighead WE, Nemeroff CB. Neurobiological and psychiatric consequences of child abuse and neglect. Dev Psychobiol. 2010;52: 671–90.

126. Ladd CO, Owens MJ, Nemeroff CB. Persistent changes in corticotropin-releasing factor neuronal systems induced by maternal deprivation. Endocrinology. 1996;137:1212–8.

127. Ladd CO, Huot RL, Thrivikraman KV, Nemeroff CB, Plotsky PM. Long-term adaptations in glucocorticoid receptor and mineralocorticoid receptor mRNA and negative feedback on the hypothalamo-pituitary-adrenal axis following neonatal maternal separation. Biol Psychiatry. 2004;55:367–75.

128. Rots NY, de Jong J, Workel JO, Levine S, Cools AR, de Kloet ER. Neonatal maternally deprived rats have as adults elevated basal pituitary-adrenal activity and enhanced susceptibility to apomorphine. J Neuroendocrinol. 1996;8:501–6.

129. Anisman H, Zaharia MD, Meaney MJ, Merali Z. Do early-life events permanently alter behavioral and hormonal responses to stressors? Int J Dev Neurosci. 1998;16:149–64.

130. Biagini G, Pich EM, Carani C, Marrama P, Agnati LF. Postnatal maternal separation during the stress hyporesponsive period enhances the adrenocortical response to novelty in adult rats by affecting feedback regulation in the CA1 hippocampal field. Int J Dev Neurosci. 1998;16:187–97.

131. van Oers HJ, de Kloet ER, Levine S. Early vs. late maternal deprivation differentially alters the endocrine and hypothalamic responses to stress. Brain Res Dev Brain Res. 1998;111:245–52.

132. Kinkead R, Gulemetova R. Neonatal maternal separation and neuroendocrine programming of the respiratory control system in rats. Biol Psychol. 2010; 84:26–38.

133. Meaney MJ, Diorio J, Francis D, Weaver S, Yau J, Chapman K, et al. Postnatal handling increases the expression of cAMP-inducible transcription factors in the rat hippocampus: the effects of thyroid hormones and serotonin. J Neurosci. 2000;20:3926–35.

134. Gardner KL, Thrivikraman KV, Lightman SL, Plotsky PM, Lowry CA. Early life experience alters behavior during social defeat: focus on serotonergic systems. Neuroscience. 2005;136:181–91.

135. Gardner KL, Hale MW, Lightman SL, Plotsky PM, Lowry CA. Adverse early life experience and social stress during adulthood interact to increase serotonin transporter mRNA expression. Brain Res. 2009;1305:47–63.

136. Gardner KL, Hale MW, Oldfield S, Lightman SL, Plotsky PM, Lowry CA. Adverse experience during early life and adulthood interact to elevate tph2 mRNA expression in serotonergic neurons within the dorsal raphe nucleus. Neuroscience. 2009;163:991–1001.

137. Quintino-Dos-Santos JW, Muller CJ, Bernabe CS, Rosa CA, Tufik S, Schenberg LC. Evidence that the periaqueductal gray matter mediates the facilitation of panic-like reactions in neonatally-isolated adult rats. PLoS One. 2014;9:e90726.

138. Dumont FS, Biancardi V, Kinkead R. Hypercapnic ventilatory response of anesthetized female rats subjected to neonatal maternal separation: insight into the origins of panic attacks? Respir Physiol Neurobiol. 2011;175:288–95.

139. Grosz HJ, Farmer BB. Pitts' and McClure's lactate-anxiety study revisited. Br J Psychiatry. 1972;120: 415–8.

140. Gorman JM, Battista D, Goetz RR, Dillon DJ, Liebowitz MR, Fyer AJ, et al. A comparison of sodium bicarbonate and sodium lactate infusion in the induction of panic attacks [published erratum appears in Arch Gen Psychiatry 1991;48:772]. Arch Gen Psychiatry. 1989;46:145–50.

141. Gorman JM, Goetz RR, Dillon D, Liebowitz MR, Fyer AJ, Davies S, et al. Sodium D-lactate infusion of panic disorder patients. Neuropsychopharmacology. 1990;3:181–9.

142. Ewaschuk JB, Naylor JM, Zello GA. D-Lactate in human and ruminant metabolism. J Nutr. 2005;135: 1619–25.
143. Dager SR, Rainey JM, Kenny MA, Artru AA, Metzger GD, Bowden DM. Central nervous system effects of lactate infusion in primates. Biol Psychiatry. 1990;27:193–204.
144. Dager SR, Marro KI, Richards TL, Metzger GD. Localized magnetic resonance spectroscopy measurement of brain lactate during intravenous lactate infusion in healthy volunteers. Life Sci. 1992;51:973–85.
145. Stein JM, Papp LA, Klein DF, Cohen S, Simon J, Ross D, et al. Exercise tolerance in panic disorder patients. Biol Psychiatry. 1992;32:281–7.
146. Ströhle A, Feller C, Onken M, Godemann F, Heinz A, Dimeo F. The acute antipanic activity of aerobic exercise. Am J Psychiatry. 2005;162:2376–8.
147. Beck JG, Ohtake PJ, Shipherd JC. Exaggerated anxiety is not unique to CO_2 in panic disorder: a comparison of hypercapnic and hypoxic challenges. J Abnorm Psychol. 1999;108:473–82.
148. Hackett PH, Roach RC. High-altitude illness. N Engl J Med. 2001;345:107–14.
149. Fagenholz PJ, Murray AF, Gutman JA, Findley JK, Harris NS. New-onset anxiety disorders at high altitude. Wilderness Environ Med. 2007;18:312–6.
150. Tweed JL, Schoenbach VJ, George LK, Blazer DG. The effects of childhood parental death and divorce on six-month history of anxiety disorders. Br J Psychiatry. 1989;154:823–8.
151. Stein MB, Walker JR, Anderson G, Hazen AL, Ross CA, Eldridge G, et al. Childhood physical and sexual abuse in patients with anxiety disorders and in a community sample. Am J Psychiatry. 1996;153:275–7.
152. Battaglia M, Pesenti-Gritti P, Medland SE, Ogliari A, Tambs K, Spatola CA. A genetically informed study of the association between childhood separation anxiety, sensitivity to CO(2), panic disorder, and the effect of childhood parental loss. Arch Gen Psychiatry. 2009;66:64–71.
153. Spatola CA, Scaini S, Pesenti-Gritti P, Medland SE, Moruzzi S, Ogliari A, et al. Gene-environment interactions in panic disorder and CO(2) sensitivity: effects of events occurring early in life. Am J Med Genet B Neuropsychiatr Genet. 2011;156B:79–88.
154. Angst J, Wicki W. The epidemiology of frequent and less frequent panic attacks. In: Montgomery SA, editor. Psychopharmacology of panic. New York: Oxford University Press; 1993. p. 7–24.
155. Lovick TA. Sex determinants of experimental panic attacks. Neurosci Biobehav Rev. 2014;46:465–71.
156. Nashold Jr BS, Wilson WP, Slaughter GS. The midbrain and pain. In: Bonica JJ, editor. International symposium on pain. New York: Raven; 1974. p. 191–6.
157. Kumar K, Toth C, Nath RK. Deep brain stimulation for intractable pain: a 15-year experience. Neurosurgery. 1997;40:736–46.
158. Bandler R, Depaulis A. Midbrain periaqueductal gray control of defensive behavior in the cat and rat.

In: Depaulis A, Bandler R, editors. The midbrain periaqueductal gray matter. New York: Plenum Press; 1991. p. 175–98.
159. Bandler R, Keay KA. Columnar organization in the midbrain periaqueductal gray and the integration of emotional expression. Prog Brain Res. 1996;107: 285–300.
160. Carrive P. Functional organization of PAG neurons controlling regional vascular beds. In: Depaulis A, Bandler R, editors. The midbrain periaqueductal gray matter: functional, anatomical, and neurochemical organization. New York: Plenum Press; 1991. p. 67–100.
161. Carrive P. The periaqueductal gray and defensive behavior: functional representation and neuronal organization. Behav Brain Res. 1993;58:27–47.
162. Keay KA, Bandler R. Periaqueductal gray. In: Paxinos G, editor. The rat nervous system. 3rd ed. San Diego: Elsevier; 2004. p. 243–57.
163. Kingsbury MA, Kelly AM, Schrock SE, Goodson JL. Mammal-like organization of the avian midbrain central gray and a reappraisal of the intercollicular nucleus. PLoS One. 2011;6:e20720.
164. Jansen AS, Farkas E, Mac SJ, Loewy AD. Local connections between the columns of the periaqueductal gray matter: a case for intrinsic neuromodulation. Brain Res. 1998;784:329–36.
165. Bittencourt AS, Carobrez AP, Zamprogno LP, Tufik S, Schenberg LC. Organization of single components of defensive behaviors within distinct columns of periaqueductal gray matter of the rat: role of N-methyl-D-aspartic acid glutamate receptors. Neuroscience. 2004;125:71–89.
166. Teixeira KV, Carobrez AP. Effects of glycine or (+/−)-3-amino-1-hydroxy-2-pyrrolidone microinjections along the rostrocaudal axis of the dorsal periaqueductal gray matter on rats' performance in the elevated plus-maze task. Behav Neurosci. 1999;113:196–203.
167. Mota-Ortiz SR, Sukikara MH, Felicio LF, Canteras NS. Afferent connections to the rostrolateral part of the periaqueductal gray: a critical region influencing the motivation drive to hunt and forage. Neural Plast. 2009; ID612698.
168. Holstege G, Kerstens L, Moes MC, Vanderhorst VG. Evidence for a periaqueductal gray-nucleus retroambiguus-spinal cord pathway in the rat. Neuroscience. 1997;80:587–98.
169. Ruiz-Torner A, Olucha-Bordonau F, Valverde-Navarro AA, Martinez-Soriano F. The chemical architecture of the rat's periaqueductal gray based on acetylcholinesterase histochemistry: a quantitative and qualitative study. J Chem Neuroanat. 2001;21: 295–312.
170. Onstott D, Mayer B, Beitz AJ. Nitric oxide synthase immunoreactive neurons anatomically define a longitudinal dorsolateral column within the midbrain periaqueductal gray of the rat: analysis using laser confocal microscopy. Brain Res. 1993;610: 317–24.

171. Bandler R, Tork I. Midbrain periaqueductal grey region in the cat has afferent and efferent connection with solitary tract nuclei. Neurosci Lett. 1987;74:1–6.

172. Lv BC, Ji GL, Huo FQ, Chen T, Li H, Li YQ. Topographical distributions of endomorphinergic pathways from nucleus tractus solitarii to periaqueductal gray in the rat. J Chem Neuroanat. 2010;39:166–74.

173. Iwata J, Ledoux JE, Reis DJ. Destruction of intrinsic neurons in the lateral hypothalamus disrupts the classical conditioning of autonomic but not behavioral emotional responses in the rat. Brain Res. 1986;368:161–6.

174. Kiser RS, Brown CA, Sanghera MK, German DC. Dorsal raphe nucleus stimulation reduces centrally elicited fearlike behavior. Brain Res. 1980;191:265–72.

175. Lovick TA. Stimulation in the ventral periaqueductal grey matter modulates the cardiovascular response evoked from the midbrain defence area in anaesthetized rats. J Physiol. 1990;91P.

176. Lovick TA. Inhibitory modulation of the cardiovascular defence response by ventrolateral periaqueductal grey matter in rats. Exp Brain Res. 1992;89:133–9.

177. Lovick TA. Influence of the dorsal and median raphe nuclei on neurons in the periaqueductal gray matter: role of 5-hydroxytryptamine. Neuroscience. 1994;59:993–1000.

178. Lovick TA, Parry DM, Stezhka VV, Lumb BM. Serotonergic transmission in the periaqueductal gray matter in relation to aversive behaviour: morphological evidence for direct modulatory effects on identified output neurons. Neuroscience. 2000;95:763–72.

179. Pobbe RL, Zangrossi Jr H. 5-HT(1A) and 5-HT(2A) receptors in the rat dorsal periaqueductal gray mediate the antipanic-like effect induced by the stimulation of serotonergic neurons in the dorsal raphe nucleus. Psychopharmacology (Berl). 2005;183:314–21.

180. Schenberg LC, Graeff FG. Role of the periaqueductal gray substance in the antianxiety action of benzodiazepines. Pharmacol Biochem Behav. 1978;9:287–95.

181. Spyer KM, Gourine AV. Chemosensory pathways in the brainstem controlling cardiorespiratory activity. Philos Trans R Soc Lond B Biol Sci. 2009;364:2603–10.

182. Finley JC, Katz DM. The central organization of carotid body afferent projections to the brainstem of the rat. Brain Res. 1992;572(1–2):108–16.

183. Loeschcke HH, Koepchen HP, Gertz KH. Über den Einfluß von Wasserstoffionenkonzentration und CO$_2$-Druck im Liquor cerebrospinalis auf die Atmung. Pflugers Arch. 1958;266:569–85.

184. Loeschcke HH. Central chemosensitivity and the reaction theory. J Physiol. 1982;332:1–24.

185. Schlaefke ME, See WR, Loeschcke HH. Ventilatory response to alterations of H+ ion concentration in small areas of the ventral medullary surface. Respir Physiol. 1970;10:198–212.

186. Schlaefke ME. Central chemosensitivity: a respiratory drive. Rev Physiol Biochem Pharmacol. 1981;90:171–244.

187. Ricardo JA, Koh ET. Anatomical evidence of direct projections from the nucleus of the solitary tract to the hypothalamus, amygdala, and other forebrain structures in the rat. Brain Res. 1978;153:1–26.

188. Luft U. Aviation physiology-the effects of altitude. In: Handbook of physiology. Washington, DC: Am. Physiol. Soc.; 1965. p. 1099–145.

189. Moosavi SH, Golestanian E, Binks AP, Lansing RW, Brown R, Banzett RB. Hypoxic and hypercapnic drives to breathe generate equivalent levels of air hunger in humans. J Appl Physiol. 2003;94:141–54.

190. Banzett RB, Lansing RW, Evans KC, Shea SA. Stimulus-response characteristics of CO$_2$-induced air hunger in normal subjects. Respir Physiol. 1996;103:19–31.

191. Banzett RB, Lansing RW, Reid MB, Adams L, Brown R. 'Air hunger' arising from increased PCO$_2$ in mechanically ventilated quadriplegics. Respir Physiol. 1989;76:53–67.

192. Gandevia SC, Killian K, McKenzie DK, Crawford M, Allen GM, Gorman RB, et al. Respiratory sensations, cardiovascular control, kinaesthesia and transcranial stimulation during paralysis in humans. J Physiol. 1993;470:85–107.

193. van den Hout MA, Boek C, van der Molen GM, Jansen A, Griez E. Rebreathing to cope with hyperventilation: experimental tests of the paper bag method. J Behav Med. 1988;11:303–10.

194. Moosavi SH, Banzett RB, Butler JP. Time course of air hunger mirrors the biphasic ventilatory response to hypoxia. J Appl Physiol. 2004;97:2098–103.

195. Spengler CM, Banzett RB, Systrom DM, Shannon DC, Shea SA. Respiratory sensations during heavy exercise in subjects without respiratory chemosensitivity. Respir Physiol. 1998;114:65–74.

196. Shea SA, Andres LP, Shannon DC, Guz A, Banzett RB. Respiratory sensations in subjects who lack a ventilatory response to CO$_2$. Respir Physiol. 1993;93:203–19.

197. Banzett RB, Mulnier HE, Murphy K, Rosen SD, Wise RJ, Adams L. Breathlessness in humans activates insular cortex. Neuroreport. 2000;11:2117–20.

198. Craig AD. How do you feel? Interoception: the sense of the physiological condition of the body. Nat Rev Neurosci. 2002;3:655–66.

199. Craig AD. Human feelings: why are some more aware than others? Trends Cogn Sci. 2004;8:239–41.

200. Casanova JP, Contreras M, Moya EA, Torrealba F, Iturriaga R. Effect of insular cortex inactivation on autonomic and behavioral responses to acute hypoxia in conscious rats. Behav Brain Res. 2013;253:60–7.

201. Prabhakar NR. Sensing hypoxia: physiology, genetics and epigenetics. J Physiol. 2013;591:2245–57.

202. Katzman MA, Struzik L, Vijay N, Coonerty-Femiano A, Mahamed S, Duffin J. Central and peripheral chemoreflexes in panic disorder. Psychiatry Res. 2002;113:181–92.

203. Corfield DR, Fink GR, Ramsay SC, Murphy K, Harty HR, Watson JD, et al. Evidence for limbic system activation during CO$_2$-stimulated breathing in man. J Physiol. 1995;488:77–84.

204. Brannan S, Liotti M, Egan G, Shade R, Madden L, Robillard R, et al. Neuroimaging of cerebral activations and deactivations associated with hypercapnia and hunger for air. Proc Natl Acad Sci U S A. 2001;98:2029–34.

205. Liotti M, Brannan S, Egan G, Shade R, Madden L, Abplanalp B, et al. Brain responses associated with consciousness of breathlessness (air hunger). Proc Natl Acad Sci U S A. 2001;98:2035–40.

206. Reiman EM, Fusselman MJ, Fox PT, Raichle ME. Neuroanatomical correlates of anticipatory anxiety. Science. 1989;243:1071–4.

207. Drevets WC, Videen TQ, MacLeod AK, Haller JW, Raichle ME. PET images of blood flow changes during anxiety: correction. Science. 1992;256:1696.

208. Benkelfat C, Bradwejn J, Meyer E, Ellenbogen M, Milot S, Gjedde A, et al. Functional neuroanatomy of CCK4-induced anxiety in normal healthy volunteers. Am J Psychiatry. 1995;152:1180–4.

209. Javanmard M, Shlik J, Kennedy SH, Vaccarino FJ, Houle S, Bradwejn J. Neuroanatomic correlates of CCK-4-induced panic attacks in healthy humans: a comparison of two time points. Biol Psychiatry. 1999;45:872–82.

210. Reiman EM, Raichle ME, Robins E, Mintun MA, Fusselman MJ, Fox PT, et al. Neuroanatomical correlates of a lactate-induced anxiety attack. Arch Gen Psychiatry. 1989;46:493–500.

211. Goossens L, Leibold N, Peeters R, Esquivel G, Knuts I, Backes W, et al. Brainstem response to hypercapnia: a symptom provocation study into the pathophysiology of panic disorder. J Psychopharmacol. 2014;28:449–56.

212. Brust RD, Corcoran AE, Richerson GB, Nattie E, Dymecki SM. Functional and developmental identification of a molecular subtype of brain serotonergic neuron specialized to regulate breathing dynamics. Cell Rep. 2014;9:2152–65.

213. Bett K, Sandkuhler J. Map of spinal neurons activated by chemical stimulation in the nucleus raphe magnus of the unanesthetized rat. Neuroscience. 1995;67:497–504.

214. Schenberg LC, Lovick TA. Attenuation of the midbrain-evoked defense reaction by selective stimulation of medullary raphe neurons in rats. Am J Physiol. 1995;269:R1378–89.

215. Beitz AJ, Clements JR, Mullett MA, Ecklund LJ. Differential origin of brainstem serotoninergic projections to the midbrain periaqueductal gray and superior colliculus of the rat. J Comp Neurol. 1986;250:498–509.

216. Fischer H, Andersson JL, Furmark T, Fredrikson M. Brain correlates of an unexpected panic attack: a human positron emission tomographic study. Neurosci Lett. 1998;251:137–40.

217. Sato M, Severinghaus JW, Basbaum AI. Medullary CO_2 chemoreceptor neuron identification by c-fos immunocytochemistry. J Appl Physiol. 1992;73: 96–100.

218. Erickson JT, Millhorn DE. Hypoxia and electrical stimulation of the carotid sinus nerve induce Fos-like immunoreactivity within catecholaminergic and serotoninergic neurons of the rat brainstem. J Comp Neurol. 1994;348:161–82.

219. Haxhiu MA, Yung K, Erokwu B, Cherniack NS. CO_2-induced c-fos expression in the CNS catecholaminergic neurons. Respir Physiol. 1996;105:35–45.

220. Hirooka Y, Polson JW, Potts PD, Dampney RA. Hypoxia-induced Fos expression in neurons projecting to the pressor region in the rostral ventrolateral medulla. Neuroscience. 1997;80:1209–24.

221. Larnicol N, Wallois F, Berquin P, Gros F, Rose D. c-fos-like immunoreactivity in the cat's neuraxis following moderate hypoxia or hypercapnia. J Physiol Paris. 1994;88:81–8.

222. Berquin P, Bodineau L, Gros F, Larnicol N. Brainstem and hypothalamic areas involved in respiratory chemoreflexes: a Fos study in adult rats. Brain Res. 2000;857:30–40.

223. Johnson PL, Fitz SD, Hollis JH, Moratalla R, Lightman SL, Shekhar A, et al. Induction of c-Fos in 'panic/defence'-related brain circuits following brief hypercarbic gas exposure. J Psychopharmacol. 2011;25:26–36.

224. Mitchell RA, Herbert DA. The effect of carbon dioxide on the membrane potential of medullary respiratory neurons. Brain Res. 1974;75:345–9.

225. Horn EM, Kramer JM, Waldrop TG. Development of hypoxia-induced Fos expression in rat caudal hypothalamic neurons. Neuroscience. 2000;99:711–20.

226. Feldman JL, Mitchell GS, Nattie EE. Breathing: rhythmicity, plasticity, chemosensitivity. Annu Rev Neurosci. 2003;26:239–66.

227. Raff H, Roarty TP. Renin, ACTH, and aldosterone during acute hypercapnia and hypoxia in conscious rats. Am J Physiol. 1988;254:R431–5.

228. Argyropoulos SV, Bailey JE, Hood SD, Kendrick AH, Rich AS, Laszlo G, et al. Inhalation of 35% CO(2) results in activation of the HPA axis in healthy volunteers. Psychoneuroendocrinology. 2002;27:715–29.

229. Kaye J, Buchanan F, Kendrick A, Johnson P, Lowry C, Bailey J, et al. Acute carbon dioxide exposure in healthy adults: evaluation of a novel means of investigating the stress response. J Neuroendocrinol. 2004;16:256–64.

230. van Duinen MA, Schruers KR, Maes M, Griez EJ. CO_2 challenge induced HPA axis activation in panic. Int J Neuropsychopharmacol. 2007;10:797–804.

231. Johnson P, Lowry C, Truitt W, Shekhar A. Disruption of GABAergic tone in the dorsomedial hypothalamus attenuates responses in a subset of serotonergic neurons in the dorsal raphe nucleus following lactate-induced panic. J Psychopharmacol. 2008;22:642–52.

232. Ryan JW, Waldrop TG. Hypoxia sensitive neurons in the caudal hypothalamus project to the periaqueductal gray. Respir Physiol. 1995;100:185–94.

233. Shams I, Avivi A, Nevo E. Oxygen and carbon dioxide fluctuations in burrows of subterranean blind

mole rats indicate tolerance to hypoxic-hypercapnic stresses. Comp Biochem Physiol A Mol Integr Physiol. 2005;142:376–82.

234. Teppema LJ, Veening JG, Berkenbosch A. Expression of c-fos in the brain stem of rats during hypercapnia. Adv Exp Med Biol. 1995;393:47–51.

235. Bodineau L, Larnicol N. Brainstem and hypothalamic areas activated by tissue hypoxia: Fos-like immunoreactivity induced by carbon monoxide inhalation in the rat. Neuroscience. 2001;108:643–53.

236. Dillon GH, Waldrop TG. In vitro responses of caudal hypothalamic neurons to hypoxia and hypercapnia. Neuroscience. 1992;51:941–50.

237. Coates EL, Li A, Nattie EE. Widespread sites of brain stem ventilatory chemoreceptors. J Appl Physiol. 1993;75:5–14.

238. Kramer JM, Nolan PC, Waldrop TG. In vitro responses of neurons in the periaqueductal gray to hypoxia and hypercapnia. Brain Res. 1999;835:197–203.

239. Kim Y, Bang H, Kim D. TASK-3, a new member of the tandem pore K(+) channel family. J Biol Chem. 2000;275:9340–7.

240. Rajan S, Wischmeyer E, Xin LG, Preisig-Muller R, Daut J, Karschin A, et al. TASK-3, a novel tandem pore domain acid-sensitive K+ channel. An extracellular histiding as pH sensor. J Biol Chem. 2000;275:16650–7.

241. Buckler KJ. TASK-like potassium channels and oxygen sensing in the carotid body. Respir Physiol Neurobiol. 2007;157:55–64.

242. Talley EM, Solorzano G, Lei Q, Kim D, Bayliss DA. CNS distribution of members of the two-pore-domain (KCNK) potassium channel family. J Neurosci. 2001;21:7491–505.

243. Bayliss DA, Talley EM, Sirois JE, Lei Q. TASK-1 is a highly modulated pH-sensitive 'leak' K(+) channel expressed in brainstem respiratory neurons. Respir Physiol. 2001;129:159–74.

244. Washburn CP, Sirois JE, Talley EM, Guyenet PG, Bayliss DA. Serotonergic raphe neurons express TASK channel transcripts and a TASK-like pH- and halothane-sensitive K+ conductance. J Neurosci. 2002;22:1256–65.

245. Washburn CP, Bayliss DA, Guyenet PG. Cardiorespiratory neurons of the rat ventrolateral medulla contain TASK-1 and TASK-3 channel mRNA. Respir Physiol Neurobiol. 2003;138:19–35.

246. Schimitel FG, de Almeida GM, Pitol DN, Armini RS, Tufik S, Schenberg LC. Evidence of a suffocation alarm system within the periaqueductal gray matter of the rat. Neuroscience. 2012;200:59–73.

247. Vianna DM, Landeira-Fernandez J, Brandão ML. Dorsolateral and ventral regions of the periaqueductal gray matter are involved in distinct types of fear. Neurosci Biobehav Rev. 2001;25:711–9.

248. Subramanian HH, Balnave RJ, Holstege G. The midbrain periaqueductal gray control of respiration. J Neurosci. 2008;28:12274–83.

249. Schenberg LC, Póvoa RMF, Costa AL, Caldellas AV, Tufik S, Bittencourt AS. Functional specializations within the tectum defense systems of the rat. Neurosci Biobehav Rev. 2005;29:1279–98.

250. Steeves JD, Jordan LM. Autoradiographic demonstration of the projections from the mesencephalic locomotor region. Brain Res. 1984;307:263–76.

251. Franchini KG, Krieger EM. Cardiovascular responses of conscious rats to carotid body chemoreceptor stimulation by intravenous KCN. J Auton Nerv Syst. 1993;42:63–9.

252. Franchini KG, Oliveira VL, Krieger EM. Hemodynamics of chemoreflex activation in unanesthetized rats. Hypertension. 1997;30:699–703.

253. Hayward LF, Von Reitzentstein M. c-Fos expression in the midbrain periaqueductal gray after chemoreceptor and baroreceptor activation. Am J Physiol Heart Circ Physiol. 2002;283:H1975–84.

254. Schimitel FG, Muller CJ, Tufik S, Schenberg LC. Evidence of a suffocation alarm system sensitive to clinically-effective treatments with the panicolytics clonazepam and fluoxetine. J Psychopharmacol. 2014;28:1184–8.

255. Spiacci Jr A, de Oliveira ST, da Silva GS, Glass ML, Schenberg LC, Garcia-Cairasco N, et al. Serotonin in the dorsal periaqueductal gray inhibits panic-like defensive behaviors in rats exposed to acute hypoxia. Neuroscience. 2015;307:191–8.

256. Suchecki D, Mozaffarian D, Gross G, Rosenfeld P, Levine S. Effects of maternal deprivation on the ACTH stress response in the infant rat. Neuroendocrinology. 1993;57:204–12.

257. Suchecki D, Rosenfeld P, Levine S. Maternal regulation of the hypothalamic-pituitary-adrenal axis in the infant rat: the roles of feeding and stroking. Brain Res Dev Brain Res. 1993;75:185–92.

258. Suchecki D, Nelson DY, Van OH, Levine S. Activation and inhibition of the hypothalamic-pituitary-adrenal axis of the neonatal rat: effects of maternal deprivation. Psychoneuroendocrinology. 1995;20:169–82.

259. Schenberg LC. Towards a translational model of panic attack. Psychol Neurosci. 2010;3:9–37.

260. Johnson PL, Lightman SL, Lowry CA. A functional subset of serotonergic neurons in the rat ventrolateral periaqueductal gray implicated in the inhibition of sympathoexcitation and panic. Ann N Y Acad Sci. 2004;1018:58–64.

261. Sachs E. On the relation of the optic thalamus to respiration, circulation, temperature, and the spleen. J Exp Med. 1911;14:408–40.

262. Tan ES. Brain-stem regions for stimulus-bound and stimulus-related respiration. Exp Neurol. 1967;17:517–28.

263. Paydarfar D, Eldridge FL. Phase resetting and dysrhythmic responses of the respiratory oscillator. Am J Physiol. 1987;252:R55–62.

264. Hayward LF, Swartz CL, Davenport PW. Respiratory response to activation or disinhibition of the dorsal periaqueductal gray in rats. J Appl Physiol. 2003;94:913–22.

265. Cohen MI. Neurogenesis of respiratory rhythm in the mammal. Physiol Rev. 1979;59:1105–73.

266. Schenberg LC, de Aguiar JC, Graeff FG. GABA modulation of the defense reaction induced by brain electrical stimulation. Physiol Behav. 1983;31:429–37.
267. Bizzi E, Libretti A, Malliani A, Zanchetti A. Reflex chemoceptive excitation of diencephalic sham rage behavior. Am J Physiol. 1961;200:923–6.
268. Hilton SM, Joels N. Facilitation of chemoreceptor reflexes during the defence reaction. J Physiol. 1965;176:20–2.
269. Guyenet PG, Koshiya N. Working model of the sympathetic chemoreflex in rats. Clin Exp Hypertens. 1995;17:167–79.
270. Hayward LF, Castellanos M, Davenport PW. Parabrachial neurons mediate dorsal periaqueductal gray evoked respiratory responses in the rat. J Appl Physiol. 2004;96:1146–54.
271. Zhang W, Hayward LF, Davenport PW. Respiratory responses elicited by rostral versus caudal dorsal periaqueductal gray stimulation in rats. Auton Neurosci. 2007;134:45–54.
272. Haibara AS, Tamashiro E, Olivan MV, Bonagamba LG, Machado BH. Involvement of the parabrachial nucleus in the pressor response to chemoreflex activation in awake rats. Auton Neurosci. 2002;101:60–7.
273. Adams DB, Baccelli G, Mancia G, Zanchetti A. Relation of cardiovascular changes in fighting to emotion and exercise. J Physiol. 1971;212:321–35.
274. Hilton SM. The defence-arousal system and its relevance for circulatory and respiratory control. J Exp Biol. 1982;100:159–74.
275. Lovick TA. Interactions between descending pathways from the dorsal and ventrolateral periaqueductal gray matter in the rat. In: Depaulis A, Bandler R, editors. The midbrain periaqueductal gray matter. New York: Plenum Press; 1991. p. 101–20.
276. Lovick TA. Midbrain and medullary regulation of defensive cardiovascular functions. Prog Brain Res. 1996;107:301–13.
277. Carobrez AP, Schenberg LC, Graeff FG. Neuroeffector mechanisms of the defense reaction in the rat. Physiol Behav. 1983;31:439–44.
278. Subramanian HH, Holstege G. Stimulation of the midbrain periaqueductal gray modulates preinspiratory neurons in the ventrolateral medulla in the in vivo rat. J Comp Neurol. 2013;521:3083–98.
279. Hilton SM. Inhibition of baroreceptor reflexes on hypothalamic stimulation. J Physiol. 1963;165:56P–7.
280. Coote JH, Hilton SM, Zbrozyna AW. The pontomedullary area integrating the defence reaction in the cat and its influence on muscle blood flow. J Physiol. 1973;229:257–74.
281. Schenberg LC, Vasquez EC, da Costa MB. Cardiac baroreflex dynamics during the defence reaction in freely moving rats. Brain Res. 1993;621:50–8.
282. Lopes LT, Patrone LG, Bicego KC, Coimbra NC, Gargaglioni LH. Periaqueductal gray matter modulates the hypercapnic ventilatory response. Pflugers Arch. 2012;464:155–66.
283. Lopes LT, Biancardi V, Vieira EB, Leite-Panissi C, Bicego KC, Gargaglioni LH. Participation of the dorsal periaqueductal grey matter in the hypoxic ventilatory response in unanaesthetized rats. Acta Physiol (Oxf). 2014;211:528–37.
284. Krout KE, Jansen AS, Loewy AD. Periaqueductal gray matter projection to the parabrachial nucleus in rat. J Comp Neurol. 1998;401:437–54.
285. Hayward LF, Castellanos M. Increased c-Fos expression in select lateral parabrachial subnuclei following chemical versus electrical stimulation of the dorsal periaqueductal gray in rats. Brain Res. 2003;974:153–66.
286. Potts JT, Rybak IA, Paton JF. Respiratory rhythm entrainment by somatic afferent stimulation. J Neurosci. 2005;25:1965–78.
287. Orwin A. 'The running treatment': a preliminary communication on a new use for an old therapy (physical activity) in the agoraphobic syndrome. Br J Psychiatry. 1973;122:175–9.
288. Sun MK, Reis DJ. Hypoxia-activated Ca2+ currents in pacemaker neurones of rat rostral ventrolateral medulla in vitro. J Physiol. 1994;476:101–16.
289. Pascual O, Morin-Surun MP, Barna B, Deavit-Saubie M, Pequignot JM, Champagnat J. Progesterone reverses the neuronal responses to hypoxia in rat nucleus tractus solitarius in vitro. J Physiol. 2002;544:511–20.
290. Peano CA, Shonis CA, Dillon GH, Waldrop TG. Hypothalamic GABAergic mechanism involved in respiratory response to hypercapnia. Brain Res Bull. 1992;28:107–13.
291. Panneton WM, Loewy AD. Projections of the carotid sinus nerve to the nucleus of the solitary tract in the cat. Brain Res. 1980;191:239–44.
292. Sandkuhler J, Herdegen T. Distinct patterns of activated neurons throughout the rat midbrain periaqueductal gray induced by chemical stimulation within its subdivisions. J Comp Neurol. 1995;357:546–53.
293. Morgan MM, Carrive P. Activation of the ventrolateral periaqueductal gray reduces locomotion but not mean arterial pressure in awake, freely moving rats. Neuroscience. 2001;102:905–10.
294. Loewy AD, Wallach JH, Kellar S. Efferent connections of the ventral medulla oblongata in the rat. Brain Res Rev. 1981;3:63–80.
295. Sandner G, Dessort D, Schmitt P, Karli P. Distribution of GABA in the periaqueductal gray matter. Effects of medial hypothalamic lesions. Brain Res. 1981;224:279–90.
296. Van den Bergh P, Wu P, Jackson IM, Lechan RM. Neurons containing a N-terminal sequence of the TRH-prohormone (preproTRH53-74) are present in a unique location of the midbrain periaqueductal gray of the rat. Brain Res. 1988;461:53–63.
297. Beitz AJ, Williams FG. Localization of putative amino acid transmitters in the PAG and their relationship to the PAG-raphe magnus pathway. The midbrain periaqueductal gray matter. New York: Plenum Press; 1991. p. 305–27.
298. Smith GST, Savery D, Marden C, Costa JJL, Averill S, Priestley JV, et al. Distribution of messenger RNAs encoding enkephalin, substance P, somatostatin, galanin, vasoactive intestinal polypeptide,

neuropeptide Y, and calcitonin gene related peptide in the midbrain periaqueductal grey in the rat. J Comp Neurol. 1994;350:23–40.

299. Mihaly E, Legradi G, Fekete C, Lechan RM. Efferent projections of ProTRH neurons in the ventrolateral periaqueductal gray. Brain Res. 2001;919:185–97.

300. Mennicken F, Hoffert C, Pelletier M, Ahmad S, O'Donnell D. Restricted distribution of galanin receptor 3 (GalR3) mRNA in the adult rat central nervous system. J Chem Neuroanat. 2002;24:257–68.

301. Harding A, Paxinos G, Halliday G. The serotonin and tachykinin systems. In: Paxinos G, editor. The rat nervous system. 3rd ed. San Diego: Elsevier; 2004. p. 1205–56.

302. Darwinkel A, Stanic D, Booth LC, May CN, Lawrence AJ, Yao ST. Distribution of orexin-1 receptor-green fluorescent protein- (OX1-GFP) expressing neurons in the mouse brain stem and pons: co-localization with tyrosine hydroxylase and neuronal nitric oxide synthase. Neuroscience. 2014;278:253–64.

303. Jenck F, Moreau JL, Martin JR. Dorsal periaqueductal gray-induced aversion as a simulation of panic anxiety: elements of face and predictive validity. Psychiatry Res. 1995;57:181–91.

304. Brandão ML, de Aguiar JC, Graeff FG. GABA mediation of the anti-aversive action of minor tranquilizers. Pharmacol Biochem Behav. 1982;16:397–402.

305. Graeff FG, Brandão ML, Audi EA, Milani H. Role of GABA in the anti-aversive action of anxiolytics. Adv Biochem Psychopharmacol. 1986;42:79–86.

306. Behbehani MM, Jiang MR, Chandler SD, Ennis M. The effect of GABA and its antagonists on midbrain periaqueductal gray neurons in the rat. Pain. 1990;40:195–204.

307. Lovick TA, Stezhka VV. Neurones in the dorsolateral periaqueductal grey matter in coronal slices of rat midbrain: electrophysiological and morphological characteristics. Exp Brain Res. 1999;124:53–8.

308. Brandao ML, Lopez-Garcia JA, Graeff FG, Roberts MH. Electrophysiological evidence for excitatory 5-HT2 and depressant 5-HT1A receptors on neurones of the rat midbrain tectum. Brain Res. 1991;556:259–66.

309. Jeong HJ, Lam K, Mitchell VA, Vaughan CW. Serotonergic modulation of neuronal activity in rat midbrain periaqueductal gray. J Neurophysiol. 2013;109:2712–9.

310. Stezhka VV, Lovick TA. Inhibitory and excitatory projections from the dorsal raphe nucleus to neurons in the dorsolateral periaqueductal gray matter in slices of midbrain maintained in vitro. Neuroscience. 1994;62:177–87.

311. Thompson RH, Canteras NS, Swanson LW. Organization of projections from the dorsomedial nucleus of the hypothalamus: a PHA-L study in the rat. J Comp Neurol. 1996;376:143–73.

312. Veening J, Buma P, Ter Horst GJ, Roeling TAP, Luiten PGM, Nieuwenhuys R. Hypothalamic projections to the PAG in the rat: topographical, immuno-electronmicroscopical and functional aspects. In: Depaulis A, Bandler R, editors. The midbrain periaqueductal gray matter. New York: Plenum Press; 1991. p. 387–415.

313. Barbaresi P. Immunocytochemical localization of substance P receptor in rat periaqueductal gray matter: a light and electron microscopic study. J Comp Neurol. 1998;398:473–90.

314. Drew GM, Mitchell VA, Vaughan CW. Postsynaptic actions of substance P on rat periaqueductal grey neurons in vitro. Neuropharmacology. 2005;49:587–95.

315. Brodin E, Rosen A, Schott E, Brodin K. Effects of sequential removal of rats from a group cage, and of individual housing of rats, on substance P, cholecystokinin and somatostatin levels in the periaqueductal grey and limbic regions. Neuropeptides. 1994;26:253–60.

316. Bassi GS, Nobre MJ, Carvalho MC, Brandão ML. Substance P injected into the dorsal periaqueductal gray causes anxiogenic effects similar to the long-term isolation as assessed by ultrasound vocalizations measurements. Behav Brain Res. 2007;182:301–7.

317. Shekhar A, DiMicco JA. Defense reaction elicited by injection of GABA antagonists and synthesis inhibitors into the posterior hypothalamus in rats. Neuropharmacology. 1987;26:407–17.

318. Shekhar A, Hingtgen JN, DiMicco JA. Selective enhancement of shock avoidance responding elicited by GABA blockade in the posterior hypothalamus of rats. Brain Res. 1987;420:118–28.

319. Shekhar A, Hingtgen JN, DiMicco JA. GABA receptors in the posterior hypothalamus regulate experimental anxiety in rats. Brain Res. 1990;512:81–8.

320. Shekhar A. GABA receptors in the region of the dorsomedial hypothalamus of rats regulate anxiety in the elevated plus-maze test. I Behav Meas Brain Res. 1993;627:9–16.

321. Shekhar A, Johnson PL, Sajdyk TJ, Fitz SD, Keim SR, Kelley PE, et al. Angiotensin-II is a putative neurotransmitter in lactate-induced panic-like responses in rats with disruption of GABAergic inhibition in the dorsomedial hypothalamus. J Neurosci. 2006;26:9205–15.

322. Johnson PL, Shekhar A. Panic-prone state induced in rats with GABA dysfunction in the dorsomedial hypothalamus is mediated by NMDA receptors. J Neurosci. 2006;26:7093–104.

323. Shekhar A. Effects of treatment with imipramine and clonazepam on an animal model of panic disorder. Biol Psychiatry. 1994;36:748–58.

324. Shekhar A, Keim SR, Simon JR, McBride WJ. Dorsomedial hypothalamic GABA dysfunction produces physiological arousal following sodium lactate infusions. Pharmacol Biochem Behav. 1996;55:249–56.

325. Swanson LW, Lind RW. Neural projections subserving the initiation of a specific motivated behavior in the rat: new projections from the subfornical organ. Brain Res. 1986;379:399–403.

326. Shekhar A, Sajdyk TS, Keim SR, Yoder KK, Sanders SK. Role of the basolateral amygdala in panic disorder. Ann N Y Acad Sci. 1999;877:747–50.

327. Shekhar A, Keim SR. LY354740, a potent group II metabotropic glutamate receptor agonist prevents lactate-induced panic-like response in panic-prone rats. Neuropharmacology. 2000;39:1139–46.

328. Sajdyk TJ, Schober DA, Gehlert DR, Shekhar A. Role of corticotropin-releasing factor and urocortin within the basolateral amygdala of rats in anxiety and panic responses. Behav Brain Res. 1999;100:207–15.

329. Sajdyk TJ, Shekhar A. Sodium lactate elicits anxiety in rats after repeated GABA receptor blockade in the basolateral amygdala. Eur J Pharmacol. 2000; 394:265–73.

330. Johnson PL, Truitt W, Fitz SD, Minick PE, Dietrich A, Sanghani S, et al. A key role for orexin in panic anxiety. Nat Med. 2010;16:111–5.

331. Antunes-Rodrigues J, Castro M, Elias LL, Valenca MM, McCann SM. Neuroendocrine control of body fluid metabolism. Physiol Rev. 2004;84:169–208.

332. Fitzsimons JT. Angiotensin, thirst, and sodium appetite. Physiol Rev. 1998;78:583–686.

333. Rodgers RJ, Ishii Y, Halford JC, Blundell JE. Orexins and appetite regulation. Neuropeptides. 2002;36:303–25.

334. Siegel JM. Narcolepsy: a key role for hypocretins (orexins). Comment. Cell. 1999;98:409–12.

335. Taheri S, Zeitzer JM, Mignot E. The role of hypocretins (orexins) in sleep regulation and narcolepsy. Ann Rev Neurosci. 2002;25:283–313.

336. Molosh AI, Johnson PL, Fitz SD, DiMicco JA, Herman JP, Shekhar A. Changes in central sodium and not osmolarity or lactate induce panic-like responses in a model of panic disorder. Neuropsychopharmacology. 2010;35:1333–47.

337. Kellner M, Wiedemann K, Holsboer F. Atrial natriuretic factor inhibits the CRH-stimulated secretion of ACTH and cortisol in man. Life Sci. 1992;50: 1835–42.

338. Tannure RM, Bittencourt AS, Schenberg LC. Short-term full kindling of the amygdala dissociates natural and periaqueductal gray-evoked flight behaviors of the rat. Behav Brain Res. 2009;199:247–56.

339. Bailey TW, DiMicco JA. Chemical stimulation of the dorsomedial hypothalamus elevates plasma ACTH in conscious rats. Am J Physiol Regul Integr Comp Physiol. 2001;280:R8–15.

340. Zaretskaia MV, Zaretsky DV, Shekhar A, DiMicco JA. Chemical stimulation of the dorsomedial hypothalamus evokes non-shivering thermogenesis in anesthetized rats. Brain Res. 2002;928:113–25.

341. Johnson PL, Truitt WA, Fitz SD, Lowry CA, Shekhar A. Neural pathways underlying lactate-induced panic. Neuropsychopharmacology. 2008;33:2093–107.

342. Olsson M, Ho HP, Annerbrink K, Thylefors J, Eriksson E. Respiratory responses to intravenous infusion of sodium lactate in male and female Wistar rats. Neuropsychopharmacology. 2002;27:85–91.

343. Herbert H, Moga MM, Saper CB. Connections of the parabrachial nucleus with the nucleus of the solitary tract and the medullary reticular formation in the rat. J Comp Neurol. 1990;293:540–80.

344. Chamberlin NL, Saper CB. Topographic organization of respiratory responses to glutamate microstimulation of the parabrachial nucleus in the rat. J Neurosci. 1994;14:6500–10.

345. Chamberlin NL. Functional organization of the parabrachial complex and intertrigeminal region in the control of breathing. Respir Physiol Neurobiol. 2004;143:115–25.

346. Kaur S, Pedersen NP, Yokota S, Hur EE, Fuller PM, Lazarus M, et al. Glutamatergic signaling from the parabrachial nucleus plays a critical role in hypercapnic arousal. J Neurosci. 2013;33:7627–40.

347. Yokota S, Kaur S, Vanderhorst VG, Saper CB, Chamberlin NL. Respiratory-related outputs of glutamatergic, hypercapnia-responsive parabrachial neurons in mice. J Comp Neurol. 2015;523: 907–20.

348. Huang GF, Besson JM, Bernard JF. Intravenous morphine depresses the transmission of noxious messages to the nucleus centralis of the amygdala. Eur J Pharmacol. 1993;236:449–56.

349. Huang GF, Besson JM, Bernard JF. Morphine depresses the transmission of noxious messages in the spino(trigemino)-ponto-amygdaloid pathway. Eur J Pharmacol. 1993;230:279–84.

350. Beitz AJ. Possible origin of glutamatergic projections to the midbrain periaqueductal grey and deep layer of the superior colliculus of the rat. Brain Res Bull. 1989;23:25–35.

351. da Silva LG, de Menezes RC, dos Santos RA, Campagnole-Santos MJ, Fontes MA. Role of periaqueductal gray on the cardiovascular response evoked by disinhibition of the dorsomedial hypothalamus. Brain Res. 2003;984:206–14.

352. da Silva LG, Menezes RC, Villela DC, Fontes MA. Excitatory amino acid receptors in the periaqueductal gray mediate the cardiovascular response evoked by activation of dorsomedial hypothalamic neurons. Neuroscience. 2006;139:1129–39.

353. Allen GV, Saper CB, Hurley KM, Cechetto DF. Organization of visceral and limbic connections in the insular cortex of the rat. J Comp Neurol. 1991;311:1–16.

354. Shipley MT, Ennis M, Rizvi TA, Behbehani MM. Topographical specificity of forebrain inputs to the midbrain periaqueductal gray: evidence for discrete longitudinally organized input columns. In: Depaulis A, Bandler R, editors. The midbrain periaqueductal gray matter. New York: Plenum Press; 1991. p. 417–48.

355. Floyd NS, Price JL, Ferry AT, Keay KA, Bandler R. Orbitomedial prefrontal cortical projections to distinct longitudinal columns of the periaqueductal gray in the rat. J Comp Neurol. 2000;422:556–78.

356. Selye H. A syndrome produced by diverse nocuous agents. Nature. 1936;138:31.

357. Selye H. Forty years of stress research: principal remaining problems and misconceptions. Can Med Assoc J. 1976;115:53–6.

358. Strohle A, Holsboer F. Stress responsive neurohormones in depression and anxiety. Pharmacopsychiatry. 2003;36 Suppl 3:S207–14.

359. Jovanovic T, Norrholm SD, Fennell JE, Keyes M, Fiallos AM, Myers KM, et al. Posttraumatic stress disorder may be associated with impaired fear inhibition: relation to symptom severity. Psychiatry Res. 2009;167:151–60.

360. Jovanovic T, Norrholm SD, Blanding NQ, Phifer JE, Weiss T, Davis M, et al. Fear potentiation is associated with hypothalamic-pituitary-adrenal axis function in PTSD. Psychoneuroendocrinology. 2010;35:846–57.

361. Jovanovic T, Norrholm SD, Blanding NQ, Davis M, Duncan E, Bradley B, et al. Impaired fear inhibition is a biomarker of PTSD but not depression. Depress Anxiety. 2010;27:244–51.

362. Keen-Rhinehart E, Michopoulos V, Toufexis DJ, Martin EI, Nair H, Ressler KJ, et al. Continuous expression of corticotropin-releasing factor in the central nucleus of the amygdala emulates the dysregulation of the stress and reproductive axes. Mol Psychiatry. 2009;14:37–50.

363. Neill JD. Effects of "stress" on serum prolactin and luteinizing hormone levels during the estrous cycle of the rat. Endocrinology. 1970;87:1192–7.

364. Siegel RA, Conforti N, Chowers I. Neural pathways mediating the prolactin secretory response to acute neurogenic stress in the male rat. Brain Res. 1980; 198:43–53.

365. Dijkstra H, Tilders FJH, Hiehle MA, Smelik PG. Hormonal reactions to fighting in rat colonies prolactin rises during defence, not during offence. Physiol Behav. 1992;51:961–8.

366. Sinha SS, Coplan JD, Pine DS, Martinez JA, Klein DF, Gorman JM. Panic induced by carbon dioxide inhalation and lack of hypothalamic-pituitary-adrenal axis activation. Psychiatry Res. 1999;86:93–8.

367. Bandelow B, Wedekind D, Pauls J, Broocks A, Hajak G, Ruther E. Salivary cortisol in panic attacks. Am J Psychiatry. 2000;157:454–6.

368. Kamilaris TC, Johnson EO, Calogero AE, Kalogeras KT, Bernardini R, Chrousos GP, et al. Cholecystokinin-octapeptide stimulates hypothalamic-pituitary-adrenal function in rats: role of corticotropin-releasing hormone. Endocrinology. 1992;130:1764–74.

369. Graeff FG, Garcia-Leal C, Del-Ben CM, Guimaraes FS. Does the panic attack activate the hypothalamic-pituitary-adrenal axis? An Acad Bras Cienc. 2005;77: 477–91.

370. De Montigny C. Cholecystokinin tetrapeptide induces panic-like attacks in healthy volunteers. Preliminary findings. Arch Gen Psychiatry. 1989;46:511–7.

371. Bradwejn J, Koszycki D, Annable L, Couetoux DT, Reines S, Karkanias C. A dose-ranging study of the behavioral and cardiovascular effects of CCK-tetrapeptide in panic disorder. Biol Psychiatry. 1992;32:903–12.

372. Abelson JL, Liberzon I. Dose response of adrenocorticotropin and cortisol to the CCK-B agonist pentagastrin. Neuropsychopharmacology. 1999;21:485–94.

373. Kobelt P, Paulitsch S, Goebel M, Stengel A, Schmidtmann M, van der Voort I, et al. Peripheral injection of CCK-8S induces Fos expression in the dorsomedial hypothalamic nucleus in rats. Brain Res. 2006;1117:109–17.

374. Shlik J, Aluoja A, Vasar V, Vasar E, Podar T, Bradwejn J. Effects of citalopram treatment on behavioural, cardiovascular and neuroendocrine response to cholecystokinin tetrapeptide challenge in patients with panic disorder. J Psychiatry Neurosci. 1997;22:332–40.

375. Depot M, Caille G, Mukherjee J, Katzman MA, Cadieux A, Bradwejn J. Acute and chronic role of 5-HT3 neuronal system on behavioral and neuroendocrine changes induced by intravenous cholecystokinin tetrapeptide administration in humans. Neuropsychopharmacology. 1999;20:177–87.

376. Chen DY, Deutsch JA, Gonzalez MF, Gu Y. The induction and suppression of c-fos expression in the rat brain by cholecystokinin and its antagonist L364,718. Neurosci Lett. 1993;149:91–4.

377. Schenberg LC, Dos Reis AM, Ferreira Povoa RM, Tufik S, Silva SR. A panic attack-like unusual stress reaction. Horm Behav. 2008;54:584–91.

378. Lim LW, Blokland A, van DM, Visser-Vandewalle V, Tan S, Vlamings R, et al. Increased plasma corticosterone levels after periaqueductal gray stimulation-induced escape reaction or panic attacks in rats. Behav Brain Res. 2011;218:301–7.

379. Armini RS, Bernabe CS, Rosa CA, Siller CA, Schimitel FG, Tufik S, et al. In a rat model of panic, corticotropin responses to dorsal periaqueductal gray stimulation depend on physical exertion. Psychoneuroendocrinology. 2015;53:136–47.

380. Beckett S, Marsden CA. Computer analysis and quantification of periaqueductal grey- induced defence behaviour. J Neurosci Methods. 1995;58: 157–61.

381. Neophytou SI, Graham M, Williams J, Aspley S, Marsden CA, Beckett SR. Strain differences to the effects of aversive frequency ultrasound on behaviour and brain topography of c-fos expression in the rat. Brain Res. 2000;854:158–64.

382. Soya H, Mukai A, Deocaris CC, Ohiwa N, Chang H, Nishijima T, et al. Threshold-like pattern of neuronal activation in the hypothalamus during treadmill running: establishment of a minimum running stress (MRS) rat model. Neurosci Res. 2007;58:341–8.

383. Pittman QJ, Blume HW, Renaud LP. Connections of the hypothalamic paraventricular nucleus with the neurohypophysis, median eminence, amygdala, lateral septum and midbrain periaqueductal gray: an electrophysiological study in the rat. Brain Res. 1981;215:15–28.

384. Ziegler DR, Edwards MR, Ulrich-Lai YM, Herman JP, Cullinan WE. Brainstem origins of glutamatergic innervation of the rat hypothalamic paraventricular nucleus. J Comp Neurol. 2012;520:2369–94.

385. Klein S, Nicolas LB, Lopez-Lopez C, Jacobson LH, McArthur SG, Grundschober C, et al. Examining face and construct validity of a noninvasive model of panic

disorder in Lister-hooded rats. Psychopharmacology (Berl). 2010;211:197–208.

386. Feldman S, Weidenfeld J. The excitatory effects of the amygdala on hypothalamo-pituitary-adrenocortical responses are mediated by hypothalamic norepinephrine, serotonin, and CRF-41. Brain Res Bull. 1998;45: 389–93.

387. Bhatnagar S, Viau V, Chu A, Soriano L, Meijer OC, Dallman MF. A cholecystokinin-mediated pathway to the paraventricular thalamus is recruited in chronically stressed rats and regulates hypothalamic-pituitary-adrenal function. J Neurosci. 2000;20:5564–73.

388. Koolhaas JM, Bartolomucci A, Buwalda B, de Boer SF, Flugge G, Korte SM, et al. Stress revisited: a critical evaluation of the stress concept. Neurosci Biobehav Rev. 2011;35:1291–301.

389. Raff H, Shinsako J, Dallman MF. Renin and ACTH responses to hypercapnia and hypoxia after chronic carotid chemodenervation. Am J Physiol. 1984;247: R412–7.

390. Raff H, Shinsako J, Keil LC, Dallman MF. Vasopressin, ACTH, and corticosteroids during hypercapnia and graded hypoxia in dogs. Am J Physiol. 1983;244:E453–8.

391. Raff H, Shinsako J, Keil LC, Dallman MF. Vasopressin, ACTH, and blood pressure during hypoxia induced at different rates. Am J Physiol. 1983;245:E489–93.

392. Raff H, Sandri RB, Segerson TP. Renin, ACTH, and adrenocortical function during hypoxia and hemorrhage in conscious rats. Am J Physiol. 1986;250:R240–4.

393. Kossowsky J, Wilhelm FH, Schneider S. Responses to voluntary hyperventilation in children with separation anxiety disorder: implications for the link to panic disorder. J Anxiety Disord. 2013;27:627–34.

394. Vargas LC, Marques TA, Schenberg LC. Micturition and defensive behaviors are controlled by distinct neural networks within the dorsal periaqueductal gray and deep gray layer of the superior colliculus of the rat. Neurosci Lett. 2000;280:45–8.

395. Schenberg LC, Marcal LPA, Seeberger F, Barros MR, Sudré ECM. L-type calcium channels selectively control the defensive behaviors induced by electrical stimulation of dorsal periaqueductal gray and overlying collicular layers. Behav Brain Res. 2000;111:175–85.

396. Goetz RR, Klein DF, Gorman JM. Consistencies between recalled panic and lactate-induced panic. Anxiety. 1994;1:31–6.

397. APA. Diagnostic and statistical manual of mental disorders. 4th-R ed. Washington, DC: American Psychiatry Association; 2000.

398. WHO. The ICD-10 classification of mental and behavioral disorders. Diagnostic criteria for research. Geneva: 1993.

399. D'Amato FR, Zanettini C, Lampis V, Coccurello R, Pascucci T, Ventura R, et al. Unstable maternal environment, separation anxiety, and heightened CO_2 sensitivity induced by gene-by-environment interplay. PLoS One. 2011;6:e18637.

400. Kinkead R, Genest SE, Gulemetova R, Lajeunesse Y, Laforest S, Drolet G, et al. Neonatal maternal separation and early life programming of the hypoxic ventilatory response in rats. Respir Physiol Neurobiol. 2005;149:313–24.

401. Genest SE, Gulemetova R, Laforest S, Drolet G, Kinkead R. Neonatal maternal separation and sex-specific plasticity of the hypoxic ventilatory response in awake rat. J Physiol. 2004;554:543–57.

402. Genest SE, Gulemetova R, Laforest S, Drolet G, Kinkead R. Neonatal maternal separation induces sex-specific augmentation of the hypercapnic ventilatory response in awake rat. J Appl Physiol. 2007;102:1416–21.

403. Francis DD, Meaney MJ. Maternal care and the development of stress responses. Curr Opin Neurobiol. 1999;9:128–34.

404. Abelson JL, Curtis GC. Hypothalamic-pituitary-adrenal axis activity in panic disorder. 24-Hour secretion of corticotropin and cortisol. Arch Gen Psychiatry. 1996;53:323–31.

405. Schreiber W, Lauer CJ, Krumrey K, Holsboer F, Krieg JC. Dysregulation of the hypothalamic-pituitary-adrenocortical system in panic disorder. Neuropsychopharmacology. 1996;15:7–15.

406. Rentesi G, Antoniou K, Marselos M, Fotopoulos A, Alboycharali J, Konstandi M. Long-term consequences of early maternal deprivation in serotonergic activity and HPA function in adult rat. Neurosci Lett. 2010;480:7–11.

407. Fournier S, Allard M, Gulemetova R, Joseph V, Kinkead R. Chronic corticosterone elevation and sex-specific augmentation of the hypoxic ventilatory response in awake rats. J Physiol. 2007;584(Pt 3):951–62.

408. Gorman JM, Coplan JD. Comorbidity of depression and panic disorder. J Clin Psychiatry. 1996;57 Suppl 10:34–41.

409. Kaufman J, Charney D. Comorbidity of mood and anxiety disorders. Depress Anxiety. 2000;12 Suppl 1:69–76.

410. Safadi G, Bradwejn J. Relationship of panic disorder to posttraumatic stress disorder. Arch Gen Psychiatry. 1995;52:76–8.

411. Koenen KC, Lyons MJ, Goldberg J, Simpson J, Williams WM, Toomey R, et al. A high risk twin study of combat-related PTSD comorbidity. Twin Res. 2003;6:218–26.

412. Nixon RD, Resick PA, Griffin MG. Panic following trauma: the etiology of acute posttraumatic arousal. J Anxiety Disord. 2004;18:193–210.

413. Hinton D, Hsia C, Um K, Otto MW. Anger-associated panic attacks in Cambodian refugees with PTSD; a multiple baseline examination of clinical data. Behav Res Ther. 2003;41:647–54.

414. Cougle JR, Feldner MT, Keough ME, Hawkins KA, Fitch KE. Comorbid panic attacks among individuals with posttraumatic stress disorder: associations with traumatic event exposure history, symptoms, and impairment. J Anxiety Disord. 2010;24: 183–8.

415. McGrath PJ, Stewart JW, Liebowitz MR, Markowitz JM, Quitkin FM, Klein DF, et al. Lactate provocation of panic attacks in depressed outpatients. Psychiatry Res. 1988;25:41–7.

416. Quintino-Dos-Santos JW, Muller CJ, Santos AMC, Tufik S, Rosa CA, Schenberg LC. Long-lasting marked inhibition of periaqueductal gray-evoked defensive behaviors in inescapably-shocked rats. Eur J Neurosci. 2014;39:275–86.

417. de Paula Soares V, Zangrossi Jr H. Involvement of 5-HT1A and 5-HT2 receptors of the dorsal periaqueductal gray in the regulation of the defensive behaviors generated by the elevated T-maze. Brain Res Bull. 2004;64:181–8.

418. Maier SF, Seligman ME. Learned helplessness: theory and evidence. J Exp Psychol. 1975;105:3–46.

419. Maier SF. Learned helplessness and animal models of depression. Prog Neuropsychopharmacol Biol Psychiatry. 1984;8:435–46.

420. Maier SF, Watkins LR. Stressor controllability, anxiety, and serotonin. Cogn Ther Res. 1998;22:595–613.

421. Maier SF, Watkins LR. Stressor controllability and learned helplessness: the roles of the dorsal raphe nucleus, serotonin, and corticotropin-releasing factor. Neurosci Biobehav Rev. 2005;29:829–41.

422. Porsolt RD, Lenegre A, McArthur RA. Pharmacological models of depression. In: Olivier B, Slangen JL, Mos J, editors. Animal models in psychopharmacology. Basle: Birkhaeuser-Verlag; 1991. p. 137–61.

423. Maier SF, Grahn RE, Kalman BA, Sutton LC, Wiertelak EP, Watkins LR. The role of the amygdala and dorsal raphe nucleus in mediating the behavioral consequences of inescapable shock. Behav Neurosci. 1993;107:377–88.

424. Amat J, Matus-Amat P, Watkins LR, Maier SF. Escapable and inescapable stress differentially and selectively alter extracellular levels of 5-HT in the ventral hippocampus and dorsal periaqueductal gray of the rat. Brain Res. 1998;797:12–22.

425. Hammack SE, Richey KJ, Watkins LR, Maier SF. Chemical lesion of the bed nucleus of the stria terminalis blocks the behavioral consequences of uncontrollable stress. Behav Neurosci. 2004;118:443–8.

426. Leshner AI, Segal M. Fornix transection blocks "learned helplessness" in rats. Behav Neural Biol. 1979;26:497–501.

427. Petty F, Sherman AD. Learned helplessness induction decreases in vivo cortical serotonin release. Pharmacol Biochem Behav. 1983;18:649–50.

428. Joca SR, Padovan CM, Guimaraes FS. Activation of post-synaptic 5-HT(1A) receptors in the dorsal hippocampus prevents learned helplessness development. Brain Res. 2003;978:177–84.

429. Joca SR, Zanelati T, Guimaraes FS. Post-stress facilitation of serotonergic, but not noradrenergic, neurotransmission in the dorsal hippocampus prevents learned helplessness development in rats. Brain Res. 2006;1087:67–74.

430. Malberg JE, Duman RS. Cell proliferation in adult hippocampus is decreased by inescapable stress: reversal by fluoxetine treatment. Neuropsychopharmacology. 2003;28:1562–71.

431. Zhou J, Li L, Tang S, Cao X, Li Z, Li W, et al. Effects of serotonin depletion on the hippocampal GR/MR and BDNF expression during the stress adaptation. Behav Brain Res. 2008;195:129–38.

432. Wu J, Kramer GL, Kram M, Steciuk M, Crawford IL, Petty F. Serotonin and learned helplessness: a regional study of 5-HT1A, 5-HT2A receptors and the serotonin transport site in rat brain. J Psychiatr Res. 1999;33:17–22.

433. Yang LM, Hu B, Xia YH, Zhang BL, Zhao H. Lateral habenula lesions improve the behavioral response in depressed rats via increasing the serotonin level in dorsal raphe nucleus. Behav Brain Res. 2008;188:84–90.

434. Lino-de-Oliveira C, De Lima TC, Carobrez AP. Dorsal periaqueductal gray matter inhibits passive coping strategy elicited by forced swimming stress in rats. Neurosci Lett. 2002;335:87–90.

435. Lino-de-Oliveira C, De Oliveira RM, Padua CA, De Lima TC, Del Bel EA, Guimaraes FS. Antidepressant treatment reduces Fos-like immunoreactivity induced by swim stress in different columns of the periaqueductal gray matter. Brain Res Bull. 2006;70:414–21.

436. Jang DP, Lee SH, Lee SY, Park CW, Cho ZH, Kim YB. Neural responses of rats in the forced swimming test: [F-18]FDG micro PET study. Behav Brain Res. 2009;203:43–7.

437. Strong PV, Christianson JP, Loughridge AB, Amat J, Maier SF, Fleshner M, et al. 5-Hydroxytryptamine 2C receptors in the dorsal striatum mediate stress-induced interference with negatively reinforced instrumental escape behavior. Neuroscience. 2011;197:132–44.

438. Krout KE, Loewy AD. Periaqueductal gray matter projections to midline and intralaminar thalamic nuclei of the rat. J Comp Neurol. 2000;424:111–41.

439. Macchi G, Bentivoglio M, Molinari M, Minciacchi D. The thalamo-caudate versus thalamo-cortical projections as studied in the cat with fluorescent retrograde double labeling. Exp Brain Res. 1984;54:225–39.

440. Kaufman J, Plotsky PM, Nemeroff CB, Charney DS. Effects of early adverse experiences on brain structure and function: clinical implications. Biol Psychiatry. 2000;48:778–90.

441. Faravelli C, Pallanti S. Recent life events and panic disorder. Am J Psychiatry. 1989;146:622–6.

442. Falsetti SA, Resnick HS. Frequency and severity of panic attack symptoms in a treatment seeking sample of trauma victims. J Trauma Stress. 1997;10:683–9.

443. Sherman AD, Sacquitne JL, Petty F. Specificity of the learned helplessness model of depression. Pharmacol Biochem Behav. 1982;16:449–54.

444. Maier SF. Exposure to the stressor environment prevents the temporal dissipation of behavioral depression/learned helplessness. Biol Psychiatry. 2001;49:763–73.

445. Yonkers KA, Pearlstein T, Rosenheck RA. Premenstrual disorders: bridging research and clinical reality. Arch Womens Ment Health. 2003;6:287–92.

446. Le Mellédo JM, Van DM, Coupland NJ, Lott P, Jhangri GS. Response to flumazenil in women with premenstrual dysphoric disorder. Am J Psychiatry. 2000;157:821–3.

447. Gorman JM, Kent J, Martinez J, Browne S, Coplan J, Papp LA. Physiological changes during carbon dioxide inhalation in patients with panic disorder, major depression, and premenstrual dysphoric disorder: evidence for a central fear mechanism. Arch Gen Psychiatry. 2001;58:125–31.

448. Kent JM, Papp LA, Martinez JM, Browne ST, Coplan JD, Klein DF, et al. Specificity of panic response to CO_2 inhalation in panic disorder: a comparison with major depression and premenstrual dysphoric disorder. Am J Psychiatry. 2001;158:58–67.

449. Lovick TA. Plasticity of GABA-A receptor subunit expression during the oestrous cycle of the rat: implications for premenstrual syndrome in women. Exp Physiol. 2006;91:655–60.

450. Lovick TA. GABA in the female brain—oestrous cycle-related changes in GABAergic function in the periaqueductal grey matter. Pharmacol Biochem Behav. 2008;90:43–50.

451. Lovick TA, Devall AJ. Progesterone withdrawal-evoked plasticity of neural function in the female periaqueductal grey matter. Neural Plast. 2009;2009:730902.

452. Lovick TA, Griffiths JL, Dunn SM, Martin IL. Changes in GABA-A receptor subunit expression in the midbrain during the oestrous cycle in Wistar rats. Neuroscience. 2005;131:397–405.

453. Griffiths J, Lovick T. Withdrawal from progesterone increases expression of alpha4, beta1, and delta GABA(A) receptor subunits in neurons in the periaqueductal gray matter in female Wistar rats. J Comp Neurol. 2005;486:89–97.

454. Griffiths JL, Lovick TA. GABAergic neurones in the rat periaqueductal grey matter express alpha4, beta1 and delta GABAA receptor subunits: plasticity of expression during the estrous cycle. Neuroscience. 2005;136:457–66.

455. Brack KE, Jeffery SM, Lovick TA. Cardiovascular and respiratory responses to a panicogenic agent in anaesthetised female Wistar rats at different stages of the oestrous cycle. Eur J Neurosci. 2006;23:3309–18.

456. Brack KE, Lovick TA. Neuronal excitability in the periaqueductal grey matter during the estrous cycle in female Wistar rats. Neuroscience. 2007;144:325–35.

457. Smith SS, Woolley CS. Cellular and molecular effects of steroid hormones on CNS excitability. Cleve Clin J Med. 2004;71 Suppl 2:S4–10.

458. Gallo MA, Smith SS. Progesterone withdrawal decreases latency to and increases duration of electrified prod burial: a possible rat model of PMS anxiety. Pharmacol Biochem Behav. 1993;46:897–904.

459. Gulinello M, Gong QH, Li X, Smith SS. Short-term exposure to a neuroactive steroid increases alpha4 GABA-A receptor subunit levels in association with increased anxiety in the female rat. Brain Res. 2001;910:55–66.

460. Gulinello M, Gong QH, Smith SS. Progesterone withdrawal increases the alpha4 subunit of the GABA-A receptor in male rats in association with anxiety and altered pharmacology—a comparison with female rats. Neuropharmacology. 2002;43:701–14.

461. Smith SS. Withdrawal properties of a neuroactive steroid: implications for GABA(A) receptor gene regulation in the brain and anxiety behavior. Steroids. 2002;67:519–28.

462. Smith SS, Gong QH, Hsu FC, Markowitz RS, Ffrench-Mullen JM, Li X. GABA-A receptor alpha4 subunit suppression prevents withdrawal properties of an endogenous steroid. Nature. 1998;392:926–30.

463. Freeman EW, Frye CA, Rickels K, Martin PA, Smith SS. Allopregnanolone levels and symptom improvement in severe premenstrual syndrome. J Clin Psychopharmacol. 2002;22:516–20.

464. Redgrave P, Dean P. Does the PAG learn about emergencies from the superior colliculus? In: Depaulis A, Bandler R, editors. The midbrain periaqueductal gray matter. New York: Plenum Press; 1991. p. 199–209.

465. King SM, Shehab S, Dean P, Redgrave P. Differential expression of fos-like immunoreactivity in the descending projections of superior colliculus after electrical stimulation in the rat. Behav Brain Res. 1996;78:131–45.

466. Canteras NS, Chiavegatto S, Valle LE, Swanson LW. Severe reduction of rat defensive behavior to a predator by discrete hypothalamic chemical lesions. Brain Res Bull. 1997;44:297–305.

467. Dielenberg RA, Hunt GE, McGregor IS. "When a rat smells a cat": the distribution of Fos immunoreactivity in rat brain following exposure to a predatory odor. Neuroscience. 2001;104:1085–97.

468. Bittencourt AS, Nakamura-Palacios EM, Mauad H, Tufik S, Schenberg LC. Organization of electrically and chemically evoked defensive behaviors within the deeper collicular layers as compared to the periaqueductal gray matter of the rat. Neuroscience. 2005;133:873–92.

469. Jenck F, Broekkamp CL, Van Delft AML. The effect of antidepressants on aversive periaqueductal gray stimulation in rats. Eur J Pharmacol. 1990;177:201–4.

470. Hogg S, Michan L, Jessa M. Prediction of anti-panic properties of escitalopram in the dorsal periaqueductal grey model of panic anxiety. Neuropharmacology. 2006;51:141–5.

471. Schenberg LC, Capucho LB, Vatanabe RO, Vargas LC. Acute effects of clomipramine and fluoxetine on dorsal periaqueductal grey-evoked unconditioned defensive behaviours of the rat. Psychopharmacology (Berl). 2002;159:138–44.

The Hippocampus and Panic Disorder: Evidence from Animal and Human Studies

3

Gisele Pereira Dias and Sandrine Thuret

Contents

G.P. Dias
Laboratory of Panic and Respiration, Institute of
Psychiatry, Federal University of Rio de Janeiro,
Rio de Janeiro, Brazil
e-mail: giseledias@ufrj.br

S. Thuret (✉)
Department of Basic and Clinical Neuroscience,
Laboratory of Adult Neurogenesis and Mental
Health, Institute of Psychiatry, Psychology and
Neuroscience, King's College London, London, UK
e-mail: sandrine.1.thuret@kcl.ac.uk

Abstract

Panic disorder (PD) is a highly incapacitating psychiatric disorder. Its wide range of somatic and psychological symptoms makes it plausible that a number of different brain structures and circuits are likely to mediate this condition. In this chapter we highlight the possible contributions of the hippocampus, a key brain region involved in the regulation of cognition (learning/memory), mood and defensive responses (fear/anxiety), for the pathophysiology of PD. This chapter will present the anatomy of the hippocampus and highlight its role in emotional regulation, so that an understanding of the involvement of the hippocampus in PD can be drawn. Evidence from both animal and human findings on this topic will be approached. Particularly, the capacity of the hippocampus to continually generate newly functional neurons throughout life, a phenomenon called adult hippocampal neurogenesis, will be pointed as part of the key future directions for the study of the neurobiological basis of PD.

Keywords

Hippocampus • Panic disorder • Fear circuitry • Anxiety • Stem cells • Animal models

3.1   Anatomy
of the Hippocampus

The hippocampus is one of the brain structures implicated in the regulation of many of the features involved in panic disorder (PD), such as those related to defensive responses to threat. Moreover, in the clinical practice with PD patients, it is very common that this condition is accompanied by another highly impairing disorder, agoraphobia. Agoraphobic patients present intense fear of being in contexts where they cannot escape or find immediate help. Thus, although normally the first panic attack occurs spontaneously (i.e., without the presence of a specific and identifiable triggering stimulus), there is often an association of the attack with the surrounding environmental (exteroceptive) and/or internal (interoceptive) cues. This latter is known to be mediated primarily by the insula but growing evidence from both human and rodent studies show that the processing of exteroceptive cues is fundamentally a function of the hippocampus, along with the processing of self-location [1], which in itself is an essential requirement for agoraphobic associations.

Before exploring the evidence on the functional contribution of the hippocampus to the development of PD-related symptomatology, it is useful to understand how the hippocampal formation is organized and integrates the emotional circuits of the limbic system.

The hippocampus is a C-shaped structure located deeply within the subcortical region of the temporal lobes, extending longitudinally along the brain. It is rolled-up in 2 *laminae*, the *cornu Ammonis* (CA; for its resemblance with Jupiter Ammon's horn [2]) and the *gyrus dentatus* (dentate gyrus, DG) [3]. It is well established that the information processing in the hippocampus occurs through its tri-synaptic circuit, starting with the glutamatergic granule cells of the DG receiving input from the entorhinal cortex (EC), via the perforant path. From the DG projections constituting the so-called Mossy fibers extend their axons to CA3, which then establishes connection with the pyramidal neurons at CA1, via the Schaffer collaterals [4]. From CA1,

the processed information is sent to other parts of the brain through projections to the subiculum and EC [5, 6].

The hippocampal formation extends dorsoventrally within the temporal lobes, an anatomic feature that appears to render functional consequences. In this sense, evidence indicates that the information processed in the dorsal hippocampus is more likely to integrate into different circuits than that processed by the ventral portion of the hippocampus. Thus, due to its projections to regions such as the dorsal septum, the mammillary complex and the lateral entorhinal cortex, the dorsal hippocampus appears to be more related to the regulation of cognitive abilities, whereas mood and anxiety are believed to be more fundamentally regulated by the circuitry of the ventral hippocampus [7–9], considering its outputs to the amygdala, hypothalamus, *nucleus accumbens* and the prefrontal cortex [10] (Fig. 3.1).

3.2   The Role
of the Hippocampus
in Emotional Regulation

It is well established that the hippocampus plays a fundamental role for cognitive abilities, such as spatial recognition and declarative memory [11–14]. The classic case of patient H.M., who had a bilateral temporal lobe resection as an intervention to reduce chronic epileptic seizures and, as a consequence of the removal of the hippocampus, started to suffer from a severe incapacity to form new factual memories [15], established the hippocampal formation as the "house of memory" in the brain. Considering the large amount of evidence that came later with hundreds of animal studies pointing for the hippocampus as one of the ultimate structures implicated in cognitive regulation, it might seem counterintuitive that this brain region also significantly participates in the regulation of mood and anxiety. However, as we shall see in Sect. 3.4, the hippocampus integrates a defensive neural network, proposed in the end of the 1980s by Gorman et al. [16] and known as the fear circuitry,

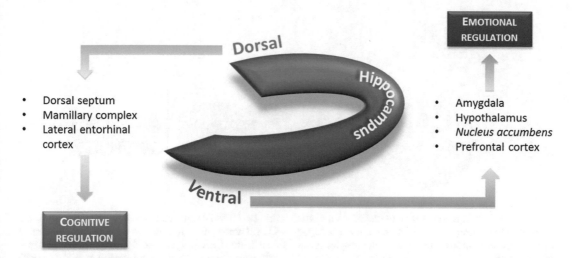

Fig. 3.1 Schematic representation of the dorsoventral division of the hippocampus. The hippocampus is a C-shaped structure that extends longitudinally along the septo-temporal axis of the brain. The dorsal hippocampus, due to its projections to structures such as the dorsal sep-tum, the mammillary complex and the lateral entorhinal cortex, is believed to be involved in cognitive abilities; the ventral hippocampus, in turn, projects to the amygdala, hypothalamus, *nucleus accumbens* and prefrontal cortex, thus being more related with emotional regulation [10]

a well-established circuit implicated in processing external and internal cues to prepare the individual for a defensive response. This network includes the amygdala, *nucleus accumbens*, hippocampus, ventromedial hypothalamus, periaqueductal gray, several brain stem and thalamic nuclei, the insular cortex, and some prefrontal regions (Box 3.1).

Box 3.1. Fear Circuitry

When a stimulus is perceived as potentially harmful for the integrity of the individual, a series of neurochemical, neuroendocrine and behavioral responses arises which, altogether, potentiate the likelihood of survival. In neurobiological terms, threatening stimuli activate the so-called fear circuitry. Fear responses are triggered through the activation of a subcortical structure, the amygdala, which receives afferents from a number of structures involved in cognitive processing, such as the sensorial cortex, the thalamus and, of our special interest, the hippocampus [16].

The hippocampal formation is, therefore, an essential component of the emotional system of the brain, playing an important functional role in the modulation of complex behavioral patterns, along with cortical and subcortical areas.

But which could be the putative roles of the hippocampus in the fear circuitry? One potential candidate is the processing of risk assessment.

When it comes to defensiveness behaviors, risk assessment emerges as a fundamental aspect of emotional regulation. This behavioral pattern is believed to aim at evaluating the odds of threat/potential danger in contrast with those of reward [17]. This defense strategy – whereby the so-called non-defensive behaviors (self-grooming, locomotion, feeding, etc.) are fundamentally inhibited [18] – has been pointed to be likely a function of the hippocampus [19] and its circuits with the septum and the amygdala [20]. The input to the hippocampus from the medial entorhinal cortex (MEC) on the spatial context of an experience, and from the lateral entorhinal cortex (LEC) on the content of an experience [21] would make it possible the delivery of information needed for risk assessment processing by the hippocampal formation, thus closely relating a

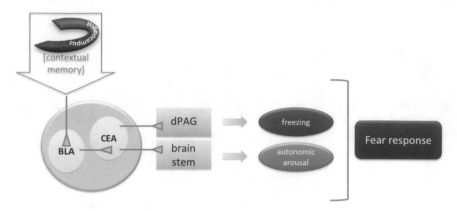

Fig. 3.2 Schematic representation of one of the potential roles of the hippocampus in the fear circuitry. The hippocampus encodes information about contexts – for example, information on the surrounding environmental stimuli associated with a previous panic attack – and projects it to the BLA, which in turn rapidly connects with the CEA. Through its connections to the dPAG and brain stem nuclei, the CEA triggers the fight/flight fear response. *BLA* basolateral amygdala, *CEA* central amygdala, *dPAG* dorsal periaqueductal gray

possible hyperactivity of this function to the threat-biased information processing observed in anxious patients. Nevertheless, some authors relate risk assessment to be a function primarily related to the mediation of danger in generalized anxiety disorder (GAD), and not in PD [20] which would be a disorder more closely related to fear- rather than anxiety- (and therefore risk assessment) related responses. Further research on this topic should help unravel if and which the contributions of the hippocampus for risk assessment in the context of panic responses are, since evidence from clinical practice largely point for the presence of such behavioral patterns among PD patients as well.

Another putative role of the hippocampus in PD – particularly in cases involving agoraphobic associations – is through its well-established mediation of contextual fear learning. With effect, the hippocampus is one of the main neural structures contributing to the fear/anxiety manifestations triggered by learned fear, most likely via its descending projections to the amygdala [22]. Maren and Fanselow [23] showed that electrolytic lesions in regions of the hippocampal formation that send projections to the amygdala eliminate pavlovian fear conditioning in situations of contextual conditioned stimuli. Phillips and Ledoux [24] also demonstrated that hippo-

campal lesions interfere in contextual fear conditioning. Anagnostaras et al. [25] reinforced this and showed further that the participation of the hippocampus in contextual fear learning is time-framed, since lesions in this structure one day after conditioning prevent fear learning, whereas hippocampal lesions 28 days after the conditioning session render no effect over the storage of contextual fear memories. In this way, animals with hippocampal lesions do not present contextual fear responses, given that they are not able to form a contextual representation necessary for the association of the context with the unconditioned stimulus and send it to the amygdala so that the expected fear responses are triggered [26]. A schematic representation of this putative role of the hippocampus within the fear circuitry is illustrated in Fig. 3.2.

Further and stronger evidence for the potential role of the hippocampus on emotional regulation, expanding its functional roles beyond those involved in cognitive processing, come from studies investigating the effects of anxiolytic drugs injected specifically to the ventral hippocampus. Indeed, growing evidence point for the notion that the functions regulated by the hippocampus are topographically distributed along its septo-temporal axis, with the dorsal hippocampus being more related to the regulation of cognition [7–9] – and the ventral hippocampus more closely regulating emotional processing [10], considering its projections to more limbic areas. According to Behrendt [27], information processed by the dorsal hippocampus would be translated into orienting and locomotor actions, whilst that processed by the ventral hippocampal formation would lead to the mapping of motivationally salient environmental information, via its projections to the ventral subiculum -ventromedial prefrontal cortex and ventral striatum. In addition, Behrendt [27] emphasizes that not only external cues would be processed by the ventral hippocampus, but also emotional information extracted from the surrounding internal physiological milieu, inducing the individual to enter a behavioral mode of emotion-guided arousal. Although the author highlights the putative links of the ventral hippocampus with appetitive behavior, considering other sources of evidence pointing for a role of this portion of the hippocampus in defensiveness, this becomes of special interest when thinking about the possible contributions of the hippocampus to PD. Indeed, if it is true that at least the ventral portion of the hippocampus is fundamental for the processing of exteroceptive and interoceptive cues, then the hippocampus could achieve a higher status among the neural substrates of PD. This is a reasonable hypothesis, considering that PD/agoraphobia mostly emerge as a consequence of aversive associations with contexts where previous panic attacks took place, as well as with bridges, tunnels and other agoraphobic stimuli (external cues), in addition with higher levels of CO_2 and other panicogenic stimuli in the blood and cerebrospinal fluid (internal cues).

Supporting this idea, a recent study has identified that microinjections of neuropeptide S – a potential novel treatment for anxiety-related illnesses such as PD – into the ventral CA1 are sufficient to reduce the anxious behavior of C57BL/6N mice [28]. The participation of the hippocampus in emotional regulation appears to be well established, however future studies are still needed for further investigation of the specific contributions of the dorsal and ventral portions of the hippocampus in this process. Indeed, some studies of the dorsal hippocampus also

showed this region to influence anxiety modulation [29–31]; the dorsoventral dichotomy [32], thus, seems to be less simple than thought, and deserves special attention from the next generation of studies aiming to unravel the particularities of the contributions of the hippocampus to emotional regulation.

3.3   Involvement of the Hippocampus in PD: Evidence from Rodent Studies

Despite the growing evidence pointing for the involvement of the hippocampal circuitry in the mediation of anxiety-related behaviors, one of the most established rodent models for PD has the hypothalamus – and not the hippocampus – at its center. The hypothalamic model is very efficient in inducing panic vulnerability in rats, by chronically disrupting the inhibition promoted by GABAergic neurons in the hypothalamus followed by intravenous injections of panicogenic stimuli, such as sodium lactate infusion [33]. The model presents robust face, predictive and construct validities and its well-conceived rationale should also provide the basis for the development of future rodent models aiming to unravel the contributions of other key brain areas involved in fear, anxiety and PD, such as the hippocampus.

At least two points should be highlighted in our pursuit to understand the contributions of the hippocampus and other brain structures in animal models of PD: (1) although the hippocampus has become a very popular structure to study due to its highly neuroplastic nature (please see Sect. 3.5), it is far from being the only brain region involved in such complex cognitive and emotional processing, such as that present in anxiety disorders, thus making it fundamental the existence of models investigating other brain areas; (2) part of the scientific literature on rodent studies more clearly distinguish the panic attack from the anxiety concept. It is worth noting here that the panic attack is a fear-related response characterized by the flight/freeze strategy mediated by the hypothalamus [33] and the dorsal periaqueductal gray to a real, *present* danger [20], not to mention the amygdala which is the hallmark structure underlying fear [19]; the anxiety concept, on the other hand, relates more to the anticipatory emotion to a *potential* threat. In the clinical practice, both emotions appear to be present in PD patients: the fear response that characterizes the panic attack itself and the anxiety presented in anticipation to the advent of a new attack or to the absence of immediate help, in the case of PD patients with agoraphobia.

But the hypothalamic rat model of PD is not the only rodent model aiming at investigating the neurobiological basis of this disorder. In this context, it has been shown, for instance, that intraperitoneal injections of lactate – which produced panic-like behavior, as seen by the induction of tachycardia and freezing response – increased the neuronal firing of neurons in the rat hippocampus [34].

Important evidence for the contribution of the hippocampus in PD have also come from studies using transgenic mice that overexpress the neurotrophin tyrosine kinase receptor type 3 (NTRK3) – a protein whose expression has been found to be altered in PD and other anxious patients [35]. TgNTRK3 mice present good: face validity, as seen by their heightened anxiety and panic-related responses; construct validity, observed by their increased density of noradrenergic neurons in the *locus coeruleus*, and finally, predictive validity, given that they respond well to diazepam in the elevated T maze [36], a paradigm for assessing panic-like behavioral patterns. These animals also present higher susceptibility to stress, as shown by their altered circadian corticosterone rhythm and more passive behaviors under certain chronic stress conditions [37], reinforcing this mouse line as an appropriate model of this disorder. Of special note, it has been recently found that these mice present hyperexcitability in the hippocampal subcircuit CA3–CA1 and that this unbalanced excitation-to-inhibition ratio in the hippocampus underlies their also increased fear memories [38].

Finally, additional evidence from animal studies for the contribution of the hippocampus in PD come from a study showing that the fear-like behavior induced by deep brain stimulation of the periaqueductal gray leads to deactivation of parvalbumin-positive interneurons in the hippocampus 12 h after the panic-like reaction observed in the open field [39]. Interestingly, these authors had previously shown that electrical stimulation of the dorsal periaqueductal gray induced an increase in the activation of cells in the CA1 and DG 2 h after the panic reaction [40]. This reinforces the notion that not only hippocampal subpopulations of neurons mediate the regulation of long-lasting anxiety- and fear-related behavioral patterns but also that they are necessary at different time frames of the fear learning processing.

3.4 Involvement of the Hippocampus in PD: Evidence from Human Studies

Neuroimaging studies have immensely contributed to our current understanding of PD [41, 42] and other anxiety disorders, especially with regard the to fear circuitry [43]. According to the original proposition by Gorman et al. [16], PD patients would present a more sensitive fear network, whose (hyper)activation would then result in the arousal of panic attacks. Some authors, however, suggest that this circuitry should be revised as to include other areas, such as the anterior cingulate [44].

Specifically with regard to the putative role of the hippocampus in mediating both trait and state anxiety in PD, data is conflicting despite a trend towards pointing the hippocampal circuits as being fundamental for our comprehension of the neurobiology of PD. In this sense, a quantitative volumetric magnetic resonance imaging (MRI) study showed that, although a bilateral reduction of the temporal lobe volume was revealed in PD patients, the amygdala-hippocampus complex (AHC) was found to be normal [45]. Similar findings had been reported by Vythilingam et al. [46] who measured the volume of the temporal lobe, hippocampus and whole brain in 13 patients with PD in comparison with 14 healthy controls, revealing that the mean volume of both left and right temporal lobes was significantly decreased in the patients' group, without changes in the hippocampal volume. The assumption that deficits in hippocampal volume appear to be a minor issue for the development of PD has also received recent support [47].

Nevertheless, other studies point for the opposite, i.e., for volume reductions of the hippocampus and the left parahippocampal gyrus in PD patients ([48]; reviewed in [49]). Future studies with larger samples and more advanced imaging techniques are, thus, needed so that a more comprehensive view of the hippocampal volume as a possible biomarker of PD can be pinpointed.

But macroscopic measures, such as the volume of a given structure, are not the only parameters implicating a certain region or network in the mediation of anxiety or other traits/states. Metabolic and functional aspects can add important evidence for understanding the contribution of the hippocampus (or of any other brain structure) for the emotional processing characteristic of PD. In this particular, changes in metabolism in hippocampal [50] and parahippocampal [51] areas have been found in PD studies using single photon emission computed tomography (SPECT). The hippocampus, along with the amygdala and insula, has also been shown to present increased activation among PD patients in a functional MRI (fMRI) study [52]. Enhanced hippocampal activation in response to a safety signal has also been revealed at baseline for medication-free PD patients with agoraphobia [53]. Furthermore, reduced binding properties of the serotonergic receptor 5-HT1A has been found in the hippocampus, amygdala, as well as in frontal and temporal cortical areas, of patients with PD [54]. Reduced binding potential of the serotonergic transporter (5-HTT) has also been reported in males with PD when compared with healthy males [55]. These data strongly support the idea that the hippocampus, especially at the metabolic and neurotransmission levels, present important components that contribute for the abnormal information processing typical of PD. Other studies, using magnetic resonance spectroscopy, also pointed for abnormalities of different neurotransmitter and metabolites in PD, highlighting the involvement of the hippocampus in this disorder (reviewed in [56]). The relevance of understand-

ing the contribution of the hippocampal formation in PD goes beyond the mapping of the neural substrates of this disorder: it is becoming increasingly relevant to comprehend how distinguished activation patterns of certain brain areas to a symptom-eliciting task can predict pharmacological and psychotherapeutic treatment outcome. In this context, a recent study has shown, among others, that activation of the hippocampus during maintenance of emotional responses to negative images during fMRI scanning of PD and GAD patients was greater in responders than non-responders to cognitive-behavioral therapy (CBT) [57]. Another study reinforced the idea of the hippocampus (particularly, of increased right hippocampal gray matter volume), along with a differential activation of other structures, as a predictor of improved CBT outcome for PD [58]. Such studies are strongly encouraged as they open important avenues for the development of more personalized and effective treatment strategies to psychiatric patients.

3.5   The Role of the Hippocampus in PD: Future Directions

Due to their remarkable plasticity – that is, to the ability to change morphologically and functionally upon environmental demands and internal signals – hippocampal circuits are the subject of innumerous studies aiming at identifying effective interventions capable to positively regulate cognition and emotion. In this context of neural plasticity, one of the most noticeable features of the hippocampus is its ability to continuously generate functional neurons throughout the lifespan of the individual.

Indeed, for many years, neuroscientists believed that once the organism was born no further neuronal cells could be added to the brain. In the 1960s, nevertheless, the technical possibility of identifying and tracking newly born neurons in the postnatal brain through autoradiography [59], and later, by the injection of the thymidine analog bromodeoxyuridine (BrdU) [60] opened a new avenue in the study of how the brain works – and how it can be modified. This represented the breakthrough for a new mentality in neurobiological sciences, based on the fundamental assumption that the adult brain could be shaped by the generation of new neurons capable to integrate into specific circuits.

Adult neurogenesis occurs in the so-called neurogenic niches – brain regions characterized by the presence of neural progenitor cells at constant self-renewal activity and holding the potentiality to differentiate into neurons in response to specific molecular signaling. In the mammalian brain, these niches are the subventricular zone (SVZ) [61, 62], adjacent to the lateral ventricles, and the subgranular zone (SGZ) of the DG in the hippocampus [63, 64]. The generation of neurons in this latter niche during postnatal life is, thus, known as adult hippocampal neurogenesis (AHN) and is of special interest for our topic on mental health.

AHN is a highly complex and intricately regulated process [65], counting on the influence of a number of different regulatory molecular signaling pathways, many of which are still largely unknown. These molecular pathways specifically target the regulation of each of AHN's stages: (1) maintenance of the pool of progenitors; (2) cell proliferation; (3) fate commitment to a neuronal phenotype; (4) acquisition of structural and functional characteristics of a mature granule neuron (maturation); (5) survival; (6) integration into pre-existing circuits. These, in turn, are believed to be the circuits upon regulation by the hippocampal formation, that is, circuits involved in cognitive functions (especially spatial and reference working memory) and mood/anxiety.

With effect, a number of papers have demonstrated that the lack of AHN induces cognitive impairment [66, 67], as well as depressive- [68] and anxiety-like [69] behavior in rodents. Despite the consistent acknowledgement in the scientific community implicating neurogenesis deficits with cognitive and mood dysfunction, not every study have succeeded in showing such association ([70]; reviewed in [71]). Nevertheless – and of special relevance for our topic on anxiety – there is more consensus in recognizing that the newly generated neurons appear to have a crucial role in the ability of the DG to distinguish similar stimuli. This process is known as pattern separation, a mechanism without which the individual loses the ability to convert similar experiences into discrete, nonoverlapping representations. This ability, in turn, is thought to be an important factor underlying the development of anxiety disorders [72].

Mostly important, it has been recently and elegantly demonstrated that AHN is a process happening not only in rodents and monkeys, but also in humans. Previous *postmortem* analysis of the human brain had already indicated the occurrence of AHN in our species [73] but not only replication was needed; functional inferences were necessary as well, so that AHN could more definitely emerge as a form of relevant neural plasticity in humans. This came with the study by Jonas Frisén's group in 2013 [74]: in an ingenious way, these researchers took advantage of the increased levels of C^{14} in the atmosphere after the nuclear bomb tests of the 1950s and 60s to hypothesize that, if adult neurogenesis was true for humans, then the *postmortem* analysis of C^{14} in the brain should show neurons that were born in years posterior to the individual's year of birth. Curiously, not only this was demonstrated to be the case, but also that: (1) this was shown primarily to happen indeed in the hippocampus; (2) through mathematical modelling, the group could show that AHN takes place in the human brain at similar rates to those found in the rodent brain. This is especially encouraging to behavioral neuroscientists, in that it gives support to the idea that understanding the mechanisms underlying AHN and behavioral change in animals might generate knowledge that is also, to great extent, applicable to humans.

One of the greatest challenges to modern neuroscience is, thus, to identify specific interventions and their respective mechanisms that are capable of upregulating AHN in health and disease. With effect, it has been demonstrated that AHN can be altered by: (1) different classes of drugs, such as hypnotics [75] and, classically, by antidepressants [76, 77]; (2) environmental factors, like the exposure to an enriched environment [78, 79] and physical exercise [80, 81]; (3)

dietary factors (reviewed in [82]), such as with regard to meal intake (caloric restriction; [83]), meal frequency (intermittent fasting; [84]), meal texture (reviewed in [85]) and meal content (reviewed in [86]; for example, polyphenol- or omega-3 fatty acids-enriched diets×high fat diets). All these three levels of interventions have been shown, in animal studies, to be able to modify mental health-related behaviors likely via the induction of AHN. This opens a new avenue for studies using PD models. In this regard, to our knowledge, only one studied aimed at associating AHN to a panic-related response [87]. Specifically, these authors showed that chronic treatment with corticosterone induced anxiety [88] and decreased AHN in rats [87], an effect that could be counteracted by the antidepressant imipramine. However, the effects of chronic treatment with corticosterone on anxiety were related to avoidance behavior, which has been associated with generalized anxiety, and not to escape behavior, a panic-related response. Besides, chronic treatment with corticosterone has failed to induce specific panic-like behaviors; it can induce generalized anxiety- [88] and depressive- [89] behavioral phenotypes, both of which can be clinically comorbid with PD but are not necessarily and specifically a feature of panic. On the other hand, higher levels of awakening cortisol have been found in PD patients [90] and higher levels of corticosterone are largely known to be associated with downregulation of AHN [89]. Thus, the study of hippocampal neurogenesis in more specific animal models of PD, which might have high levels of corticosterone as one of their biomarkers of stress, may render invaluable responses on the mechanisms by which the hippocampus may contribute to anxiety in PD, as well as serve as a platform for the screening of potentially effective interventions modifying AHN and panic-related behaviors.

But animal models are not the only way to study the putative roles of AHN in PD. The recent advent of patient-specific induced pluripotent stem cells (iPSCs) (reviewed in [91]) has immensely added to the toolkit of scientists aiming to unravel the cellular and molecular substrates of potentially any disease with a strong genetic component, which is the case of a multitude of mental illnesses, including PD [92]. The use of these cells offers a variety of relevant advantages, such as: (1) the fact that it is not only human, but patient-specific tissue; (2) collecting samples for the generation of iPSCs is a non-invasive procedure; (3) it allows for *in vivo* correlation of the cellular/molecular findings with those from the same patients in neuropsychological assessment and neuroimaging evaluation. Of special interest for our topic, it has been recently demonstrated that DG granule neurons can be generated from human pluripotent stem cells [93]. Subsequently, a next frontier in the study of the neurobiology of PD, particularly with regard to the contributions of the hippocampus, could be the generation of DG neurons from iPSCs of PD patients.

3.6 Conclusion

The lifetime prevalence of PD is estimated to reach 3.7 %, with panic attacks reaching alarming 22.7 % [94], thus posing the study of the biopsychosocial features related to this impairing disorder as an important challenge for contemporary science.

This chapter discussed empirical evidence on the putative roles of the hippocampal formation for the threat-biased emotional processing characteristic of PD. Although, as described, the literature comprises growing evidence for a contribution of the hippocampus in PD and other anxious states, it is noteworthy that the complex cognitive and emotional processing of any neuropsychiatric illness results from abnormal functioning of neural circuits encompassing a number of brain regions, each contributing for different aspects of the disease. The future of our understanding of the neurobiological basis of PD, and of other psychiatric disorders, lies thus on the active integration of knowledge not only about the brain regions involved, but also on their functional connectivity. This is a dynamic result of intricate genetic and epigenetic factors regulating neurochemical, endocrine and behavioral systems. Interestingly, if on the one hand these factors are at the very basis of psychiatric disorder, on the other they also hold promise to be modifiable targets for effective pharmacological, psychotherapeutic/environmental and dietary interventions.

References

1. Witter M, Canto CB, Couey JJ, Koganezawa N, O'Reilly KC. Architecture of spatial circuits in the hippocampal region. Philos Trans R Soc Lond B Biol Sci. 2013;369(1635):20120515.
2. Pearce J. Ammon's horn and the hippocampus. J Neurol Neurosurg Psychiatry. 2001;71(3):351.
3. Destrieux C, Bourry D, Velut S. Surgical anatomy of the hippocampus. Neurochirurgie. 2013;59(4–5): 149–58.
4. Szirmai I, Buzsáki G, Kamondi A. 120 years of hippocampal Schaffer collaterals. Hippocampus. 2012; 22(7):1508–16.
5. Amaral D, Lavenex P. Hippocampal neuroanatomy. In: Andersen P, Morris R, Amaral D, Bliss T, O'Keefe J, editors. The hippocampus book. New York: Oxford University Press; 2007. p. 37–114.
6. Doetsch F, Hen R. Young and excitable: the function of new neurons in the adult mammalian brain. Curr Opin Neurobiol. 2005;15(1):121–8.
7. Fanselow M, Dong H. Are the dorsal and ventral hippocampus functionally distinct structures? Neuron. 2010;65:7–19.
8. Kheirbek M, Hen R. Dorsal vs ventral hippocampal neurogenesis: implications for cognition and mood. Neuropsychopharmacology. 2011;36(1):373–4.
9. Tanti A, Rainer Q, Minier F, Surget A, Belzung C. Differential environmental regulation of neurogenesis along the septo-temporal axis of the hippocampus. Neuropharmacology. 2012;63(3):374–84.
10. Sahay A, Hen R. Adult hippocampal neurogenesis in depression. Nat Neurosci. 2007;10(9):1110–5.
11. Squire L. Memory and the hippocampus: a synthesis from findings with rats, monkeys, and humans. Psychol Rev. 1992;99:195–231.
12. Squire L, Ojemann JG, Miezin FM, Petersen SE, Videen TO, Raichle ME. Activation of the hippocampus in normal humans: a functional anatomical study of memory. Proc Natl Acad Sci USA. 1992;89:1837–41.
13. Squire L. The hippocampus and spatial memory. Trends Neurosci. 1993;16:56–7.
14. Kosaki Y, Lin TC, Horne MR, Pearce JM, Gilroy KE. The role of the hippocampus in passive and active spatial learning. Hippocampus. 2014;24(12): 1633–52.
15. Squire L. The legacy of patient H.M. for neuroscience. Neuron. 2009;61(1):6–9.
16. Gorman J, Liebowitz MR, Fyer AJ, Stein J. A neuroanatomical hypothesis for panic disorder. Am J Psychiatry. 1989;146(2):148–61.
17. Blanchard D, Blanchard RJ. Ethoexperimental approaches to the biology of emotion. Annu Rev Psychol. 1988;39:43–68.
18. Blanchard R, Nikulina JN, Sakai RR, McKittrick C, McEwen B, Blanchard DC. Behavioral and endocrine change following chronic predatory stress. Physiol Behav. 1998;63(4):561–9.
19. Morris R. Stress and the hippocampus. In: Andersen P, Morris R, Amaral D, Bliss T, O'Keefe J, editors. The hippocampus book. New York: Oxford University Press; 2007. p. 751–68.
20. Shuhama R, Del-Ben CM, Loureiro SR, Graeff FG. Animal defense strategies and anxiety disorders. An Acad Bras Ciênc. 2007;79(1):97–109.
21. Knierim J, Neunuebel JP, Deshmukh SS. Functional correlates of the lateral and medial entorhinal cortex: objects, path integration and local-global reference frames. Philos Trans R Soc Lond B Biol Sci. 2013;369(1635):20130369.
22. Anagnostaras S, Craske MG, Fanselow MS. Anxiety: at the intersection of genes and experience. Nat Neurosci. 1999;2(9):780–2.
23. Maren S, Fanselow MS. Synaptic plasticity in the basolateral amygdala induced by hippocampal formation stimulation in vivo. J Neurosci. 1995;15(11):7548–64.
24. Phillips R, Ledoux J. Differential contribution of amygdala and hippocampus to cued and contextual fear conditioning. Behav Neurosci. 1992;106(2):274–85.
25. Anagnostaras S, Maren S, Fanselow MS. Temporally graded retrograde amnesia of contextual fear after hippocampal damage in rats: within-subjects examination. J Neurosci. 1999;19(3):1106–14.
26. Ledoux J. The emotional brain: the mysterious underpinnings of emotional life. New York: Phoenix; 1999.
27. Behrendt R. Situationally appropriate behavior: translating situations into appetitive behavior modes. Rev Neurosci. 2013;24(6):577–606.
28. Dine J, Ionescu IA, Stepan J, Yen YC, Holsboer F, Landgraf R, et al. Identification of a role for the ventral hippocampus in neuropeptide S-elicited anxiolysis. PLoS One. 2013;8(3):e60219.
29. Gonzalez L, Ouagazzal AM, File SE. Stimulation of benzodiazepine receptors in the dorsal hippocampus and median raphe reveals differential GABAergic control in two animal test of anxiety. Eur J Neurosci. 1998;10(12):3673–80.
30. Nazar M, Jessa M, Płaźnik A. Benzodiazepine-GABAA receptor complex ligands in two models of anxiety. J Neural Transm. 1997;104(6–7):733–46.
31. Rezayat M, Roohbakhsh A, Zarrindast MR, Massoudi R, Djahanguiri B. Cholecystokinin and GABA interaction in the dorsal hippocampus of rats in the elevated plus-maze test of anxiety. Physiol Behav. 2005;84(5):775–82.
32. Strange B, Witter MP, Lein ES, Moser EI. Functional organization of the hippocampal longitudinal axis. Nat Rev Neurosci. 2014;15(10):655–69.
33. Johnson P, Shekhar A. An animal model of panic vulnerability with chronic disinhibition of the dorsomedial/

periformical hypothalamus. Physiol Behav. 2012; 107(5):686–98.

34. Bergold P, Pinkhasova V, Syed M, Kao HY, Jozwicka A, Zhao N, et al. Production of panic-like symptoms by lactate is associated with increased neural firing and oxidation of brain redox in the rat hippocampus. Neurosci Lett. 2009;453(3):219–24.

35. Muiños-Gimeno MGM, Kagerbauer B, Martín-Santos R, Navinés R, Alonso P, et al. Allele variants in functional MicroRNA target sites of the neurotrophin-3 receptor gene (NTRK3) as susceptibility factors for anxiety disorders. Hum Mutat. 2009;30(7):1062–71.

36. Dierssen M, Gratacòs M, Sahún I, Martín M, Gallego X, Amador-Arjona A, et al. Transgenic mice overexpressing the full-length neurotrophin receptor TrkC exhibit increased catecholaminergic neuron density in specific brain areas and increased anxiety-like behavior and panic reaction. Neurobiol Dis. 2006;24(2): 403–18.

37. Amador-Arjona A, Delgado-Morales R, Belda X, Gagliano H, Gallego X, Keck ME, et al. Susceptibility to stress in transgenic mice overexpressing TrkC, a model of panic disorder. J Psychiatr Res. 2010;44(3): 157–67.

38. Santos M, D'Amico D, Spadoni O, Amador-Arjona A, Stork O, Dierssen M. Hippocampal hyperexcitability underlies enhanced fear memories in TgNTRK3, a panic disorder mouse model. J Neurosci. 2013;33(38): 15259–71.

39. Temel Y, Blokland A, Lim LW. Deactivation of the parvalbumin-positive interneurons in the hippocampus after fear-like behaviour following electrical stimulation of the dorsolateral periaqueductal gray of rats. Behav Brain Res. 2012;233(2):322–5.

40. Lim L, Temel Y, Visser-Vandewalle V, Blokland A, Steinbusch H. Fos immunoreactivity in the rat forebrain induced by electrical stimulation of the dorsolateral periaqueductal gray matter. J Chem Neuroanat. 2009;38(2):83–96.

41. de Carvalho M, Dias GP, Cosci F. de-Melo-Neto VL, Bevilaqua MC, Gardino PF, et al. Current findings of fMRI in panic disorder: contributions for the fear neurocircuitry and CBT effects. Expert Rev Neurother. 2010;10(2):291–303.

42. de Carvalho M, Rozenthal M, Nardi AE. The fear circuitry in panic disorder and its modulation by cognitive-behaviour therapy interventions. World J Biol Psychiatry. 2010;11(2 Pt 2):188–98.

43. Shin L, Liberzon I. The neurocircuitry of fear, stress, and anxiety disorders. Neuropsychopharmacology. 2010;35(1):169–91.

44. Dresler T, Guhn A, Tupak SV, Ehlis AC, Herrmann MJ, Fallgatter AJ, et al. Revise the revised? New dimensions of the neuroanatomical hypothesis of panic disorder. J Neural Transm. 2013;120(1):3–29.

45. Sobanski T, Wagner G, Peikert G, Gruhn U, Schluttig K, Sauer H, et al. Temporal and right frontal lobe alterations in panic disorder: a quantitative volumetric

and voxel-based morphometric MRI study. Psychol Med. 2010;40(11):1879–86.

46. Vythilingam M, Anderson ER, Goddard A, Woods SW, Staib LH, Charney DS, et al. Temporal lobe volume in panic disorder—a quantitative magnetic resonance imaging study. Psychiatry Res. 2000;99(2):75–82.

47. Lai C. Hippocampal and subcortical alterations of first-episode, medication-naïve major depressive disorder with panic disorder patients. J Neuropsychiatry Clin Neurosci. 2014;26(2):142–9.

48. Massana G, Serra-Grabulosa JM, Salgado-Pineda P, Gastó C, Junqué C, Massana J, et al. Parahippocampal gray matter density in panic disorder: a voxel-based morphometric study. Am J Psychiatry. 2003;160(3):566–8.

49. Del Casale A, Serata D, Rapinesi C, Kotzalidis GD, Angeletti G, Tatarelli R, et al. Structural neuroimaging in patients with panic disorder: findings and limitations of recent studies. Psychiatr Danub. 2013; 25(2):108–14.

50. Sakai Y, Kumano H, Nishikawa M, Sakano Y, Kaiya H, Imabayashi E, et al. Cerebral glucose metabolism associated with a fear network in panic disorder. Neuroreport. 2005;16(9):927–31.

51. Koh K, Kang JI, Lee JD, Lee YJ. Shared neural activity in panic disorder and undifferentiated somatoform disorder compared with healthy controls. J Clin Psychiatry. 2010;71(12):1576–81.

52. Wittmann A, Schlagenhauf F, John T, Guhn A, Rehbein H, Siegmund A, et al. A new paradigm (Westphal-Paradigm) to study the neural correlates of panic disorder with agoraphobia. Eur Arch Psychiatry Clin Neurosci. 2011;261(3):185–94.

53. Lueken U, Straube B, Konrad C, Wittchen HU, Ströhle A, Wittmann A, et al. Neural substrates of treatment response to cognitive-behavioral therapy in panic disorder with agoraphobia. Am J Psychiatry. 2013;170(11):1345–55.

54. Nash J, Sargent PA, Rabiner EA, Hood SD, Argyropoulos SV, Potokar JP, et al. Serotonin 5-HT1A receptor binding in people with panic disorder: positron emission tomography study. Br J Psychiatry. 2008;193(3):229–34.

55. Maron E, Tõru I, Hirvonen J, Tuominen L, Lumme V, Vasar V, et al. Gender differences in brain serotonin transporter availability in panic disorder. J Psychopharmacol. 2011;25(7):952–9.

56. Pannekoek J, van der Werff SJ, Stein DJ, van der Wee NJ. Advances in the neuroimaging of panic disorder. Hum Psychopharmacol. 2013;28(6):608–11.

57. Ball T, Stein MB, Ramsawh HJ, Campbell-Sills L, Paulus MP. Single-subject anxiety treatment outcome prediction using functional neuroimaging. Neuropsychopharmacology. 2014;39(5):1254–61.

58. Reinecke A, Thilo K, Filippini N, Croft A, Harmer CJ. Predicting rapid response to cognitive-behavioural treatment for panic disorder: the role of hippocampus, insula, and dorsolateral prefrontal cortex. Behav Res Ther. 2014;S0005-7967(14):00120-X.

59. Altman J. Autoradiographic investigation of cell proliferation in the brains of rats and cats. Anat Rec. 1963;145(4):573–91.

60. Nowakowski R, Lewin SB, Miller MW. Bromodeoxyuridine immunohistochemical determination of the lengths of the cell cycle and the DNA-synthetic phase for an anatomically defined population. J Neurocytol. 1989;18(3):311–8.

61. Petreanu L, Alvarez-Buylla A. Maturation and death of adult-born olfactory bulb granule neurons: role of olfaction. J Neurosci. 2002;22(14):6106–13.

62. Ernst A, Alkass K, Bernard S, Salehpour M, Perl S, Tisdale J, et al. Neurogenesis in the striatum of the adult human brain. Cell. 2014;156(5):1072–83.

63. Alvarez-Buylla A, Lim DA. For the long run: maintaining germinal niches in the adult brain. Neuron. 2004;41(5):683–6.

64. van Praag H, Schinder AF, Christie BR, Toni N, Palmer TD, Gage FH. Functional neurogenesis in the adult hippocampus. Nature. 2002;415(6875):1030–4.

65. Mu Y, Lee SW, Gage FH. Signalling in adult neurogenesis. Curr Opin Neurobiol. 2010;20:416–23.

66. Rola R, Raber J, Rizk A, Otsuka S, VandenBerg SR, Morhardt DR, et al. Radiation-induced impairment of hippocampal neurogenesis is associated with cognitive deficits in young mice. Exp Neurol. 2004; 188(2):316–30.

67. Villeda S, Luo J, Mosher KI, Zou B, Britschgi M, Bieri G, et al. The ageing systemic milieu negatively regulates neurogenesis and cognitive function. Nature. 2011;477(7362):90–4.

68. Snyder J, Soumier A, Brewer M, Pickel J, Cameron HA. Adult hippocampal neurogenesis buffers stress responses and depressive behaviour. Nature. 2011; 476(7361):458–61.

69. Revest J, Dupret D, Koehl M, Funk-Reiter C, Grosjean N, Piazza PV, et al. Adult hippocampal neurogenesis is involved in anxiety-related behaviors. Mol Psychiatry. 2009;14(10):959–67.

70. Groves J, Leslie I, Huang GJ, McHugh SB, Taylor A, Mott R, et al. Ablating adult neurogenesis in the rat has no effect on spatial processing: evidence from a novel pharmacogenetic model. PLoS Genet. 2013;9(9):e1003718.

71. Tanti A, Belzung C. Hippocampal neurogenesis: a biomarker for depression or antidepressant effects? Methodological considerations and perspectives for future research. Cell Tissue Res. 2013;354(1): 203–19.

72. Sahay A, Scobie KN, Hill AS, O'Carroll CM, Kheirbek MA, Burghardt NS, et al. Increasing adult hippocampal neurogenesis is sufficient to improve pattern separation. Nature. 2011;472(7344):466–70.

73. Eriksson P, Perfilieva E, Björk-Eriksson T, Alborn AM, Nordborg C, Peterson DA, et al. Neurogenesis in the adult human hippocampus. Nat Med. 1998;4(11):1313–7.

74. Spalding K, Bergmann O, Alkass K, Bernard S, Salehpour M, Huttner HB, et al. Dynamics of hippo-campal neurogenesis in adult humans. Cell. 2013;153(6):1219–27.

75. Methippara M, Bashir T, Suntsova N, Szymusiak R, McGinty D. Hippocampal adult neurogenesis is enhanced by chronic eszopiclone treatment in rats. J Sleep Res. 2010;19(3):384–93.

76. Su X, Li XY, Banasr M, Duman RS. Eszopiclone and fluoxetine enhance the survival of newborn neurons in the adult rat hippocampus. Int J Neuropsychopharmacol. 2009;12(10):1421–8.

77. Mateus-Pinheiro A, Pinto L, Bessa JM, Morais M, Alves ND, Monteiro S, et al. Sustained remission from depressive-like behavior depends on hippocampal neurogenesis. Transl Psychiatry. 2013;3:e210.

78. Monteiro B, Moreira FA, Massensini AR, Moraes MF, Pereira GS. Enriched environment increases neurogenesis and improves social memory persistence in socially isolated adult mice. Hippocampus. 2014;24(2):239–48.

79. Hattori S, Hashimoto R, Miyakawa T, Yamanaka H, Maeno H, Wada K, et al. Enriched environments influence depression-related behavior in adult mice and the survival of newborn cells in their hippocampi. Behav Brain Res. 2007;180(1):69–76.

80. Farioli-Vecchioli S, Mattera A, Micheli L, Ceccarelli M, Leonardi L, Saraulli D, et al. Running rescues defective adult neurogenesis by shortening the length of the cell cycle of neural stem and progenitor cells. Stem Cells. 2014;32(7):1968–82.

81. Winocur G, Wojtowicz JM, Huang J, Tannock IF. Physical exercise prevents suppression of hippocampal neurogenesis and reduces cognitive impairment in chemotherapy-treated rats. Psychopharmacology (Berl). 2014;231(11):2311–20.

82. Murphy T, Dias GP, Thuret S. Effects of diet on brain plasticity in animal and human studies: mind the gap. Neural Plast. 2014;2014:563160.

83. Park J, Glass Z, Sayed K, Michurina TV, Lazutkin A, Mineyeva O, et al. Calorie restriction alleviates the age-related decrease in neural progenitor cell division in the aging brain. Eur J Neurosci. 2013;37(12):1987–93.

84. Mattson M, Duan W, Guo Z. Meal size and frequency affect neuronal plasticity and vulnerability to disease: cellular and molecular mechanisms. J Neurochem. 2003;84(3):417–31.

85. Zainuddin M, Thuret S. Nutrition, adult hippocampal neurogenesis and mental health. Br Med Bull. 2012;103(1):89–114.

86. Dias G, Cavegn N, Nix A, do Nascimento Bevilaqua MC, Stangl D, Zainuddin MS, et al. The role of dietary polyphenols on adult hippocampal neurogenesis: molecular mechanisms and behavioural effects on depression and anxiety. Oxid Med Cell Longev. 2012;2012:541971.

87. Diniz L, dos Santos TB, Britto LR, Céspedes IC, Garcia MC, Spadari-Bratfisch RC, et al. Effects of chronic treatment with corticosterone and imipramine on fos immunoreactivity and adult hippocampal neurogenesis. Behav Brain Res. 2013;238:170–7.

88. Diniz L, Dos Reis BB, de Castro GM, Medalha CC, Viana MB. Effects of chronic corticosterone and imipramine administration on panic and anxiety-related responses. Braz J Med Biol Res. 2011; 44(10):1048–53.

89. Murray F, Smith DW, Hutson PH. Chronic low dose corticosterone exposure decreased hippocampal cell proliferation, volume and induced anxiety and depression like behaviours in mice. Eur J Pharmacol. 2008;583(1):115–27.

90. Vreeburg S, Zitman FG, van Pelt J, Derijk RH, Verhagen JC, van Dyck R, et al. Salivary cortisol levels in persons with and without different anxiety disorders. Psychosom Med. 2010;72(4):340–7.

91. Cocks G, Curran S, Gami P, Uwanogho D, Jeffries AR, Kathuria A, et al. The utility of patient specific induced pluripotent stem cells for the modelling of Autistic Spectrum Disorders. Psychopharmacology (Berl). 2013;231(6):1079–88.

92. Gregersen N, Dahl HA, Buttenschøn HN, Nyegaard M, Hedemand A, Als TD, et al. A genome-wide study of panic disorder suggests the amiloride-sensitive cation channel 1 as a candidate gene. Eur J Hum Genet. 2012;20(1):84–90.

93. Yu D, Marchetto MC, Gage FH. How to make a hippocampal dentate gyrus granule neuron. Development. 2014;141(12):2366–75.

94. Kessler R, Chiu WT, Jin R, Ruscio AM, Shear K, Walters EE. The epidemiology of panic attacks, panic disorder, and agoraphobia in the National Comorbidity Survey Replication. Arch Gen Psychiatry. 2006;63(4): 415–24.

Panic Disorder, Is It Really a Mental Disorder? From Body Functions to the Homeostatic Brain

4

Giampaolo Perna, Giuseppe Iannone, Tatiana Torti, and Daniela Caldirola

Contents

Abstract

Panic disorder (PD), is characterized by repeated PAs (i.e. abrupt surges of anxiety and fear accompanied by physical – e.g. pounding heart, sweating, trembling, etc. – and cognitive – e.g. fear of dying, fear of losing control, etc. – symptoms that usually reach their peak within 10 min), and major changes in behavior or persistent anxiety over having further attacks for at least 1 month. Since PD can be treated with psychotropic drugs and/or psychotherapy it has been commonly considered a mental disorder. However, recent evidence indicates that patients with PD exhibit subclinical anomalies in the respiratory, cardiac, and balance systems. In addition, apart from reducing panic symptoms, many antipanic pharmacotherapies (e.g. SSRIs) improve the functioning of the

G. Perna (✉)
Department of Clinical Neurosciences, Villa San Benedetto Menni, Hermanas Hospitalarias, FoRiPsi, Albese con Cassano, Italy

Department of Psychiatry and Neuropsychology, Maastricht University, Maastricht, The Netherlands

Department of Psychiatry and Behavioral Sciences, Leonard Miller School of Medicine, University of Miami, Miami, FL, USA

AIAMC (Italian Association for Behavioural Analysis, Modification and Behavioural and Cognitive Therapies), Milan, Italy
e-mail: pernagp@gmail.com

G. Iannone • D. Caldirola
Department of Clinical Neurosciences, Villa San Benedetto Menni, Hermanas Hospitalarias, FoRiPsi, Albese con Cassano, Italy
e-mail: g.iannone@alumni.maastrichtuniversity.nl; caldiroladaniela@gmail.com

T. Torti
Department of Clinical Neurosciences, Villa San Benedetto Menni, Hermanas Hospitalarias, FoRiPsi, Albese con Cassano, Italy

AIAMC (Italian Association for Behavioural Analysis, Modification and Behavioural and Cognitive Therapies), Milan, Italy
e-mail: tatiana.torti@gmail.com

© Springer International Publishing Switzerland 2016
A.E. Nardi, R.C.R. Freire (eds.), *Panic Disorder*, DOI 10.1007/978-3-319-12538-1_4

abovementioned systems. Therefore, some authors believe PAs may be real alarms arising from transient instability of homeostatic body functions. The idea PD is a mere psychiatric disease may be challenged by acknowledging a paramount role also to aberrant homeostatic functioning. This might pave the way to a more integrated approach of treating PD.

Keywords

Panic disorder • Homeostasis • Respiration • Cardiac system • Balance • Vision • Photosensitivity

4.1 Introduction

According to the latest edition of the Diagnostic and Statistical Manual of Mental Disorders (DSM-5) panic disorder (PD) is an anxiety disorder characterized by recurrent unexpected panic attacks (PAs), consisting of physical and cognitive symptoms such as palpitations, dyspnea, dizziness, derealisation, fear of losing control, and fear of dying, that surge abruptly and that reach a peak within minutes, provoking intense fear or discomfort. Beyond the PAs themselves, which are the hallmark of the disorder, another key feature of PD is fear (1 month at least) of having future PAs, which can lead to important maladaptive changes in behaviour (i.e. anticipatory anxiety and phobic avoidance of places and situations where an attack has occurred or where patients believe it may occur). PD frequently occurs in comorbidity with other mental disorders, such as agoraphobia, major depression disorder, and bipolar disorder, it can be chronic and disabling, can cause distress and impair quality of life [1]. PD has a lifetime prevalence of approximately 3.5% in the general population and 5–8% in primary care settings [2–4]. Etiology of PD has not been fully unfolded; however research suggests interaction of genetic predisposition and specific environmental factors. PD is considered a mental disorder because: (1) PAs are deemed false alarms. It has been more than 20 years since germinal theories postulated the

existence of hypersensitive alarm systems in PD. According to Klein PAs occur when the suffocation alarm system is erroneously triggered [5]; Gorman's neuroanatomical model postulates PAs are conditioned fear responses mediated by an overly sensitive fear network [6, 7]; the three-alarms (true, false, learned) theory deems PAs as the results of both spontaneous firing of the fear system and conditioning processes to internal or external cues [8]. Clark et al. [9] consider PAs catastrophic misinterpretations of harmless bodily sensations. These theories share the assumption that alarms are false because patients with PD are physically healthy. (2) PD is treated with psychotropic drugs and/or psychotherapies. Withal there has been some debate whether to consider PD just a mental disorder.

Beside cognitive symptoms, patients with PD often complain of several somatic symptoms including respiratory difficulties, irregular heartbeat, dizziness, and photophobia. Usually physicians and psychiatrists after conducting standard procedures (such as physical and/or clinical tests) reassure patients that their bodies function perfectly, and they ascribe the somatic symptoms entirely to anxiety. Conversely numerous scientific findings suggest that patients with PD may suffer from subclinical abnormal organic systems functioning (e.g. in the cardio-respiratory and the balance systems), which may be associated with hyperreactivity to hypercapnic and hypoxic inhalations, subclinical autonomic hyperreactivity, and space and motion discomfort.

In the next paragraphs we examine evidence that patients with PD may manifest subtle physiological functions abnormalities as well as reduced adaptability to changes, and that their brains are more akin to react when these systems are stimulated. In particular, we will focus on three main systems: the cardiac, the respiratory, and the balance systems. We hypothesize that PAs are true alarms signaling aberrant functioning of one or more of these body systems. Finally we speculate that patients with PD are physiologically different from healthy subjects and therefore PD may be reconsidered from two non mutually-exclusive perspectives (a somatic one and a psychological one) and that patients may benefit from therapeutic approaches that simultaneously take into account both facets of the disorder.

4.2   Respiration in Panic Disorder

Breathing is involved in the phenomenology and the biological mechanisms of PD. Both experimental and clinical observations provide strong evidence that a connection between panic and respiration does exist [10, 11]. Respiratory symptoms are very frequent among patients with PD [12], both during spontaneous PAs and during daily-life [13], and constitute a hallmark of PAs and PD. Further information corroborates this correlation. First, an association between PD and hyperventilation is known. Indeed, several studies reported low resting partial pressure of carbon dioxide (CO_2) in patients with PD [14, 15]. Also up to 40% of patients with PD suffers from hyperventilation and both patients with PAs and patients with hyperventilation syndrome exhibit similar respiratory symptoms, such as dyspnea [16]. Hence it might be that hyperventilation causes PAs in patients suffering from PD. However chronic hyperventilation (and hypocapnia) has been reported in less than 50% of the patients with PD and it does not seem to be unique of PD. In fact it is also prevalent in patients suffering from other anxiety disorders and it may just reflect background anxiety [17]. Hyperventilation seems to induce anxiety, but not PAs, in patients with PD [18]. Panic precedes hyperventilation, as it is the case during hypercapnic challenges [19], and a review claimed that a causal role of hyperventilation in the etiology of PD seems unlikely [11]. Recently Meuret et al. [20] examined changes in respiration, heart rate, and skin conductance level 60 min before and 10 min after PAs in individuals suffering from PD. The authors observed important patterns of autonomic and respiratory irregularity that were not detected by the patients, as early as 47 min before panic onset. Respiratory changes, such as decreased tidal volume and pCO_2 increase, characterized the final minutes preceding a PA. These findings suggest that hypoventilation precedes

Box 4.1. Panic Disorder

Respiration:
- Respiratory symptoms and respiratory diseases are very frequent.
- Hypersensitivity to hypercapnic gas-mixture inhalation.
- Baseline respiratory irregularity (e.g. chronic hyperventilation).
- Respiratory PD subtype.

Cardiac system:
- Comorbidity with cardiac disorders.
- Comorbidity with coronary artery disease.
- Decreased cardiac vagal function.
- Higher exercise avoidance and worse cardiopulmonary performance.

Balance and vision:
- High prevalence of vestibular symptoms.
- Balance control mainly relies on non-vestibular cues, especially visual cues.

Photosensitivity:
- Lowered threshold of tolerance to light.
- Photophobic behavior.

PAs. Given that during the PA, tidal volume (and heart rate) increased and pCO_2 dropped, suggesting hyperventilation, it might plausible to assume that hyperventilation is merely a consequence (and probably a compensatory mechanism) rather than a cause of PAs. Skin conductance levels raised in the hour preceding PAs as well as during the attacks. These changes were largely absent when an attack did not occur. These findings suggest that unexpected PAs are preceded by significant autonomic irregularities and invite to rethink the classic nosotaxy between situational and spontaneous PAs.

Second, an association between PD and respiratory diseases was found. Goodwin, Pine [21] documented an association between self-reported respiratory diseases and increased likelihood of PAs among adults in the general population, also when controlling for differences in sociodemographic characteristics, physical illnesses, and comorbid mental disorders (adjusted OR = 1,7–95 % CI) and concluded that this association is specific to PAs. Up to 40 % of patients with PD have a history of respiratory disease, in particular asthma and bronchitis [11]. PD prevalence in patients suffering from chronic obstructive pulmonary disease is higher both than in people suffering from other mental disorders (i.e. obsessive-compulsive, depressive, and eating disorders) and than in healthy controls [22, 23]. The nature of this association remains largely unknown. It may be that either PAs precede the onset of respiratory disease or that respiratory or lung disease lead to the development of PAs. Alternatively a third factor (e.g. cigarette smoking) may increase co-occurrence of respiratory disease and PAs. Future studies that identify genetic and/or environmental routes of transmission are warranted to determine the specific mechanisms of this association, albeit mounting evidence indicates that respiratory illness precedes PD.

Third, patients with PD showed behavioural and respiratory hypersensitivity to hypercapnic gas-mixture inhalation [24, 25] and subclinical abnormalities in respiratory patterns. Klein [5] speculated that PAs are "false suffocations alarms" resulting from carbon dioxide hypersensitivity that induces the brain's suffocation monitor to erroneously signal lack of air and breathlessness. This might lead to respiratory distress, hyperventilation and, eventually, to a PA. Preter and Klein [26] further developed the suffocation false alarm theory and proposed that endogenous opioidergic regulation dysfunctions increase suffocation sensitivity, separation anxiety, and PAs. Just recently, the same authors demonstrated an association between endogenous opioid system deficiency, panic-like suffocation sensitivity, and childhood parental loss. Finally, this theory reshaped the role of the amygdala (and in general of the fear system) in PD. This is consistent with findings of patients with focal bilateral amygdale lesions who also exhibited PAs in response to CO_2 inhalation [27]. Hence, an alternative alarm system beyond the amygdala is likely to underlie hypersensitivity to CO_2. On

the top of that, absence of hypothalamic pituitary adrenal activation during PAs contrasts the assumption that panic is a manifestation of a hypersensitive fear system [28]. However both normal and increased hormonal activity has been reported in PD [29]. It might be that HPA-axis abnormal function may render some individuals more vulnerable to PD (and to psychopathology in general) over time, perhaps by increasing vulnerability to future stressors.

Children suffering from anxiety disorders exhibit greater changes in somatic symptoms after CO_2 inhalation and those who developed panic symptoms manifest respiratory rate increases in response to CO_2 breathing and elevated mean tidal volume levels, and higher respiratory rate variability during room-air breathing [30]. Behavioral hyperreactivity to CO_2 has been found in healthy first-degree relatives of patients with PD. Finally, even mentally healthy individuals with first-degree relatives suffering from PD experience behavioral and respiratory hyperactivity in response to CO_2 administration when compared to controls without such a family history [31, 32].

Our group investigated the breath-by-breath complexity of respiration dynamics in patients with PD and in healthy controls. We found that patients exhibited greater baseline respiratory irregularity, which may be a vulnerability factor to PAs [33]. We also found higher respiratory irregularity in children of patients with PD when compared to children of psychiatrically healthy parents, even when children with anxiety disorders were excluded. Hence irregular breathing may represent a risk marker of familial vulnerability to PD that is independent of "state" effects and it may anticipate PD [34]. In addition, subjects with PD show increased respiratory variability during mild physical activity even in the absence of full blown respiratory diseases [24, 35].

In a recent meta-analytic review our research group compared baseline respiratory and hematic parameters related to the respiratory function in subjects with PD and in healthy controls. We found higher baseline mean minute ventilation (MV), and lower end-tidal partial pressure (et-pCO$_2$) and venous pCO$_2$ in subjects with PD, which indicated a condition of baseline hyperventilation. In addition, reduced HCO$_3^-$ (bicarbonate ion) and PO$_4^-$ (phosphate ion) venous concentrations suggest that hyperventilation may be chronic and not just related to higher anxiety states during the respiratory assessment. Finally preliminary indication of higher MV, respiration rate (RR) and tidal volume (TV) variability, higher RR and TV irregularity, and higher rate of sighs and apneas in the respiratory patterns of these subjects was found [14]. Whether these respiratory abnormalities are peculiar to PD or whether they are common to other anxiety disorders is currently debated. On the one hand it is known that anxiety influences respiration [36], therefore respiratory abnormalities may be not specific to PD but they might be common also in subjects suffering from other anxiety disorders. Previous research on baseline respiratory parameters (especially mean RR, TV, MV, and et-pCO$_2$) in subjects with PD and in subjects suffering from other anxiety disorders yielded conflicting results [37–39]. Just recently in a meta-analysis our research team observed significant differences between the baseline respiratory parameters in subjects with PD and in subjects with social phobia (SP) or generalized anxiety disorder (GAD). We found significantly lower mean end-tidal partial pressure of CO$_2$ (et-pCO$_2$) in subjects with PD than in those with SP or GAD, and higher mean respiratory rate, lower venous et-pCO$_2$ and HCO$_3^-$ concentration in subjects with PD than in those with SP. These findings suggested that subjects with PD have a condition of baseline hyperventilation when compared to subjects with SP or GAD. Hematic variables suggested that the hyperventilation may be chronic. Conversely, we did not find significant differences in respiratory abnormalities between subjects with SP or GAD and healthy controls [15]. These results support the idea that baseline respiratory abnormalities are specific to PD. It is still not clear whether hyperventilation may arise from an intrinsic malfunction of the respiratory system or rather from panic-related anxiety. Chronic hyperventilation in PD indicates

that state anxiety during respiratory assessment does not fully explain these respiratory abnormalities. Yet, patients with PD exhibit specific emotional, cognitive and behavioral characteristics, such as fear of somatic sensations, panic-specific beliefs, and phobic/protective behaviors. Given that these features are not completely captured by trait/state anxiety measurements, they may specifically influence respiration in this population.

Several hypotheses have been advanced to elucidate the origin of respiratory irregularity in PD. It is known that complex regulatory systems modulate respiration. Many brainstem regions containing CO$_2$-sensitive neurons are implicated in regulating both ventilation and panic [40, 41]. Hence it is plausible to hypothesize that some overlap exist between neurons that serve respiration and those that elicit panic. Protopopescu et al. [42] found increased brainstem gray matter volume in the ventral and dorsal midbrain, and in the rostral pons of patients with PD compared with healthy controls. In a recent review, Perna et al. [43] also suggested that the brainstem volume is larger in patients with PD and that aberrant functioning of the brainstem serotonergic system (i.e. altered serotonergic receptors and 5-HT-transporter bindings) is likely to be involved in panic modulation as well as in (cardio)respiratory activity. The pre-Bötzinger complex is a cluster of brainstem neurons that contains the basic circuits for respiratory rhythm and pattern generation [44]. Breathing irregularities may arise from an intrinsic deranged activity of the pre-Bötzinger complex neurons, that fail to adequately cope with external stimuli and that would underlie respiratory phenomena, such as sighs and gasps, and/or from compensatory-like responses to abnormal central and/or peripheral signals (the brainstem communicates with both sensory afferents from central and peripherical chemoreceptors and from pulmonary/chest wall receptors). In addition, more rostral areas such as the hypothalamus, cerebellum and cortex, modulate respiration across different physiological states [45, 46].

The limbic circuit regulates respiration during arousal and emotional states. The dorsal

periaqueductal gray seems to regulate unconditioned defensive responses to proximal threats,
including physical stimuli, therefore it might be
implicated in PD [47]. Respiratory irregularity in
patients with PD may derive from abnormalities
in these brain centers. Finally, a wider and more
general dysfunction across the homeostatic systems, including the respiratory system, the cardiac
system, and the balance system may underlie such
respiratory irregularity. For instance the parabrachial nucleus in the brainstem filters and orga-
nizes interoceptive stimuli from our basic
homeostatic functions, and maintains a representation of internal stability and bodily well-being
[48]. Given that basic physiologic systems work
in concert with mutual modulation, we speculated
that breathing irregularities may be related to per-
turbations of the cardiovascular system and the
balance system and that the respiratory irregularity in patients with PD might originate from a
more general dysfunction of the brainstem circuits that modulate homeostatic functions. Within
this framework, PAs may represent a primal emotion arising from those specific brain circuits that
process bodily sensations and perceptions that are
related to homeostatic functions. Functional failure in these centers might account for the emergence of PAs [25]. In conclusion homeostatic
dysfunctions (in particular respiratory dysfunctions) may be crucial in the pathophysiology of PD
and trigger neuroanatomical networks involved
in PAs. Maddock [49] claimed that excessive
response to lactate in the brain may underlie these
abnormal respiratory findings. Similarly, Esquivel
et al. [45] maintained that much of the connection
between panic and respiration might be explained
by altered acid-base levels in the brain: acute
brain acidosis may be linked to PD and PAs might
represent a defensive response to subtle potentially
threatening acid-base alteration. Administration
of panicogenic substances (e.g. lactate or CO_2)
would activate the centers that govern brain pH
and evoke spontaneous PAs. It is plausible to
assume that the degree of brain acidosis is relevant to the panic symptoms induced by CO_2 inhalation and it may reflect an underlying metabolic
disturbance (see next paragraphs). Whether higher
respiratory irregularity is a consequence or rather

a trait marker vulnerability of PD is a question
that deserves a conclusive and definite answer.
So far, preliminary findings supported the latter
hypothesis.

Smoking has been seen as a risk factor for the
first occurrence of PAs and the onset of PD [50,
51]. Withal the biological mechanisms underlying the link between PAs and smoking are largely
unknown. Our team investigated the effects of
smoking on respiratory irregularity in patients
with PD. When compared with healthy controls,
both smoker and non-smoker patients exhibited
greater respiratory irregularity but smoker
patients showed higher irregularities than non-
smoker patients. On the contrary, smoking did
not influence the regularity of respiratory patterns in healthy subjects. Overall, smokers had
more severe PAs than nonsmokers [52]. It might
be that smoking impairs respiration in patients
with PD, and influences the onset and/or maintenance of the disorder. Abelson et al. [53] claimed
that respiration irregularity in PD is influenced by
neither doxapram-induced hyperventilation
(doxapram is a respiratory stimulant) nor cognitive manipulation but it appears to have intrinsic
and stable features. However replication studies
with a greater number of participants necessitate
to yield more robust and convincing evidence and
to confirm behavioral and/or respiratory hyperreactivity both at baseline and following CO_2
administration in patients with PD.

Fourth, there has been a long scientific debate
on whether PD might be categorized in different
subtypes according to specific clusters of symptoms. Several investigators speculated about the
existence of a respiratory and a non-respiratory
PD. Klein [5] hypothesized a connection between
the respiratory system and PD. Ley [54] also postulated the existence of a subcategory of PD with
dyspnea, heart palpitations, terror, and a strong
desire to flee, as predominant symptoms.
Similarly Briggs et al. [13] contemplated the
existence of a subgroup of PD characterized by
more respiratory symptoms (e.g. shortness of
breath, feelings of choking, etc.), higher occurrence of spontaneous PAs, and better response to
imipramine, whereas the non-respiratory group
seems to suffer more from situational PAs and to

respond better to alprazolam. More recently Song et al. [55] found that patients with the respiratory subtype exhibited earlier onset of PD, more severe clinical symptoms, higher fear of respiratory symptoms, higher co-occurrence of agoraphobia, and better response to SSRIs, when compared to patients with non-respiratory subtype. Those with predominant respiratory symptoms were also more sensitive to CO_2 challenges, exhibited higher familial prevalence of PD and earlier onset of the disorder, as well as more previous depressive episodes [56].

35 % CO_2 inhalation induced dyspnea in patients with PD [22]. We found that patients with PD who reacted to 35 % CO_2 were characterized by greater baseline pattern of tidal volume and inspiratory drive compared to patients who did not react to 35 % CO_2 [34]. Meuret et al. [57] identified three dimensions of panic: the cardiorespiratory (with palpitations, shortness of breath, choking, chest pain, numbness, fear of dying) the autonomic/somatic (with sweating, trembling, nausea, chills/hot flashes, and dizziness) and the cognitive (with derealisation, fear of going crazy or losing control). A systematic review by Niccolai et al. [24] indicated that patients with PD exhibit an abnormal breathing pattern when compared to controls, during rest/baseline, challenge, and recovery conditions, and that respiratory variability may be a possible endophenotype of PD. Roberson-Nay et al. [58] compared the respiratory and non-respiratory panic to investigate whether these subtypes represent a single disorder, whether they differed on a quantitative (i.e. severity) or rather qualitative (i.e. distinct patterns) dimension from one another. Their results suggested that the two panic subtypes had distinct symptom profiles that differed in severity and that respiratory panic represented the most severe form of the disorder. In sum, sufficient scientific evidence supports that patients with PD with prominent respiratory symptoms constitute a discrete subgroup with more severe symptoms. Finally, from a review by it emerged that hypersensitivity to CO_2 might be a valid marker of the respiratory PD subtype [2]. We believe it is plausible to distinguish between respiratory and non-respiratory PD. Genetic and neuroimaging

studies are encouraged to provide a definite answer on whether it is appropriate to separate PD into distinct subcategories.

In conclusion, the respiratory system of individuals suffering from PD appears to be more unstable and sensitive than subjects without PD. It is pivotal to verify to which extent the respiratory abnormalities arise from intrinsic malfunctions of the respiratory system.

4.3   Cardiovascular System and PD

Heart palpitations or a racing heart, and chest pain or discomfort are among the core somatic symptoms that characterize PAs. Subjects who experience PAs are often concerned they may suffer from a cardiac disease or that they are dying of a heart attack and repeatedly refer emergency room. Of these individuals, more than 20 % is not diagnosed with a cardiac problem but rather with PD [59]. Although routinely cardiac examination yields negative result, the symptoms these subjects experience feel so real that reassurance from the doctors is not enough to convince them their cardiovascular system performs normally. Most of these individuals believe there might be some undetected abnormality in their cardiovascular system [60]. As a results they tend to worry about future attacks and develop anticipatory anxiety, which negatively impacts on their quality of life and daily functioning. Also, frequent special consultation results in increased health care consumption and costs [61]. A link between PD and cardiac disorders (CDs) emerged. Some historical studies suggested an association between PD, arrhythmias, sudden cardiac death or idiopathic cardiomyopathy [62–64]. Several reports found a relationship between PD and coronary artery disease (CAD). PD prevalence ranged from 11 % to 53 % in patients with documented CAD who visited emergency rooms (ERs) or outpatient cardiology clinics [65–67], especially in those with atypical chest pain [65]. Chance of suffering from CAD was 26 % in patients with PD who referred to ERs for chest pain [68]. Conversely, other studies failed to find an association between PD and CAD in patients presenting with chest pain to ERs or cardiology settings [69–71]. Such discrepancy may arise from methodological limitations of the studies, such as low sample sizes, lack of standardized tests for cardiac diagnoses and/or clinician-reported questionnaires to evaluate PD, and/or comorbid psychiatric disorders associated with cardiac risk.

Although many studies did not find an association between PD and CAD [72, 73], others that used more robust methodological criteria (e.g. inclusion of subjects with primary diagnosis of PD performed by a clinical interview and/or structured clinician-administered interview, diagnosis of CDs obtained by medical examination, sensitive standardized tests, and standardized criteria) support a cross-sectional association between current PD and CAD (prevalence across studies ranged from 4.7–21 %) [74–76]. This suggests that past history of PD may also be relevant for the occurrence of CAD, even in the absence of current panic symptoms at the time of CAD diagnosis. Thus, lack of investigating lifetime PD in some studies may have hampered the identification of a relationship between PD and CDs. In addition, in older subjects with cardiovascular diseases (CVDs) higher rates of subthreshold panic-phobic symptoms emerged and this suggests a possible association between CVDs and panic even in the absence of a full-blown PD [77]. Finally, a nationwide population-based study by examined prospectively the relationship between PD and acute myocardial infarction risk within 1 year of follow-up and found that almost 5 % of patients with PD experienced an acute myocardial infarction episode within a year, compared with less than 3 % in the comparison cohort [78]. The association persisted also when controlling for hypertension, coronary heart diseases, and age and the authors concluded that PD may represent an independent risk factor for developing acute myocardial infarction. These results confirm previous findings of a large cohort study based on the US National Managed Cara Database [79]. They showed that patients with PD (about 40,000 subjects) had a nearly twofold increased risk for subsequent CAD (AMI, unstable angina or angina pectoris) than subjects without PD (about 40,000 subjects), even after adjusting for age at entry in the cohort, smoking, obesity, use of angiotensin converting enzyme inhibitors, beta blockers, diuretics, and statins. The average time span between diagnosis of PD and incident diagnosis of a CDs event was about 1.5 years. Both direct (e.g. physiological alterations) and indirect (e.g. unhealthy behaviors) mechanisms may underlie the association between panic and cardiac risk. Many studies described imbalanced autonomic regulation and reduced heart rate variability (lower parasympathetic activity and higher sympathetic/parasympathetic ratio) in patients with PD. Defective neuronal noradrenaline reuptake in the heart may contribute to adverse cardiac events in subjects with PD by augmenting the sympathetic cardiac firing [80]. Important associations between PD and increased arterial stiffness (that predicts cardiovascular mortality), poor cardiovascular fitness and several factors negatively affecting the endothelial function, and the development of atherosclerosis, (such as increased homocysteine levels, platelet aggregation, lower levels of nitric oxide, lipid pattern abnormalities, and higher inflammatory indexes) emerged [81–83]. In subjects with PD, inhalation of 35 % CO_2–65 % O_2 gas mixture (which is known to induce PAs), provoked transient myocardial ischemia at least in "high risk" CAD patients [84]. Niccolai et al. [85] found that heart rate in response to the 35 % CO_2 challenge was higher in subjects with PD than controls. Moreover patients needed more time to recover, and showed increased respiratory parameters variability. Taken together these findings suggest that PAs are disturbing events for the cardiorespiratory system that may contribute to increase cardiac risk in this population over time. As we discussed in the previous chapter, subjects with PD have also irregular respiratory patterns [33, 39] and baseline hyperventilation [14]. Since respiration influences the autonomic regulation of cardiac activity [86] and coronary vasospasm can be precipitated by hyperventilation [87, 88], such peculiar respiratory features may increase vulnerability to CAD.

The QT interval indicates the time the ventricular myocardium needs to depolarize and repolarize. A lengthened QT interval is a marker for ventricular tachyarrhythmia and a risk factor for sudden death. QT variability is elevated in patients with PD than in healthy controls [89]. QT interval variability is closely related to HRV. Decreased heart rate and heart period variability in patients with PD compared to normal control subjects, suggests decreased cardiac vagal function [90, 91]. Increased QT variability in combination with decreased HRV may significantly increase the risk for cardiovascular morbidity and sudden death [92, 93]. Both HRV and QT interval can be influenced by sympathetic mechanisms. Considering that PAs are associated with several autonomic symptoms, including chest pain, heart pounding, tachycardia, and shortness of breath, such enhanced autonomic activity can result in a significant increase in QT variability [94]. Increased QT in patients with PD may be due to increased sympathetic activity which may put them at a greater risk for significant cardiovascular events and sudden death. Finally, some studies performing surface electrocardiogram in subjects with PD found that the QT interval augmented even more during hyperventilation challenges [95]. Patients with PD also manifest increased dispersion of the QT and P-wave [96, 97], which indicates higher regional heterogeneity of ventricular repolarization and atrial depolarization, respectively, and are considered indicators of arrhythmia and sudden death risk [98–100].

The physiologic alterations found in PD patients are consistent with the idea that abnormal regulation of the body homeostatic functions may be involved in the pathophysiology of PD [25, 43] and confer a peculiar vulnerability to medical diseases, including CDs [25, 101]. Lastly, behavioral risk factors, such as smoking [102, 103] (which is known to be associated with increased cardiovascular risk) and physical exercise avoidance [103] may contribute to cardiac morbidity in PD. Cigarette smoking prevalence is high in patients with PD when compared with both healthy controls and subjects with other anxiety disorders [50]. Patients with PD exhibit higher exercise avoidance and worse cardiopul-

monary exercise performance when compared to healthy controls. We investigated cardiorespiratory fitness levels in subjects with PD and in a group of age-, gender-, weight-, and physical activity levels-matched healthy controls. We also investigated whether psychological variables (i.e. state and trait anxiety, fear of physical sensations, and fear of autonomic arousal) influenced cardiorespiratory responses and perceived exertion of patients PD during a submaximal exercise test. We found that although patients had poorer cardiorespiratory fitness and spent more effort during physical exercise, this was not related to the psychological variables examined but it might be related to a diminished ability of the cardiac system to efficiently respond to physical efforts [104]. This might contribute to a sedentary lifestyle and to increased cardiovascular risk in the long run [103, 105].

Taking into account the available data and given the involvement of the cardio-respiratory system in the pathophysiology of PD, a more in-depth investigation of the association between panic and CDs is strongly recommended. Therefore research, preferably longitudinal, is encouraged to confirm the association between both full-blown and subthreshold panic and CAD, to investigate the association between panic and other CDs, such as arrhythmias, and hypertension and to investigate whether specific subgroups of subjects with PD (e.g. those with biological subclinical risk factors and/or higher cardio-respiratory instability and symptoms) might be at higher risk of cardiovascular disease. Unfortunately PD is often under-recognized in cardiological care settings and diagnosing PD may result in failure to recognize cardiac diseases [70]. Hence a better understanding of this relationship between may contribute to improve treatment and prevention of both PD and CDs.

4.4   Panic Disorder and the Balance System

Prevalence of vestibular symptoms, such as vertigo, instability, and lightheadedness is higher in individuals suffering from PD, when compared to control populations [106]. Jacob [107] noticed that 75 % of patients with PD manifest postural instability. Up to one third of patients with PD and agoraphobia with chronic dizziness suffer from peripheral vestibular alterations [108]. Stambolieva et al. assessed postural instability by static posturographic tests of standing on stable and foam surfaces with open and closed eyes in 30 patients with PD and in 30 sex- and age-matched healthy controls. They noticed that almost 84 % of the patients experienced dizziness and imbalance, both during and between PAs. No differences of sway velocity (an indicator of postural stability) emerged between the groups while standing on both surfaces with open eyes. However the sway velocity of the patients with PD was higher when compared to controls while standing with closed eyes both on the stable and the foam surfaces. The authors concluded that visual information plays a more important role in maintaining postural stability when sensory conflict exists [109].

Finally, peripheral vestibular disorders seem to be more prevalent in patients with PD and agoraphobia than in patients with PD alone [110]. In a double-blind, random, cross-over study Perna et al. [111] compared a group of patients with PD, with and without agoraphobia, and a group of sex- and age-matched healthy controls who underwent static posturography in three conditions (eyes open, eyes closed and neck extension) and the 35 % CO_2 challenge. Symptomatological reactivity to CO_2 correlated with balance system dysfunction in patients only in the eyes-closed condition. Up to 42 % of patients with PD (compared to 0–5 % in controls) exhibited aberrant balance system functioning, which correlated with agoraphobic avoidance. The authors concluded that the balance system seems to influence the psychobiological mechanisms underlying agoraphobic avoidance and therefore plays a role in the behavioural features of PD.

Staab et al. suggested three patterns of illness in patients with clinical syndrome of what they labeled *psychogenic dizziness*, which is characterized by vague and elusive physical symptoms of vertigo and lightheadedness, in the absence of objective clinical tests abnormalities. The first one is a psychogenic pattern, in which anxiety is the sole cause of the disorder. The second one is an otogenic pattern, in which a neurotologic condition triggers the development of anxiety [112]. The mismatch theory of Furman and Jacob [113] also suggested that vestibular dysfunctions may provoke space and motion discomfort and elicit anxiety in patients with PD. The third one is an interactive pattern in which a neurotologic condition is responsible for the onset of dizziness but also exacerbates preexisting or prodromal anxiety or panic symptoms. Therefore the connection between vestibular manifestations and anxiety disorders seems to be bidirectional: on the one hand vestibular disorders can trigger anxiety and on the other hand anxiety symptoms can trigger vestibular symptoms. This implies that at least some anomalies of the balance system can be ascribed to psychological factors rather than to vestibular dysfunctions.

Neuroanatomical/neurophysiological circuitries that might account for the relationships between the vestibular system and PD encompass: (1) connections between the locus coeruleus and the lateral vestibular nucleus [114], (2) vestibular inputs to the raphae nuclei [115], (3) serotonergic influence on the vestibular system [116]. Finally vestibular–respiratory connections have been proposed (4). Vestibular nuclei project to and receive from the caudal parabrachial nucleus and the subparabrachial nucleus (i.e. the Kölliker-Fuse) [117, 118]. A review of anatomic and physiologic studies demonstrated direct connections between the vestibular nuclei and the brainstem regions that influence both sympathetic and parasympathetic activity [119]. A combination of autonomic, vestibular, and limbic information both in the brainstem and forebrain areas seems to regulate balance control. In particular, the parabrachial nucleus (PBN), which is located in the pons, influences cardiovascular, respiratory, and autonomic responses and transmits interoceptive information, as suggested by chemical stimulation and lesion studies [120–122]. Respiration and the balance system are intertwined and such connections may partly explain the association between hyperventilation and postural instability in patients with PD (who are in fact in a chronic state of hyperventilation), in patients with vestibular diseases, and in healthy controls. These neuroanatomical connections indicate that respiration and the balance system are intertwined. In subjects with dizziness, hyperventilation may induce nystagmus and reveal vestibular dysfunction [123]. Accordingly, we believe chronic hyperventilation contributes to postural instability and might aggravate dizziness in subjects with PD.

The PBN mediates both awareness of, and affective and emotional responses to intrinsic and extrinsic stimuli that alter the sense of physiological well-being (i.e. interoception) [124–127] and, in concert with the central amygdala, it also plays a role in recognizing 'innate danger stimuli' [128]. Because interconnections between the PBN, the limbic system and the prefrontal cortex seem to be involved in the development and expression of PD, it has been proposed that these structures may also be a substrate for the co-occurrence of balance disorders and PD [48].

In conclusion, patients with PD seem to present subclinical abnormalities in their balance system. According to Balaban [129] such vestibular instability may be linked to changes in homeostatic autonomic responses which in turn would trigger affective and emotional responses, including panic.

4.4.1 Balance and the Visual System in PD

Maintaining proper balance and posture partly depends on visual information. Scientific literature identified that the visual system has two visual pathways referred to as central (or focal/parvocellular) and peripheral (or ambient/magnocellular). Visual information from central and peripheral visual fields have complementary roles and might differently affects postural control. The central vision enables object identification and recognition, it works largely consciously and in isolation and enables direct visual impact on objects that appear straight in front of the visual field. The peripheral vision underlies the perception of self-motion and body stance in the environment, and the ability to scan surroundings

for changing conditions and to begin fast responses. It is generally considered to be involved in motion processing, posture, movement and balance and in orienting and defensive reactions to visual motion, involving short latency postural adjustments as well as head and eye movements. Finally peripheral vision also recognizes harmful threats coming up alongside [130].

In patients with PD balance control seems to rely mainly on non-vestibular cues, such as proprioceptive and especially visual cues. In these individuals vision is perhaps the sensory systems most strongly associated with postural balance in particular when sensory conflict exists or when the other sensory systems are compromised or damaged (e.g. while standing on an unstable surface, such as a foam platform, in broad spaces, heights, crowded places, or in eyes closed conditions). Impaired visual inputs enhance sensory conflict and may lead to adulterated proprioceptive information and balance impairment which in turn may elicit panic symptoms. There is evidence that in patients with PD, anxiety and discomfort may arise when visual information is inaccurate as a result of hyperexcitability of the locus coeruleus and the vestibular brain nuclei, which would result in higher postural instability [109, 111]. Our group found that patients with PD and agoraphobia manifest postural instability during peripheral visual stimulation whereas controls did not. Conversely, the two groups showed similar patterns of postural instability during central visual stimulation [104]. Hence we concluded that patients with PD, especially those with comorbid agoraphobia, might be hypersensitive to the influence of peripheral visual system on balance. Such higher sensitivity is perhaps linked to a more active visual alarm system that scans the environment for possible threats. Connections between visual, vestibular and limbic areas may increase postural sway when the visual environment is changing in an uncertain way, such as during motion in the peripheral visual field [131]. Indeed, the relationship between state anxiety and postural instability during peripheral stimulation supports the idea that these situations are emotionally relevant for patients with PD. Such hypersensitivity might be specific of PD and arise from multiple sources. It might be an idiosyncratic perceptual habit that, in concert with PAs, might lead to consequent agoraphobia. Alternatively, it might follow PD and agoraphobia, which would act as "disrupting" factors on balance control systems by vestibular brainstem-limbic connections. Also, panic-phobic conditions might involve activation of complex alarm systems including interoceptive conditioning processes linked to destabilizing visual stimuli and operant learning processes related to the avoidance of visual experiences and provoking discomfort in everyday life [8].

In conclusion, even though patients with PD report normal baseline sway, they seem unable to effectively counterbalance somatosensory and/or vestibular information during irrelevant visual stimuli and increase postural sway more than control subjects when exposed to visual sensory conflict. Such increase is unlikely to result from general destabilization, given that they tend to sway synchronously with the optic stimulus [132]. Patients may sway excessively also when standing on a fixed platform with a static visual scene, and this suggests generalized balance abnormality rather than a problem limited to sensory integration in situations involving sensory conflict. Alternatively, it might be that these individuals are less capable of ignoring the misleading visual information (i.e., they depend more on visual cues). This increased reliance on vision may be reflected in complaints of disequilibrium in complex moving visual environments. Whether such visual dependence hinges on trait- or state-like enhanced vigilance needs to be determined.

Many complications emerge when examining patients with vestibular dysfunctions and comorbid psychiatric disorders [133]. Multidisciplinary evaluation of these patients is warranted, unfortunately psychiatrists usually do not refer patients with PD for otoneurologic evaluations as they refer them to cardiologic evaluations for symptoms like palpitations. On the other hand, it is also rare that vestibular specialists refer patients to a psychiatrist to exclude the diagnosis of PD, which may be present in addition to the organic vestibular disease. Exploratory research indicates that adequate treatment of vestibular abnormalities in patients suffering from PD is possible. Jacob et al. [134] claimed that 8–12 weeks of vestibular rehabilitation alone was sufficient to reduce anxiety and avoidance in patients suffering from PD and from vestibular abnormalities and that 4 weeks of cognitive-behavioral therapy produced little benefit prior to vestibular rehabilitation. Teggi et al. [110] found that early vestibular rehabilitation had even greater beneficial effects on anxiety than on the balance function.

Pharmacotherapy is another viable treatment option. Mezzasalma et al. [108] evaluated the efficacy and effectiveness of imipramine on the treatment of comorbid chronic dizziness and PD in nine patients with a diagnosis of PD with agoraphobia. The authors found peripheral vestibular alteration in about one third of patients with PD and agoraphobia with chronic dizziness. After a 3-month treatment with imipramine, patients exhibited a significant decrease in anxiety, dizziness, quality of life, and PD severity. Preliminary results indicate reduced anxious symptoms and impairment due to dizziness after 12 weeks of fluoxetine treatment in patients with vestibular dysfunction (e.g. dizziness that included vertigo, motion sickness, nausea, and anxiety) without anxiety disorders [135]. Our team assessed posturography in a small group of patients with PD and agoraphobia who were partial responders to pharmacotherapy and cognitive-behavioral therapy, before and after 10 weeks of vestibular rehabilitation treatment consisting of stimulation of the peripheral visual field in concomitance with a series of head/body movement patients had to perform. Following the vestibular rehabilitation patients exhibited improved balance performance as well as diminished agoraphobic symptoms (unpublished study). In light of this, it is central that both patients and clinicians acknowledge that PD and vestibular anomalies can occur separately or in comorbidity. Clinicians should be advised to conduct exhaustive medical investigation in order to convey the most accurate diagnosis to their patients, to inform them about the causes of their symptoms, and about the treatment options, which encompasses specific rehabilitation protocols addressed to re-establish proper functioning of the vestibular system and decrease anxiety-and dizziness-related symptomatology, as well as SSRIs, TCAs, and CBT.

4.5   Photosensitivity and PD

Photosensitivity (i.e. an abnormally high sensitivity to light exposure) seems to contribute both to the etiopathogenesis of and response to therapy in PD [136]. Indeed patients with PD have a lowered threshold of tolerance to light and develop the tendency to adopt photophobic behavior when compared to healthy controls (e.g. to protect themselves from light by wearing sunglasses and/or by avoiding to go out during daytime) [137]. Indeed they score significantly higher on photophobia (light avoidance) and significantly lower on photophilia (light pursuit) symptoms, as measured by the Photosensitivity Assessment Questionnaire (PAQ). Interestingly, photosensitivity seems not to be linked to the mere presence of a diagnosed PD, in fact it correlated with the panic-agoraphobic spectrum, regardless of diagnosis, within both clinical and healthy populations. These results suggest that photosensitivity may belong to the core of PD, regardless of the presence of current active symptomatology. More precisely, photophobia may be integrated in the panic-agoraphobic spectrum, as it seems to run in parallel to other panic spectrum dimensions [138]. Photophobia has been associated with agoraphobia as well. Light frequently elicits anxious/panic symptoms also in these patients [139].

Light sensitivity may be correlated with subclinical autonomic system dysfunctions in patients with PD. Research indicates anatomical links between the amygdala, which exerts a key role in anxiety, and the Edinger-Westphal nucleus, a midbrain center that controls pupil movement, lens accommodation, and eyes convergence [140]. The specific associations between light exposure and the neurotransmitters involved in PD are largely uninvestigated. For instance, it might be interesting to explore to which extent serotonin or melatonin (both of which have been involved in PD) synthesis is activated or deactivated by light. Finally, whether photosensitivity is a *state* characteristic secondary to the active disease or a *trait* predisposing to the full-blown disorder is currently debated although research tends to favor the former hypothesis on the basis that photophobia in PD renormalized after cognitive behavior therapy treatment [141].

4.6   Conclusions and Clinical Implications

Patients with PD often complain of somatic symptoms such as abnormalities in the cardiorespiratory and vestibular systems. They exhibit poorer physical fitness [103, 104], higher respiratory variability during mild physical activity, higher respiratory dysfunctions/breathing patterns irregularities, when compared with subjects without PD [14, 24, 33]. Even unexpected PAs have been associated with significant autonomic and respiratory instability that largely preceded panic onset [20]. Patients often manifest a less efficient cardiovascular system [101] and higher cardiovascular risk [142]. Finally subclinical abnormalities in the balance system, as well as higher photosensitivity, are also common [33, 111, 138].

Taken together these findings partly contradict the assumption that PAs are just false alarms. PAs may be real alarms and reflect reduced adaptability to changes, and true homeostatic instability, which may sustain the experience of anticipatory anxiety and phobic avoidance and increase vulnerability to panic. For these reasons we feel confident to state that patients with PD are physiologically different from healthy subjects and that cardiac, respiratory, and balance symptoms may be the outcome of the inability of these systems to relate with the environment.

In addition mounting evidence suggests that antipanic medications are not merely psychotropic medications. For instance, the selective serotonin re-uptake inhibitors (SSRIs), which are considered the first choice drug treatment of PD, also act on the respiratory, cardiovascular and balance systems. Asymptomatic patients with PD (with and without agoraphobia) with no pulmonary diseases exhibited improved lung function following administration of antipanic drugs (i.e. paroxetine, imipramine, and clonazepam) when compared to those in the washout period [143]. These results indicate that anti-panic drugs ameliorate pulmonary function in this population.

In line with these findings, Perna et al. [31] found that patients with PD manifested abnormal values for many dynamic lung volumes (i.e. peak expiratory flow rate, expiratory flow at 75 % of vital capacity, and maximum mid-expiratory flow rate) when compared to a group of healthy controls. The authors believe that such functional abnormalities indicate subclinical obstruction of lung airways, which are perhaps relevant to the mechanisms related to PD.

In rats a significant increase in baseline respiratory rate was found after 5 and 15 weeks of treatment with paroxetine. In particular following 15 weeks of treatment the rats exhibited reduced respiratory rate in response to CO_2 exposure. These results indicate that the regulation of respiration may be an important factor for the paroxetine antipanic effect [144]. Lungs are the main reservoir of the serotonin transporter, which is the main target of the SSRIs, and this may explain the favourable effect of the SSRIs on respiratory irregularities as well as in pulmonary arterial hypertension. Paroxetine treatment at 20 mg/day for 4 weeks increased heart rate variability and total parasympathetic activity and decreased total sympathetic activity in patients with PD [145]. These results indicate that paroxetine (and arguably SSRIs in general) may protect against or even decrease cardiovascular morbidity and mortality in patients with PD. Finally, we investigated the effects of a 6-week treatment with citalopram on the balance system function in 15 patients with PD, with or without agoraphobia, who underwent static posturography on days 0 and 42. After 6 weeks of treatment with citalopram there was a significant decrease of four out of six posturography measures in eyes-closed and neck extension conditions [146]. This study suggests that serotonergic modulation can improve the balance system function in patients with PD, especially when visual information is lacking.

We believe SSRIs might exert their anti-panic effect also by reducing homeostatic dysfunctions in patients with PD. Besides cognitive-behavior psychotherapy, other non-pharmacologic treatments, such as breathing therapies, and physical exercise, can help to normalize homeostatic dysfunctions in PD. Indeed breathing therapies [147] and aerobic physical exercise reduce panic symptoms, therefore they might represent valid adjunctive treatment options for PD [148]. In addition, preliminary data suggest that vestibular rehabilitation might also be beneficial for patients with PD (especially in those with comorbid agoraphobia). Although we cannot exclude that these somatic treatments act by increasing the perceived sense of control in patients with PD, they might improve panic symptoms also via their positive effect on the homeostatic body functions. A recent meta-analysis indicated that combining exposure, relaxation training, and breathing retraining is more efficacious than cognitive therapy alone, suggesting that somatic interventions are more effective than the mental ones [149]. Taken together, these findings reinforce the assumption that PAs may represent real alarms. Homeostatic dysfunctions may represent a candidate endophenotype of panic vulnerability and underlie maintenance of defensive active mechanisms such as anticipatory anxiety and phobic avoidance. Many patients complain of somatic and cognitive symptoms even when not experiencing a PA. Therefore, treatment should focus not only on reducing PAs occurrence but also aim at reaching a thorough state of physical well being. Anti-panic drugs posology might be adjusted until a full sensation of physical wellbeing is reached. Physical exercise could also be recommended as an additional intervention to potentiate cardiovascular fitness. We believe diagnosis and treatment of patients with PD must be approached with a multidisciplinary evaluation, and therefore all treatment options must be considered. Since cardiorespiratory and vestibular symptoms can aggravate psychiatric symptoms and psychiatric disorders can complicate even further the evaluation of patients with somatic complaints, every therapeutic option available must be carefully considered. Hopefully in the future PD treatment will involve interventions that address not only the core symptoms but also focus on harmonizing respiration and balance, and increase heart rate variability. Regaining physical wellbeing will help patients overcoming anticipatory anxiety and agoraphobia more rapidly and effectively. In conclusion, the assumption that PD is merely a mental disorder seems unsatisfactory. We believe PD should be integrated in a broader vision that acknowledges a central role to the homeostatic systems in the etiopathogenesis of the disorder. Actually the somatic complains of our patients might reflect real bodily dysfunctions. In the future we are convinced that patients with PD may benefit from multidisciplinary evaluation, therefore it is crucial that mental health professionals, vestibular specialists, cardiac specialists, and lung specialists mutually exchange information in order to assess the contingent co-occurrence of PD and its related somatic diseases. This will allow a better diagnosis and lay the foundation for more adequate treatment options that may encompass pharmacological, psychological and somatic treatments, and physical exercise, in a joint effort to decrease panic symptomatology on the one hand and to re-establish proper functioning of the homeostatic systems, on the other hand. We trust this approach will favor not only remission but recovery (which encompasses both symptoms remission and more functional aspects of the patients' wellbeing, such as social functionality and quality of life), and avoid relapse in the future.

References

1. Association AP. Diagnostic and statistical manual of mental disorders. 5th ed. Arlington, VA: American Psychiatric Publishing; 2013.
2. Amaral JM, Spadaro PT, Pereira VM, Silva AC, Nardi AE. The carbon dioxide challenge test in panic disorder: a systematic review of preclinical and clinical research. Rev Bras Psiquiatr. 2013;35(3):318–31. doi:10.1590/1516-4446-2012-1045.
3. Kessler RC, Petukhova M, Sampson NA, Zaslavsky AM, Wittchen HU. Twelve-month and lifetime prevalence and lifetime morbid risk of anxiety and mood disorders in the United States. Int J Methods Psychiatr Res. 2012;21(3):169–84. doi:10.1002/mpr.1359.
4. Goodwin RD, Faravelli C, Rosi S, Cosci F, Truglia E, de Graaf R, et al. The epidemiology of panic disorder and agoraphobia in Europe. Eur Neuropsychopharmacol. 2005;15(4):435–43. doi:10.1016/j.euroneuro.2005.04.006.
5. Klein DF. False suffocation alarms, spontaneous panics, and related conditions. An integrative hypothesis. Arch Gen Psychiatry. 1993;50(4): 306–17.
6. Gorman JM, Liebowitz MR, Fyer AJ, Stein J. A neuroanatomical hypothesis for panic disorder. Am J Psychiatry. 1989;146(2):148–61.
7. Gorman JM, Kent JM, Sullivan GM, Coplan JD. Neuroanatomical hypothesis of panic disorder, revised. Am J Psychiatry. 2000;157(4):493–505.
8. Bouton ME, Mineka S, Barlow DH. A modern learning theory perspective on the etiology of panic disorder. Psychol Rev. 2001;108(1):4–32.
9. Clark DM, Salkovskis PM, Ost LG, Breitholtz E, Koehler KA, Westling BE, et al. Misinterpretation of body sensations in panic disorder. J Consult Clin Psychol. 1997;65(2):203–13.
10. Bellodi L, Perna G. The panic respiration connection. Milan, Italy: Medical Media; 1998.
11. Davidson Jr, Nutt D. Anxiety and respiration. In: Disorders eA, editor. Anxiety disorders. Oxford: Blackwell Science Ltd.; 2003.
12. Schruers K, Griez E. The effects of tianeptine or paroxetine on 35 % CO_2 provoked panic in panic disorder. J Psychopharmacol. 2004;18(4):553–8. doi:10.1177/0269881104047283.
13. Briggs AC, Stretch DD, Brandon S. Subtyping of panic disorder by symptom profile. Br J Psychiatry. 1993;163:201–9.
14. Grassi M, Caldirola D, Vanni G, Guerriero G, Piccinni M, Valchera A, et al. Baseline respiratory parameters in panic disorder: a meta-analysis. J Affect Disord. 2013;146(2):158–73. doi:10.1016/j.jad.2012.08.034.
15. Grassi M, Caldirola D, Di Chiaro NV, Riva A, Dacco S, Pompili M, et al. Are respiratory abnormalities specific for panic disorder? A meta-analysis. Neuropsychobiology. 2014;70(1):52–60. doi:10.1159/000364830.
16. Cowley DS, Roy-Byrne PP. Hyperventilation and panic disorder. Am J Med. 1987;83(5):929–37.
17. van den Hout MA, Hoekstra R, Arntz A, Christiaanse M, Ranschaert W, Schouten E. Hyperventilation is not diagnostically specific to panic patients. Psychosom Med. 1992;54(2):182–91.
18. Griez E, Zandbergen J, Lousberg H, van den Hout M. Effects of low pulmonary CO_2 on panic anxiety. Compr Psychiatry. 1988;29(5):490–7.
19. Gorman JM, Papp LA, Martinez J, Goetz RR, Hollander E, Liebowitz MR, et al. High-dose carbon dioxide challenge test in anxiety disorder patients. Biol Psychiatry. 1990;28(9):743–57.

20. Meuret AE, Rosenfield D, Wilhelm FH, Zhou E, Conrad A, Ritz T, et al. Do unexpected panic attacks occur spontaneously? Biol Psychiatry. 2011;70(10): 985–91. doi:10.1016/j.biopsych.2011.05.027.

21. Goodwin RD, Pine DS. Respiratory disease and panic attacks among adults in the United States. Chest. 2002;122(2):645–50.

22. Perna G, Battaglia M, Garberi A, Arancio C, Bertani A, Bellodi L. Carbon dioxide/oxygen challenge test in panic disorder. Psychiatry Res. 1994;52(2):159–71.

23. Spinhoven P, Ros M, Westgeest A, Van der Does AJ. The prevalence of respiratory disorders in panic disorder, major depressive disorder and V-code patients. Behav Res Ther. 1994;32(6):647–9.

24. Niccolai V, van Duinen MA, Griez EJ. Respiratory patterns in panic disorder reviewed: a focus on biological challenge tests. Acta Psychiatr Scand. 2009;120(3): 167–77. doi:10.1111/j.1600-0447.2009.01408.x.

25. Perna G, Caldirola D, Bellodi L. Panic disorder: from respiration to the homeostatic brain. Acta Neuropsychiatr. 2004;16:57–67.

26. Preter M, Klein DF. Panic, suffocation false alarms, separation anxiety and endogenous opioids. Prog Neuropsychopharmacol Biol Psychiatry. 2008;32(3): 603–12. doi:10.1016/j.pnpbp.2007.07.029.

27. Feinstein JS, Buzza C, Hurlemann R, Follmer RL, Dahdaleh NS, Coryell WH, et al. Fear and panic in humans with bilateral amygdala damage. Nat Neurosci. 2013;16(3):270–2. doi:10.1038/nn.3323.

28. Preter M, Klein DF. Lifelong opioidergic vulnerability through early life separation: A recent extension of the false suffocation alarm theory of panic disorder. Neurosci Biobehav Rev. 2014;46:345–51. doi:10.1016/j.neubiorev.2014.03.025.

29. Faravelli C, Lo Sauro C, Godini L, Lelli L, Benni L, Pietrini F, et al. Childhood stressful events, HPA axis and anxiety disorders. World J Psychiatry. 2012;2(1):13–25. doi:10.5498/wjp.v2.i1.13.

30. Pine DS, Klein RG, Coplan JD, Papp LA, Hoven CW, Martinez J, et al. Differential carbon dioxide sensitivity in childhood anxiety disorders and nonill comparison group. Arch Gen Psychiatry. 2000; 57(10):960–7.

31. Perna G, Cocchi S, Bertani A, Arancio C, Bellodi L. Sensitivity to 35 % CO_2 in healthy first-degree relatives of patients with panic disorder. Am J Psychiatry. 1995;152(4):623–5.

32. Coryell W, Pine D, Fyer A, Klein D. Anxiety responses to CO_2 inhalation in subjects at high-risk for panic disorder. J Affect Disord. 2006;92(1):63–70. doi:10.1016/j.jad.2005.12.045.

33. Caldirola D, Bellodi L, Caumo A, Migliarese G, Perna G. Approximate entropy of respiratory patterns in panic disorder. Am J Psychiatry. 2004;161(1):79–87.

34. Perna G, Bertani A, Caldirola D, Gabriele A, Cocchi S, Bellodi L. Antipanic drug modulation of 35 % CO_2 hyperreactivity and short-term treatment outcome. J Clin Psychopharmacol. 2002;22(3):300–8.

35. Nardi AE. Panic disorder is closely associated with respiratory obstructive illnesses. Am J Respir Crit Care Med. 2009;179(3):256–7. doi:10.1164/ajrccm.179.3.256.

36. Homma I, Masaoka Y. Breathing rhythms and emotions. Exp Physiol. 2008;93(9):1011–21. doi:10.1113/expphysiol.2008.042424.

37. Munjack DJ, Brown RA, Cabe DD, McDowell DE, Baltazar PL. A naturalistic follow-up of panic patients after short-term pharmacologic treatment. J Clin Psychopharmacol. 1993;13(2):156–8.

38. Wilhelm FH, Gerlach AL, Roth WT. Slow recovery from voluntary hyperventilation in panic disorder. Psychosom Med. 2001;63(4):638–49.

39. Wilhelm FH, Gevirtz R, Roth WT. Respiratory dysregulation in anxiety, functional cardiac, and pain disorders. Assessment, phenomenology, and treatment. Behav Modif. 2001;25(4):513–45.

40. Putnam RW, Filosa JA, Ritucci NA. Cellular mechanisms involved in CO(2) and acid signaling in chemosensitive neurons. Am J Physiol Cell Physiol. 2004;287(6):C1493–526. doi:10.1152/ajpcell.00282.2004.

41. Bailey JE, Argyropoulos SV, Lightman SL, Nutt DJ. Does the brain noradrenaline network mediate the effects of the CO_2 challenge? J Psychopharmacol. 2003;17(3):252–9.

42. Protopopescu X, Pan H, Tuescher O, Cloitre M, Goldstein M, Engelien A, et al. Increased brainstem volume in panic disorder: a voxel-based morphometric study. Neuroreport. 2006;17(4):361–3. doi:10.1097/01.wnr.0000203354.80438.1.

43. Perna G, Guerriero G, Brambilla P, Caldirola D. Panic and the brainstem: clues from neuroimaging studies. CNS Neurol Disord Drug Targets. 2014;13(6):1049–56.

44. Remmers JE. Central neuron control of breathing. In: Altose MD, Kawakami Y, editors. Control of breathing in health and disease (lung biology in health and disease). New York: Marcel Dekker; 1999.

45. Esquivel G, Schruers KR, Maddock RJ, Colasanti A, Griez EJ. Acids in the brain: a factor in panic? J Psychopharmacol. 2010;24(5):639–47. doi:10.1177/0269881109104847.

46. Horn EM, Waldrop TG. Suprapontine control of respiration. Respir Physiol. 1998;114(3):201–11.

47. Roncon CM, Biesdorf C, Santana RG, Zangrossi Jr H, Graeff FG, Audi EA. The panicolytic-like effect of fluoxetine in the elevated T-maze is mediated by serotonin-induced activation of endogenous opioids in the dorsal periaqueductal grey. J Psychopharmacol. 2012;26(4):525–31. doi:10.1177/0269881111434619.

48. Balaban CD, Thayer JF. Neurological bases for balance-anxiety links. J Anxiety Disord. 2001;15(1–2):53–79.

49. Maddock RJ. The lactic acid response to alkalosis in panic disorder: an integrative review. J Neuropsychiatry Clin Neurosci. 2001;13(1):22–34.

50. Knuts IJ, Cosci F, Esquivel G, Goossens L, van Duinen M, Bareman M, et al. Cigarette smoking and 35 % CO(2) induced panic in panic disorder patients. J Affect Disord. 2010;124(1-2):215–8. doi:10.1016/j.jad.2009.10.012.

51. Breslau N, Klein DF. Smoking and panic attacks: an epidemiologic investigation. Arch Gen Psychiatry. 1999;56(12):1141–7.

52. Caldirola D, Bellodi L, Cammino S, Perna G. Smoking and respiratory irregularity in panic disorder. Biol Psychiatry. 2004;56(6):393–8. doi:10.1016/j.biopsych.2004.06.013.

53. Abelson JL, Weg JG, Nesse RM, Curtis GC. Persistent respiratory irregularity in patients with panic disorder. Biol Psychiatry. 2001;49(7):588–95.

54. Ley R. The many faces of Pan: psychological and physiological differences among three types of panic attacks. Behav Res Ther. 1992;30(4):347–57.

55. Song HM, Kim JH, Heo JY, Yu BH. Clinical characteristics of the respiratory subtype in panic disorder patients. Psychiatry Investig. 2014;11(4):412–8. doi:10.4306/pi.2014.11.4.412.

56. Nardi AE, Valenca AM, Mezzasalma MA, Lopes FL, Nascimento I, Veras AB, et al. 35 % carbon dioxide and breath-holding challenge tests in panic disorder: a comparison with spontaneous panic attacks. Depress Anxiety. 2006;23(4):236–44. doi:10.1002/da.20165.

57. Meuret AE, White KS, Ritz T, Roth WT, Hofmann SG, Brown TA. Panic attack symptom dimensions and their relationship to illness characteristics in panic disorder. J Psychiatr Res. 2006;40(6):520–7. doi:10.1016/j.jpsychires.2005.09.006.

58. Roberson-Nay R, Latendresse SJ, Kendler KS. A latent class approach to the external validation of respiratory and non-respiratory panic subtypes. Psychol Med. 2012;42(3):461–74. doi:10.1017/S0033291711001425.

59. Soares-Filho GL, Machado S, Arias-Carrion O, Santulli G, Mesquita CT, Cosci F, et al. Myocardial perfusion imaging study of CO(2)-induced panic attack. Am J Cardiol. 2014;113(2):384–8. doi:10.1016/j.amjcard.2013.09.035.

60. Starcevic V, Berle D. Cognitive specificity of anxiety disorders: a review of selected key constructs. Depress Anxiety. 2006;23(2):51–61. doi:10.1002/da.20145.

61. Batelaan N, De Graaf R, Van Balkom A, Vollebergh W, Beekman A. Thresholds for health and thresholds for illness: panic disorder versus subthreshold panic disorder. Psychol Med. 2007;37(2):247–56. doi:10.1017/S0033291706009007.

62. Katon WJ. Chest pain, cardiac disease, and panic disorder. J Clin Psychiatry. 1990;51(Suppl):27–30. discussion 50–3.

63. Coryell W, Noyes Jr R, House JD. Mortality among outpatients with anxiety disorders. Am J Psychiatry. 1986;143(4):508–10.

64. Chignon JM, Lepine JP, Ades J. Panic disorder in cardiac outpatients. Am J Psychiatry. 1993;150(5):780–5.

65. Fleet R, Lavoie K, Beitman BD. Is panic disorder associated with coronary artery disease? A critical review of the literature. J Psychosom Res. 2000;48(4–5):347–56.

66. Jeejeebhoy FM, Dorian P, Newman DM. Panic disorder and the heart: a cardiology perspective. J Psychosom Res. 2000;48(4–5):393–403.

67. Huffman JC, Pollack MH. Predicting panic disorder among patients with chest pain: an analysis of the literature. Psychosomatics. 2003;44(3):222–36. doi:10.1176/appi.psy.44.3.222.

68. Lynch P, Galbraith KM. Panic in the emergency room. Can J Psychiatry. 2003;48(6):361–6.

69. Katerndahl D. Panic plaques: panic disorder & coronary artery disease in patients with chest pain. J Am Board Fam Pract. 2004;17(2):114–26.

70. Katerndahl DA. The association between panic disorder and coronary artery disease among primary care patients presenting with chest pain: an updated literature review. Prim Care Companion J Clin Psychiatry. 2008;10(4):276–85.

71. Katerndahl DA. Chest pain and its importance in patients with panic disorder: an updated literature review. Prim Care Companion J Clin Psychiatry. 2008;10(5):376–83.

72. Bunevicius A, Staniute M, Brozaitiene J, Pop VJ, Neverauskas J, Bunevicius R. Screening for anxiety disorders in patients with coronary artery disease. Health Qual Life Outcomes. 2013;11:37. doi:10.1186/1477-7525-11-37.

73. Frasure-Smith N, Lesperance F. Depression and anxiety as predictors of 2-year cardiac events in patients with stable coronary artery disease. Arch Gen Psychiatry. 2008;65(1):62–71. doi:10.1001/archgenpsychiatry.2007.4.

74. Parker GB, Owen CA, Brotchie HL, Hyett MP. The impact of differing anxiety disorders on outcome following an acute coronary syndrome: time to start worrying? Depress Anxiety. 2010;27(3):302–9. doi:10.1002/da.20602.

75. Todaro JF, Shen BJ, Raffa SD, Tilkemeier PL, Niaura R. Prevalence of anxiety disorders in men and women with established coronary heart disease. J Cardiopulm Rehabil Prev. 2007;27(2):86–91. doi:10.1097/01.HCR.0000265036.24157.e7.

76. Dammen T, Arnesen H, Ekeberg O, Friis S. Psychological factors, pain attribution and medical morbidity in chest-pain patients with and without coronary artery disease. Gen Hosp Psychiatry. 2004;26(6):463–9. doi:10.1016/j.genhosppsych.2004.08.004.

77. Grenier S, Potvin O, Hudon C, Boyer R, Preville M, Desjardins L, et al. Twelve-month prevalence and correlates of subthreshold and threshold anxiety in community-dwelling older adults with cardiovascular diseases. J Affect Disord. 2012;136(3):724–32. doi:10.1016/j.jad.2011.09.052.

78. Chen YH, Tsai SY, Lee HC, Lin HC. Increased risk of acute myocardial infarction for patients with panic disorder: a nationwide population-based study.

Psychosom Med. 2009;71(7):798–804. doi:10.1097/PSY.0b013e3181ad55e3.

79. Gomez-Caminero A, Blumentals WA, Russo LJ, Brown RR, Castilla-Puentes R. Does panic disorder increase the risk of coronary heart disease? A cohort study of a national managed care database. Psychosom Med. 2005;67(5):688–91.

80. Alvarenga ME, Richards JC, Lambert G, Esler MD. Psychophysiological mechanisms in panic disorder: a correlative analysis of noradrenaline spillover, neuronal noradrenaline reuptake, power spectral analysis of heart rate variability, and psychological variables. Psychosom Med. 2006;68(1):8–16. doi:10.1097/01.psy.0000195872.00987.db.

81. Yapislar H, Aydogan S, Ozum U. Biological understanding of the cardiovascular risk associated with major depression and panic disorder is important. Int J Psychiatry Clin Pract. 2012;16(1):27–32. doi:10.3109/13651501.2011.620127.

82. Jakovljevic M, Reiner Z, Milicic D. Mental disorders, treatment response, mortality and serum cholesterol: a new holistic look at old data. Psychiatr Danub. 2007;19(4):270–81.

83. Hoge EA, Brandstetter K, Moshier S, Pollack MH, Wong KK, Simon NM. Broad spectrum of cytokine abnormalities in panic disorder and posttraumatic stress disorder. Depress Anxiety. 2009;26(5):447–55. doi:10.1002/da.20564.

84. Fleet R, Lesperance F, Arsenault A, Gregoire J, Lavoie K, Laurin C, et al. Myocardial perfusion study of panic attacks in patients with coronary artery disease. Am J Cardiol. 2005;96(8):1064–8. doi:10.1016/j.amjcard.2005.06.035.

85. Niccolai V, van Duinen MA, Griez EJ. Objective and subjective measures in recovery from a 35 % carbon dioxide challenge. Can J Psychiatry. 2008;53(11):737–44.

86. Guyenet PG, Stornetta RL, Abbott SB, Depuy SD, Fortuna MG, Kanbar R. Central CO_2 chemoreception and integrated neural mechanisms of cardiovascular and respiratory control. J Appl Physiol. 2010;108(4):995–1002. doi:10.1152/japplphysiol.00712.2009.

87. Girotti LA, Crosatto JR, Messuti H, Kaski JC, Dyszel E, Rivas CA, et al. The hyperventilation test as a method for developing successful therapy in Prinzmetal's angina. Am J Cardiol. 1982;49(4):834–41.

88. Fujii H, Yasue H, Okumura K, Matsuyama K, Morikami Y, Miyagi H, et al. Hyperventilation-induced simultaneous multivessel coronary spasm in patients with variant angina: an echocardiographic and arteriographic study. J Am Coll Cardiol. 1988;12(5):1184–92.

89. Yeragani VK, Pohl R, Jampala VC, Balon R, Ramesh C, Srinivasan K. Increased QT variability in patients with panic disorder and depression. Psychiatry Res. 2000;93(3):225–35.

90. Yeragani VK, Sobolewski E, Igel G, Johnson C, Jampala VC, Kay J, et al. Decreased heart-period variability in patients with panic disorder: a study of Holter ECG records. Psychiatry Res. 1998;78(1–2):89–99.

91. Yeragani VK, Pohl R, Berger R, Balon R, Ramesh C, Glitz D, et al. Decreased heart rate variability in panic disorder patients: a study of power-spectral analysis of heart rate. Psychiatry Res. 1993; 46(1):89–103.

92. Molgaard H, Sorensen KE, Bjerregaard P. Attenuated 24-h heart rate variability in apparently healthy subjects, subsequently suffering sudden cardiac death. Clin Auton Res. 1991;1(3):233–7.

93. Bigger Jr JT, Fleiss JL, Rolnitzky LM, Steinman RC. Frequency domain measures of heart period variability to assess risk late after myocardial infarction. J Am Coll Cardiol. 1993;21(3):729–36.

94. Yeragani VK, Pohl R, Jampala VC, Balon R, Kay J, Igel G. Effect of posture and isoproterenol on beat-to-beat heart rate and QT variability. Neuropsychobiology. 2000;41(3):113–23. doi:26642.

95. Sullivan GM, Kent JM, Kleber M, Martinez JM, Yeragani VK, Gorman JM. Effects of hyperventilation on heart rate and QT variability in panic disorder pre- and post-treatment. Psychiatry Res. 2004;125(1):29–39. doi:10.1016/j.psychres.2003.10.002.

96. Atmaca M, Yavuzkir M, Izci F, Gurok MG, Adiyaman S. QT wave dispersion in patients with panic disorder. Neurosci Bull. 2012;28(3):247–52.

97. Yavuzkir M, Atmaca M, Dagli N, Balin M, Karaca I, Mermi O, et al. P-wave dispersion in panic disorder. Psychosom Med. 2007;69(4):344–7. doi:10.1097/PSY.0b013e3180616900.

98. Tomaselli GF, Beuckelmann DJ, Calkins HG, Berger RD, Kessler PD, Lawrence JH, et al. Sudden cardiac death in heart failure. The role of abnormal repolarization. Circulation. 1994;90(5):2534–9.

99. Manttari M, Oikarinen L, Manninen V, Viitasalo M. QT dispersion as a risk factor for sudden cardiac death and fatal myocardial infarction in a coronary risk population. Heart. 1997;78(3):268–72.

100. Dilaveris PE, Gialafos EJ, Andrikopoulos GK, Richter DJ, Papanikolaou V, Poralis K, et al. Clinical and electrocardiographic predictors of recurrent atrial fibrillation. Pacing Clin Electrophysiol. 2000;23(3):352–8.

101. Fisher AJ, Woodward SH. Cardiac stability at differing levels of temporal analysis in panic disorder, post-traumatic stress disorder, and healthy controls. Psychophysiology. 2014;51(1):80–7. doi:10.1111/psyp.12148.

102. Cosci F, Knuts IJ, Abrams K, Griez EJ, Schruers KR. Cigarette smoking and panic: a critical review of the literature. J Clin Psychiatry. 2010;71(5):606–15. doi:10.4088/JCP.08r04523blu.

103. Muotri RW, Bernik MA. Panic disorder and exercise avoidance. Rev Bras Psiquiatr. 2014;36(1):68–75. doi:10.1590/1516-4446-2012-1012.

104. Caldirola D, Namia C, Micieli W, Carminati C, Bellodi L, Perna G. Cardiorespiratory response to physical exercise and psychological variables in

panic disorder. Rev Bras Psiquiatr. 2011;33(4): 385–9.

105. Sardinha A, Araujo CG, Soares-Filho GL, Nardi AE. Anxiety, panic disorder and coronary artery disease: issues concerning physical exercise and cognitive behavioral therapy. Expert Rev Cardiovasc Ther. 2011;9(2):165–75. doi:10.1586/erc.10.170.

106. Furman JM, Redfern MS, Jacob RG. Vestibulo-ocular function in anxiety disorders. J Vestib Res. 2006;16(4–5):209–15.

107. Jacob RG. Panic disorder and the vestibular system. Psychiatr Clin North Am. 1988;11(2):361–74.

108. Mezzasalma MA, Mathias Kde V, Nascimento I, Valenca AM, Nardi AE. Imipramine for vestibular dysfunction in panic disorder: a prospective case series. Arq Neuropsiquiatr. 2011;69(2A):196–201.

109. Stambolieva K, Angov G. Balance control in quiet upright standing in patients with panic disorder. Eur Arch Otorhinolaryngol. 2010;267(11):1695–9. doi:10.1007/s00405-010-1303-2.

110. Teggi R, Caldirola D, Fabiano B, Recanati P, Bussi M. Rehabilitation after acute vestibular disorders. J Laryngol Otol. 2009;123(4):397–402. doi:10.1017/S0022215108002983.

111. Perna G, Dario A, Caldirola D, Stefania B, Cesarani A, Bellodi L. Panic disorder: the role of the balance system. J Psychiatr Res. 2001;35(5):279–86.

112. Staab JP, Ruckenstein MJ. Which comes first? Psychogenic dizziness versus otogenic anxiety. Laryngoscope. 2003;113(10):1714–8.

113. Furman JM, Jacob RG. A clinical taxonomy of dizziness and anxiety in the otoneurological setting. J Anxiety Disord. 2001;15(1–2):9–26.

114. Schuerger RJ, Balaban CD. Immunohistochemical demonstration of regionally selective projections from locus coeruleus to the vestibular nuclei in rats. Exp Brain Res. 1993;92(3):351–9.

115. Yates BJ, Goto T, Kerman I, Bolton PS. Responses of caudal medullary raphe neurons to natural vestibular stimulation. J Neurophysiol. 1993;70(3): 938–46.

116. Licata F, Li Volsi G, Maugeri G, Santangelo F. Excitatory and inhibitory effects of 5-hydroxytryptamine on the firing rate of medial vestibular nucleus neurons in the rat. Neurosci Lett. 1993;154(1–2):195–8.

117. Balaban CD. Projections from the parabrachial nucleus to the vestibular nuclei: potential substrates for autonomic and limbic influences on vestibular responses. Brain Res. 2004;996(1):126–37.

118. Balaban CD. Vestibular nucleus projections to the parabrachial nucleus in rabbits: implications for vestibular influences on the autonomic nervous system. Exp Brain Res. 1996;108(3):367–81.

119. Balaban CD, Beryozkin G. Vestibular nucleus projections to nucleus tractus solitarius and the dorsal motor nucleus of the vagus nerve: potential substrates for vestibulo-autonomic interactions. Exp Brain Res. 1994;98(2):200–12.

120. Reilly S, Trifunovic R. Lateral parabrachial nucleus lesions in the rat: neophobia and conditioned taste aversion. Brain Res Bull. 2001;55(3):359–66.

121. Hayward LF, Felder RB. Lateral parabrachial nucleus modulates baroreflex regulation of sympathetic nerve activity. Am J Physiol. 1998;274(5 Pt 2):R1274–82.

122. Chamberlin NL, Saper CB. Topographic organization of respiratory responses to glutamate microstimulation of the parabrachial nucleus in the rat. J Neurosci. 1994;14(11 Pt 1):6500–10.

123. Huh YE, Kim JS. Bedside evaluation of dizzy patients. J Clin Neurol. 2013;9(4):203–13. doi:10.3988/jcn.2013.9.4.203.

124. Craig AD. How do you feel? Interoception: the sense of the physiological condition of the body. Nat Rev Neurosci. 2002;3(8):655–66. doi:10.1038/nrn894.

125. Craig AD. How do you feel—now? The anterior insula and human awareness. Nat Rev Neurosci. 2009;10(1):59–70. doi:10.1038/nrn2555.

126. Gauriau C, Bernard JF. Pain pathways and parabrachial circuits in the rat. Exp Physiol. 2002;87(2): 251–8.

127. Zylka MJ. Nonpeptidergic circuits feel your pain. Neuron. 2005;47(6):771–2. doi:10.1016/j.neuron.2005.09.003.

128. Fanselow MS. Neural organization of the defensive behavior system responsible for fear. Psychon Bull Rev. 1994;1(4):429–38. doi:10.3758/BF03210947.

129. Balaban CD. Vestibular autonomic regulation (including motion sickness and the mechanism of vomiting). Curr Opin Neurol. 1999;12(1):29–33.

130. Berencsi A, Ishihara M, Imanaka K. The functional role of central and peripheral vision in the control of posture. Hum Mov Sci. 2005;24(5–6):689–709. doi:10.1016/j.humov.2005.10.014.

131. Balaban CD. Neural substrates linking balance control and anxiety. Physiol Behav. 2002;77(4–5): 469–75.

132. Redfern MS, Yardley L, Bronstein AM. Visual influences on balance. J Anxiety Disord. 2001;15(1–2): 81–94.

133. Szirmai A, Kisely M, Nagy G, Nedeczky Z, Szabados EM, Toth A. Panic disorder in otoneurological experience. Int Tinnitus J. 2005;11(1): 77–80.

134. Jacob RG, Whitney SL, Detweiler-Shostak G, Furman JM. Vestibular rehabilitation for patients with agoraphobia and vestibular dysfunction: a pilot study. J Anxiety Disord. 2001;15(1–2):131–46.

135. Simon NM, Parker SW, Wernick-Robinson M, Oppenheimer JE, Hoge EA, Worthington JJ, et al. Fluoxetine for vestibular dysfunction and anxiety: a prospective pilot study. Psychosomatics. 2005; 46(4):334–9. doi:10.1176/appi.psy.46.4.334.

136. Palazzo L, Clemente R, Bersani G. Disturbo di panico e stagionalità: ipotesi di una sottotipizzazione del disturbo sulla base del suo andamento stagionale. Giorn Ital Psicopat. 2005;11:235–6.

137. Lelliott P, Marks I, McNamee G, Tobena A. Onset of panic disorder with agoraphobia. Toward an integrated model. Arch Gen Psychiatry. 1989;46(11): 1000–4.

138. Bossini L, Frank E, Campinoti G, Valdagno M, Caterini C, Castrogiovanni P, et al. Photosensitivity and panic-agoraphobic spectrum: a pilot study. Riv Psichiatr. 2013;48(2):108–12. doi:10.1708/1272.14034.

139. Kellner M, Wiedemann K, Zihl J. Illumination perception in photophobic patients suffering from panic disorder with agoraphobia. Acta Psychiatr Scand. 1997;96(1):72–4.

140. Bitsios P, Szabadi E, Bradshaw CM. The inhibition of the pupillary light reflex by the threat of an electric shock: a potential laboratory model of human anxiety. J Psychopharmacol. 1996;10(4):279–87. doi:10.1177/026988119601000404.

141. Bossini L, Fagiolini A, Valdagno M, Padula L, Hofkens T, Castrogiovanni P. Photosensitivity in panic disorder. Depress Anxiety. 2009;26(1):E34–6. doi:10.1002/da.20477.

142. Alici H, Ercan S, Bulbul F, Alici D, Alpak G, Davutoglu V. Circadian blood pressure variation in normotensive patients with panic disorder. Angiology. 2014;65(8):747–9. doi:10.1177/0003319713512172.

143. Nascimento I, de Melo-Neto VL, Valenca AM, Lops FL, Freire RC, Cassabian LA, et al. Antipanic drugs and pulmonary function in panic disorder patients. Rev Psiq Clín. 2009;36:123–9.

144. Olsson M, Annerbrink K, Bengtsson F, Hedner J, Eriksson E. Paroxetine influences respiration in rats: implications for the treatment of panic disorder. Eur Neuropsychopharmacol. 2004;14(1):29–37.

145. Tucker P, Adamson P, Miranda Jr R, Scarborough A, Williams D, Groff J, et al. Paroxetine increases heart rate variability in panic disorder. J Clin Psychopharmacol. 1997;17(5):370–6.

146. Perna G, Alpini D, Caldirola D, Raponi G, Cesarani A, Bellodi L. Serotonergic modulation of the balance system in panic disorder: an open study. Depress Anxiety. 2003;17(2):101–6. doi:10.1002/da.10092.

147. Meuret AE, Wilhelm FH, Ritz T, Roth WT. Feedback of end-tidal pCO_2 as a therapeutic approach for panic disorder. J Psychiatr Res. 2008;42(7):560–8. doi:10.1016/j.jpsychires.2007.06.005.

148. Broocks A, Bandelow B, Pekrun G, George A, Meyer T, Bartmann U, et al. Comparison of aerobic exercise, clomipramine, and placebo in the treatment of panic disorder. Am J Psychiatry. 1998;155(5): 603–9.

149. Sanchez-Meca J, Rosa-Alcazar AI, Marin-Martinez F, Gomez-Conesa A. Psychological treatment of panic disorder with or without agoraphobia: a meta-analysis. Clin Psychol Rev. 2010;30(1):37–50. doi:10.1016/j.cpr.2009.08.011.

Staging of Panic Disorder: Implications for Neurobiology and Treatment

5

Fiammetta Cosci

Contents

Abstract

The staging model of panic includes the following stages. Stage 1 in which subclinical symptoms of agoraphobia or social phobia or generalized anxiety disorder or hypochondriasis are present. Stage 2 characterized by the acute manifestations of agoraphobia or social phobia or generalized anxiety disorder or hypochondriasis. Panic Disorder (PD) with worsening of anxiety and hypochondriacal symptoms characterizes stage 3 together with demoralization or major depression. Chronic PD and agoraphobia or social phobia or generalized anxiety disorder or hypochondriasis together with increased liability to major depression may occur at stage 4. This staging model is applicable in clinical practice. In a substantial proportion of patients with PD a prodromal phase and, despite successful treatment, residual symptoms can be identified. Both prodromes and residual symptoms allow to monitor the evolution of the disorder during recovery via the rollback phenomenon. The different stages of PD and the steps of the rollback have a correspondence in its neurobiology and in its treatment. The translation of staging in the neurobiology of panic identifies different phases in the development of PD which involve the amygdala, the hippocampus, and the medial/orbital prefrontal cortex. The treatment implications, although still too disregarded, emphasize the importance to consider residual symptoms as the final target

F. Cosci (✉)
Department of Health Sciences,
University of Florence, Florence, Italy
e-mail: fiammetta.cosci@unifi.it

© Springer International Publishing Switzerland 2016
A.E. Nardi, R.C.R. Freire (eds.), *Panic Disorder*, DOI 10.1007/978-3-319-12538-1_5

of the therapy. In addition, psychotherapy (mainly cognitive-behavioral therapy) has shown good effectiveness while sequential or stage-oriented treatments have shown promising results.

Keywords
Staging • Stage • Panic • Panic disorder • Neurobiology • Treatment

5.1   Introduction

Current emphasis in psychiatry is on cross-sectional assessment of symptoms resulting in diagnostic criteria and on comorbidity. The use of diagnostic criteria is derived from the traditional method of clinical medicine, in which they provide operating specifications for making a clinical decision about the existence of a particular disease [1]. However, clinicians usually evaluate, in their daily practice, also issues that do not simply apply to the severity of the disorder, such as its longitudinal development; social support and adaptation; the resilience and reaction to previous conflicts, threats, or losses; the motivation and compliance with the treatment; the pre-morbid personality and potential abnormal personality traits.

Over the time, a clinical reasoning which goes through a series of "transfer stations" [2], where potential connections between presenting symptoms and pathophysiological processes are drawn and which are amenable to longitudinal verification and modification as long as therapeutic goals are achieved [3], has been proposed. Notwithstanding this, a lack of attention to the longitudinal development of the disorders has been maintained and has apparently deprived the clinical process of a number of important "transfer stations".

The creator of this innovative and emerging staging approach for assessment was Feinstein who introduced in 1987 the term "clinimetrics" [1]. Clinimetrics indicates a domain concerned with indices, rating scales, and other expressions that are used to describe or measure symptoms, physical signs, and other distinctly clinical phenomena in medicine. Further, it provides a home for a number of clinical phenomena which do not find

room in customary clinical taxonomy, such as types, severity, and sequence of symptoms; rate of progression in illness (staging); severity of comorbidity; problems of functional capacity; reasons for medical decision; and other aspects of daily life, such as well-being, distress, and mental pain [1, 4–6]. In recent years, there have been several exemplifications of this approach in research on mood and anxiety disorders, reviewed by Fava et al. [7].

Although current diagnostic entities (e.g., the Diagnostic and Statistical Manual of Mental Disorders – DSM) are based on clinimetric principles, their use is still strongly influenced by psychometric models [3, 8]. This means that the severity of each disorder is determined by the number of symptoms and not by their intensity or quality, to the same extent that a score in a self-rating scale depends on the number of symptoms that are scored as positive [8]. As a consequence, the preferential target of therapy tends to become the syndrome resulting from a certain number of symptoms (which may be of mild intensity and of doubtful impact on quality of life), instead of individual symptoms that may be incapacitating for the patient [3]. Moreover, clinicians may find some difficulties in formulating a treatment plan for people who, for instance, do not reach the known diagnostic threshold. It might be sometimes difficult to disentangle whether the symptoms the patient is complaining about should be addressed for treatment or whether the conditions need to be addressed in an integrated way, and if so, by which professional figures.

5.2   The Staging Model of Panic Disorder

In 1993, Fava and Kellner [9] proposed to use in psychiatry a staging method that allows characterizing a disorder according to seriousness, extension, features, and longitudinal development following standardized and well-defined models. They published a seminal paper suggesting a staging model for schizophrenia, mood, and panic disorder (PD). The core concept was that these psychiatric disorders develop according to main stages. The first stage usually involves the

Table 5.1 Stages of panic disorder with agoraphobia according to Fava and Kellner [9]

Stages	
1	Prodromal or predisposing: anxiety sensitivity, health anxiety, harm avoidance and dependence
2	Agoraphobia of mild or moderate severity (DSM IIIR) (APA, 1987)
3	Panic disorder, acute phase (DSM IIIR) (APA, 1987). Worsening of agoraphobia, anxiety and hypochondriacal fears and beliefs
4	Panic disorder, chronic phase (longer than 6 months). Increased liability to major depression

Table 5.2 Staging of panic disorder according to Cosci and Fava [12]

Stages	
1	Prodromal phase: subclinical symptoms of agoraphobia and/or social phobia and/or generalized anxiety disorder and/or hypochondriasis
2	Acute manifestations of agoraphobia and/or social phobia and/or generalized anxiety disorder and/or hypochondriasis
3	Panic disorder with worsening of anxiety and hypochondriacal symptoms. Demoralization and/or major depression may occur
4	Chronic panic disorder and agoraphobia and/or social phobia and/or generalized anxiety disorder and/or hypochondriasis (in attenuated or persistent form). Increased liability to major depression

presence of predisposing factors (e.g., genetic vulnerabilities, pre-morbid personality, lack of psychological well-being); the second stage is characterized by the acute symptoms; the third includes the residual symptoms; the fourth implies sub-chronic symptoms; and the fifth stage, when present, is characterized by the chronic illness [9].

Regarding PD, Fava and Kellner [9] described a staging model, suggesting that, in a substantial proportion of patients, anxiety sensitivity, health anxiety, harm avoidance and dependence are the prodromes of the disorder (stage 1). In the same staging model agoraphobia precedes the first panic attack (stage 2), and acute manifestations of panic disorder, worsened by agoraphobia, hypochondriacal fears and beliefs (stage 3), can be followed by a stage 4 characterized by the chronic manifestations of PD and increased liability to major depression (see Table 5.1).

This four-stage model was consistent with symptomatic patterns of improvement upon behavioral treatment of panic disorder as well as with the rollback phenomenon upon drug treatment. However, since, at least in some patients, the first panic attack apparently occurred without conspicuous prodromal symptoms, while anticipatory anxiety, phobic avoidance and hypochondriasis may develop subsequently, Sheehan and Sheehan [10] outlined a different staging process: stage 1 (subpanic) characterized by panic attacks with limited symptoms; stage 2 (panic); stage 3 (hypochondriasis); stage 4 (single phobia, that is the setting in which panic occurs); stage 5 (social phobia); stage 6 (agoraphobia); stage 7 (depression).

Over the time, the staging model of panic has been updated [11, 12]. According to the most recent revision [12], prodromal phase (stage 1) may include subclinical symptoms of agoraphobia and/or social phobia and/or generalized anxiety disorder and/or hypochondriasis; stage 2 is characterized by the acute manifestations of agoraphobia and/or social phobia and/or generalized anxiety disorder and/or hypochondriasis; PD with worsening of anxiety and hypochondriacal symptoms characterizes stage 3 together with demoralization and/or major depression; chronic PD and agoraphobia and/or social phobia and/or generalized anxiety disorder and/or hypochondriasis together with increased liability to major depression may occur at stage 4 (see Table 5.2).

This staging model includes agoraphobia, social anxiety, generalized anxiety, hypochondriasis in stages 1, 2, and 4 suggesting to consider agoraphobia, social anxiety, generalized anxiety, or hypochondriasis as a stage of development of PD. Thus, PD seems to become an aspecific clinical manifestation in the frame of other anxiety disorders. A confirmation of such view comes from Kessler et al. [13] who found that isolated panic attacks are quite common and significantly comorbid with other DSM-IV disorders [14].

A growing literature has supported the existence of the above mentioned clinical stages in the longitudinal development of panic.

Roy-Byrne and Cowley [15] observed that, despite the availability of effective anti-panic treatments, PD remains a chronic illness and the presence of agoraphobia, major depression, and personality disorder predict a poor outcome. In the general population of the Epidemiologic Catchment Area study, the prodromal period was about 10–15 years long. Panic attacks occurring in the year before the first interview and the perception that one is a "nervous person" were strong predictors of the onset of PD [16]. According to Keller and colleagues [17], the probability of having a panic-free interval by 12 months was 0.68 for subjects with PD and 0.55 for those with PD with agoraphobia [17]. O'Rourke et al. [18] almost confirmed these results observing that at 12 month follow-up, 23 (33.8 %) PD patients had recovered and remained well, 12 reported relapses and remissions, 19 were still improved short of full recovery, and 14 had persistent PD. The strongest predictor of sustained recovery was good clinical status between 6 and 12 months from baseline, while personality dysfunction was the most important characteristic of patients with persistent PD. Over a 15–60 month period of follow up, Cowley et al. [19] found that only 10 % of the PD patients were asymptomatic; the strongest predictors of overall improvement were avoidance coping for outcome at 12 months and Axis I comorbidity for outcome at the time of the follow-up evaluation. Finally, 23 % of 132 PD with agoraphobia patients treated with behavioral methods based on exposure homework and followed for a median period of 8 years had a relapse. The estimated cumulative percentage of patients remaining in remission was: 93.1 for at least 2 years, 82.4 after 5 years, 78.8 after 7 years, and 62.1 after 10 years. The presence of a personality disorder and the pretreatment level of depressed mood indicated a worse prognosis while patients who completely overcame agoraphobic avoidance had a better outcome. Two additional risk factors involved the use of psychotropic drugs: those who were still taking benzodiazepines at the end of exposure therapy had a less favorable outcome than those who were drug free; and patients under antidepressant treatment before starting the behavioral therapy had a worse outcome than those antidepressant-free [20].

5.3 Subclinical Symptoms

Prodromes can be identified with the early symptoms and signs of a disease. The prodromal phase connotes a time interval between the onset of prodromal symptoms and the onset of the characteristic manifestations of the fully developed illness. Residual symptoms have been identified with the persistent symptoms and signs despite apparent remission or recovery. Of course, in any chronic and recurring medical illness, subclinical fluctuations, either in terms of symptomatology and laboratory markers, may occur [21].

Detre and Jarecki [22] provided a model for relating prodromal and residual symptomatology in psychiatric illness: the "rollback phenomenon". According to this phenomenon, as the illness remits, it progressively recapitulates even though in a reverse order many of the stages and symptoms that were seen during the time it developed [21]. According to the rollback model, there is also a temporal relationship between the time of development of a disorder and the duration of the phase of recovery. Moreover, there appears to be a relationship between residual and prodromal symptomatology since certain prodromal symptoms may persist and progress to become prodromes of relapse.

Fava et al. reviewed the literature on several psychiatric disorders and, among them PD [21, 23, 24], paving the way for the characterization of the phenomenological development of these illnesses [25] and the application of sequential treatment [3].

Regarding PD with agoraphobia, Fava and Mangelli [24] illustrated prodromes starting from the Klein's model [26] who observed that when patients are suddenly struck by the first panic attack, they develop persistent anticipatory anxiety and hypochondriacal fears, leading to avoidant behaviour and agoraphobia. This view has been supported over time [27, 28]. However, also different sequences of events have been observed. Wittchen et al. [28] pointed out that 45.1% of individuals at risk with PD and 76.5% of those at risk for panic attacks did not develop agoraphobia (AG). Faravelli et al. [29] re-interviewed 41 subjects with a lifetime history of AG 4 years after the first evaluation and found that 12 cases had the original diagnosis of agoraphobia without a history of panic attacks, 29 had PD with agoraphobia, 2 no longer met the criterion for agoraphobia turning into social phobia or into specific phobia. In 2010, Wittchen et al. [30] confirmed that there is no empirical evidence which unequivocally demonstrates that agoraphobia is temporally primarily and exclusively a function of panic attack (PA) or panic-like features. According to this body of scientific evidences, the DSM 5 [31] introduced agoraphobia as a clinically significant disorder that exists independently from panic.

Some authors also found generalized anxiety to be prodromal of panic [27]; others observed that agoraphobic avoidance, generalized anxiety, hypochondriacal fears and beliefs occur before the first panic attack [32–34]; further works showed that some patients have prodromal depression, anxiety, or avoidance [35, 36].

Since 1990s, PD has been recognized as a chronic illness with little spontaneous improvement, high rates of relapse after remission, and longer episodes when agoraphobia is a part of the constellation of symptoms [17]. Thus, residual symptoms have been found extremely common and encompassing phobic and anxiety disturbances, social impairment, and dependence.

An interesting brief report by Fava et al. [37] assessed psychological well-being and residual symptoms in a sample of 30 patients who had recovered from PD with agoraphobia and 30 controls. Remitted patients displayed significantly more psychological distress (i.e., anxiety, depression, somatic symptoms) than controls; the most common residual symptoms were generalized anxiety, somatic anxiety, low self-esteem, agoraphobia, and hypochondriasis. Patients with PD also showed less psychological and physical well-being than controls. On the other hand, Corominas et al. [38] evaluated a clinical sample

of 64 PD outpatients comparing those who developed residual symptoms at 1-year of follow-up with those who did not. Surprisingly, the two groups did not differ in terms of achieved improvement measured via the frequency of panic attacks, the degree of avoidant behavior, the depressive symptoms, or the anxious symptoms. Those who developed residual symptoms had higher rates in history of anxious disorders in childhood, pretreatment panic attacks, presence of depersonalization and derealization symptoms during the panic attacks, comorbidity (particularly with simple phobia) than those who did not develop residual symptoms. In addition, the likelihood of developing residual symptoms was higher in patients who reported dyspnea as the main symptom during panic attacks.

In a study conducted by Marchesi et al. [39], 65 PD patients were followed over 12 months and randomly treated with paroxetine or citalopram. A complete remission was achieved by only 47.6% of the subjects, whereas in the remaining patients, that is those who did not reach a complete remission, limited symptom of panic attacks, anticipatory anxiety, phobic avoidance, and depression were present at the end of the study.

Several studies have addressed the issue of the sequence of improvement of symptoms in patients with PD upon behavioral [40–43] or pharmacological [44–47] treatment. A seminal paper was proposed by Fava et al. in 1991 [40] in which they specifically investigated the rollback phenomenon in 25 PD patients who received 12 sessions of in vivo exposure. After the first six sessions, agoraphobia was significantly improved while 12 sessions of treatment yielded the disappearance of panic attacks and a further reduction of agoraphobic avoidance. The phenomenological sequence observed retrospectively for the prodromal symptoms of PD was: phobic avoidance and hypochondriasis which leaded to panic and, thereafter, to more phobic avoidance and hypochondriasis. On the other hand, the rollback phenomenon observed was: a decrease in avoidance by exposure which seemed to improve agoraphobia and panic, with eventual disappearance of panic in majority of patients, whereas agoraphobia persisted at least to a lesser degree. Prodromal symptoms of PD with agoraphobia thus tend to become residual symptoms, which, in turn, may progress to prodromal symptoms of relapse.

Other authors also found interesting results. Bouchard et al. [48] observed that cognitive changes precede changes in the level of panic apprehension both when treated with cognitive restructuring or exposure. Hoffart et al. [43] administered an integrated cognitive and behavioral model of agoraphobic avoidance to patients with PD and agoraphobia and found a feedback loop of effects during treatment: the anxiety elicited by bodily sensations influenced catastrophic beliefs and such beliefs influenced avoidant behavior. A reduction of avoidance, in turn, decreased a fear of bodily sensations. Thus, avoidant behavior seems to be maintained by cognitive appraisal, while avoidance maintained anxiety conditioned to bodily sensations.

The role of subclinical symptoms has been further increased by the development of the Diagnostic Criteria for Psychosomatic Research (DCPR) [49, 50]. Fava and Wise [51] suggested to modify the DSM-IV category [14] concerned with psychological factors affecting medical conditions into an expanded category of psychological factors affecting either identified or feared medical conditions. They proposed a new section with 6 most frequent DCPR syndromes. Among them, the DCPR health anxiety, disease phobia, persistent somatization, demoralization, and irritable mood offer interesting specifiers for subclinical symptoms. For instance, health anxiety, that encompasses nonspecific dimensions of abnormal illness and somatic amplification that readily respond to appropriate reassurance, may be the prodromal or the residual symptom of PD. Fava et al. [52] found a significant association between PD and DCPR syndromes of health anxiety, disease phobia, patterns of somatization, and irritable mood.

Unfortunately, the DSM 5 [31] did not agree to this proposal although DCPR are extremely important, being good predictors of impaired psychosocial functioning in medically ill people [53], and having high sensitive instruments in detecting sub-threshold psychological distress as well as sub-threshold psychiatric comorbidity [54].

5.4 Psychological Development of Panic and Neurobiology

A longitudinal view of PD, encompassing prodromal and residual symptoms, can find interesting links to the neurobiology of panic.

In 1989, Gorman et al. [55] articulated a neuroanatomical hypothesis of PD positing that a panic attack stems from loci in the brainstem that involve serotonergic and noradrenergic transmission and respiratory control, that anticipatory anxiety arises after the kindling of limbic area structures, and that phobic avoidance is a function of pre-cortical activation. The hypothesis asserted that medication exerts its therapeutic effect by normalizing brainstem activity in patients with panic disorder, whereas cognitive behavioural therapy works at the cortical level. However, this original idea has been surpassed because it was almost completely divorced from research that has mapped out the neuroanatomical basis for fear. For this reason, Gorman et al. in 2000 [56] proposed a revised neuroanatomical hypothesis of PD. According to this model, the sensory input for the conditioned stimulus runs through the anterior thalamus to the lateral nucleus of the amygdala and is then transferred to the central nucleus of the amygdala which stands as the central point for dissemination of information that coordinates autonomic and behavioural responses. Efferents of the central nucleus of the amygdala have many targets: the parabrachial nucleus, producing an increase in respiratory rate; the lateral nucleus of the hypothalamus, activating the sympathetic nervous system and causing autonomic arousal and sympathetic discharge; the locus coeruleus, resulting in an increase in norepinephrine release and contributing to increase blood pressure, heart rate, and the behavioral fear response; and the paraventricular nucleus of the hypothalamus, causing an increase in the release of adrenocorticoids. A projection from the central nucleus of the amygdala to the periaqueductal gray region would be responsible for additional behavioural responses, including phobic avoidance. There are also important reciprocal connections between the amygdala and the sensory thalamus, prefrontal cortex, insula, and primary somatosensory cortex. Although the amygdala receives direct sensory input from brainstem structures and the sensory thalamus enabling a rapid response to potentially threatening stimuli, it also receives afferents from cortical regions involved in the processing and evaluation of sensory information. Potentially, a neurocognitive deficit in these cortical processing pathways could result in the misinterpretation of sensory information known to be a hallmark of panic disorder, leading to an inappropriate activation of the "fear network" via misguided excitatory input to the amygdala. It is thus conceivable that the misinterpretation of sensory information, which in clinical practice can be represented by prodromal health anxiety/hypochondriacal beliefs and fear, may kindle the fear network inducing a panic attack.

In this framework, Gorman et al. [56] also hypothesised the possible neurobiological mechanisms of Selective Serotonin Reuptake Inhibitors (SSRIs) as leading pharmacological treatment of PD. They observed that serotonergic neurons originate in the brainstem raphe region and project widely throughout the entire central nervous system. Three of these projections are of particular relevance to an understanding of the SSRI antipanic effect. First, the projection of serotonin (5-HT) neurons to the locus coeruleus is generally inhibitory, such that the greater the activity of the serotonergic neurons in the raphe, the smaller the activity of the noradrenergic neurons in the locus coeruleus. This suggests that SSRIs, by increasing serotonergic activity in the brain, have a secondary effect of decreasing noradrenergic activity leading to a decrease in many of the cardiovascular symptoms associated with panic attacks, including tachycardia and increased diastolic blood pressure. Second, the projection of the raphe neurons to the periaqueductal gray region appears to modify defense/escape behaviors. Thus, the serotonergic projections from the

dorsal raphe nuclei play a role in modifying defense/escape responses by means of their inhibitory influence on the periaqueductal gray region. Third, long-term treatment with SSRIs may reduce hypothalamic release of corticotropin-releasing factor (CRF). CRF, which initiates the cascade of events that leads to adrenal cortical production of cortisol, is also a neurotransmitter in the central nervous system and has been shown to increase fear. According to Gorman et al. [56] equally intriguing was the possibility that SSRIs, by increasing serotonergic activity, have an effect on the central nucleus of the amygdala itself, this may be a prime site for the anxiolytic action of the SSRIs, whereby an increase in 5-HT inhibits excitatory cortical and thalamic inputs from activating the amygdala. In addition to their psychic effects, drugs such as SSRIs may eliminate most of the troubling physical effects that may occur during panic by affecting heart rate, blood pressure, breathing rate, and glucocorticoid release. This would lead to a secondary decrease in anticipatory anxiety as a patient recognizes that the seemingly life-threatening physical manifestations of panic have been blocked. It is not uncommon, particularly in the early stages of medication treatment of a patient with PD, to hear "I sometimes feel as if the attack is coming on but then nothing happens. My thoughts don't seem to be able to cause a panic attack anymore". It is thus conceivable that SSRIs induce a rollback phenomenon in which pharmacological treatments first reduce the severity and frequency of panic attacks and then, indirectly, ameliorate anticipatory anxiety and avoidance.

In 2005, Ninan et al. [57] interestingly matched the sequence of events in a prototypical panic attack and what follows with potential neurobiological alterations. They made the example of a young lady experiencing an unexpected panic attack while driving on the highway. Some confluence of events triggers the amygdala, the central command switch and activates a fixed action patterns of responses in her brain and body. Thus, she pulls over the side of the road paralyzed with fear and, after a few minutes when the worst is over, she gathers up her courage and slowly drives to the safety of her home.

The terrifying experience leaves her an emotion memory (i.e., a strengthening of synapses in the lateral nucleus of the amygdala that represents the experience). Subsequent experiences, either anticipated or actual, that match components of that emotional memory, now trigger the anxiety response. The conventional memory system also remembers the panic attack. Explicit memory involves the hippocampus which is crucial for the autobiographical memory of the attack. Moreover, the involvement of hippocampus has also a role in recording the context in which the attack occurred. Thus, for instance, the highway is now associated with panic attack and driving may elevate the risk of a further panic attack. Although not previously connected with fear, driving is now associated with vigilance, anxiety, arousal. The anticipation of highway drive may become dysphoric and the situation may be therefore avoided. The functional anatomy of such avoidance is the medial/orbital prefrontal cortex and its reciprocal connections with the amygdala. Excessive activation of the amygdala decreases prefrontal activity, which, in turn, reduces the inhibitory control of amygdala. Thus, the learning of new information that may counter the initial association is impaired and avoidance becomes lasting. Once again, this matched sequence of events of a panic attack and the corresponding neurobiological alterations might identify different stages of the development of PD which are the mirror of what happens during the rollback.

In 2008, Graeff and Del-Ben [58] further clarified the neuroanatomical model of panic focusing on the role of 5-HT as enhancer of inhibitory avoidance in the forebrain and inhibitor of the one-way escape in the midbrain periaqueductal gray (PAG). Indeed, experimental studies led to the association of escape with PD; functional neuroimaging show activation of the insula and upper brain stem (including the PAG), as well as deactivation of the Anterior Cingulated Cortex (ACC) during experimental panic attacks; and voxel-based morphometric analyses of brain magnetic resonance images suggest an increase of grey matter volume in the insula and upper brain stem, and a decrease in the ACC of panic

patients as compared to healthy controls. Since the insula and the ACC are thought to translate interoceptive stimulation into feeling, and panic patients overestimate bodily signals, they are a likely neural substrate of interoceptive supersensitivity, and a possible site of action of both drug and cognitive behavior therapy. As a complement, antidepressants seem also to prevent panic attacks by enhancing 5-HT inhibition in the PAG.

5.5   Psychological Development of Panic and Treatment

The staging model of psychiatric disorders has been strongly related to treatment in unipolar depression, bipolar disorder, and schizophrenia. On one hand, it has been observed that it allows to recognize a disorder early enough to treat it precociously, that is making early intervention and prevention. On the other hand, it has been observed that it may help in identifying possible therapeutic strategies for treatment resistant cases. Unfortunately, current therapeutic models for PD [59, 60] disregard staging as well as subclinical symptomatology. Yet, there should be more emphasis on treatment outcome, especially on long-term outcome because of the chronic nature of PD. Disappearance of residual symptoms, upon abatement of panic attacks, should be the final target of therapy since they constitute a substantial risk of relapse. Adequate treatments of enduring effects should become of paramount importance including, at least in some patients, long-lasting treatment, sequential, or stage-oriented combination of different therapeutic modalities.

In this framework, Shear et al. [61] proposed to use a better system to monitor the patient's progress because, if residual symptoms are identified clearly and the progress of treatment can be mapped, clinicians may appropriately increase or possibly decrease medication dosage.

For a long time, the literature has suggested to increase the rates of remission combining pharmacological and psychological treatments. More recently, evidence seems not to strongly agree with it. While in the 1990s antidepressants plus exposure in vivo were proposed to treat PD [62]; in 2000s, the combination of psychotherapy and antidepressant showed higher effectiveness than antidepressant alone but no differences over psychotherapy alone [63]. Similarly, psychotherapy plus benzodiazepines did not provide a significant advantage over psychotherapy alone [64]. Thus, for those PD patients who have access to appropriate behavior therapy services, psychotherapy alone might be the most favorable intervention. In terms of differential effectiveness of various forms of psychotherapy, only behavior therapy and cognitive-behavior therapy were homogeneously effective while brief psychodynamic and cognitive-interpersonal therapy produced mixed results [63].

When appropriate behavior therapy services are not available or when the patient does not want to engage in a psychological path, pharmacological treatment remains the only option. In this case, the choice of the treatment should be done on the basis of a flourishing literature which strongly suggests that there is no adequate evidence of SSRIs higher effectiveness if compared to tricyclic antidepressants or benzodiazepines [65, 66]. Benzodiazepine, in particular clonazepam, diazepam, alprazolam, lorazepam [67], are a possible first-line treatment in a PD patient who do not have a comorbid depressive disorder and are not prone to addiction [67, 68]. However, it should be noted that alprazolam has been related to continuous and high-dose use; thus, it should be carefully used or simply avoided [69]

Different treatment approaches have been also proposed. Some authors evaluated the effects of a sequential treatment, that is a planned sequential administration of different therapies based on specific effects induced by each therapy that provide additional benefits in the course of time.

Goldstein [70] administered a 8-week treatment program, starting with alprazolam and switching gradually to imipramine, to 6 PD patients. Five completed the treatment program and, among them, 4 had no panic attacks by the end of the first week of treatment and maintained the improvement throughout the shift to imipramine. Mavissakalian [71] proposed 8 weeks of treatment with imipramine followed by 8 weeks

of treatment with imipramine combined with behavior therapy to 38 PD with agoraphobia patients. Of them, 63% responded markedly to the sequential treatment; most of the improvement in panic occurred during the first 8 weeks, whereas improvement in severity, anxiety, depression, and phobias, continued to be significant between mid-treatment and end of study. Further analyses revealed that improvement in phobic anxiety and avoidance in the first 8 weeks of treatment, rather than improvement in panic, predicted the final outcome. De Beurs et al. [72] compared fluvoxamine plus exposure, psychological panic management plus exposure, and exposure alone and found that the combination of fluvoxamine and exposure demonstrated efficacy superior than that of other treatments at the end of the trial. However, these advantages faded at a 2-year naturalistic follow up [73].

Finally, 63 patients with a primary diagnosis of PD who had residual symptoms (i.e., panic attacks, anticipatory anxiety, phobic avoidance) despite being on a stable dose of medications for at least 4 months, were treated with CBT using a group format by means of 12 sessions over 4 months. Significant reductions in symptoms were evident for all outcome measures (i.e., frequency of panic attacks, agoraphobia, anticipatory anxiety) across treatment, with maintenance of these gains at 1-year follow-up. At least a 50% reduction in symptoms was achieved by 78% of the sample for ratings of agoraphobia, 62% for anticipatory anxiety, and 49% for the Hamilton Anxiety Scale score. Fully 81% of the sample achieved a panic-free status, and 64% met criteria for remission. The presence of dysthymia, generalized anxiety disorder, or social phobia at pre-treatment was associated with a lower likelihood of remission [74]. This study provides further evidence for the efficacy of CBT as a next-step strategy for patients who fail to respond adequately to pharmacotherapy for PD. The improvement was also maintained despite overall reductions in medication use, indicating that CBT can be used as a strategy for medication discontinuation, with longer-term maintenance of treatment gains. Indeed, persistent post-withdrawal disorders induced, for instance, by

Table 5.3 Staging of levels of treatment resistance in panic disorder according to Cosci and Fava [12]

Stages	
0	No history of failure to respond to therapeutic trial
1	Failure of at least one adequate therapeutic trial (either pharmacological or psychological)
2	Failure of at least two adequate therapeutic trials, including at least one psychological
3	Failure of three or more adequate therapeutic trials, including at least one concerned with psychotherapy
4	Failure of three or more adequate therapeutic (either pharmacological or psychological) trials, including at least one concerned with psychotherapy/pharmacotherapy combination

paroxetine can be successfully treated with a specific cognitive behavioral therapy which includes explanatory therapy, monitoring of emergent symptoms, homework exposure for avoidance patterns, lifestyle modifications, techniques of decreasing abnormal reactivity to the social environment, and well-being therapy [75].

An interesting proposal of stage-oriented therapy comes from Fava et al. [76]. They administered well-being therapy or CBT of residual symptoms to three PDA outpatients, one major depressive disorder, four social phobia, one generalized anxiety disorder, and one obsessive compulsive disorder patient. Both treatments were associated with decrease in residual symptoms but a significant advantage of well-being therapy over cognitive behavioral strategies was observed when the residual symptoms at the second assessment (after treatment) of the two groups were compared with the initial measurements as covariates [76]. This study has obvious limitations, nonetheless, it provides important clinical insights since a novel, and specific psychotherapeutic technique addressed to increasing well-being [77] was found to be significantly associated with a decrease in residual symptoms in patients with mood and anxiety disorders. More recently, a treatment-resistant panic disorder patient having difficulties in automatic thought identification with the distress oriented approach of cognitive therapy, showed to improve when the wellbeing therapy was administered [78].

Finally, considering that treatment resistance in PD is a growing and emerging issue [79], a staging levels of treatment resistance was proposed by Cosci and Fava [12] (see Table 5.3).

5.6 Conclusions

Despite the relative paucity of research on psychological development of PD, the reports summarized in this chapter address important clinical issues that deserve further study.

The prodromal period seems to be about 10–15 years long, the perception that one is a "nervous person" [16] or the presence of specific personality characteristics are strong predictors of the onset of panic disorder. The most common prodromes are depressed mood, illness phobia, distress and avoidance of closed spaces, excessive worries, negative affectivity, anxiety sensitivity, health anxiety or fear of disease, separation anxiety. The phenomenological clinical sequence of PD with agoraphobia is: phobic avoidance and hypochondriasis leading to panic, which, in turn, leads to more phobic avoidance and hypochondriasis. On the other hand, the rollback phenomenon is: a decrease in avoidance by exposure, which improves agoraphobia and panic, with eventual disappearance of panic, whereas agoraphobia persists although to a less degree. Prodromal symptoms of PD with agoraphobia thus tend to become residual symptoms which, in turn, may progress to prodromal symptoms of relapse [40]. Alternatively, the rollback might be characterized by anxiety elicited by bodily sensations which influences catastrophic beliefs which, in turn, influence avoidant behavior [43].

The translation of staging in the neurobiology of panic identifies different phases in the development of PD. They are the mirror of the steps observed during the rollback. The experience of a panic attack triggers the amygdala which activates patterns of responses in the brain and body. The terrifying experience leaves an emotion memory because of which subsequent anticipated or actual experience, that match components of that emotional memory, will trigger the anxiety response. The hippocampus is also involved and records the context in which the attack occurred so that some details of the threatening experience are now associated with panic attack. The anticipation of such details may become dysphoric, and the situation may be, therefore, avoided. The functional anatomy of such avoidance is the medial/orbital prefrontal cortex and its reciprocal connections with the amygdala [57].

The treatment implications of the longitudinal model of PD, although still too disregarded, emphasize the importance to consider residual symptoms as the final target of the therapy and offer adequate treatments of enduring effects. In this regard, psychotherapy (mainly cognitive-behavioral therapy) [63] has shown the highest effectiveness and sequential or stage-oriented treatments have shown promising results.

References

1. Feinstein AR. Clinimetrics. New Haven: Yale University Press; 1987.
2. Feinstein AR. Basic biomedical science and the destruction of the pathophysiological bridge from bench to bedside. Am J Med. 1999;107:461–7. doi:10.1016/S0002-9343(99)00264-8.
3. Fava GA, Tomba E. New modalities of assessment and treatment planning in depression: the sequential approach. CNS Drugs. 2010;24:453–65. doi:10.2165/11531580-000000000-00000.
4. Cosci F, Fava GA. New clinical strategies of assessment of comorbidity associated with substance use disorders. Clin Psychol Rev. 2011;31:418–27. doi:10.1016/j.cpr.2010.11.004.
5. Fava GA, Tomba E, Sonino N. Clinimetrics: the science of clinical measurements. Int J Clin Pract. 2012;66:11–5. doi:10.1111/j.1742-1241.2011.02825.x.
6. Tossani E. The concept of mental pain. Psychother Psychosom. 2013;82:67–73. doi:10.1159/000343003.
7. Fava GA, Rafanelli C, Tomba E. The clinical process in psychiatry: a clinimetric approach. J Clin Psychiatry. 2012;73(2):177–84. doi:10.4088/JCP.10r06444.
8. Bech P. Fifty years with the Hamilton scales for anxiety and depression. A tribute to Max Hamilton. Psychother Psychosom. 2009;78:202–11. doi:10.1159/000214441.
9. Fava GA, Kellner R. Staging: a neglected dimension in psychiatric classification. Acta Psychiatr Scand. 1993;87:225–30.   doi:10.1111/j.1600-0447.1993.tb03362.x.
10. Sheehan DV, Sheehan KH. The classification of anxiety and hysterical states. Part II. Toward a more heuristic classification. J Clin Psychopharmacol. 1982;2:386–93.
11. Fava GA, Rafanelli C, Tossani E, Grandi S. Agoraphobia is a disease: a tribute to Sir Martin Roth. Psychother Psychosom. 2008;77:133–8. doi:10.1159/000116606.
12. Cosci F, Fava GA. Staging of mental disorders: systematic review. Psychother Psychosom. 2013;82:20–34. doi:10.1159/000342243.
13. Kessler RC, Chiu WT, Jin R, Ruscio AM, Shear K, Walters EE. The epidemiology of panic attacks, panic disorder, and agoraphobia in the National Comorbidity Survey Replication. Arch Gen Psychiatry. 2006;63:415–24. doi:10.1001/archpsyc.63.4.415.
14. American Psychiatric Association. Diagnostic and statistical manual of mental disorders (DSM IV). 4th ed. Washington: APA; 1994.
15. Roy-Byrne PP, Cowley DS. Course and outcome in panic disorder: a review of recent follow-up studies. Anxiety. 1994–1995;1:151–60. doi:10.1002/anxi.3070010402.
16. Eaton WW, Badawi M, Melton B. Prodromes and precursors: epidemiologic data for primary prevention of disorders with slow onset. Am J Psychiatry. 1995;152:967–72.
17. Keller MB, Yonkers KA, Warshaw MG, Pratt LA, Gollan JK, Massion AO, et al. Remission and relapse in subjects with panic disorder and panic with agoraphobia: a prospective short-interval naturalistic follow-up. J Nerv Ment Dis. 1994;182:290–6. doi:10.1097/00005053-199405000-00007.
18. O'Rourke D, Fahy TJ, Brophy J, Prescott P. The Galway Study of Panic Disorder: III. Outcome at 5 to 6 years. Br J Psychiatry. 1996;168:462–9. doi:10.1192/bjp.168.4.462.
19. Cowley DS, Flick SN, Roy-Byrne PP. Long-term course and outcome in panic disorder: a naturalistic follow-up study. Anxiety. 1996;2:13–21. doi:10.1002/(sici)1522-7154(1996)2:1<13::aid-anxi2>3.0.co;2-e.
20. Fava GA, Rafanelli C, Grandi S, Conti S, Ruini C, Mangelli L, et al. Long-term outcome of panic disorder with agoraphobia treated by exposure. Psychol Med. 2001;31:891–8.
21. Fava GA. Subclinical symptoms in mood disorders: pathophysiological and therapeutic implications. Psychol Med. 1999;29:47–61.
22. Detre TP, Jarecki HG. Modern psychiatric treatment. Philadelphia: Lippincott; 1971.
23. Fava GA, Kellner R. Prodromal symptoms in affective disorder. Am J Psychiatry. 1991;148:823–30.
24. Fava GA, Mangelli L. Subclinical symptoms of panic disorder: new insights into pathophysiology and treatment. Psychother Psychosom. 1999;68:281–9. doi:10.1159/000012345.
25. Fava GA, Tomba E, Grandi S. The road to recovery from depression—don't drive today with yesterday's map. Psychother Psychosom. 2007;76:260–5. doi:10.1159/000104701.
26. Klein DF. Anxiety reconceptualized. In: Klein DF, Rabkin J, editors. Anxiety: new research and changing concepts. New York: Raven; 1981. p. 235–63.

27. Garvey MJ, Tuason VB. The relationship of panic disorder to agoraphobia. Compr Psychiatry. 1984; 25(5):529–31. doi:10.1016/0010-440X(84)90052-X.
28. Wittchen HU, Nocon A, Beesdo K, Pine DS, Hofler M, Lieb R, et al. Agoraphobia and panic. Prospective-longitudinal relations suggest a rethinking of diagnostic concepts. Psychother Psychosom. 2008;77:147–57. doi:10.1159/000116608.
29. Faravelli C, Cosci F, Rotella F, Faravelli L, Catena Dell'osso M. Agoraphobia between panic and phobias: clinical epidemiology from the Sesto Fiorentino Study. Compr Psychiatry. 2008;49:283–7. doi: 10.1016/j.comppsych.2007.12.001.
30. Wittchen HU, Gloster AT, Beesdo-Baum K, Fava GA, Craske MG. Agoraphobia: a review of the diagnostic classificatory position and criteria. Depress Anxiety. 2010;27:113–33. doi:10.1002/da.20646.
31. American Psychiatric Association. Diagnostic and statistical manual of mental disorders. 5th ed. Arlington: American Psychiatric Publishing; 2013.
32. Fava GA, Grandi S, Canestrari R. Prodromal symptoms in panic disorder with agoraphobia. Am J Psychiatry. 1988;145:1564–7.
33. Fava GA, Grandi S, Rafanelli C, Canestrari R. Prodromal symptoms in panic disorder with agoraphobia: a replication study. J Affect Disord. 1992; 26:85–8.
34. Rudaz M, Craske MG, Becker ES, Ledermann T, Margraf J. Health anxiety and fear of fear in panic disorder and agoraphobia vs. social phobia: a prospective longitudinal study. Depress Anxiety. 2010;27: 404–11. doi:10.1002/da.20645.
35. Hayward C, Killen JD, Kraemer HC, Taylor CB. Predictors of panic attacks in adolescents. J Am Acad Child Adolesc Psychiatry. 2000;39:207–14. doi:10.1097/00004583-200002000-00021.
36. Lelliott P, Marks I, McNamee G, Tobeña A. Onset of panic disorder with agoraphobia. Toward an integrated model. Arch Gen Psychiatry. 1989;46:1000–4. doi:10.1001/archpsyc.1989.01810110042006.
37. Fava GA, Rafanelli C, Ottolini F, Ruini C, Cazzaro M, Grandi S. Psychological well-being and residual symptoms in remitted patients with panic disorder and agoraphobia. J Affect Disord. 2001;65:185–90.
38. Corominas A, Guerrero T, Vallejo J. Residual symptoms and comorbidity in panic disorder. Eur Psychiatry. 2002;17:399–406. doi:10.1016/S0924-9338 (02)00693-4.
39. Marchesi C, De Panfilis C, Cantoni A, Giannelli MR, Maggini C. Effect of pharmacological treatment on temperament and character in panic disorder. Psychiatry Res. 2008;158:147–54. doi:10.1016/j. psychres.2006.08.009.
40. Fava GA, Grandi S, Canestrari R, Grasso P, Pesarin F. Mechanisms of change of panic attacks with exposure treatment of agoraphobia. J Affect Disord. 1991;22:65–71.
41. Stanley MA, Beck JG, Averill PM, Baldwin LE, Deagle 3rd EA, Stadler JG. Patterns of change during cognitive behavioral treatment for panic disorder. J Nerv Ment Dis. 1996;184:567–72.
42. Noda Y, Nakano Y, Lee K, Ogawa S, Kinoshita Y, Funayama T, et al. Sensitization of catastrophic cognition in cognitive-behavioral therapy for panic disorder. BMC Psychiatry. 2007;7:70. doi:10.1186/ 1471-244X-7-70.
43. Hoffart A, Sexton H, Hedley LM, Martinsen EW. Mechanisms of change in cognitive therapy for panic disorder with agoraphobia. J Behav Ther Exp Psychiatry. 2008;39:262–75. doi:10.1016/j.jbtep.2007.07.006.
44. Rifkin A. The sequence of improvement of the symptoms encountered in patients with panic disorder. Compr Psychiatry. 1991;32:559–60.
45. Deltito JA, Argyle N, Buller R, Nutzinger D, Ottosson JO, Brandon S, et al. The sequence of improvement of the symptoms encountered in patients with panic disorder. Compr Psychiatry. 1991;32:120–9. doi:10.1016/0010-440X(91)90003-U.
46. Mavissakalian MR. Phenomenology of panic attacks: responsiveness of individual symptoms to imipramine. J Clin Psychopharmacol. 1996;16:233–7.
47. Liebowitz MR, Asnis G, Mangano R, Tzanis E. A double-blind, placebo-controlled, parallel-group, flexible-dose study of venlafaxine extended release capsules in adult outpatients with panic disorder. J Clin Psychiatry. 2009;70:550–61. doi:10.4088/ JCP.08m04238.
48. Bouchard S, Gauthier J, Nouwen A, Ivers H, Vallières A, Simard S, et al. Temporal relationship between dysfunctional beliefs, self-efficacy and panic apprehension in the treatment of panic disorder with agoraphobia. J Behav Ther Exp Psychiatry. 2007;38:275–92. doi:10.1016/j.jbtep.2006.08.002.
49. Fava GA, Freyberger HJ, Bech P, Christodoulou G, Sensky T, Theorell T, et al. Diagnostic criteria for use in psychosomatic research. Psychother Psychosom. 1995;63:1–8. doi:10.1159/000288931.
50. Wise TH. Diagnostic criteria for psychosomatic research are necessary for DSM V. Psychother Psychosom. 2009;78:330–2. doi:10.1159/000235735.
51. Fava GA, Wise TN. Psychological factors affecting either identified or feared medical conditions: a solution for somatoform disorders. Am J Psychiatry. 2007;164:1002–3.
52. Fava GA, Porcelli P, Rafanelli C, Mangelli L, Grandi S. The spectrum of anxiety disorders in the medically ill. J Clin Psychiatry. 2010;71:910–4. doi:10.4088/ JCP.10m06000blu.
53. Porcelli P, Bellomo A, Quartesan R, Altamura M, Iuso S, Ciannameo I, et al. Psychosocial functioning in consultation-liaison psychiatry patients: influence of psychosomatic syndromes, psychopathology and somatization. Psychother Psychosom. 2009;78:352–8. doi:10.1159/000235739.
54. Ferrari S, Galeazzi GM, Mackinnon A, Rigatelli M. Frequent attenders in primary care: impact of medical, psychiatric and psychosomatic diagnoses. Psychother Psychosom. 2008;77:306–14. doi:10.1159/ 000142523.
55. Gorman JM, Liebowitz MR, Fyer AJ, Stein J. A neuroanatomical hypothesis for panic disorder. Am J Psychiatry. 1989;146:148–61.

56. Gorman JM, Kent JM, Sullivan GM, Coplan JD. Neuroanatomical hypothesis of panic disorder, revised. Am J Psychiatry. 2000;157:493–505.
57. Ninan PT, Dunlop BW. Neurobiology and etiology of panic disorder. J Clin Psychiatry. 2005;66:3–7.
58. Graeff FG, Del-Ben CM. Neurobiology of panic disorder: from animal models to brain neuroimaging. Neurosci Biobehav Rev. 2008;32:1326–35. doi:10.1016/j.neubiorev.2008.05.017.
59. American Psychiatric Association. Practice guideline for the treatment of patients with panic disorder. 2nd ed. Washington: APA; 2009.
60. Royal Australian and New Zealand College of Psychiatrists Clinical Practice Guidelines Team for Panic Disorder and Agoraphobia. Australian and New Zealand clinical practice guidelines for the treatment of panic disorder and agoraphobia. Aust N Z J Psychiatry. 2003;37:641–56. doi:10.1111/j.1440-1614.2003.01254.x.
61. Shear MK, Clark D, Feske U. The road to recovery in panic disorder: response, remission, and relapse. J Clin Psychiatry. 1998;8:4–8.
62. van Balkom AJ, Bakker A, Spinhoven P, Blaauw BM, Smeenk S, Ruesink B. A meta-analysis of the treatment of panic disorder with or without agoraphobia: a comparison of psychopharmacological, cognitive-behavioral, and combination treatments. J Nerv Ment Dis. 1997;185:510–6.
63. Furukawa TA, Watanabe N, Churchill R. Combined psychotherapy plus antidepressants for panic disorder with or without agoraphobia. Cochrane Database Syst Rev. 2007;1:CD004364. doi:10.1002/14651858.CD004364.pub2.
64. Watanabe N, Churchill R, Furukawa TA. Combined psychotherapy plus benzodiazepines for panic disorder. Cochrane Database Syst Rev. 2009;1:CD005335. doi:10.1002/14651858.CD005335.pub2.
65. Berney P, Halperin D, Tango R, Daeniker-Dayer I, Schulz P. A major change of prescribing pattern in absence of adequate evidence: benzodiazepines versus newer antidepressants in anxiety disorders. Psychopharmacol Bull. 2008;41:39–47.
66. Offidani E, Guidi J, Tomba E, Fava GA. Efficacy and tolerability of benzodiazepines versus antidepressants in anxiety disorders: a systematic review and meta-analysis. Psychother Psychosom. 2013;82:355–62. doi:10.1159/000353198.
67. Freire RC, Machado S, Arias-Carrión O, Nardi AE. Current pharmacological interventions in panic disorder. CNS Neurol Disord Drug Targets. 2014;13:1057–65. doi:10.2174/1871527313666140612125028.
68. Starcevic V. The reappraisal of benzodiazepines in the treatment of anxiety and related disorders. Expert Rev Neurother. 2014;14:1275–86. doi:10.1586/14737175.2014.963057.
69. Cosci F, Guidi J, Balon R, Fava GA. Clinical Methodology Matters in Epidemiology: Not All Benzodiazepines Are the Same. Psychother Psychosom. 2015;84(5):262–4. doi:10.1159/000437201.
70. Goldstein S. Sequential treatment of panic disorder with alprazolam and imipramine. Am J Psychiatry. 1986;143:1634.
71. Mavissakalian M. Sequential combination of imipramine and self-directed exposure in the treatment of panic disorder with agoraphobia. J Clin Psychiatry. 1990;51:184–8.
72. de Beurs E, van Balkom AJ, Lange A, Koele P, van Dyck R. Treatment of panic disorder with agoraphobia: comparison of fluvoxamine, placebo, and psychological panic management combined with exposure and of exposure in vivo alone. Am J Psychiatry. 1995;152:683–91.
73. de Beurs E, van Balkom AJ, Van Dyck R, Lange A. Long-term outcome of pharmacological and psychological treatment for panic disorder with agoraphobia: a 2-year naturalistic follow-up. Acta Psychiatr Scand. 1999;99:59–67. doi:10.1111/j.1600-0447.1999.tb05385.x.
74. Heldt E, Manfro GG, Kipper L, Blaya C, Isolan L, Otto MW. One-year follow-up of pharmacotherapy-resistant patients with panic disorder treated with cognitive-behavior therapy: outcome and predictors of remission. Behav Res Ther. 2006;44:657–65. doi:10.1016/j.brat.2005.05.003.
75. Belaise C, Gatti A, Chouinard VA, Chouinard G. Persistent postwithdrawal disorders induced by paroxetine, a selective serotonin reuptake inhibitor, and treated with specific cognitive behavioral therapy. Psychother Psychosom. 2014;83:247–8. doi:10.1159/000362317.
76. Fava GA, Rafanelli C, Cazzaro M, Conti S, Grandi S. Well-being therapy. A novel psychotherapeutic approach for residual symptoms of affective disorders. Psychol Med. 1998;28:475–80.
77. Ryff CD, Singer B. Psychological well-being: meaning, measurement, and implications for psychotherapy research. Psychother Psychosom. 1996;65:14–23.
78. Cosci F. Well-Being Therapy in a Patient with Panic Disorder Who Failed to Respond to Paroxetine and Cognitive Behavior Therapy. Psychother Psychosom. 2015;84(5):318–9. doi:10.1159/000430789.
79. Sanderson WC, Bruce TJ. Causes and management of treatment-resistant panic disorder and agoraphobia: a survey of expert therapists. Cogn Behav Pract. 2007;14:26–35. doi:10.1016/j.cbpra.2006.04.020.

Panic Disorder Respiratory Subtype

6

Morena Mourao Zugliani,
Rafael Christophe R. Freire,
and Antonio Egidio Nardi

Contents

M.M. Zugliani • R.C.R. Freire (✉) • A.E. Nardi
Laboratory of Panic and Respiration, Institute of
Psychiatry, Federal University of Rio de Janeiro,
Rio de Janeiro, Brazil
e-mail: morezugli@gmail.com;
rafaelcrfreire@gmail.com;
antonioenardi@gmail.com

Abstract

Research into panic disorder (PD) has long indicated the possibility of distinct PD subtypes, such as respiratory, cardiovascular or gastrointestinal, based on the symptoms experienced during a panic attack (PA). In this chapter, we will elaborate on the new developments concerning the respiratory subtype (RS) of panic disorder (PD) since its first description, presenting psychopathological features, diagnostic criteria, genetic and physiopathological hypotheses, as well as therapeutic and prognostic characteristics. Evidence drawn from the available literature indicates a greater incidence of family history of PD, as well as higher comorbidity rates for disorders of anxiety and depression, among RS patients in comparison with patients of the non-respiratory subtype (NRS). RS patients were also more sensitive to CO_2, hyperventilation and caffeine. These patients were clearly distinguished from the NRS patients by certain characteristics, such as the heightened sensitivity to CO_2 and the higher incidence of a family history of PD. Nonetheless, it was not possible to demonstrate differential responses to pharmacological treatment and cognitive behavioral therapy across the subtypes. RS patients seem to respond faster than NRS to pharmacological treatment with antidepressants and

© Springer International Publishing Switzerland 2016
A.E. Nardi, R.C.R. Freire (eds.), *Panic Disorder*, DOI 10.1007/978-3-319-12538-1_6

benzodiazepines, but more studies are needed to confirm this finding.

Keywords

Caffeine • Hypercapnia • Comorbidity • Hyperventilation • Nocturnal panic • Dyspnea • Respiratory diseases • Respiration

6.1 Introduction

Panic Disorder (PD) can be defined as experiencing unexpected recurring episodes of panic attacks (PA) along with concerns about having other PA, whereby the subject often worries about the consequences of PA or suffers significant behavioral changes related to the PA [1]. According to the DSM-5, a panic attack may be expected or unexpected, and is defined as being a sudden episode of intense anxiety and fear with a number of symptoms [1]. However, the heterogeneity among panic symptoms indicates that there may be distinct PD subtypes [2, 3] such as respiratory, cardiovascular or gastrointestinal; as proposed by Klein [4] and others [5], based on the most prominent symptoms occurring during a typical PA. Distinct groups of symptoms suffered by PD patients may have an association with specific clinical courses, sensitivity to respiratory tests, and efficacy of pharmacological treatment [6]. Moreover, the diverse clinical presentations of PD may reflect the distinct pathways producing PD [7].

Although there are a number of similarities between PA and common fear reactions, there are also distinct psychopathological and neurobiological differences [8]. For instance, in PA there is a marked feeling of air hunger, which is not normally associated with fear reactions that result of external danger. Another important difference, demonstrated clinically and in PD challenge studies [8], is that PA fail to cause activation of the HPA (hypothalamic-pituitary-adrenal) axis. Furthermore, a great deal of research has been done on the connection between panic disorder and the respiratory system [4, 9–11]. Some of these studies have indicated that PD patients may suffer from subclinical respiratory abnormalities [11–14], as well as the fact that patients with

respiratory diseases are prone to developing panic disorder and agoraphobia [15–17]. There are also findings which point to a link between abnormalities in the respiratory control centers and the physiopathology of PD [4, 18]. Klein [4] proposed that spontaneous PA happen when a lack of useful air is signaled by the brain suffocation monitor, thereby acting as a hyper-sensitive alarm system against suffocation. However, such a dysfunction would cause an individual to be susceptible to PA as episodes of "false suffocation alarms" [4]. Several studies indicate that patients suffering distinct respiratory symptoms differ in their responses to respiratory and non-respiratory challenges, when compared to PD patients without such prominent symptoms [19–22].

In the studies of Briggs et al. [6], Roberson-Nay et al. [2, 3] and other authors [7, 19] there is a great deal of evidence to indicate that patients with prominent respiratory symptoms during PA may represent a distinct PD subtype which is stable through time. Briggs and colleagues studied 1034 PD patients' accounts of their most recent severe PA [6] and separated the patients into two groups, based on the presence of prominent respiratory symptoms. Patients considered as being RS showed at least four of the following symptoms: Fear of death, pain/discomfort in the chest, breathlessness, paresthesias, and the sensation of choking. Those considered as suffering more natural PA, who also seemed to respond better to antidepressants, belonged to the RS group. Those having more situational PA, who showed a better response to benzodiazepines, were part of the NRS group [6].

In this chapter, the psychopathology, demographic features, clinical features, psychological factors, neurobiological factors and treatment of the respiratory subtype were reviewed. The most relevant findings regarding the RS since its first description in 1993 have been summarized [6].

6.2 Psychopathology

In 1993, Briggs et al. [6] studied the psychopathology of PD and discovered a frequent incidence of respiratory symptoms. A principal component analysis separated two symptom groups: the non-respiratory symptom group, which included 8 PD symptoms; and the respiratory symptom group, which included fear of death, breathlessness, paresthesias, pain/discomfort in the chest and a sensation of choking. The NRS group incorporated most symptoms of PD and an overall severity factor was taken into consideration. A different PD subtype was determined by the symptoms from the second group, and another subtype was thus defined through the absence of such symptoms. When at least four of the five symptoms from the respiratory symptom group were present, the RS was defined; otherwise the RS was absent [6]. Bandelow et al. [23] performed an oblique principal component cluster analysis on a sample of 330 PD patients from 14 centers in six different countries and his findings were almost identical to those of Briggs et al. [6]. It was shown that there was a latent tendency towards symptoms such as fear of death, pain/discomfort in the chest, tingling or numbness, dyspnea and choking/suffocation. Patients showing at least four of these symptoms suffered fewer situational PA and appeared to have episodes of a more spontaneous nature [23].

Factor and group analyses were performed by Shioiri et al. [24] among a Japanese sample of 207 PD patents and a distinct RS was not found. The respiratory symptoms were not together, they were divided into two different groups [24]. Also, in the study from Rees et al. [25] the five respiratory symptoms were not grouped together. The authors made a principal component analysis of a sample of 153 PD patients and identified five components of somatic symptoms. Component one included choking and shortness of breath, component five consisted of chest pain and numbness. The other PD somatic symptoms were distributed across the other three components [25]. Two studies [26, 27] have shown that cardio-respiratory symptoms were frequently present among Spanish PD patients. In both studies the factor analysis grouped together the symptoms breathlessness/dyspnea, fear of death and pain/discomfort in the chest, although in the study of Marquez et al. [27] the choking sensation was also included in this cluster. In the study by Segui et al. [26] trembling, palpitations and paresthesias were added to the cardio-respiratory cluster along with fear of dying, chest pain and dyspnea.

Three factors were recently identified in a later study [28] by factor analysis: the cognitive, with fear of going mad or suffering loss of control and derealization; the autonomic/somatic, including dizziness, perspiration, tremors, nausea, chills/hot flushes; and cardio-respiratory, involving palpitations, pain in the chest, breathlessness, choking, sensations of numbness and fear of death. The cardio-respiratory dimension was found to be associated with agoraphobic avoidance and severity of panic in multiple regression analyses, while those suffering from interference in daily life, mainly with panic preoccupation, were linked to autonomic/somatic and cognitive subscales [28].

A multi-centric study [29] conducted in Europe indicated than the lifetime prevalence of PA with "sensation of shortness of breath" (PASB) was 6.77%, while the lifetime prevalence of PA without this symptom was 3.14%, the 12-month prevalence of these PA were 2.26% and 1% respectively. The prevalence of PASB was significantly higher in Spain and Italy, compared to France, Belgium, Germany and the Netherlands. The PASB was also associated to "any chronic physical condition" and to high use of health services due to mental health problems [29].

Roberson-Nay and Kendler [2] examined panic symptoms across four samples from epidemiologic studies and one sample from a clinical study with the aim of determining if patients with PD have a tendency to co-vary within distinct subgroups as a function of symptomatic similarity. Examination of panic symptoms from the Epidemiological Catchment Area (ECA), Virginia Adult Twin Study of Psychiatric and Substance Use Disorders (VATSPSUD) and the Cross-National Collaborative Panic Study (CNCPS) revealed two very distinct groups. The first group was differentiated by its prominent respiratory symptoms as well as higher incidence across a number of other symptoms of PD, while the other group was characterized by its low endorsement rates in regard to respiratory symptoms of panic but high endorsement of non-respiratory.

The analysis of samples from the National Comorbidity Study (NCS) and the National Epidemiologic Survey on Alcohol and Related Conditions (NESARC) were also in agreement with the findings from Briggs et al. [6], with only slight differences [2].

With the exceptions of the studies from Bandelow et al. [23] and Roberson-Nay and Kendler [2], other studies involving factor analysis [24–28] failed to reproduce the precise respiratory symptom distinction described by Briggs et al. [6], although there were still many similarities. These disparities may occur as a result of uneven sampling or reduced numbers within the sample. A further possibility in relation to elucidation for this phenomenon is that certain factors such as cross-cultural samples or hereditary influences could have had an effect on the prevalence of panic symptoms in different samples, affecting the PD subtyping.

6.3 Demographic and Clinical Features

It has also been shown in studies of Demographic and clinical features that RS PD patients, according to the criteria of Briggs et al. [6] have specific demographics and clinical aspects.

Some research indicated that in PD patients in the RS there is a higher age of onset [19, 30], although other studies indicated the opposite [21, 31], while another recent study found no difference between the subtypes regarding age of onset [3]. Studies comparing the RS with the NRS have not found any significant difference in regard to gender, occupation, marital status or education [3, 21].

Several studies indicated a higher degree of family history of PD for RS patients, compared to NRS patients [21, 32, 33] (Table 6.1).

Previously, some studies indicated that the NRS had greater levels of comorbidity with major depression [32, 33] and found no differences between the subtypes regarding comorbidity with agoraphobia and other anxiety disorders [31, 34, 35]. A recent analysis of a large epidemiologic sample (NESARC) demonstrated that,

compared to NRS patients, RS patients showed greater comorbidity with agoraphobia, specific phobias, social or generalized anxiety disorder and major depression [3]. In the same study [3], an analysis of a clinical sample (CNCPS) identified only specific phobias and major depression as comorbidities associated with the RS (Table 6.1).

Higher rates of alcohol consumption [21] and cigarette smoking [19] for RS patients were found by some authors, when compared to the NRS, although other studies failed to show these disparities [3, 34–36]. Another study [37] recently discovered a negative connection between the RS and neuroticism, while other PD subtypes were not correlated to this personality trait.

It is not clear if the RS is associated only with spontaneous PA [6, 23], both spontaneous and situational PA [20] or if there are no differences between the subtypes regarding the types of PA [3].

There is also controversy as to whether or not the RS correlates with nocturnal panic attacks (NPA). These occur, usually after 2–3 h of sleep, when the subject suddenly wakes up with anxiety, a sensation of fear and physical symptoms [38]. During these attacks, marked respiratory symptoms such as breathlessness or dyspnea [38], chest pains, a sensation of choking, paresthesias and fear of death arise [39]. 49–69 % of PD sufferers experience this kind of episode [40–42]. It is therefore common for PD patients to develop anticipatory anxiety [39], sleep-onset insomnia and phobic avoidance of sleep secondary to the NPA [38]. Sarisoy et al. [40] also found patients with NPA showing marked respiratory symptoms. Said patients presented substantially higher levels of chest pain or distress, the sensation of choking, paresthesias, dizziness and fear of loss of control or going mad than those patients without NPA [40]. Despite evidence indicating a correlation between NPA and RS, two more recent research papers [41, 42] failed to demonstrate this correlation.

Compared to the NRS, patients in the RS exhibited higher scores on the Clinical Global Impression (CGI) scale [31], a lengthier period of illness, more severe panic and phobic symptoms [19].

Table 6.1 Clinical variations from the respiratory subtype to the non-respiratory subtype

Evidence type	RS	NRS	Reference
Family history of PD (%)	62.1–75.7	28.6–35.3	Freire et al. [21], Nardi et al. [32], Nardi et al. [33]
Comorbidity			
With agoraphobia (OR/95 % CI)	2.16 (1.51–3.11)**		Roberson-Nay et al. [3][a]
With GAD (OR/95 % CI)	3.31 (2.16–5.08)**		Roberson-Nay et al. [3][a]
With SAD (OR/95 % CI)	1.78 (1.29–2.49)**		Roberson-Nay et al. [3][a]
With specific phobia (OR/95 % CI)	1.88 (1.40–2.54)**		Roberson-Nay et al. [3][a]
With major depression (OR/95 % CI)	2.00 (1.16–3.45)*		Roberson-Nay et al. [3][a]
Scores on scales of PD severity			
Clinical Global Impression (median)	5**	4**	Valenca et al. [31]
Anxiety Sensitivity Index (mean/SD)	35.1 (13.2)*	29.5 (13.2)*	Onur et al. [34]
Panic-Agoraphobia Spectrum Scale (mean/SD)	65.2 (14.3)**	55.6 (17.2)**	Onur et al. [34]
WHOQOL[b] (mean/SD)	64.3 (15.2)*	48.1 (19.9)*	De-Melo-Neto [43]

PD panic disorder, *SAD* social anxiety disorder, *GAD* generalized anxiety disorder, *RS* respiratory subtype, *NRS* non-respiratory subtype, *OR* odd ratio, *95 % CI* 95 % confidence interval, *SD* standard deviation, *WHOQOL* World Health Organization Quality of Life scale
*P ≤ 0.05
** P ≤ 0.01
[a]Only the National Epidemiologic Survey on Alcohol and Related Conditions (NESARC) sample
[b]The difference between the two subtypes was only in the psychological domain

Onur et al.[34] detected no subtype distinctions in scores on the Panic Disorder Severity Scale (PDSS) and the Panic and Agoraphobia Scale (PAS). However, on the Anxiety Sensitivity Index (ASI) and the Panic-Agoraphobia Spectrum Scale (PAS-SR), patients in the RS still recorded higher scores. This dissimilarity between the subtypes lay among a number of domains of the PAS-SR, such as "panic-like symptoms", "agoraphobia", "reassurance orientation" and "separation sensitivity". These domains also appear to be statistically discriminative of RS and the NRS [34]. A study [43] using the World Health Organization Quality of Life Scale (WHOQOL) showed higher scores in the psychological domain for the RS, compared to NRS patients (Table 6.1).

Two epidemiological studies [3, 29] indicated that RS patients have a tendency to seek more psychosocial treatment and are often given prescription medicine for symptoms of panic. These findings may explain why there are so many differences between clinical sample and epidemiological studies regarding the RS.

6.4   Psychological Factors

Behavioral sensitization following near-drowning or suffocation incidents may have a significant role in the psychobiology of PD. Research indicates that 19.3–33 % of PD patients have a history of traumatic suffocation (TSH), while only 6.7 % of "normal" controls had TSH [44]. Near-drowning, torture or rape involving suffocation or choking were some of the many types of traumatic suffocation episode that were reported [44]. In comparing PD patients with and without this background, the authors found a higher prevalence of respiratory symptoms and nocturnal PA in those with TSH, and more symptoms of agoraphobia or cardiovascular symptoms in those patients without such a background [44]. In this comparison no contrasts were presented concerning childhood separation anxiety or familial history of PD [44]. It was discovered in another study that PD patients who had suffered torture suffocation in the past, showed a higher incidence of depression and respiratory symptoms than other patients [45]. TSH patients also presented more posttraumatic stress symptoms on CAPS, the Clinician-Administered Posttraumatic Stress Disorder Scale, than those without THS, although this was not statistically significant [45].

Bouwer and Stein [44] put forward the hypothesis that an actual suffocation experience could augment the sensitivity of the suffocation alarm, which would subsequently be more easily activated.

6.5   Biological Factors

6.5.1   Oxidative Stress and Inflammation

There is mounting evidence to indicate that oxidative stress is involved in the development of neuropsychiatric disorders, including schizophrenia, bipolar disorder, depression and PD [46]. The free radicals involved in oxidative stress have short half-lives but may be indirectly evaluated through measurement of the activity of certain antioxidant enzymes like superoxide dismutase (SOD) and catalase or glutathione peroxidase (GSH-Px). The study from Ozdemir et al. [46] indicated that SOD and GSH-Px activity was significantly lower in PD patients compared to healthy subjects, but there were no significant differences in the comparison between the RS and NRS. The authors concluded that there is a high level of oxidative stress in PD, regardless of the subtype [46].

There is also evidence of comorbidity between PD and immunological diseases such as asthma, in which there is an increase of T lymphocytes [46]. Adenosine deaminase (ADA) is an enzyme which is important to the maturation and function of T lymphocytes, as well as being an indicator of cellular immunity. It has been suggested that increased plasma activity of this enzyme occurs in diseases of an inflammatory nature involving a cell-mediated immune response. Ozdemir et al. [46] found higher ADA activity in PD patients in comparison with controls. The ADA activity was not significantly different when comparing RS and NRS. High ADA activity indicates that inflammatory processes are present in PD patients and may increase oxidative stress [46].

6.5.2   Genetics

Some studies [21, 32, 33] indicated that RS patients have a higher frequency of familial history of PD compared to NRS, suggesting that genetic factors may play a role in the physiopathology of respiratory PD.

It is also important to study the role of steroid hormones such as progesterone in PD patients with respiratory symptoms because they may have an effect on breathing [35]. Hormonal changes directly influence the regulation of breathing due to changes in the state of excitability of the respiratory center [35]. Progesterone influences ventilatory control while its impact on breath stimulation among healthy male subjects has also been demonstrated [35]. Two progesterone receptor isoforms (A and B) mediated all the principal actions of progesterone [35]. The progesterone receptor gene is located on chromosome 11q22-23, and consists of a number of polymorphic regions, including an ALU insertion polymorphism in intron 7 (PROGINS) and a single nucleotide polymorphism at position +331 (G331A) in the promoter region [35]. Pirildar et al. [35] compared RS patients with healthy controls and discovered a trend towards a statistically significant difference (P=0.06) in the PROGINS. The healthy subjects were significantly different from the RS and NRS patients regarding the G331A, but no differences were found in the comparison between the RS and NRS [35].

Currently there is insufficient data to determine what role genetics may play in the RS and further study is required.

6.5.3   Respiratory Dysfunction

Asmundson and Stein [47] carried out research on pulmonary function in patients with PD and did not find any respiratory damage, although patients with a low forced expiratory flow, at 50% of a forced expired vital capacity (FEF 50%), were different to patients with high FEF 50%, in regard to panic symptoms. Compared to patients who presented a high FEF 50%, during a PA, those with low FEF 50% had a higher number of strong respiratory and cognitive symptoms, such as breathlessness, a sensation of smothering, giddiness or feeling unsteady, lethargy or tingling sensations, fear of death, loss of control or going mad [47]. It was hypothesized that PD patients with marked respiratory and cognitive symptoms may characterize a different group of PD patients with early manifestation of obstructive pulmonary disease [47]. Pfaltz et al. [48] found no differences between PD patients and healthy subjects regarding respiratory frequency, volumes and irregularities in ambulatory monitoring. However, the severity of respiratory symptoms in PD patients was positively correlated with the breath time and the variability of the breath time [48]. There was also a negative correlation between the intensity of respiratory symptoms and an index of rapid shallow breathing [48].

Moynihan and Gevirtz [49] discovered that RS patients showed lower partial pressure of end-tidal CO_2 (PETCO$_2$) (35.14, SD=3.89 mmHg) compared to NRS (39.27, SD=3.33 mmHg) and healthy subjects (39.43, SD=2.72 mmHg). The PETCO$_2$ is a measure of the quantity of CO2 in exhaled air as an indication of the partial pressure of CO_2 in the arterial blood. The low PETCO$_2$ in RS patients indicates that these patients hyperventilate and eliminate more CO_2 than the subjects in the other two groups [49].

6.5.4   Diagnostic Challenge Tests

Biber and Alkin [19] found that single-breath 35% CO_2 inhalation induced PA in 79% of patients in the RS, while just 48% of those in the NRS suffered PA in this respiratory challenge. Valenca et al. [31] also found that 93.7% of the RS patients had PA during a double-breath 35% CO_2 inhalation, while only 43.4% of the NRS had PA, indicating increased sensitivity to CO_2

among RS patients. A study with a challenge of 5% carbon dioxide rebreathing [50], had higher numbers of RS patients terminating the procedure voluntarily, compared to the NRS patients. There was also higher respiratory frequency and a greater incidence of suffocation sensations in RS patients, than the other subtype [50]. The respiratory ratio, which is a dimensional construct based on the respiratory subtype, was also positively correlated to carbon dioxide sensitivity [51]. Freire et al. [51] found that the respiratory ratio could predict CO_2-induced PA with a sensitivity of 67.7% and a specificity of 65.5%, using a cutoff of 0.437 in the respiratory ratio score. The idea of two marked PD subtypes was reinforced by the differential responses to CO_2 and these findings are also compatible with the theory of false suffocation alarm [19]. Patients in the RS could be oversensitive to CO_2, and any trivial increase in levels of carbon dioxide may be readily misinterpreted as a lack of useful air, causing severe respiratory symptoms as well as other symptoms of PA (Table 6.2).

In a study by Nardi et al. [52] 62.0% of the patients who responded with PA to 35% CO_2 inhalation and breath holding test were in the RS group. Only 30.8% of the non-responders were RS patients [52]. In another study from the same group [53], patients who responded to the double-breath 35% CO_2 inhalation presented more respiratory symptoms – chest pain/discomfort, shortness of breath, paresthesias and feelings of choking – in a typical spontaneous PA compared to non-responders (Table 6.2).

In another study by Nardi et al. [20], PD patients were submitted to two panic provoking challenges with a 1-week interval: a double-breath 35% CO_2 inhalation and the ingestion of 480 mg of caffeine. In the CO_2 test, 61.4% of the patients suffered PA, while in the caffeine test only 45.8% had PA. Those who had PA in the caffeine challenge were also sensitive to CO_2, and 76.3% of these patients were in the RS group. Among patients who did not respond to either of the two tests only 37.5% were in the RS group [20] (Table 6.2).

It was demonstrated in two studies [22, 54] that of those who had PA in the hyperventilation

Table 6.2 Sensitivity to respiratory and non-respiratory challenge tests

Author	Year	N	N respiratory subtype	N non-respiratory subtype	Method	OR	95% CI lower	95% CI upper	P
Abrams	2006	33	10	23	Standardized 5% CO_2 rebreathing challenge	2.267	0.452	11.349	0.319
Biber	1999	51	28	23	Single breath 35% CO_2/65% O2 mixture	4.000	1.182	13.525	0.025
Freire	2008	117	66	51	35% CO_2 double-breath challenge	15.500	4.367	55.011	<0.001
					Hyperventilation test	0.648	0.273	1.537	0.325
Valença	2002	27	16	11	35% CO_2 double-breath challenge	18.0	1.722	188.090	0.015
Nardi	2006a	76	39	37	35% CO_2 inhalation double-breath	3.671	1.337	10.077	0.011
					Breath holding	12.056	4.090	35.528	<0.001
Nardi	2004a	85	52	33	Hyperventilation test	0.847	0.323	2.218	0.736
					Breath holding	1.238	0.435	3.517	0.688
Nardi	2007	83	41	42	Caffeine 480 mg	8.861	3.266	24.037	<0.001
					35% CO_2 inhalation double breath	2.197	0.888	5.431	0.088
Nardi	2006b	91	31	60	35% CO_2 inhalation double-breath	1.222	0.475	3.138	0.676
Nardi	2004b	88	51	37	Hyperventilation test	4.727	1.894	11.794	<0.001

95% CI *95% confidence interval*, N number of subjects, OR odd ratio

challenge, 75.0–75.6% were RS patients. Also 70.0% of the patients sensitive to the breath holding test were in the RS group [22]. In patients sensitive to both tests 72.0% were in the RS group, in those not sensitive to any of these tests only 37.5% were in the RS group [55] (Table 6.2).

Several other methods are used to provoke PA in PD patients; one of these methods is lactate infusion. Massana et al. [56] found that during the infusion, those with the cardiorespiratory subtype had tachycardia and localized sweating, while the "pseudoneurological" PD patients had bradycardia and generalized sweating.

Studies have shown that hypoxic challenge tests also provoke PA in PD patients, similarly to carbon dioxide [30]. RS patients did not differ from NRS patients regarding the level of anxiety and panic symptoms produced by the challenge. However, the RS group showed more ventilatory irregularities and lower $PETCO_2$ than the other group [30].

Several studies demonstrated that RS patients are more prone to panic attacks in challenge tests. Regarding sensitivity to CO_2, these patients were more responsive to the 5% rebreathing challenge [50], the 35% single-breath challenge [19], the 35% double-breath challenge [21, 31, 52] and the breath holding challenge [22, 52], compared to NRS patients. The RS patients were also more sensitive to the hyperventilation challenge test

[21] and to the caffeine challenge test [20]. The differential responses to diagnostic challenge tests seem to be the most relevant feature of the RS (Table 6.2).

6.6   Treatment

Briggs et al. [6] conducted a multi-centric, randomized, controlled and double-blind medication trial in 1034 PD patients. At the beginning of the trial patients had been free of all medication for at least 1 week, and they were randomized to receive placebo, imipramine or alprazolam. The treatment lasted for 8 weeks and the clinicians were allowed to adjust the dosage according to response or adverse effects. Alprazolam acted faster than imipramine, although at the endpoint both active drugs were significantly superior to the placebo. Among RS patients, those who received imipramine improved more than those who received alprazolam. Among NRS patients, those who received alprazolam showed greater improvement [6]. The higher efficacy of serotonergic medications may indicate that the physiopathology of the RS may be linked to a serotonin imbalance. Nardi et al. [32] treated 118 PD patients with nortriptyline, with a dosage from 50 mg to 150 mg per day, for 52 weeks. The RS patients improved faster than the NRS patients, and in week 8 there was a statistically significant difference in the outcome measures, nevertheless by week 52 both groups had improved equally [32]. In a 3-year follow-up study [33] there was also a significantly faster response in the RS patients compared to the NRS patients, both being treated with clonazepam in doses from 1 mg to 4 mg per day. At the endpoint, 3 years later, there were no differences between these two groups regarding the efficacy of the treatment [33].

Taylor et al. [57] conducted a study on the treatment of PD with cognitive behavioral therapy (CBT) with 22 un-medicated patients who were submitted to 10 weekly CBT sessions, with all patients improving. There were no significant differences between those with prominent respiratory symptoms and those without prominent respiratory symptoms. At a 3-month follow-up, the previously obtained improvement was maintained [57]. A promising new treatment [58] is capnometry-assisted breathing therapy, which uses a feedback system based on the $PETCO_2$. This treatment consists of: educating patients on the role of breathing in PD; correcting problematic respiratory patterns; having them perform different breathing maneuvers with capnometer feedback to experience how changes in breathing affect physiology, symptoms, and mood; teaching them to control $PETCO_2$ level and respiratory rate; and having them practice breathing exercises daily [58]. PD patients, regardless of the subtype, showed significant improvement with a decrease in PDSS scores after five weekly treatment sessions [58].

These studies indicate that imipramine, alprazolam, nortriptyline and clonazepam are effective medications in the treatment of all PD patients; nevertheless tricyclic antidepressants may be more effective than benzodiazepines in RS patients. Trials with newer drugs are needed to ascertain if these medications produce different improvements in RS and NRS patients. The CBT has proved itself equally effective in RS and NRS patients.

6.7   Discussion

In the last 20 years the PD RS subtype has been extensively studied, and although some findings were replicated in more than one study, there were also some controversial findings regarding the PD subtypes.

There were significant differences in the prevalence of respiratory symptoms in samples from different studies, indicating that the prevalence of the RS may vary from one population to the other. There is evidence of higher prevalence of choking in Caucasian Hispanic compared to Caucasian non-Hispanic PD patients [59], also, African Americans have greater fear of dying and fear of insanity, as well as a higher incidence of tingling sensations than European Americans [60], indicating that ethnicity may play a role in the unbalanced distribution of PD symptoms across populations. Cultural factors cannot be ruled out, but the authors believe that the distinct gene pools from the studied populations are the main reason for differences in the prevalence of PD symptoms. The increased risk of PD in RS family members also indicates that genetics play an important role in the physiopathology of this PD subtype. Unfortunately only one study [35] addressed the genetic differences between the RS and NRS and the results were inconclusive. The genes implicated in anxiety and respiratory function should both be investigated in future studies to identify which of these genes may be responsible for the RS.

The RS patients were also more sensitive than NRS patients to challenge tests with CO_2 inhalation, hyperventilation, breath holding and caffeine. The differential responses to diagnostic challenges were consistent across several studies, indicating that respiratory system dysfunctions may contribute to the physiopathology of the RS subtype. Recent studies indicated that twin siblings and other family members share carbon dioxide sensitivity [61–63], once again indicating the importance of genetic factors as mediators of this phenomenon. Given that RS is associated with CO_2 sensitivity and both features are influenced by genetic factors, it is reasonable to assume that they share at least some genetic characteristics.

RS patients had high comorbidity with agoraphobia, social anxiety disorder, generalized anxiety disorder, specific phobias and major depression. These patients also had higher scores on clinical scales and lower neuroticism scores, compared to NRS. These differences in the clinical presentations of PD, with greater comorbidity and more intense PD symptoms, raise the question of whether the RS could predict a worse response to pharmacological and psychological treatments compared to the NRS. With the exception of the study from Briggs et al. [6] in which RS responded better to treatment with imipramine and NRS responded better to alprazolam, the other studies failed to demonstrate differential responses to pharmacological treatment and CBT across the subtypes. RS patients seem to respond faster than NRS to pharmacological treatment with antidepressants and benzodiazepines, although further studies are required to confirm this finding.

6.8 Conclusion

The definition of the RS made by Briggs et al. [6] has been confirmed by two other high quality studies [2, 23], but there continues to be no consensus concerning the definition of the respiratory subtype of panic disorder. This problem may be due to methodological variability between studies and small sample sizes. The authors believe that the respiratory subtype should not be defined exclusively based on a symptomatological profile, but also based on other features such as respiratory challenge profiles. Incorporating new criteria would increase the validity of the respiratory subtype, and it would become a useful tool for research on panic disorder.

References

1. APA. Diagnostic and statistical manual of mental disorders. 5th ed. Arlington, VA: American Psychiatric Publishing; 2013.
2. Roberson-Nay R, Kendler KS. Panic disorder and its subtypes: a comprehensive analysis of panic symptom heterogeneity using epidemiological and treatment seeking samples. Psychol Med. 2011;41(11):2411–21.
3. Roberson-Nay R, Latendresse SJ, Kendler KS. A latent class approach to the external validation of respiratory and non-respiratory panic subtypes. Psychol Med. 2012;42(3):461–74.
4. Klein DF. False suffocation alarms, spontaneous panics, and related conditions. An integrative hypothesis. Arch Gen Psychiatry. 1993;50(4):306–17.
5. Aronson TA, Logue CM. Phenomenology of panic attacks: a descriptive study of panic disorder patients' self-reports. J Clin Psychiatry. 1988;49(1):8–13.
6. Briggs AC, Stretch DD, Brandon S. Subtyping of panic disorder by symptom profile. Br J Psychiatry. 1993;163:201–9.
7. Freire RC, Perna G, Nardi AE. Panic disorder respiratory subtype: psychopathology, laboratory challenge tests, and response to treatment. Harv Rev Psychiatry. 2010;18(4):220–9.
8. Klein DF. Panic developments. Rev Bras Psiquiatr. 2012;34 Suppl 1:S1–2.
9. Freire RC, Nardi AE. Panic disorder and the respiratory system: clinical subtype and challenge tests. Rev Bras Psiquiatr. 2012;34 Suppl 1:S32–41.
10. Nardi AE, Freire RC, Zin WA. Panic disorder and control of breathing. Respir Physiol Neurobiol. 2009;167(1):133–43.
11. Gorman JM, Fyer MR, Goetz R, Askanazi J, Liebowitz MR, Fyer AJ, et al. Ventilatory physiology of patients with panic disorder. Arch Gen Psychiatry. 1988;45(1):31–9.

12. Stein MB, Millar TW, Larsen DK, Kryger MH. Irregular breathing during sleep in patients with panic disorder. Am J Psychiatry. 1995;152(8):1168–73.
13. Martinez JM, Papp LA, Coplan JD, Anderson DE, Mueller CM, Klein DF, et al. Ambulatory monitoring of respiration in anxiety. Anxiety. 1996;2(6):296–302.
14. Abelson JL, Weg JG, Nesse RM, Curtis GC. Persistent respiratory irregularity in patients with panic disorder. Biol Psychiatry. 2001;49(7):588–95.
15. Hasler G, Gergen PJ, Kleinbaum DG, Ajdacic V, Gamma A, Eich D, et al. Asthma and panic in young adults—a 20-year prospective community study. Am J Respir Crit Care Med. 2005;171(11):1224–30.
16. Nascimento I, Nardi AE, Valenca AM, Lopes FL, Mezzasalma MA, Nascentes R, et al. Psychiatric disorders in asthmatic outpatients. Psychiatry Res. 2002;110(1):73–80.
17. Scott KM, Von Korff M, Ormel J, Zhang MY, Bruffaerts R, Alonso J, et al. Mental disorders among adults with asthma: results from the World Mental Health Survey. Gen Hosp Psychiatry. 2007;29(2):123–33.
18. Gorman JM, Kent JM, Sullivan GM, Coplan JD. Neuroanatomical hypothesis of panic disorder, revised. Am J Psychiatry. 2000;157(4):493–505.
19. Biber B, Alkin T. Panic disorder subtypes: differential responses to CO_2 challenge. Am J Psychiatry. 1999;156(5):739–44.
20. Nardi AE, Valenca AM, Lopes FL, De-Melo-Neto VL, Freire RC, Veras AB, et al. Caffeine and 35% carbon dioxide challenge tests in panic disorder. Hum Psychopharmacol Clin Exp. 2007;22(4):231–40.
21. Freire RC, Lopes FL, Valenca AM, Nascimento I, Veras AB, Mezzasalma MA, et al. Panic disorder respiratory subtype: a comparison between responses to hyperventilation and CO_2 challenge tests. Psychiatry Res. 2008;157(1–3):307–10.
22. Nardi AE, Valenca AM, Lopes FL, Nascimento I, Mezzasalma MA, Zin WA. Clinical features of panic patients sensitive to hyperventilation or breath-holding methods for inducing panic attacks. Braz J Med Biol Res. 2004;37(2):251–7.
23. Bandelow B, Amering M, Benkert O, Marks I, Nardi AE, Osterheider M, et al. Cardio-respiratory and other symptom clusters in panic disorder. Anxiety. 1996;2(2):99–101.
24. Shioiri T, Someya T, Murashita J, Takahashi S. The symptom structure of panic disorder: a trial using factor and cluster analysis. Acta Psychiatr Scand. 1996;93(2):80–6.
25. Rees CS, Richards JC, Smith LM. Symptom clusters in panic disorder. Aust J Psychol. 1998;50(1):19–24.
26. Segui J, Salvador-Carulla L, Garcia L, Canet J, Ortiz M, Farre JM. Semiology and subtyping of panic disorders. Acta Psychiatr Scand. 1998;97(4):272–7.
27. Marquez M, Segui J, Garcia L, Canet J, Ortiz M. Is panic disorder with psychosensorial symptoms (depersonalization-derealization) a more severe clinical subtype? J Nerv Ment Dis. 2001;189(5):332–5.
28. Meuret AE, White KS, Ritz T, Roth WT, Hofmann SG, Brown TA. Panic attack symptom dimensions and their relationship to illness characteristics in panic disorder. J Psychiatr Res. 2006;40(6):520–7.
29. Fullana MA, Vilagut G, Ortega N, Bruffaerts R, de Girolamo G, de Graaf R, et al. Prevalence and correlates of respiratory and non-respiratory panic attacks in the general population. J Affect Disord. 2011;131(1–3):330–8.
30. Beck JG, Shipherd JC, Ohtake P. Do panic symptom profiles influence response to a hypoxic challenge in patients with panic disorder? A preliminary report. Psychosom Med. 2000;62(5):678–83.
31. Valenca AM, Nardi AE, Nascimento I, Zin WA, Versiani M. Respiratory panic disorder subtype and sensitivity to the carbon dioxide challenge test. Braz J Med Biol Res. 2002;35(7):783–8.
32. Nardi AE, Nascimento I, Valenca AM, Lopes FL, Mezzasalma MA, Zin WA, et al. Respiratory panic disorder subtype: acute and long-term response to nortriptyline, a noradrenergic tricyclic antidepressant. Psychiatry Res. 2003;120(3):283–93.
33. Nardi AE, Valenca AM, Nascimento I, Lopes FL, Mezzasalma MA, Freire RC, et al. A three-year follow-up study of patients with the respiratory subtype of panic disorder after treatment with clonazepam. Psychiatry Res. 2005;137(1–2):61–70.
34. Onur E, Alkin T, Tural U. Panic disorder subtypes: further clinical differences. Depress Anxiety. 2007;24(7):479–86.
35. Pirildar S, Bayraktar E, Berdeli A, Kucuk O, Alkin T, Kose T. Progesterone receptor gene polymorphism in panic disorder: associations with agoraphobia and respiratory subtype of panic disorder. Klin Psikofarmakol B. 2010;20(2):153–9.
36. Freire RC, Nardi AE. Are patients with panic disorder respiratory subtype more vulnerable to tobacco, alcohol or illicit drug use? Rev Psiquiatr Clin. 2013;40(4):135–8.
37. Kristensen AS, Mortensen EL, Mors O. The association between bodily anxiety symptom dimensions and the scales of the Revised NEO Personality Inventory and the Temperament and Character Inventory. Compr Psychiatry. 2009;50(1):38–47.
38. Lepola U, Koponen H, Leinonen E. Sleep in panic disorders. J Psychosom Res. 1994;38:105–11.
39. Lopes FL, Nardi AE, Nascimento I, Valenca AM, Zin WA. Nocturnal panic attacks. Arq Neuropsiquiatr. 2002;60(3B):717–20.
40. Sarisoy G, Boke O, Arik AC, Sahin AR. Panic disorder with nocturnal panic attacks: symptoms and comorbidities. Eur Psychiatry. 2008;23(3):195–200.
41. Lopes FL, Nardi AE, Nascimento I, Valenca AM, Mezzasalma MA, Freire RC, et al. Diurnal panic attacks with and without nocturnal panic attacks: are there some phenomenological differences? Rev Bras Psiquiatr. 2005;27(3):216–21.
42. Freire RC, Valenca AM, Nascimento I, Lopes FL, Mezzasalma MA, Zin WA, et al. Clinical features of

respiratory and nocturnal panic disorder subtypes. Psychiatry Res. 2007;152(2–3):287–91.

43. De-Melo-Neto VL, King AL, Valenca AM, da Rocha Freire RC, Nardi AE. Respiratory and non-respiratory panic disorder subtypes: clinical and quality of life comparisons. Rev Port Pneumol. 2009;15(5):859–74.

44. Bouwer C, Stein DJ. Association of panic disorder with a history of traumatic suffocation. Am J Psychiatry. 1997;154(11):1566–70.

45. Bouwer C, Stein D. Panic disorder following torture by suffocation is associated with predominantly respiratory symptoms. Psychol Med. 1999;29(1):233–6.

46. Ozdemir O, Selvi Y, Ozkol H, Tuluce Y, Besiroglu L, Aydin A. Comparison of superoxide dismutase, glutathione peroxidase and adenosine deaminase activities between respiratory and nocturnal subtypes of patients with panic disorder. Neuropsychobiology. 2012;66(4): 244–51.

47. Asmundson GJG, Stein MB. A preliminary-analysis of pulmonary-function in panic disorder—implications for the dyspnea-fear theory. J Anxiety Disord. 1994;8(1):63–9.

48. Pfaltz MC, Michael T, Grossman P, Blechert J, Wilhelm FH. Respiratory pathophysiology of panic disorder: an ambulatory monitoring study. Psychosom Med. 2009;71(8):869–76.

49. Moynihan JE, Gevirtz RN. Respiratory and cognitive subtypes of panic—preliminary validation of Ley's model. Behav Modif. 2001;25(4):555–83.

50. Abrams K, Rassovsky Y, Kushner MG. Evidence for respiratory and nonrespiratory subtypes in panic disorder. Depress Anxiety. 2006;23(8):474–81.

51. Freire RC, Nascimento I, Valenca AM, Lopes FL, Mezzasalma MA, de Melo Neto VL, et al. The panic disorder respiratory ratio: a dimensional approach to the respiratory subtype. Rev Bras Psiquiatr. 2013; 35(1):57–62.

52. Nardi AE, Valenca AM, Mezzasalma MA, Lopes FL, Nascimento I, Veras AB, et al. 35% carbon dioxide and breath-holding challenge tests in panic disorder: a comparison with spontaneous panic attacks. Depress Anxiety. 2006;23(4):236–44.

53. Nardi A, Valenca A, Lopes F, Nascimento I, Veras A, Freire R, et al. Psychopathological profile of 35%

CO$_2$ challenge test-induced panic attacks: a comparison with spontaneous panic attacks. Compr Psychiatry. 2006;47(3):209–14.

54. Nardi AE, Lopes FL, Valenca AM, Nascimento I, Mezzasalma MA, Zin WA. Psychopathological description of hyperventilation-induced panic attacks: a comparison with spontaneous panic attacks. Psychopathology. 2004;37(1):29–35.

55. Nardi A, Valenca A, Mezzasalma M, Levy S, Lopes F, Nascimento I, et al. Comparison between hyperventilation and breath-holding in panic disorder: patients responsive and non-responsive to both tests. Psychiatry Res. 2006;142(2–3):201–8.

56. Massana J, Risueno JAL, Masana G, Marcos T, Gonzalez L, Otero A. Subtyping of panic disorder patients with bradycardia. Eur Psychiatry. 2001; 16(2):109–14.

57. Taylor S, Woody S, Koch WJ, Mclean PD, Anderson KW. Suffocation false alarms and efficacy of cognitive behavioral therapy for panic disorder. Behav Ther. 1996;27(1):115–26.

58. Meuret AE, Wilhelm FH, Ritz T, Roth WT. Feedback of end-tidal pCO$_2$ as a therapeutic approach for panic disorder. J Psychiatr Res. 2008;42(7):560–8.

59. Hollifield M, Finley MR, Skipper B. Panic disorder phenomenology in urban self-identified Caucasian-Non-Hispanics and Caucasian-Hispanics. Depress Anxiety. 2003;18(1):7–17.

60. Smith LC, Friedman S, Nevid J. Clinical and sociocultural differences in African American and European American patients with panic disorder and agoraphobia. J Nerv Ment Dis. 1999;187(9):549–60.

61. Bellodi L, Perna G, Caldirola D, Arancio C, Bertani A, Di BD. CO$_2$-induced panic attacks: a twin study. Am J Psychiatry. 1998;155(9):1184–8.

62. Perna G, Cocchi S, Bertani A, Arancio C, Bellodi L. Sensitivity to 35% CO$_2$ in healthy first-degree relatives of patients with panic disorder. Am J Psychiatry. 1995;152(4):623–5.

63. Battaglia M, Pesenti-Gritti P, Spatola CA, Ogliari A, Tambs K. A twin study of the common vulnerability between heightened sensitivity to hypercapnia and panic disorder. Am J Med Genet B Neuropsychiatr Genet. 2008;147B(5):586–93.

Lifelong Opioidergic Vulnerability Through Early Life Separation: A Recent Extension of the False Suffocation Alarm Theory of Panic Disorder

7

Maurice Preter

Contents

M. Preter (✉)
Department of Psychiatry, College of Physicians
and Surgeons, Columbia University,
New York, NY, USA

Department of Neurology, Mount Sinai School
of Medicine, New York, NY, USA
e-mail: mp2285@cumc.columbia.edu

Abstract

Suffocation-False Alarm Theory (Klein, Arch Gen Psychiatry 50:306–317, 1993) postulates the existence of an evolved physiologic suffocation alarm system that monitors information about potential suffocation. Panic attacks maladaptively occur when the alarm is erroneously triggered. The expanded Suffocation-False Alarm Theory (Preter and Klein, Biol Psychiatry 32(3):603–612, 2008) hypothesizes that endogenous opioidergic dysregulation may underlie the respiratory pathophysiology and suffocation sensitivity in panic disorder. Opioidergic dysregulation increases sensitivity to CO_2, separation distress and panic attacks. That sudden loss, bereavement and childhood separation anxiety are also antecedents of "spontaneous" panic requires an integrative explanation. Our work unveiling the lifelong endogenous opioid system impairing effects of childhood parental loss (CPL) and parental separation in non-ill, normal adults opens a new experimental, investigatory area.

Keywords

Affective neuroscience • Childhood parental loss (CPL) • Endogenous opioids • Panic disorder pathophysiology • Expanded Suffocation-False Alarm Theory • Panic disorder comorbidity

7.1 Introduction

I briefly reference previous material on Klein's suffocation false alarm theory (SFA) of panic disorder [1] and its amplification in 2008 [2]. I then discuss a recent finding showing a fundamental difference in endogenous opioid reactivity to a naloxone challenge in psychiatrically and medically healthy adults, depending on whether or not they had experienced childhood separation and parental loss.

In 1993, Klein published the original SFA theory of panic disorder [1], attempting to integrate the multiplicity of apparently unrelated clinical and laboratory observations. We posited "that a physiologic misinterpretation by a suffocation monitor misfires an evolved suffocation alarm system. This produces sudden respiratory distress followed swiftly by a brief hyperventilation, panic, and the urge to flee. Carbon dioxide hypersensitivity is seen as due to the deranged suffocation alarm monitor. If other indicators of potential suffocation provoke panic, this theoretical extension is supported." In the original paper, we tested "the theory by examining Ondine's curse as the physiologic and pharmacologic converse of panic disorder, splitting panic in terms of symptomatology and challenge studies, reevaluating the role of hyperventilation, and reinterpreting the contagiousness of sighing and yawning, as well as mass hysteria" [1]. Original SFA focused on relating the observed lactate and carbon dioxide hypersensitivity in panic disorder to a putative dysfunction in a hypothesized suffocation alarm. At the time, the underlying pathophysiology that might connect the apparent disparate "phenomena of panic during relaxation and sleep, late luteal phase dysphoric disorder, pregnancy, childbirth, pulmonary disease, separation anxiety, and treatment", was unknown.

Over the intervening decades, much data evolved linking Separation Anxiety, Panic, and respiratory dysfunction. This suggested a potential missing link:

"Klein and Fink [3] posited a developmental pathophysiological link between [clinical levels of] separation anxiety and PD and subsequent agoraphobia, since 50 % of hospitalized agoraphobics

reported severe early separation anxiety that often prevented school attendance. Further, panic, in this group, was frequently precipitated by bereavement, or separation […]" [2]. This was also noted by others, e.g., Faravelli and Pallanti [4], Kaunonen et al. [5], Milrod et al. [6].

We noted that "[p]atients highly comorbid for multiple anxiety disorders are particularly likely to recall childhood SAD [7]", and that "[c]laims that separation anxiety equivalently antecedes other anxious states [8] may be due to diagnostically ambiguous limited symptom attacks and the unreliability of the questionnaire method". We concluded that "in the only controlled, long-term, direct, blind, clinical interview follow-up of separation-anxious, school-phobic children, the only significant finding was an increased PD rate" [2].

Adding further support, Battaglia et al. [9] showed that "[s]eparation anxiety correlates with increased familial loading and early onset of PD." More recently (2012), as part of their series of brilliant twin studies, Robertson-Nay et al. demonstrated that, "[childhood] separation anxiety disorder and adult onset panic attacks share a common genetic diathesis".

7.2 Panic and Comorbid Conditions

There has been a regained awareness of the relevance of panic states to other clinical contexts, as Freud suggested in his pioneering statement [10].

In the US, nearly half of panic patients are initially seen in the medical emergency room of a hospital. They may undergo extensive diagnostic medical procedures, such as MRI scans of the brain for headaches, or coronary angiograms for chest pain. Cardiovascular symptoms, particularly pseudo-anginal chest pain resembling a heart attack are the most common symptoms in these PD patients' experience. Accordingly, 25 % or more of outpatients seen by a cardiologist have a current diagnosis of PD [11]. Since primary complaints are of distress, rather than anxiety, they have been termed "non-fearful" panics. This seemingly oxymoronic term nevertheless indicates that fearfulness is not essential to panic disorder [12].

Migraine and other chronic headaches are highly comorbid with panic disorder [13, 14]. Having PD increases the risk of migraine four-fold, and vice versa. This bidirectionality suggests that the migraine-panic association is unlikely to be merely coincidental and that shared environmental and familial factors are involved [13]. In a longitudinal study, separation and parental loss early in life increased the risk of both headaches and psychiatric morbidity, mainly anxiety and depression, in adulthood [15].

Panic disorder is also comorbid with other somatic pain syndromes [16]. In a cross-sectional survey of 1219 female veterans studying the prevalence and frequency of mastalgia, women reporting frequent mastalgia were much more likely to have comorbid panic disorder [OR 7.1], but also post-traumatic stress disorder, mood disorders, and other somatic pain syndromes, such as fibromyalgia, chronic pelvic pain or irritable bowel syndrome.

Although panic attacks as they occur in panic disorder often prominently feature air hunger, other panic subtypes have prominent vestibular symptoms and unspecific dizziness/lightheadedness. True vertigo was historically recognized as a common presentation of panic disorder [17, 18]. The current rigid distinction between psychiatry and neurology [19] interferes with proper assessment [20].

In Mandarin Chinese, *tou yun* refers to the disabling sensation of a constant state of movement of oneself or one's surroundings. This dizziness (note that *tou yun* also describes vertigo), is probably the most common expression of panic disorder in Chinese patients [21], so by sheer numbers, this may well be the most prevalent panic subtype worldwide.

Taken together, the various comorbidities of panic disorder and untreated sequelae massively impact people's quality of life [22].

One prominent characteristic of the panic attack and subthreshold panic-related anxiety is respiratory dysregulation and chaotic breathing. This can be experimentally reproduced in adult panic sufferers, but also in children with separation anxiety disorder [2, 23, 24].

Air hunger and chronic sighing outside of the acute attack are hallmarks of panic that rarely occurs under acute, external-threat initiated fear [1, 25]. Increasing hypercapnia is a more salient indicator of potential suffocation than hypoxia, but hypoxia also serves this alarm function. "Beck et al. [26, 27] showed that panic patients respond with increased panic symptoms not only to CO_2 inhalation, but also to normocapnic hypoxia, as predicted by SFA." [2]. Unsurprisingly, numerous studies found that panic disorder and lung disease commonly occur together [28–37]. More specifically, we wrote, "early lung disease, including asthma and COPD may predispose to PD [28, 38–42], or present solely with panic symptoms [43, 44]. Asthma and PD are both characterized by acute episodes, salient respiratory symptoms and anxiety with avoidance of situations related to acute attacks [1, 37]. There is a significantly higher (6.5–24%) prevalence of PD in asthmatics [36, 45, 46] than the 1–3% reported in the general population [47, 48]. Perna et al. [41] found a significantly higher prevalence of PD, sporadic panic attacks, and social phobia in asthmatics than the general population. In 90% of asthmatics with PD, asthma appeared first. Panic symptomatology during the asthmatic attack predicted longer hospitalizations in asthmatic patients [49–51]." [2].

The recent amplification of SFA centers on the observation that both separation anxiety and suffocation sensitivity are under endogenous opioidergic control. We amplified the SFA theory by suggesting that PD may be due to an episodic functional endogenous opioid deficit [25]. The following is a necessarily brief explanation.

The endogenous opioid system was discovered in the early 1970s. Electrical stimulation the periaqueductal gray [52] produced analgesia that was reversed by naloxone, suggesting an endogenous opioid system. Opioid molecules are among the oldest evolved signaling substances., functioning in many physiological processes e.g., pain perception, respiration [53]. Dyspnea is modulated by central and peripheral opioid levels in both rodents and humans [54].

In mice, exposure to intermittent, severe hypoxia prolonged survival during subsequent lethal suffocation [55]. This effect was blocked by naloxone, implying that endogenous opioids increase adaptability to hypoxic environments. Opioid receptors, including 'non-conventional' ones, can be found throughout the respiratory tract. Nebulized morphine is an outstanding treatment for chronic dyspnea ([56–59].

In our 2008 paper, we summarized data from developmental psychobiology and neuroanatomy that point to a possible link between separation and the endogenous opioid system, as follows:

"Following birth, mammalian infants cannot survive independently. Survival requires reliable distress signaling mechanisms to elicit parental care and retrieval. Distress vocalizations (DVs) are a primitive form of audio-vocal communication [60]. A common neuroanatomy subserving DVs may be shared by all mammals, although substantial functional variations depend on the ontogenetic niche. The latter [61] signifies the ecological and social legacies ("the inherited environment") in which a given set of genes develops. For instance, isolated altricial (developmentally immature) infants do not emit DVs compared to other species, since it is not likely they will stray from the nest [62].

Immature human infants practically never get lost for their first 6 months. Despite frequent maternal absence, separation anxiety in humans develops only after their motor system matures. Young rats are not specifically attached to their mother, i.e. any mother will do as heater or feeder. Only once mobile do they socially bond, but their responses do not compare with the vigor seen in other species. Rats also differ from other species, including primates, dogs and chicks in their greater DV suppression by benzodiazepines [62–64]. Since benzodiazepines differentially alleviate anticipatory anxiety, social isolation in young rodents, as compared to many other mammals, may activate anxiety mechanisms other than separation distress." Thus, Panksepp emphasizes that when using cross-species analogies, it is important to keep in mind that the type and degree of social separation distress depends on ecological and developmental parameters [62].

The developmental phase of separation anxiety serves as a biologic leash for the increasingly mobile, but helpless infant who continually checks for the mother's presence, becomes acutely distressed on discovering her absence, and immediately attempts to elicit retrieval by crying. In humans, separation anxiety usually wanes around age four when the now verbally skilled child can successfully elicit care even from non-relatives.

Using electrical brain stimulation (ESB), DVs have been elicited in many species from homologous areas, including the midbrain, dorsomedial thalamus, ventral septum, preoptic area, and the bed nucleus striae terminalis (BNST). In some higher species, one can obtain separation calls by stimulating the central amygdala and dorsomedial hypothalamus. All these sites have high opioid receptor densities and figure heavily in sexual and maternal behaviors [60]. Cortically, electrical stimulation of the rostral cingulate gyrus in monkeys consistently elicits distress calls [65, 66]. The cingulate cortex, found exclusively in mammals, is particularly well developed in humans and contains high densities of opioid receptors [67].

Naloxone-blockable opioid agonists reduce isolation-induced distress vocalizations (DVs) across mammalian species [68–71]. In beagles, imipramine, the classic anti-panic agent, and morphine were the only psychotropic drug that yielded specific DV reduction at nonsedating doses [64, 71].

Naloxone given to guinea pigs and young chicks [71] increased baseline vocalizations (by 600 %), but only when the animals were in a group, since isolates already emitted maximum DVs.

Kalin et al. [69] studied opioid modulation of separation distress in primates, showing morphine (0.1 mg/kg) significantly decreased separation distress vocalizations without changes in autonomic and hormonal activation. Naloxone (0.1 mg/kg) blocked this effect. Sympathetic blockade using the α(2) agonist, clonidine, and the β adrenergic antagonist, propranolol, had no specific effect on separation-induced "coos" in infant rhesus monkeys [2, 72].

7.3   Testing the Panic-Suffocation-False Alarm-Endogenous Opioid Connection

"Panic Disorder is unique among psychiatric disorders in that its salient component, the panic attack, can be reliably incited in laboratory settings by specific chemical challenges as well as having challenges specifically blocked by anti-panic agents, e.g. imipramine. We can experimentally turn panic on and off, producing trenchant causally related data rather than inferences from naturalistic data." [2]. Specifically, sodium lactate infusions and CO_2 inhalation regularly produce panic attacks in patients with panic disorder [73–75]. However, while normal controls or patients with other anxiety disorders rarely show such reactivity (i.e., progress to a full-blown panic attack) [1], higher concentrations of inhaled CO_2 are highly aversive and can produce respiratory panic symptomatology in a dose-dependent fashion [76–78].

Both spontaneous and lactate induced panic attacks in panic patients produce air hunger and marked, objective increases in tidal volume (Vt) [79, 80]. Since sodium lactate infusion causes a metabolic alkalosis, a compensatory decrease in ventilation would be expected. This would homeostatically buffer blood pH, by increasing CO_2 retention. However, the converse actually occurs indicating a specific lactate stimulating effect on respiration.

The usual response of healthy control subjects to a sodium lactate infusion is a minor, but definite increase in Vt [81]. The lesser tidal volume response in lactate challenged normal subjects may be due to buffering by their intact endogenous opioid system.

An open pilot study showed that naloxone infusion (ranging from an initial 0.5 mg/kg to a maximum of 2 mg/kg) followed by lactate (N + L), caused significant tidal volume increments similar to those observed during clinical and lactate induced panic attacks in 8 of 12 normal subjects, supporting the hypothesis that opioidergic deficiency might be necessary for lactate to produce a marked increase in tidal volume in normal subjects [82].

Based on these initial findings, and cognizant that previous experiments using smaller doses of both intravenous and oral opioid blocking agents had shown little results [83, 84] we decided to conduct a controlled, randomized experimental study to investigate whether high-dose naloxone, an intravenous opioid receptor antagonist, could change the regularly resistant normal controls to become more sensitive to intravenous lactate as a respiratory stimulus to tidal volume increment. Study design and statistical analysis are detailed elsewhere [85], but in addition to the usual standard recruitment procedure for healthy research subjects, "eligible volunteers were further interviewed about potentially significant individual and family antecedents and comorbidities· of panic, such as near-suffocation, pulmonary disease, and migraine headache. Recent and childhood loss and separation events (parental divorce or death, childhood abuse) were specifically reviewed." [85].

Results showed that "[n]ormal subjects, usually relatively insensitive to the TV effects of lactate infusion, in this study, given opioid antagonist pretreatment, developed TV and RR increments resembling those occurring in both spontaneous clinical panic attacks and in panic patients who panic during lactate infusions [73–75]. The hypothesis that a functioning endogenous opioid system buffers normal subjects from the behavioral and physiological effects of lactate is consonant with these results."

The most interesting aspect of this study is that "for the first time the prolonged physiological effects of actual separations and losses during childhood, i.e. parental death, parental separation or divorce, on the endogenous opioid system of healthy adults have been objectively, experimentally shown. Presence or absence of childhood parental loss (CPL) antecedents determined the response to the naloxone-lactate probe." [85].

In these carefully screened medically and psychiatrically healthy subjects, a history of childhood parental loss or separation decreased the naloxone + lactate effect, implying that there was an antecedent decrement in opioidergic activity, so that the naloxone had nothing to block. It was the subjects that had not suffered such separation events that showed the expectable tidal volume increment (hyperventilation) used as outcome measure.

"The import of these findings is that analyses attempting to relate CPL to other baseline variables may well fail since CPL impact may be specific to challenges to the endogenous opioid system." [85]. Also it implies that separation-induced, baseline opioidergic deficiency, while it may not be sufficient to induce overt disease, confers lifetime vulnerability even in healthy adults. Whether it is longitudinally, or cross-sectionally relevant to somatic pain syndromes such as migraine, and to opiate abuse ought to be determined.

Again, we emphasize that these CPL effects were apparent in a 'normal' sample. "In a society where divorce rates approach 50%, the results raise the question whether current psychiatric classification and diagnostic scales are sensitive enough to detect the effects of childhood parental loss. This applies as well to developing societies like China, where the massive migration of mostly young individuals from the countryside to urban areas has left approximately 30 million small children behind [85, 86]."

It is known that early maternal separation is a risk factor for adult anxious-depressive and borderline psychopathology [87–89]. "However, its detrimental long-term effect is not limited to psychiatric illness [90]. Using criteria similar to ours, the Adverse Childhood Experiences (ACE) Study, a CDC supported prospective cohort study of 16,908 adults found a significant relationship between CPL and premature death in adulthood [91]. Retrospective (e.g., [92, 93]) and prospective longitudinal data [15, 94–96], link family disruption, physical abuse, separation and maternal loss in early life to chronic physical pain in adulthood. It should be explored whether the NL vs. SL probe has a differential effect on pain perception and physiological pain measures, and whether the presence of childhood parental loss antecedents modifies this interaction. Unfortunately, our exploratory pain measure was limited to a single item, and in retrospect, was clearly inadequate.

CPL as related to childhood separation anxiety, adult panic disorder (PD) and suffocation hypersensitivity was studied by Battaglia et al. [97]. In a large sample of twins from Norway, CPL accounted in no small part for "the covariation between separation anxiety in childhood, hypersensitivity to CO_2 (as indexed by the anxiety response to a 35% CO_2/65% O_2 mixture), and PD in adulthood". Note that in Battaglia's study, CPL increased reactivity to the 35% CO_2 probe." [85] [...]

"Testing the specificity of the naloxone-lactate model of clinical panic requires double-blind investigation whether specific anti-panic drugs, but not panic irrelevant drugs, block this effect." If found, this has practical and heuristic implications.

First, there is currently no specific, screening method for testing putative anti-panic drugs except by the experimental treatment of panic disorder patients. The naloxone + lactate effect in normal humans may afford such a screening method, and may be extended to preclinical studies. Second, these data offer heuristic support for the theory that an opioidergic dysfunction is the pathophysiological mechanism underlying panic disorder. If so, the appropriateness of opioidergic therapeutic agents comes into question. The use of morphine or other simple agonists would probably be rejected for fear of inducing addiction, although the evidence for addiction during indicated medical treatment is slim. However, recent work with opioidergic mixed agonist-antagonists [98, 99], e.g. buprenorphine, may be relevant. The concern about addiction would be mitigated by the fact that higher doses become receptor blockers rather than agonists. Positive results would foster investigations into basic molecular mechanisms. For instance, we note that the dose of naloxone used in our study (2 mg/kg) substantially exceeds that needed for μ opioid receptor (MOR) blockade [100], suggesting a role for the δ opioid receptor (DOR). This could spark interest in the development of specific DOR agonists suitable for human use. Currently, such agents have not been developed, although agents suitable for animal use are available." [85].

Since the publication of our paper, exciting new work in panic disorder has emerged, notably from Brazil. Appropriately, the "First World Symposium On Translational Models Of Panic Disorder", was held in Vitoria, E.S., in November of 2012. Moreira et al. [101] present a thoughtful review of the use of rodents in panic disorder research. Graeff [102], studying an animal model of panic disorder found that the inhibitory action of serotonin is connected with activation of endogenous opioids in the periaqueductal gray (PAG). Schenberg and colleagues [103, 104] suggest "[the PAG] harbors an anoxia-sensitive suffocation alarm system". Activation precipitates panic attacks and potentiates the subject's responses to hypercapnia. Notably, the resemblance of these effects to panic disorder was supported by their pharmacological parallel to panic disorder treatment. This model was also supported by demonstrating a lack of stress hormone release during DPAG stimulation thus paralleling panic disorder [105, 106]. The utility of opioidergic mixed agonist-antagonists in animal models of panic disorder and in treatment refractory patients would seem promising.

The lack of hypothalamic pituitary adrenal (HPA) activation during the panic attack, as it occurs in panic disorder, is a striking peculiarity since it contradicts the belief that the panic is an expression of a hypersensitive fear mechanism that stimulates the HPA anti-stress response, supposedly reactive to all dangers. This is usually understood as dependent on hyper-responsiveness of the amygdala. For instance, both Stein [107], and Gorman et al. [108] neglect or dismiss the incongruity of the lack of HPA response, claiming a supposed amygdala-based hypersensitive fear system as central to panic.

Further damage to the amygdalocentric fear system theory is provided by Feinstein et al. [109] who studied three patients with amygdala damage produced by Urbach-Wiethe syndrome. It is worth extensive citation:

"A substantial body of evidence has emphasized the importance of the amygdala in fear [...]. In animals, amygdala-restricted manipulations interfere with the acquisition, expression and recall of conditioned fear and other forms of fear and anxiety-related behaviors. In humans, focal bilateral amygdala lesions are extraordinarily rare, and such cases have been crucial for understanding the role of the human amygdala in fear. [...] The most intensively studied case is patient SM, whose amygdala damage stems from Urbach-Wiethe disease [...] Previous studies have shown that patient SM does not condition to aversive stimuli [...], fails to recognize fearful faces [...] and demonstrates a marked absence of fear during exposure to a variety of fear-provoking stimuli, including life-threatening traumatic events [...]. Patients with similar lesions have largely yielded similar results [...].

One stimulus not previously tested in humans with amygdala damage is CO_2 inhalation. Inhaling CO_2 stimulates breathing and can provoke both air hunger and fear [...] Furthermore, CO_2 can trigger panic attacks, especially in patients with panic disorder [...]. Recent work in mice found that the amygdala directly detects CO_2 and acidosis to produce fear behaviors [...]. Thus, we hypothesized that bilateral amygdala lesions would reduce CO_2-evoked fear in humans.

In contrast with our prediction, patient SM reported fear in response to a 35 % CO_2 inhalation challenge. To the best of our knowledge, this was the first time patient SM experienced fear in any setting, laboratory or otherwise, since childhood [...]. To further explore this issue, we tested two additional patients (AM and BG), monozygotic twin sisters with focal bilateral amygdala lesions resulting from Urbach-Wiethe disease [...] As with patient SM, both patients also reported experiencing fear during the CO_2 challenge." [109].

These startling observations affirm that the reaction to carbon dioxide must be due to an alternative alarm system, such as has been proposed for possible suffocation. We have suggested that under conditions of threatened asphyxia the activation of the HPA system would produce a counterproductive hyperoxidative state and is therefore inactivated. This is in keeping with the observation that the tachycardia during panic is produced by vagal withdrawal rather than a counterproductive sympathetic oxidative surge [2].

In conclusion, we objectively, experimentally showed a physiological link between endogenous opioid system deficiency and panic-like suffocation sensitivity in healthy adults. This is consonant with the expanded Suffocation-False Alarm theory of panic suggesting an episodic functional endogenous opioid deficit [25, 110]. The specificity of the naloxone + lactate model of clinical panic should be tested using specific anti-panic components, possibly including opioidergic mixed agonist-antagonists such as buprenorphine. If specific, the naloxone + lactate effect in normal humans affords a screening method for testing putative anti-panic drugs which is currently not available. This could obviate the experimental treatment of panic disorder patients in drug development.

Our data also show for the first time that actual separations and losses during childhood, such parental death, parental separation or divorce (CPL), effect lifelong alterations in the physiological reactivity of the endogenous opioid system of healthy adults.

This result encourages epigenetic inquiry into the effects of CPL on endogenous opioid systems, and the role of these systems in resilience, but also in the chronification of symptoms. A second step would be to explore similarities and differences between CPL effects and those of "subthreshold" events (such as maternal rejection; e.g. [111, 112]). Finally, a redefinition of what constitutes a (truly) healthy control (without antecedents of CPL) in clinical research protocols may be called for.

References

1. Klein DF. False suffocation alarms, spontaneous panics, and related conditions. An integrative hypothesis. Arch Gen Psychiatry. 1993;50:306–17.
2. Preter M, Klein DF. Panic, suffocation false alarms, separation anxiety and endogenous opioids. Prog Neuropsychopharmacol Biol Psychiatry. 2008;32(3):603–12.
3. Klein DF, Fink M. Psychiatric reaction patterns to imipramine. Am J Psychiatry. 1962;119:432–8.
4. Faravelli C, Pallanti S. Recent life events and panic disorder. Am J Psychiatry. 1989;146:622–6.
5. Kaunonen M, Paivi AK, Paunonen M, Erjanti H. Death in the Finnish family: experiences of spousal bereavement. Int J Nurs Pract. 2000;6:127–34.
6. Milrod B, Leon AC, Shear MK. Can interpersonal loss precipitate panic disorder? Am J Psychiatry. 2004;161:758–9.
7. Lipsitz JD et al. Childhood separation anxiety disorder in patients with adult anxiety disorders. Am J Psychiatry. 1994;151:927–9.
8. van der Molen GM, van den Hout MA, van Dieren AC, Griez E. Childhood separation anxiety and adult-onset panic disorders. J Anxiety Disord. 1989;3:97–106.
9. Battaglia M et al. Age at onset of panic disorder: influence of familial liability to the disease and of childhood separation anxiety disorder. Am J Psychiatry. 1995;152:1362–4.
10. Freud, S. [1895].Über die Berechtigung, von der Neurasthenie einen bestimmten Symptomenkomplex als "Angst-Neurose" abzutrennen. Neurol. Zbl., XIV, p. 50-66; G.W., I, p. 315–342; On the grounds for detaching a particular syndrome from neurasthenia under the description "anxiety neurosis." SE, 3: 1895;85–115.
11. Ballenger JC. Treatment of panic disorder in the general medical setting. J Psychosom Res. 1998;44(1):5–15.
12. Beitman BD, Kushner M, Lamberti JW, Mukerji V. Panic disorder without fear in patients with angiographically normal coronary arteries. J Nerv Ment Dis. 1990;178:307–12.
13. Breslau N, Schultz LR, Stewart WF, Lipton R, Welch KM. Headache types and panic disorder: directionality and specificity. Neurology. 2001;56:350–4.
14. Hamelsky SW, Lipton RB. Psychiatric comorbidity of migraine. Headache. 2006;46(9):1327–33.
15. Fearon P, Hotopf M. Relation between headache in childhood and physical and psychiatric symptoms in adulthood: national birth cohort study. BMJ. 2001;322(7295):1145.
16. Johnson KM, Bradley KA, Bush K, Gardella C, Dobie DJ, Laya MB. Frequency of mastalgia among women veterans. Association with psychiatric conditions and unexplained pain syndromes. J Gen Intern Med. 2006;21 Suppl 3:S70–5.
17. Benedikt M. Ueber Platzschwindel. Allg Wien Med Zig. 1870;15:488–90.
18. Frommberger UH, Tettenborn B, Buller R, Benkert O. Panic disorder in patients with dizziness. Arch Intern Med Mar. 1994;154(5):590–1.
19. Staab JP. Chronic dizziness: the interface between psychiatry and neuro-otology. Curr Opin Neurol. 2006;19(1):41–8.
20. Preter M, Bursztajn HJ. Crisis and opportunity—the DSM-V and its neurology quandary. Asian J Psychiatry. 2009;2(4):143.

21. Park L, Hinton D. Dizziness and panic in China: associated sensations of zang fu organ disequilibrium. Cult Med Psychiatry. 2002;26(2):225–57.

22. Sareen J, Jacobi F, Cox BJ, Belik SL, Clara I, Stein MB. Disability and poor quality of life associated with comorbid anxiety disorders and physical conditions. Arch Intern Med. 2006;166(19): 2109–16.

23. Pine DS et al. Differential carbon dioxide sensitivity in childhood anxiety disorders and nonill comparison group. Arch Gen Psychiatry. 2000;57:960–7.

24. Pine DS et al. Response to 5 % carbon dioxide in children and adolescents: relationship to panic disorder in parents and anxiety disorders in subjects. Arch Gen Psychiatry. 2005;62:73–80.

25. Preter M, Klein DF. Panic disorder and the suffocation false alarm theory: current state of knowledge and further implications for neurobiologic theory testing. In: Bellodi L, Perna G, editors. The panic respiration connection. Milan: MDM Medical Media; 1998.

26. Beck JG, Ohtake PJ, Shipherd JC. Exaggerated anxiety is not unique to CO_2 in panic disorder: a comparison of hypercapnic and hypoxic challenges. J Abnorm Psychol. 1999;108:473–82.

27. Beck JG, Shipherd JC, Ohtake P. Do panic symptom profiles influence response to a hypoxic challenge in patients with panic disorder? A preliminary report. Psychosom Med. 2000;62:678–83.

28. Goodwin RD, Eaton WW. Asthma and the risk of panic attacks among adults in the community. Psychol Med. 2003;33:879–85.

29. Goodwin RD, Fergusson DM, Horwood LJ. Asthma and depressive and anxiety disorders among young persons in the community. Psychol Med. 2004;34: 1465–74.

30. Katon WJ, Richardson L, Lozano P, McCauley E. The relationship of asthma and anxiety disorders. Psychosom Med. 2004;66:349–55.

31. Klein DF. Asthma and psychiatric illness. JAMA. 2001;285:881–2.

32. Nascimento I et al. Psychiatric disorders in asthmatic outpatients. Psychiatry Res. 2002;110:73–80.

33. Roy-Byrne PP, Craske MG, Stein MB. Panic disorder. Lancet. 2006;368:1023–32.

34. Valença AM, Falcão R, Freire RC, Nascimento I, Nascentes R, Zin WA, et al. The relationship between the severity of asthma and comorbidities with anxiety and depressive disorders. Rev Bras Psiquiatr. 2006;28:206–8.

35. Wingate BJ, Hansen-Flaschen J. Anxiety and depression in advanced lung disease. Clin Chest Med. 1997;18:495–505.

36. Yellowlees PM, Haynes S, Potts N, Ruffin RE. Psychiatric morbidity in patients with life-threatening asthma: initial report of a controlled study. Med J Aust. 1988;149:246–9.

37. Yellowlees PM, Kalucy RS. Psychobiological aspects of asthma and the consequent research implications. Chest. 1990;97:628–34.

38. Craske MG, Poulton R, Tsao JC, Plotkin D. Paths to panic disorder/agoraphobia: an exploratory analysis from age 3 to 21 in an unselected birth cohort. J Am Acad Child Adolesc Psychiatry. 2001;40: 556–63.

39. Hasler G, Gergen PJ, Kleinbaum DG, Ajdacic V, Gamma A, Eich D, et al. Asthma and panic in young adults: a 20-year prospective community study. Am J Respir Crit Care Med. 2005;171:1224–30.

40. Karajgi B, Rifkin A, Doddi S, Kolli R. The prevalence of anxiety disorders in patients with chronic obstructive pulmonary disease. Am J Psychiatry. 1990;147:200–1.

41. Perna G, Bertani A, Politi E, Colombo G, Bellodi L. Asthma and panic attacks. Biol Psychiatry. 1997;42:625–30.

42. Verburg K, Griez E, Meijer J, Pols H. Respiratory disorders as a possible predisposing factor for panic disorder. J Affect Disord. 1995;33:129–34.

43. Edlund MJ, McNamara ME, Millman RP. Sleep apnea and panic attacks. Compr Psychiatry. 1991;32: 130–2.

44. Sietsema KE, Simon JI, Wasserman K. Pulmonary hypertension presenting as a panic disorder. Chest. 1987;91:910–2.

45. Goodwin RD, Messineo K, Bregante A, Hoven CW, Kairam R. Prevalence of probable mental disorders among pediatric asthma patients in an inner-city clinic. J Asthma. 2005;42:643–7.

46. Shavitt RG, Gentil V, Mandetta R. The association of panic/agoraphobia and asthma. Contributing factors and clinical implications. Gen Hosp Psychiatry. 1992;14:420–3.

47. Kessler RC, Chiu WT, Jin R, Ruscio AM, Shear K, Walters EE. The epidemiology of panic attacks, panic disorder, and agoraphobia in the National Comorbidity Survey Replication. Arch Gen Psychiatry. 2006;63:415–24.

48. Weissman MM. The epidemiology of anxiety disorders: rates, risks and familial patterns. J Psychiatr Res. 1988;22 Suppl 1:99–114.

49. Baron C, Lamarre A, Veilleux P, Ducharme G, Spier S, Lapierre JG. Psychomaintenance of childhood asthma: a study of 34 children. J Asthma. 1986; 23:69–79.

50. Brooks CM, Richards Jr JM, Bailey WC, Martin B, Windsor RA, Soong SJ. Subjective symptomatology of asthma in an outpatient population. Psychosom Med. 1989;51:102–8.

51. Jurenec GS. Identification of subgroups of childhood asthmatics: a review. J Asthma. 1988;25:15–25.

52. Mayer DJ, Wolfle TL, Akil H, Carder B, Liebeskind JC. Analgesia from electrical stimulation in the brainstem of the rat. Science. 1971;174:1351–4.

53. Stefano GB, Scharrer B, Smith EM, Hughes Jr TK, Magazine HI, Bilfinger TV, et al. Opioid and opiate immunoregulatory processes. Crit Rev Immunol. 1996;16:109–44.

54. Santiago TV, Edelman NH. Opioids and breathing. J Appl Physiol. 1985;59:1675–85.

55. Mayfield KP, D'Alecy LG. Role of endogenous opioid peptides in the acute adaptation to hypoxia. Brain Res. 1992;582:226–31.

56. Baydur A. Nebulized morphine: a convenient and safe alternative to dyspnea relief? Chest. 2004; 125:363–5.

57. Mahler DA. Understanding mechanisms and documenting plausibility of palliative interventions for dyspnea. Curr Opin Support Palliat Care. 2011; 5(2):71–6.

58. Zebraski SE, Kochenash SM, Raffa RB. Lung opioid receptors: pharmacology and possible target for nebulized morphine in dyspnea. Life Sci. 2000; 66:2221–31.

59. Bruera E, Sala R, Spruyt O, Palmer JL, Zhang T, Willey J. Nebulized versus subcutaneous morphine for patients with cancer dyspnea: a preliminary study. J Pain Symptom Manage 2005;29:613–8.

60. Panksepp J. Affective neuroscience: the foundations of human and animal emotions. New York: Oxford University Press; 1998.

61. West MJ, King AP. Settling nature and nurture into an ontogenetic niche. Dev Psychobiol 1987;20:549–62.

62. Panksepp J, Newman JD, Insel TR. Critical conceptual issues in the analysis of separation-distress systems of the brain. In: Strongman KT, editor. International review of studies on emotion, vol. 2. New York: Wiley; 1992.

63. Kalin NH, Shelton SE, Barksdale CM. Separation distress in infant rhesus monkeys: effects of diazepam and Ro 15-1788. Brain Res. 1987;408:192–8.

64. Scott JP. Effects of psychotropic drugs in separation distress in dogs. Amsterdam: Proc IX Congress ECNP Exc Med; 1974.

65. Jurgens U, Ploog D. Cerebral representation of vocalization in the squirrel monkey. Exp Brain Res. 1970;10:532–54.

66. Ploog D. Neurobiology of primate audio-vocal behavior. Brain Res. 1981;228:35–61.

67. Wise SP, Herkenham M. Opiate receptor distribution in the cerebral cortex of the Rhesus monkey. Science. 1982;218:387–9.

68. Hofer MA, Shair H. Ultrasonic vocalization during social interaction and isolation in 2-weeek-old rats. Dev Psychobiol. 1978;11:495–504.

69. Kalin NH, Shelton SE, Barksdale CM. Opiate modulation of separation-induced distress in non-human primates. Brain Res. 1988;440:285–92.

70. Kehoe P, Blass EM. Opioid-mediation of separation distress in 10-day-old rats: reversal of stress with maternal stimuli. Dev Psychobiol. 1986;19: 385–98.

71. Panksepp J, Herman B, Conner R, Bishop P, Scott JP. The biology of social attachments: opiates alleviate separation distress. Biol Psychiatry. 1978;13: 607–18.

72. Kalin NH, Shelton SE. Effects of clonidine and propranolol on separation-induced distress in infant rhesus monkeys. Brain Res. 1988;470:289–95.

73. Gorman JM, Askanazi J, Liebowitz MR, Fyer AJ, Stein J, Kinney JM, et al. Response to hyperventilation in a group of patients with panic disorder. Am J Psychiatry. 1984;141:857–61.

74. Liebowitz MR, Fyer AJ, Gorman JM, Dillon D, Appleby IL, Levy G, et al. Lactate provocation of panic attacks: I. Clinical and behavioral findings. Arch Gen Psychiatry. 1984;41:764–70.

75. Papp LA, Klein DF, Gorman JM. Carbon dioxide hypersensitivity, hyperventilation, and panic disorder. Am J Psychiatry. 1993;150:1149–57.

76. Esquivel G, Schruers KR, Maddock RJ, Colasanti A, Griez EJ. Acids in the brain: a factor in panic? J Psychopharmacol. 2010;24(5):639–47.

77. Griez EJ, Colasanti A, van Diest R, Salamon E, Schruers K. Carbon dioxide inhalation induces dose-dependent and age-related negative affectivity. PLoS One. 2007;2(10):e987.

78. Leibold NK, Viechtbauer W, Goossens L, De Cort K, Griez EJ, Myin-Germeys I, et al. Carbon dioxide inhalation as a human experimental model of panic: the relationship between emotions and cardiovascular physiology. Biol Psychol. 2013;94(2):331–40.

79. Goetz RR, Klein DF, Gully D, Kahn J, Liebowitz M, Fyer A, et al. Panic attacks during placebo procedures in the laboratory: physiology and symptomatology. Arch Gen Psychiatry. 1993;50:280–5.

80. Martinez JM, Papp LA, Coplan JD, Anderson DE, Mueller CM, Klein DF, et al. Ambulatory monitoring of respiration in anxiety. Anxiety. 1996;2: 296–302.

81. Liebowitz MR, Gorman JM, Fyer AJ, Levitt M, Dillon D, Levy G, et al. Lactate provocation of panic attacks. II. Biochemical and physiological findings. Arch Gen Psychiatry. 1985;42:709–19.

82. Sinha SS, Goetz RR, Klein DF. Physiological and behavioral effects of naloxone and lactate in normal volunteers with relevance to the pathophysiology of panic disorder. Psychiatry Res. 2007;149:309–14.

83. Esquivel G, Fernández-Torre O, Schruers KR, Wijnhoven LL, Griez EJ. The effects of opioid receptor blockade on experimental panic provocation with CO_2. J Psychopharmacol. 2009;23(8): 975–8.

84. Liebowitz MR, Gorman JM, Fyer AJ, Dillon DJ, Klein DF. Effects of naloxone on patients with panic attacks. Am J Psychiatry. 1984;141(8):995–7.

85. Preter M, Lee SH, Petkova E, Vannucci M, Kim S, Klein DF. Controlled cross-over study in normal subjects of naloxone-preceding-lactate infusions; respiratory and subjective responses: relationship to endogenous opioid system, suffocation false alarm theory and childhood parental loss. Psychol Med. 2011;41(2):385–93.

86. Liu Z, Li X, Ge X. Left too early: the effects of age at separation from parents on Chinese rural children's symptoms of anxiety and depression. Am J Public Health. 2009;99:2049–54.

87. Bandelow B, Spath C, Tichauer GA, Broocks A, Hajak G, Ruther E. Early traumatic life events, parental attitudes, family history, and birth risk factors in patients with panic disorder. Compr Psychiatry. 2002;43:269–78.

88. Crawford TN, Cohen PR, Chen H, Anglin DM, Ehrensaft M. Early maternal separation and the trajectory of borderline personality disorder symptoms. Dev Psychopathol. 2009;21(3):1013–30.

89. Kendler KS, Neale MC, Kessler RC, Heath AC, Eaves LJ. Childhood parental loss and adult psychopathology in women. A twin study perspective. Arch Gen Psychiatry. 1992;49:109–16.

90. Shonkoff JP, Boyce WT, McEwen BS. Neuroscience, molecular biology, and the childhood roots of health disparities: building a new framework for health promotion and disease prevention. JAMA. 2009; 301:2252–9.

91. Brown DW, Anda RF, Tiemeier H, Felitti VJ, Edwards VJ, Croft JB, et al. Adverse childhood experiences and the risk of premature mortality. Am J Prev Med. 2009;37:389–96.

92. Juang KD, Wang SJ, Fuh JL, Lu SR, Chen YS. Association between adolescent chronic daily headache and childhood adversity: a community-based study. Cephalalgia. 2004;24:54–9.

93. Kopec JA, Sayre EC. Stressful experiences in childhood and chronic back pain in the general population. Clin J Pain. 2005;21:478–83.

94. Harter MC, Conway KP, Merikangas KR. Associations between anxiety disorders and physical illness. Eur Arch Psychiatry Clin Neurosci. 2003; 253:313–20.

95. Jones GT, Power C, Macfarlane GJ. Adverse events in childhood and chronic widespread pain in adult life: results from the 1958 British Birth Cohort Study. Pain. 2009;143:92–6.

96. Katerndahl DA. Chest pain and its importance in patients with panic disorder: an updated literature review. Prim Care Companion J Clin Psychiatry. 2008;10:376–83.

97. Battaglia M, Pesenti-Gritti P, Medland SE, Ogliari A, Tambs K, Spatola CA. A genetically informed study of the association between childhood separation anxiety, sensitivity to CO(2), panic disorder, and the effect of childhood parental loss. Arch Gen Psychiatry. 2009;66:64–71.

98. Gerra G, Leonardi C, D'Amore A, Strepparola G, Fagetti R, Assi C, et al. Buprenorphine treatment outcome in dually diagnosed heroin dependent patients: a retrospective study. Prog Neuropsychopharmacol Biol Psychiatry. 2006;30:265–72.

99. Wallen MC, Lorman WJ, Gosciniak JL. Combined buprenorphine and chlonidine for short-term opiate detoxification: patient perspectives. J Addict Dis. 2006;25:23–31.

100. Sluka KA, Deacon M, Stibal A, Strissel S, Terpstra A. Spinal blockade of opioid receptors prevents the analgesia produced by TENS in arthritic rats. J Pharmacol Exp Ther. 1999;289:840–6.

101. Moreira FA, Gobira PH, Viana TG, Vicente MA, Zangrossi H, Graeff FG. Modeling panic disorder in rodents. Cell Tissue Res. 2013;354:119–25.

102. Graeff FG. New perspective on the pathophysiology of panic: merging serotonin and opioids in the periaqueductal gray. Braz J Med Biol Res. 2012; 45(4):366–75.

103. Schenberg LC, Schimitel FG, Armini Rde S, Bernabe CS, Rosa CA, Tufik S, et al. Translational approach to studying panic disorder in rats: hits and misses. Neurosci Biobehav Rev. 2014;46(Pt 3): 472–96.

104. Schimitel FG, De Almeida GM, Pitol DN, Armini RS, Tufik S, Schenberg LC. Evidence of a suffocation alarm system within the periaqueductal gray matter of the rat. Neuroscience. 2012;200:59–73.

105. Armini R, Bernabe CS, Rosa CA, Siller CA, Schimitel FG, Tufik S, et al. In a rat model of panic, corticotropin responses to dorsal periaqueductal gray stimulation depend on physical exertion. Psychoneuroendocrinology. 2015;53:136–47.

106. Schenberg LC, Dos Reis AM, Ferreira Póvoa RM, Tufik S, Silva SR. A panic attack-like unusual stress reaction. Horm Behav. 2008;54(5):584–91.

107. Stein G. Panic disorder: the psychobiology of external treat and introceptive distress. CNS Spectrums. 2008;13(1):26–30.

108. Gorman JM, Kent JM, Sullivan GM, Coplan JD. Neuroanatomical hypothesis of panic disorder, revised. Am J Psychiatry. 2000;157(4):493–505.

109. Feinstein JS, Buzza C, Hurlemann R, Follmer RL, Dahdaleh NS, Coryell WH, et al. Fear and panic in humans with bilateral amygdala damage. Nat Neurosci. 2013;16(3):270–2.

110. Preter M, Klein DF. Lifelong opioidergic vulnerability through early life separation: a recent extension of the false suffocation alarm theory of panic disorder. Neurosci Biobehav Rev. 2014;46(Pt 3): 345–51.

111. Eisenberger NI, Lieberman MD. Why rejection hurts: a common neural alarm system for physical and social pain. Trends Cogn Sci. 2004;8(7):294–300.

112. Hsu DT, Sanford BJ, Meyers KK, Love TM, Hazlett KE, Walker SJ, et al. It still hurts: altered endogenous opioid activity in the brain during social rejection and acceptance in major depressive disorder. Mol Psychiatry. 2015;20(2):193–200.

Circadian Rhythm in Panic Disorder

8

Michelle Levitan and Marcelo Papelbaum

Contents

Abstract

The relationship between panic disorder and sleep problems has been studied, and possible explanations for this association are discussed in this chapter. So far, the results of polysomnographic studies in PD patients are inconclusive, but seem to suggest that patients with PD have impaired initiation and maintenance of sleep. The presence of nocturnal panic attacks induce an intense fear of sleep, leading to anticipatory anxiety and sleep avoidance, resulting in secondary insomnia and facilitating the development of new panic attacks. Other hypotheses, as the co-occurrence of depression, cortisol levels and anxiety sensitivity are also raised as mechanisms related to sleep problems in PD patients. Treatments are available, as the cognitive behavioral treatment and some novel treatments that may improve panic attacks as well as insomnia.

Keywords
Panic disorder • Sleep • Nocturnal panic attacks • Sleep treatment

M. Levitan (✉)
Laboratory of Panic and Respiration, Institute of Psychiatry, Federal University of Rio de Janeiro, Rio de Janeiro, Brazil
e-mail: milevitan@gmail.com

M. Papelbaum
State Institute of Diabetes and Endocrinology of Rio de Janeiro, Rio de Janeiro, Brazil
e-mail: marcelo@papelbaum.com

8.1 Introduction

The relationship between psychiatric disorders and sleep problems has been widely studied. Indeed, sleep complains integrate such an important feature of psychiatric disorders, being, for instance, part of the diagnostic criteria for depression and generalized anxiety disorder. Specifically for panic disorder (PD), the presence of sleep disturbance is very common, with 68–77 % of the patients complaining about trouble with sleep [1]. However, there is a paucity of research regarding the comorbidity between this disorder and sleep disturbance.

To evaluate sleep quality, subjective reports and polysomnographic data are often used. The clinical interview and scales are important to identify sleep complaints and accordingly, refer to a more accurate exam, as a polysomnography, that provides sleep cycles and help identifying any alteration. However, because PD is not directly associated to sleep problems, health professionals tend to spend less time evaluating its occurrence and impact.

Possible explanations for the association between PD and sleep difficulties are discussed in this chapter, focusing mostly in insomnia, the most common sleep disorder in PD. A potential link between both disorders relies on the occurrence of nocturnal panic attacks (NPA), that happen repeatedly in 20–45 % of PD patients [2, 3] and appear to predispose patients to be fearful and to stay awake to avoid their recurrence. Other association hypotheses are discussed, as the physiological arousal, increased base levels of cortisol [4] or high anxiety sensitivity. At last, the comorbid depression [3] as an intermediate between sleep complaints and PD is addressed.

8.2 Sleep Cycles

For a better understanding, sleep architecture stages are exposed according to the Committee on Sleep Medicine and Research [5] and differential diagnosis are made between sleep problems (Table 8.1). Overall, sleep is divided in two main types: rapid eye movement (REM) and non-REM (NREM), the latter divided in four stages. A sleep episode begins in the first stage of NREM sleep through the four stage, and ends up at REM sleep. Individuals do not remain in REM sleep the rest of the night but, rather, cycle between stages of NREM and REM throughout the night.

Stage 1 is easily interrupted by a disruptive noise. Brain activity on the electroencephalogram in stage 1 transitions from wakefulness (marked by rhythmic alpha waves) to low-voltage, mixed-frequency waves, in which alpha waves are associated with a wakeful relaxation state. It constitutes 2–5 % of the total sleep.

An individual in stage 2 sleep requires more intense stimuli than in stage 1 to awaken. In this stage, brain activity on an EEG shows relatively low-voltage, mixed-frequency activity characterized by the presence of sleep spindles and K-complexes. It lasts 45–55 % of total sleep.

Sleep stages 3 and 4 are collectively referred to as slow-wave sleep, in which the EEG shows increased high-voltage, slow-wave activity. It lasts 10–15 % of sleep. The REM stage is defined by the presence of desynchronized (low-voltage,

Table 8.1 Differential diagnosis between nocturnal panic and sleep complaints Institute of Medicine (US) Committee on Sleep Medicine and Research

Disorder	Definition	Cycle of sleep
Nocturnal panic attack	Waking up in a state of horror	Late stage II or early stage III sleep
Insomnia	Trouble in falling or staying asleep	Before sleep begins
Nightmares	Distressing dream that forces awakenings; the person usually remembers the episode	Occur largely in REM sleep, after hours of sleep
Night terrors	Cause feelings of terror; for the sleeper is hard to awaken	Occur mostly in stage IV
Obstructive sleep apnea	Upper airway obstruction occur repeatedly during sleep	Occurs in stage I and II and REM sleep

Sleep Disorders and Sleep Deprivation: An Unmet Public Health Problem. Washington (DC): National Academies Press (US); 2006. 3, Extent and Health Consequences of Chronic Sleep Loss and Sleep Disorders [5]

mixed-frequency) brain wave activity and bursts of rapid eye movements, in which dreaming is most often associated.

8.3   Sleep Quality Studies

Irregular cycling sleep is associated with sleep disorders. Subjective reports of PD patients evidenced that 68 % of the sample described difficulties in falling asleep and 77 % reported disturbed sleep [6]. Higher percentages of sleep complaints in PD patients compared to healthy subjects were found in other study, especially middle night insomnia (67 % vs 23 %) and late night insomnia (67 % vs 31 %) [7].

So far, the results of polysomnographic studies in PD patients are inconclusive, but seem to suggest that patients with PD have impaired initiation and maintenance of sleep, characterized by increased sleep latency and increased time awake after sleep onset [3, 8]. Some authors also observed a decrease in sleep efficiency, total sleep time and amount of non-REM sleep in stage 4 compared to healthy controls, whereas others identified an increased percentage of non-REM sleep stage 1 in PD patients [3, 9, 10]. Regarding patients with repeated NPAs, more severe subjective sleep complaints were reported when compared to those with only diurnal panic attacks, but no differences were found on electroencephalogram indices or polysomnographic parameters [11, 12].

Commentary must be made on the possible co-occurrence of PD with another sleep disorder: the obstructive sleep apnea (OSA). The OSA may cause NPAs symptoms [13], and treating one condition with continuous positive airway pressure (CPAP), could impact on the prognosis of the other one [14, 15]. Thus, in addition to electroencephalogram alterations, respiratory parameters might be able to guide the accurate diagnosis and treatment of sleep problems, and impact the clinical course of panic symptoms.

8.4   Mechanism of Association Between Panic and Sleep Disorders

8.4.1   Obstructive Sleep Apnea Syndrome

Studies investigating the prevalence of both disorders in clinical samples are scarce, however attention must be drawn to this comorbidity because both conditions are relative common, with sleep apnea prevalence ranging between 2–4 % on population-based studies [16]. In addition, authors suggest that the presence of sleep apnea could be associated to higher prevalence of psychiatric comorbid conditions [17].

Patients with PD have greater respiratory variability than comparison subjects [18]. This trait marker may predispose patients to CO_2-induced panic attack [19]. In addition, respiratory events during sleep, in the presence o comorbid OSA, may trigger panic attacks. In this way, treatment results indicate that the vast majority of PD with OSA patients experienced a diminution or disappearance of panic attacks with CPAP, besides a decrease in the use of alprazolam, what might suggest that CPAP works as a treatment for diurnal and NPAs alone too [17]. These data seem to be in accordance with an influential etiological hypothesis for PD, the Klein's suffocation alarm theory [20], in which panic attacks would be a result of inappropriate activation of an alarm system by the brain that signals lack of air [18, 20].

8.4.2   Nocturnal Panic Attacks

NPAs refer to waking up in a state of panic, defined as an abrupt and discrete period of intense fear and discomfort accompanied by cognitive and physical symptoms of arousal [21]. They are NREM-related events, usually emerging from late stage 2 or early stage 3 of sleep [3] and differ

Fig. 8.1 Nocturnal
panic attacks cycle

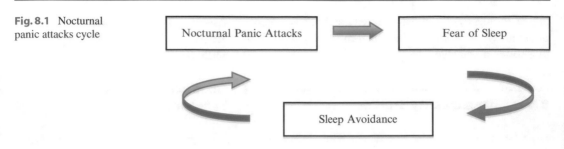

from sleep problems regarding clinical features and stage of occurrence.

The association between PD and sleep complaints is often attributed to NPAs, that may occur in 33–69 % of PD patients [7]. Due to the nocturnal occurrence, they induce an intense fear of sleep, leading to anticipatory anxiety and sleep avoidance, resulting in secondary insomnia and facilitating the development of new NPAs [22] (Fig. 8.1). This perspective has its foundation in the cognitive behavioral model, in which learned fearfulness of bodily sensations leads to reactivity changes in bodily state. Indeed, slight changes in heartbeat and sweat have been observed preceding NPAs [23].

In a polysomnographic study, Stein et al. [3] found that non-depressed PD patients presented regular sleep stages indistinguishable from the healthy group. An interesting finding of this study was that healthy individuals presented brief arousals ten times every hour. The authors point out that for PD patients, each of these arousals represents an opportunity to become aware of his sensations and end up interpreting them as dangerous, possibly leading to a NPA [24].

Some authors suggest that NPAs would be a more severe subtype of PD. This hypothesis is based in studies that found that patients who had NPAs reported more fears and a more severe history of psychopathology than those who only had daytime panic attacks [25]. Other studies also found and a greater severity of symptoms, compared to diurnal panic attacks, especially chest pain, nausea and discomfort [26]. Regarding sleep quality, patients with NPAs did not differ significantly from those with diurnal panic attacks, except a tendency to report a more severe self-rating of past and current sleep difficulties with sleep onset [24].

8.4.3 Depression

Survey data on sleep and psychiatric disorders in 14,915 subjects using questionnaires evidenced that insomnia is the most frequent sleep disorder in psychiatric disorders [27]. However, the natural course of it incidence tend to differ according to the type of psychopathology. Specifically, insomnia tended to precede the onset of mood disorders in 40 % of cases, whereas in the case of anxiety disorders, in 80 % of the time it appeared either at the same time or following the onset of anxious symptoms [27]. Therefore, sleep disorders and psychiatric disorders seem to interact in multiple ways [27].

Previous research has evidenced association between sleep complaints and comorbid depressive symptomatology in patients with PD. There are some possible explanations for this relationship. First, the occurrence of sleep problems might be associated directly to the comorbid depression, commonly coexisted with PD. In fact, depression occurs in 50–65 % of PD patients. A second possibility is that comorbid depression may be more common in the presence of a severe form of PD that features with sleep problems [28]. Lastly, insomnia itself, as a part of PD symptoms might be a risk factor for depression comorbidity. Indeed, longitudinal studies with individuals with insomnia demonstrated an increase in the likelihood of developing major depression, especially when the sleep disorder is persistent [29, 30]. Moreover, insomnia might be an independent risk factor for the development of PD. In this way, results of the epidemiological catchment area revealed that individuals with insomnia were five times more likely to have a panic attack and being at risk of developing PD than individuals without insomnia. For these subjects, insomnia persisted after the remission of the mental disorder [31].

8.4.4   The Cortisol and Panic Disorder

The hypothalamic-pituitary-adrenal (HPA) axis is frequently referred as disturbed in PD [32]. HPA regulates cortisol secretion and is under the influence of various factors such as sleep, that exerts an inhibitory effect on cortisol secretion [34]. During nocturnal awakenings there is a transient elevation of cortisol levels, followed by temporary inhibition of cortisol secretion [33].

Studies have pointed out the importance of stress hormone regulation in anxiety disorders [32, 33]. Patients with PD and controls participated in a study in which salivary and urinary cortisol levels determined in 2-h spans during 3 consecutive days. The more severe group of clinical patients presented salivary cortisol, nocturnal and urinary cortisol levels significantly higher than the controls in the whole group of PD patients. Other studies found elevated baseline plasma levels of cortisol in PD patients [34, 35]. The results suggest that cortisol elevations are more pronounced during the night, mainly in severe ill PD patients [32].

Despite the results, the authors still cannot affirm whether increased cortisol levels in PD patients only reflect the chronic stress due to panic attacks and anticipatory anxiety or are the expression of a neurobiological defect involved in the pathogenesis of panic attacks [32].

8.4.5   Anxiety Sensitivity

Anxiety sensitivity (AS) is a dispositional feature associated to a excessive fear of anxiety-related sensations based on beliefs that these sensations are harmful [36]. The AS is considered a main factor in the development and maintenance of PD. Studies indicate that the association between AS and PD is largely attributable to the somatic aspects of AS [37], considered not the fear of the general anxiety, but the fear of physical sensations that are associated to the panic related-psychology [38]. Based on these data, a possible explanation for sleep complaints and PD would rely on a selective attention to fear of anxiety and physical sensations that would occur at night while trying to fall asleep [39]. Studies show an association between AS and sleep disturbance and suggest that AS may contribute to initial insomnia in PD [39].

This hypothesis lead authors the suggestion to include insomnia-targeted components in the cognitive behavioral therapy (CBT) interventions, which are effective to reduce AS in PD patients [40] and may interfere with a possible development of sleep problems. Besides CBT techniques as interoceptive exposure, pharmacotherapy seems to decrease AS levels well [41].

8.5   Treatment

Mellman et al. points out that there is overlap between interventions that target sleep disturbances and those that are used in treating anxiety disorders [2]. Overlapping approaches include medications and cognitive behavioral strategies that target worry, tension, and maladaptive cognitions. Optimal sequencing or integration of treatments targeting anxiety and sleep disturbance were not fully investigated.

Some authors suggest that once treatment is initiated at the earliest phase, sleep problems in anxious patients are considered secondary symptoms that improve with the other symptomatology. Clinical experience shows that most patients with comorbid anxiety and insomnia seek treatment months to years after the initial presentation of symptoms. At this point, it is necessary to treat both symptoms [42].

However, others state that in this comorbid treatment, insomnia should be treated concurrently with, but independently of the anxiety disorder *per se*. Many authors criticize the idea that the clinician should wait to evaluate whether the insomnia improves as a consequence of the anxiety disorder treatment. Clinical experience has shown that without targeted insomnia treatment, insomnia frequently persists [43].

8.5.1 Pharmacological Treatment

Specific treatments for sleep disturbance are available; however studies with PD and these conditions are scarce and hardly ever address this complaint. In fact, most studies related to treatment of anxiety disorders with comorbid insomnia were conducted with generalized anxiety disorder and post-traumatic stress disorder.

Despite well-established pharmacological treatments for panic symptoms, persistence of sleep complaints is common and, therefore, discontinuation of medication for comorbid chronic insomnia proves to be a difficult task. Benzodiazepines (BDZ) represent an advance by offering safer alternatives that demonstrated long-term benefit, but they also showed problems of occasional dependence and residual side effects. Nevertheless, attempts to slow tapering chronic use of BDZ in asymptomatic PD patients should be made. Nardi et al. [44] showed successful discontinuation of 3-years of more use of clonazepam in patients with PD over the course of 4 months with mild and transient withdrawal symptoms. However, although panic attacks might be remitted, the re-occurrence of sleep problems on the follow-up is common. In a study that intervened in the long-term use of benzodiazepines for the treatment of chronic insomnia, efforts to discontinue medication resulted in a high rate of return to the hypnotic over the 2 years of follow-up [43].

In the case of antidepressants, the improvements of PD symptoms, commonly, do not reflect into benefits of sleep complaints. Small but significant change of the subjective quality of sleep might be noted, objective sleep parameters such as total sleep time and sleep onset latency remain unchanged [45]. Indeed, Todder et al. [46] found improvement of subjectively quality of sleep and persistence of objective parameter alterations in patients with PD treated with escitalopram. Also, the study did not found a correlation between clinical improvement and the changes in the sleep quality, suggesting that the mechanism associated to improvement of the sleep might be non-specific related to the reduction of panic attacks. In this way, antidepressants with sedative side effects might be helpful in patients with anxiety who also report significant comorbid insomnia [46].

Mirtazapine is a noradrenergic and specific serotoninergic antidepressant with known sedative effect. It has been studied in some uncontrolled and head-to-head studies (against active treatments), evidencing efficacy in the remission of panic symptoms [47]. However, specific sleep parameters were not evaluated. Nevertheless, two case reports showed remission of insomnia with mirtazapine added onto selective serotonin reuptake inhibitors treatment [47, 48]. In this way, it is worth mentioning the use of antidepressant trazodone (serotonin antagonist and reuptake inhibitor) as a sleep-inducing medication, especially at lower doses [49]. Although, only two past studies showed limited or negative benefit on PD, the use of trazodone as an add-on therapy to mitigate sleep problems must be kept in mind, especially when BDZ use should be avoided [50, 51].

Another possibility is agomelatine, an antidepressant with the novel mechanism of being a selective melatonergic MT1/MT2 receptor agonist with serotonin 5-HT2c and 5-HT2b receptor antagonist activities [52]. Although still not fully elucidated, it seems that its melatonergic agonism and its 5-HT2c antagonism could act synergistically in the restoration of disrupted circadian rhythms [52]. So far, few studies evaluated the efficacy of agomelatine in treating anxiety disorders and sleep symptoms, mostly in generalized anxiety disorder. Overall, the use of agomelatine was associated with a larger improvement in subjective sleep symptoms, including getting of to sleep, quality of sleep and sleep awakening [54]. Independently of the reduction of the anxiety symptoms. Fornaro et al. [55] discussed a possible benefit of agomelatine in diminishing the need for benzodiazepines, which are highly prescribed in patients with PD [53]. This effect of agomelatine could be a protective factor in reducing the risk of benzodiazepines abuse or dependence, with the benefit of improving sleep, being an off-label option in the treatment of PD, especially when the usual serotonin reuptake inhibitors are not well-tolerated [56].

Finally, regarding the comorbidity between PD and OSA, consideration must be made regarding the pharmacologic treatment of panic symptoms. In a PD patient with OSA symptoms such as sleep apnea episodes, if the clinician does not recognize the OSA, he could prescribe benzodiazepines. In this case the benzodiazepine could suppress the tonus of the upper airway respiratory muscles, making sleep apnea worse [15, 17]. On the other hand, when specific treatment for OSA is initiated, PD patients could have their symptoms diminished or disappeared with CPAP, a treatment that can suppress upper airway collapse during sleep with a pneumatic splint [17, 57].

8.5.2   Psychological Treatment

CBT is a short-term, multi-component psychotherapy, currently considered the treatment of choice for insomnia [58] and PD [59]. When these conditions co-occur, a CBT protocol that addresses both difficulties: panic attacks and insomnia, is highly recommended [22]. Based on the most intense difficulty for the patient (panic attack or insomnia), the treatment may begin with one target at a time or manage both at the same time. This decision will depend on the patient impairment and the therapist management. The sessions are divided into goals and stages, adjusted to each patient; bellow are some concise outlines [58–60].

- Step 1: Sleep and PD education: Provides information about anxiety and sleep. The therapists help the patients to identify dysfunctional beliefs about panic attacks and sleep and correct them.
- Step 2: Breathing and relaxation techniques: Helps the patients to drift off to sleep as well as deactivate the hyper stimulation of the autonomic nervous system. It is possible that NPAs patients will be too sensitive to a relaxation state similar to the one felt during sleep, avoiding this exercise. This should be accomplished when the patient feels ready.
- Step 3: Behavioral changes: This behavioral phase use stimulus control and sleep restriction strategies to regularize the patient's sleep/wake schedule and eliminate sleep incompatible behaviors in an effort to force the development of an efficient consolidated sleep pattern. Some sleep hygiene orientations (sleep rules) are given: (1) To choose a wake up time; (2) not to entertain with activities while in bed; (3) not to stay in bed if not to sleep; (4) avoid daytime napping; (5) try not to worry with problems in bed and (6) go to bed when sleepy.
- Step 4: Cognition restructuring: Targets misappraisals of bodily sensations as threatening or dangerous, being the technique most used in CBT. The therapist helps the patient to identify and correct negative thoughts by evaluating evidence for and against them. Patients with insomnia often develop erroneous thoughts that worsen their difficulties, such as "I'll never get to sleep tonight;" "I'll be a wreck tomorrow;" "I'll get sick unless I sleep eight hours a night". By the same way NPAs patients tend to anticipate the consequences of their sensations, ultimately believing that that they will have a heart attack or die while sleeping.

8.6   Conclusions

The importance of sleep in PD should not be underestimated. Nearly 80 % of PD patients complain about disturbed sleep. In fact, this relationship is so intimate that a subtype of PD with nocturnal symptoms is recognized. Indeed, due to the sudden arousal without an obvious trigger, NPAs are considered a severe form of PD that leads to anticipatory anxiety and sleep avoidance.

Several hypotheses attempt to explain the mechanisms related to the association between PD and sleep difficulties. Regardless, in clinical practice, investigation of sleep problems in patients with PD is of extreme importance. Firstly, it can provide symptom relief to the patient and, secondly, the alleviation of sleep disturbance can have a positive impact on panic symptoms. Additional, although not apparently related to the panic symptoms, clinical problems such should also be investigated, especially because treatment of one condition could affect the other one. Therefore, asking family members about patient's sleep or submitting patient to a polysomnography could help in the exclusion of differential diagnosis.

Although pharmacological treatment for PD is effective, remission of panic symptoms is commonly associated with persistence of sleep complaints, which makes it difficult to discontinue medication. In this way, the review of different pharmacological strategies seen in this chapter, also showed the need of other nonpharmacological tools, such as the ones presented based on CBT, to address issues related to irrational thoughts, avoidance and sleep hygiene, considered main components for the patient's improvement.

References

1. Papadimitriou GN, Linkowski P. Sleep disturbance in anxiety disorders. Int Rev Psychiatry. 2005;17(4):229–36.
2. Mellman TA. Sleep and anxiety disorders. Psychiatr Clin North Am. 2006;29(4):1047–58.
3. Stein MB, Enns MW, Kryger MH. Sleep in nondepressed patients with panic disorder: II. Polysomnographic assessment of sleep architecture and sleep continuity. J Affect Disord. 1993;28(1):1–6.
4. Bonnet MH, Arand DL. Hyperarousal and insomnia: state of the science. Sleep Med Rev. 2010;14(1):9–15.
5. Institute of Medicine (US) Committee on Sleep Medicine and Research; Colten HR, Altevogt BM, editors. Sleep disorders and sleep deprivation: an unmet public health problem. Washington, DC: National Academies Press (US); 2006. 3, Extent and Health Consequences of Chronic Sleep Loss and Sleep Disorders. Available from: http://www.ncbi.nlm.nih.gov/books/NBK19961.
6. Sheehan DV, Ballenger J, Jacobsen G. Treatment of endogenous anxiety with phobic, hysterical and hypochondriacal symptoms. Arch Gen Psychiatry. 1980;37:51–9.
7. Mellman TA, Uhde TW. Sleep in panic and generalised anxiety disorder. In: Ballanger J, editor. Neurobiology of panic disorder. New York, NY: Alan R. Liss; 1990. p. 365–6.
8. Staner L. Sleep and anxiety disorders. Dialogues Clin Neurosci. 2003;5(3):249–58.
9. Sloan EP, Natarajan M, Baker B, Dorian P, Mironov D, Barr A, et al. Nocturnal and daytime panic attacks—comparison of sleep architecture, heart rate variability, and response to sodium lactate challenge. Biol Psychiatry. 1999;45:1313–20.
10. Arriaga F, Paiva T, Matos-Pires A, et al. The sleep of non-depressed patients with panic disorder: a comparison with normal controls. Acta Psychiatr Scand. 1996;93(3):191–4.
11. Uhde TW, Cortese BM, Vedeniapin A. Anxiety and sleep problems: emerging concepts and theoretical treatment implications. Curr Psychiatry Rep. 2009;11(4):269–76.
12. Landry P, Marchand L, Mainguy N, Marchand A, Montplaisir J. Electroencephalography during sleep of patients with nocturnal panic disorder. J Nerv Ment Dis. 2002;190(8):559–62.
13. Edlund MJ, McNamara ME, Millman RP. Sleep apnea and panic attacks. Compr Psychiatry. 1991;32:130–2.
14. Hanly P, Powles P. Hypnotics should never be used in patients with sleep apnea. J Psychosom Res. 1993;37 Suppl 1:59–65.
15. Takaesu Y, Inoue Y, Komada Y, Kagimura T, Iimori M. Effects of nasal continuous positive airway pressure on panic disorder comorbid with obstructive sleep apnea syndrome. Sleep Med. 2012;13(2):156–60.
16. Lee W, Nagubadi S, Kryger MH, Mokhlesi B. Epidemiology of obstructive sleep apnea: a population-based perspective. Expert Rev Respir Med. 2008;2(3):349–64.
17. Sharafkhaneh A, Giray N, Richardson P, Young T, Hirshkowitz M. Sleep. 2005;28(11):1405–11.
18. Martinez JM, Kent JM, Coplan JD, Browne ST, Papp LA, Sullivan GM, et al. Respiratory variability in panic disorder. Depress Anxiety. 2001;14(4):232–327.
19. Freire RC, Nardi AE. Panic disorder and the respiratory system: clinical subtype and challenge tests. Rev Bras Psiquiatr. 2012;34 Suppl 1:S32–41.
20. Klein DF. False suffocation alarms, spontaneous panics, and related conditions: an integrative hypothesis. Arch Gen Psychiatry. 1993;50(4):306–17.
21. American Psychiatric Association. Diagnostic and statistical manual of mental disorders, 4th ed. Text Revision (DSM-IV-TR). Arlington, VA: American Psychiatric Publishing; 2000.
22. Craske MG, Tsao JC. Assessment and treatment of nocturnal panic attacks. Sleep Med Rev. 2005;9(3):173–84.
23. Roy-Byrne P, Mellman T, Uhde T. Biological findings in panic disorder: neuroendocrine and sleep- related abnormalities. J Anxiety Dis. 1988;2:17–29.
24. Craske MG, Lang AJ, Mystkowski JL, Zucker BG, Bystritsky A, Yan-Go F. Does nocturnal panic represent a more severe form of panic disorder? J Nerv Ment Dis. 2002;190(9):611–8.
25. Labbate LA, Pollack MH, Otto MW, Langenauer S, Rosenbaum JF. Sleep panic attacks: an association with childhood anxiety and adult psychopathology. Biol Psychiatry. 1994;36:57–60.
26. Craske MG, Barlow DH. Nocturnal panic. J Nerv Ment Dis. 1989;177:160–7.
27. Ohayon MM, Roth T. Place of chronic insomnia in the course of depressive and anxiety disorders. J Psychiatr Res. 2003;37(1):9–15.

28. Mellman TA. Sleep and anxiety disorders. Sleep Med Clin. 2008; 261–8.

29. Roberts RE, Shema SJ, Kaplan GA, Strawbridge WJ. Sleep complaints and depression in an aging cohort: a prospective perspective. Am J Psychiatry. 2000;157: 81–8.

30. Ford DE, Kamerow DB. Epidemiologic study of sleep disturbances and psychiatric disorders. An opportunity for prevention? JAMA. 1989;262:1479–84.

31. Breslau N, Roth T, Rosenthal L, Andreski P. Sleep disturbance and psychiatric disorders: a longitudinal epidemiological study of young adults. Biol Psychiatry. 1996;39:411–8.

32. Bandelow B, Wedekind D, Sandvoss V, Broocks A, Hajak G, Pauls J, et al. Diurnal variation of cortisol in panic disorder. Psychiatry Res. 2000;95:245–50.

33. Balbo M, Leproult R, Van Cauter E. Impact of sleep and its disturbances on hypothalamo-pituitary-adrenal axis activity. Int J Endocrinol. 2010;vol. 2010, Article ID 759234, 16 pages, 2010. doi:10.1155/2010/759234.

34. Nesse RM, Cameron OG, Curtis GC, McCann DS, Huber Smith MJ. Adrenergic function in patients with panic anxiety. Arch Gen Psychiatry. 1984; 41:771–6.

35. Goldstein S, Halbreich U, Asnis G, Endicott J, Alvir J. The hypothalamic-pituitary-adrenal system in panic disorder. Am J Psychiatry. 1987;144:1320–3.

36. Reiss S, Peterson RA, Gursky DM, McNally RJ. Anxiety sensitivity, anxiety frequency and the predictions of fearfulness. Behav Res Ther. 1986;24:1–8.

37. McNally RJ. Anxiety sensitivity and panic disorder. Biol Psychiatry. 2002;52(10):938–46.

38. Deacon B, Abramowitz J. Anxiety sensitivity and its dimensions across the anxiety disorders. J Anxiety Disord. 2006;20(7):837–57.

39. Hoge EA, Marques L, Wechsler RS, Lasky AK, Delong HR, Jacoby RJ, et al. The role of anxiety sensitivity in sleep disturbance in panic disorder. J Anxiety Disord. 2011;25(4):536–8.

40. Smits JA, Berry AC, Tart CD, Powers MB. The efficacy of cognitive-behavioral interventions for reducing anxiety sensitivity: a meta-analytic review. Behav Res Ther. 2008;46:1047–54.

41. Simon NM, Otto MW, Smits JA, Nicolaou DC, Reese HE, Pollack MH. Changes in anxiety sensitivity with pharmacotherapy for panic disorder. J Psychiatr Res. 2004;38:491–5.

42. Marcks BA, Weisberg RB. Co-occurrence of insomnia and anxiety disorders: a review of the literature. Am J Lifestyle Med. 2009;3:300–9.

43. Morin CM. Insomnia: psychological assessment and management. New York: Guillford Press; 1993.

44. Nardi AE, Freire RC, Valença AM, Amrein R, de Cerqueira AC, Lopes FL, et al. Tapering clonazepam in patients with panic disorder after at least 3 years of treatment. J Clin Psychopharmacol. 2010;30(3): 290–3.

45. Cervena K, Matousek M, Prasko J, Brunovsky M, Paskova B. Sleep disturbances in patients treated for panic disorder. Sleep Med. 2005;6(2):149–53.

46. Todder D, Baune BT. Quality of sleep in escitalopram-treated female patients with panic disorder. Hum Psychopharmacol. 2010;25(2):167–73.

47. Milan Pavlovic Z. Remission of panic disorder with mirtazapine augmentation of paroxetine: a case report. Prim Care Companion J Clin Psychiatry. 2007;9(5):396.

48. Uguz F. Low dose mirtazapine added to selective serotonin reuptake inhibitors in pregnant women with major depression or panic disorder including symptoms of severe nausea, insomnia and decreased appetite: three cases. J Matern Fetal Neonatal Med. 2013;26(11):1066–8.

49. Bossini L, Casolaro I, Koukouna D, Cecchini F, Fagiolini A. Off label uses of trazodone: a review. Expert Opin Pharmacother. 2012;13(12):1707–17.

50. Charney DS, Woods SW, Goodman WK, Rifkin B, Kinch M, Aiken B, et al. Drug treatment of panic disorder: the comparative efficacy of imipramine, alprazolam, and trazodone. J Clin Psychiatry. 1986;47(12): 5806.

51. Serretti A, Chiesa A, Calati R, Perna G, Bellodi L, De Ronchi D. Novel antidepressants and panic disorder: evidence beyond current guidelines. Neuropsychobiology. 2011;63(1):1–7.

52. McAllister-Williams RH, Baldwin DS, Haddad PM, Bazire S. The use of antidepressants in clinical practice: focus on agomelatine. Hum Psychopharmacol. 2010;25(2):95–102.

53. Levitan MN, Papelbaum M, Nardi AE. A review of preliminary observations on agomelatine in the treatment of anxiety disorders. J Clin Psychol. 2012; 68(4):397–402.

54. Stein DJ, Ahokas AA, de Bodinat C. Efficacy of agomelatine in generalized anxiety disorder: a randomized, double-blind, placebo-controlled study. J Clin Psychopharmacol. 2008;28(5):561–6.

55. Fornaro M. Agomelatine in the treatment of panic disorder. Prog Neuropsychopharmacol Biol Psychiatry. 2011;35(1):286–7.

56. Millan MJ, Brocco M, Gobert A, Dekeyne A. Anxiolytic properties of agomelatine, an antidepressant with melatoninergic and serotonergic properties: role of 5-HT2C receptor blockade. Psychopharmacology (Berl). 2005;177:448–58.

57. American Thoracic Society. Indications and standards for use of nasalcontinuous positive airway pressure (CPAP) in sleep apnea syndromes. American Thoracic Society. Official statement adopted March 1944. Am J Respir Crit Care Med. 1994;150:1738–45.

58. Mitchell MD, Gehrman P, Perlis M, Umscheid CA. Comparative effectiveness of cognitive behavioral therapy for insomnia: a systematic review. BMC Fam Pract. 2012;13:40.

59. Barlow DH, Craske MG. Mastery of your anxiety and panic II. San Antonio, TX: The Psychological Corporation; 1994.

60. Edinger JD, Hoelscher TJ, Marsh GR, Ionescu-Pioggia M, Lipper S. A cognitive-behavioral therapy for sleep maintenance insomnia in older adults. Psychol Aging. 1992;7:282–9.

Some Genetic Aspects of Panic Disorder

9

Fabiana Leão Lopes

Contents

Abstract

Panic disorder (PD) is a multifactorial disease and despite being the anxiety disorder with higher heritability, the underlying genetic basis of this disorder remains poorly elucidated. Several candidate genes have been described so far, but they generally are characterized by small sample sizes, have small effects, lack replication and translational models. Initial attempts to perform genome-wide association studies (GWAS) in PD did not lead to significant results nor were confirmed in subsequent studies. These facts serve to call attention to the PD, as other psychiatric disorders, is likely to be a multigenic and heterogeneous disease, with small-effect alleles that do not reach genome-wide significant results. Moreover, even presenting a polygenic basis, future genetic studies for PD should comprise large-scale multicenter studies under international collaboration in order to obtain representative samples. Still, the techniques of next generation sequencing – which are already dominating the field in other psychiatric disorders – aim to reveal the common and rare genetic variants associated with the PD. With a lifetime prevalence of approximately 4 % and outlined endophenotypes (i.e. carbon dioxide sensitivity), a better understanding of the genetic basis and biological mechanisms underlying PD is very important.

F.L. Lopes (✉)
Laboratory of Panic and Respiration, Institute of
Psychiatry, Federal University of Rio de Janeiro,
Rio de Janeiro, Brazil
e-mail: lopes.fabiana@gmail.com

© Springer International Publishing Switzerland 2016
A.E. Nardi, R.C.R. Freire (eds.), *Panic Disorder*, DOI 10.1007/978-3-319-12538-1_9

Keywords

Panic disorder • Single nucleotide polymorphism • Genome-wide association study • Genetic association studies • Inheritance patterns

9.1   Introduction

Panic disorder (PD) is a multifactorial disease likely involving biological, psychological, survival-related evolutionary factors and the interaction of all of the above factors. To understand the contribution of genetic factors in the etiology of a condition of interest, some resources from clinical genetics are frequently used, like family studies, twin studies, adoption studies and segregation studies. For instance, family studies of PD support a pattern of familial aggregation indicating that the disease runs in families. In a comprehensive meta-analysis [1] the results showed a significant association between PD in the probands and in their first-degree relatives (summary odds ratio of the studies taken into account: 5.0; 95% CI: 3.0–8.2). The estimated relative risk to siblings of PD probands is five- to tenfold higher than the population risk, and the heritability of PD is 0.48 [1].

Twin studies comprise monozygotic and dizygotic ones. One could assume that the correlation (C) should be 1 or 100 % for monozygotics (MZ) and 0.5 or 50 % for dizygotics (DZ). But what is generally found is a different condition (p.ex., height – C_{MZ}: 0.94 and C_{DZ}: 0.44; bipolar disorder – C_{MZ}: 0.79 and C_{DZ}: 0.24; measles – C_{MZ}: 0.95 and C_{DZ}: 0.87) [2] . The difference of concordance between monozygotic and dizygotic twins is used to estimate the heritability. Indeed, when the difference is large means that exists a great role of genes in determining the condition (as we can observe in both examples of height and bipolar disorder). Hettema et al. [1] in their meta-analysis pointed out the estimate heritability for PD to be 43 %. For illustration, Fig. 9.1 depicts the estimate heritability of psychiatric diseases [3]. In addition, Kendler and Myers [4] have recently highlighted that disorders that optimally index the genetic liability to externalizing and internalizing disorders in the general population differ significantly in relation to gender. PD accounted for the largest sex difference in their review, demonstrating a much better index for genetic risk in women than in men (heritability in males: 21 %; heritability in females: 96 %). See Fig. 9.1.

Segregation studies also consistently confirm a genetic component to transmission of PD, but they had not supported any specific mode of inheritance. While some studies have suggested a dominant transmission pattern, others have found support for both dominant and recessive modes. In this way, segregation studies in PD have not found a pattern of inheritance according to Mendelian rules, indicating that PD, as like the whole psychiatric disorders, are complex genetic disorders raised by an interaction of many factors such as polygenic and environmental factors.

The great majority of molecular studies conducted on PD still consist of linkage and candidate gene association studies. A summary of the main findings, advantages and pitfalls are detailed elsewhere. Unlike other psychiatric diseases – schizophrenia, bipolar disorder, major depression disorder, autism and attention deficit hyperactive disorder – genetics research in PD is still at the beginning. One of the barriers leading to this scenario could be the complex nature of PD phenotype. First of all, we have to remember that PD is characterized by increased pattern of comorbidities with medical and psychiatric conditions. Second, a decade of research trying to highlight the commonalities and differences between panic and fear has not reached a consensus yet. The main repercussion that comes out is a fuzzy phenotype definition. And last, but not least, we currently have disparities in funding and advocacy leading the field of panic disorder to be behind other mental health diseases [5].

9.2   Linkage Studies

Many linkage methods dating from the mid-1900 were performed to analyze genetic data from populations with rare traits. However, rare traits are usually influenced by a major gene or high

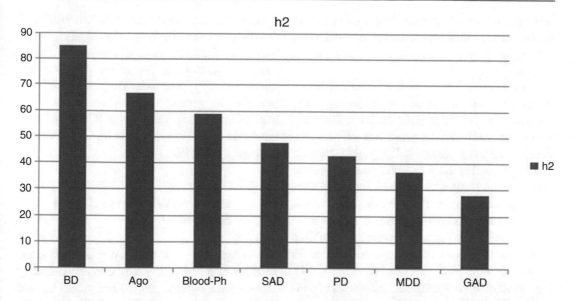

Fig. 9.1 Heritability estimates (h²) for mood and anxiety disorders. Adapted from Bienvenu et al. [3], Stein et al. [25] and Kendler et al. [26]. *BD* Bipolar Disorder, *Ago* Agoraphobia, *Blood-Ph* Blood Injury phobia, *SAD* Social Anxiety Disorder, *PD* Panic Disorder, *MDD* Major Depression Disorder, *GAD* Generalized Anxiety Disorder

penetrant condition what does not seem to occur in complex disorders like PD. Therefore, the detection sensitivity of linkage studies conducted in PD is low, due to the small individual effect of single genes. Moreover, these studies give an approximate chromosomal location of the gene related to an inherited phenotype. Using affected pedigrees, the rationale is if a marker significantly co-segregates with the disease then the region around this marker contains genes conferring a disease risk. These studies use the logarithm of the odds (LOD), which is based on the number of estimates of recombination frequency. A LOD score greater than 3.0 is considered *significant* (indicating that a thousand to one in favor of genetic linkage) and a LOD score greater than 1.9 is considered *suggestive* [6].

Webb et al. [7] performed a meta-analysis of Genome Wide Linkage Scan on independent samples of Neuroticism and Anxiety (two studies of panic disorder and one using a 'broad anxiety' definition). The data comprised 5341 families encompassing 15,529 individuals (the anxiety sample consisted of 718 subjects from 162 families). Rank based genome scan was used to analyze each trait separately and combined. They presented the results for 10 CM intervals and unweighted genome scan meta-analysis (GSMA) approach. The analysis for anxiety identified nominal significance ($p < 0.05$) for regions on chromosomes 1, 2, 5, 9, 11, 15, 16 and 22. Chromosome 1 was correlated to both phenotypes, being nominally significant. The authors hypothesized that this region harbors genes broadly underlying anxiety susceptibility. Linkage evidence for a broad phenotype "panic disorder + bipolar disorder" has been obtained on chromosomes 2q (lodscore = 4.6) and 12 (lodscore = 3.6), under a dominant model of inheritance [27]. Descriptions of linkage on chromosome 2p, 2q, 9p, 12 and 15q (near GABA-A receptor subunitor genes) among families with PD have been consistently found on the scientific literature [8].

9.3 Candidate Gene Association Studies

This approach has been the major source of investigations and the available published data in PD. These studies are based on association paradigm using case-control and/or family based analysis. Although they have stronger power than linkage studies to identify loci associated to complex diseases, there are important pitfalls that should be taken into account when facing those studies.

So far, candidate genes have been identified on the basis of our current knowledge regarding the pathophysiology of panic disorder. There are more than 360 candidate genes currently investigated to be associated with PD [9, 10]. Despite this huge number, only a small number has been replicated and only a handful has dealt with sample sizes higher than 200 subjects. In addition, the results are nominally significant or even present a trend but are often uncorrected for multiple comparisons. Out of 364 genes, only about 56 presented with positive associations [10, 11]. Due to all of these conditions, the results of candidate genes association studies in PD seem not to be robust and indicate that most of the candidate studies are not likely to have a main role in the susceptibility to this disorder.

The genes that have been more consistently studied are cholecystokinin (CCK), cholecystokinin B receptor (CCKBR), 5-hydroxytryptamine receptor 2A (HTR2A), solute carrier family 6 (neurotransmitter transporter), member 4 (SLC6A4), adenosine A2a receptor (ADORA2A), catechol-O-methyltransferase (COMT) and monoamine oxidase A (MAO). The most recent data is for transmembrane protein 132D (TMEM132D) [12], acid-sensing (proton gated) ion channel 1 (ASIC1) [13], gamma-aminobutyric acid (GABA) A receptor, alpha 5 (GABRA5) and gamma-aminobutyric acid (GABA) A receptor, beta3 (GABRB3) [14]. The function of TMEM132D still remains to be confirmed but there is suggestion of its involvement in neuronal sprouting and brain connectivity [12]. The amiloride-sensitive cation channel 2 gene (ACCN2) is the human ortholog of Asic1a and is highly expressed in the amygdala [13]. In a case-control study, the authors observed an increased effect of ACCN2 alleles on early-onset PD and on the respiratory subtype of PD. In this way, they speculate that ACCN2 variants may lead to PD risk by lowering the threshold for amygdala sensing of acidosis [13]. The study of Hodges et al. [14] found additional support for association

between PD and the region of the GABRB3 and GABRA5 genes in both United States and Sardinian families, and a number of novel sequence variants in the region of these two genes among PD probands. They also found support for the genetic contribution to GABRB3 expression, which may suggest a regulatory mechanism for the hypothesized GABAergic dysregulation in PD [14].

9.4 Genome Wide Association Studies

The first genome-wide association study (GWAS) in PD was a Japanese study conducted among 200 subjects with PD and 200 controls [15]. All of the included subjects had a Japanese ascendant. None of the initial nominal association findings were replicated in the subsequent study, with a sample size comprising 558 cases and 566 controls [16]. A meta-analysis of these two Japanese studies (718 cases/1717 controls) followed by a replication analysis (329 cases/861 controls) found no significant genome-wide results [17]. Evidence emerging from this study support the involvement of the previous candidate gene NPYSR (4q31.3-32; p = 6.4 × 10^{-4}). In a genome-wide scan using microsatellite markers among PD patients and controls, from the isolated population of the Faroe Islands, Gregerson et al. [18] found evidence for the amiloride-sensitive cation channel 1 (ACCN1) – located on chromosome 17q11.2-q12 – as a potential candidate gene for PD. Further analyses of the ACCN1 gene using single-nucleotide polymorphisms (SNPs) revealed significant association with PD in an extended Faroese case-control sample. However, the authors were not able to replicate the findings in a larger and independent Danish case-control sample and concluded that possible risk alleles associated with PD in the isolated population are not those ones involved in the development of PD in a larger outbred population. A Japanese study evaluated genome-wide copy number variation association in PD [19] and detected positive results for duplications in the peri-centromeric region in the chromosome 16p11.2. Nevertheless, the association level was borderline, the results were not replicated, translational validation still lacks and though, additional confirmation studies are needed [20].

Among a sample consisting of 1001 European American bipolar cases and 1034 controls, a GWAS was conducted throughout bipolar related phenotypes. Thus, in the panic attack-adjusted analysis, top-ranking SNPs included rs599845 in a protein-coding region of SGOL1. Although this result has never been associated with PD before, it had already been linked to temperament traits in bipolar disorder. The results in this study did not reach genome-wide significance [21].

The most robust results are for the transmembrane protein gene 1342D located on chromosome 12. This GWAS was initially conducted in MaxPlanck and expanded with the Consortium of Panic, totaling 2678 PD cases and 3262 controls. A correlation with European ancestry was detected and the results were not replicated in the Japanese population. The results involving the transmembrane protein gene 1342D were further validated in animal models with greater anxiety associated with increased expression of mRNA TMEM in the anterior cingulate cortex. In humans, associations with increased expression in the frontal cortex in postmortem brains have been reported [12].

9.5   Other Genetics Studies

A study conducted by the Spanish team analyzed the involvement and regulation of micro-RNAs (mi-RNA) in candidate genes for PD. They used 712 SNPs covering 325 regions of mi-RNAs in a sample of 203 subjects with PD and 341 controls. The strongest associations occurred for two SNP: rs6502892 tagging miR-22 ($p < .0002$) and rs11763020 tagging miR-339 ($p < .00008$), although such associations have not survived after multiple corrections. Replications in Finnish and Estonian samples did not support these associations. Functional studies have shown that miR-22 regulates the four candidate genes of PD – BDNF, HTR2C, MAOA and RGS2 [22]. Though preliminary, such data may call attention

to further studies of mi-RNAs and/or the involvement of regulatory regions in the etiopathological basis of PD.

Epigenetic studies are also scarce in the field of PD. Domschke et al. [23] investigated DNA methylation patterns in the regions of glutamate decarboxylases – GAD1 and GAD2 promoter and GAD1 intron 2 – to be associated with PD. To this end, 65 subjects with PD and 65 controls were analyzed. PD patients exhibited significantly lower average GAD1 methylation than healthy controls, particularly at three CpG sites in the promoter as well as in intron 2. The authors point to a potentially compensatory role of GAD1 gene hypomethylation in PD probably mediating the influence of negative life events.

Recently, Sasaki's team [24] performed a pathway analysis from the SNPs data genotyped for GWAS and converted to genes associations. They performed three different types of pathway analysis and each one showed that those pathways related to immunity had the strongest association with PD. The authors focused on and investigated HLA-B and HLA-DRB1 as candidate susceptibility genes for PD. Therefore, 744 PD subjects and 1418 controls were typed for HLA-B and HLA-DRB1. It came out that patients with PD were significantly more likely to carry HLA-DRB1*13:02 ($p = 2.50 \times 10^{-4}$, odds ratio = 1.49). This study provided initial evidence that genes involved in immune related pathways are associated with PD.

9.6   Conclusion

Studies of PD molecular genetics are at a preliminary stage compared to other pathologies with greater investment in the field of research and development, such as schizophrenia, depression and bipolar disorder. Despite being the anxiety disorder with higher heritability, the underlying genetic basis of PD remains poorly elucidated. Several candidate genes have been reported presenting, in some way, a disappointing result. Small sample size (N < 200, largely) with only small effects that fail replication have been almost the rule than exception, dominating thus the scientific literature of such studies. Initial attempts to perform GWAS in PD did not lead to significant results nor were confirmed in subsequent studies. These facts serve to call attention to the PD, as in other psychiatric disorders, is likely to be a multigenic and heterogeneous disease, with small-effect alleles that does not reach genome-wide significant results. Moreover, even presenting likely a polygenic basis future delineations for panic disorder comprise larger-scale multicenter studies under international collaboration in order to obtain a representative sample. Still, the techniques of next generation sequencing – which are already dominating the field in other psychiatric disorders – aim to reveal the common and rare genetic variants associated with the PD. With a lifetime frequency of approximately 4% and outlined endophenotypes (i.e. carbon dioxide sensitivity), a better understanding of the genetic pathways interactions and biological mechanisms underlying PD became very important.

References

1. Hettema JM, Neale MC, Kendler KS. A review and meta-analysis of the genetic epidemiology of anxiety disorders. Am J Psychiatry. 2001;158:1568–78.
2. Thomas A. Thrive in genetics. Oxford: Oxford University Press; 2013. Quantitative genetics, p. 97.
3. Bienvenu OJ, Davydow DS, Kendler KS. Psychiatric 'diseases' versus behavioral disorders and degree of genetic influence. Psychol Med. 2011;41(1):33–40.
4. Kendler KS, Myers J. The boundaries of internalizing and externalizing genetic spectra in men and women. Psychol Med. 2014;44:647–55.
5. Smoller JW. Who's afraid of anxiety genetics? Biol Psychiatry. 2011;69:506–7.
6. Lander ES, Schork NJ. Genetic dissection of complex traits. Science. 1994;265(5181):2037–48.
7. Webb BT, Guo AY, Maher BS, Zhao Z, van den Oord EJ, Kendler KS, et al. Meta-analyses of genome-wide linkage scans of anxiety-related phenotypes. Eur J Hum Genet. 2012;20(10):1078–84.
8. Fyer AJ, Hamilton SP, Durner M, Haghighi F, Heiman GA, Costa R, et al. A third-pass genome scan in panic disorder: evidence for multiple susceptibility loci. Biol Psychiatry. 2006;60(4):388–401.
9. Maron E, Lang A, Tasa G, Liivlaid L, Tõru I, Must A, et al. Associations between serotonin-related gene polymorphisms and panic disorder. Int J Neuropsychopharmacol. 2005;8(2):261–6.
10. McGrath LM, Weill S, Robinson EB, MacRae R, Smoller JW. Bringing a developmental perspective to anxiety genetics. Dev Psychopathol. 2012;24(4):1179–93.
11. Maron E, Hettema JM, Shlik J. Advances in molecular genetics of panic disorder. Mol Psychiatry. 2010;15(7):681–701.
12. Erhardt A, Akula N, Schumacher J, Czamara D, Karbalai N, Müller-Myhsok B, et al. Replication and meta-analysis of TMEM132D gene variants in panic disorder. Transl Psychiatry. 2012;2:e156.
13. Smoller JW, Gallagher PJ, Duncan LE, McGrath LM, Haddad SA, Holmes AJ, et al. The human ortholog of acid-sensing ion channel gene ASIC1a is associated with panic disorder and amygdala structure and function. Biol Psychiatry. 2014;76(11):902–10.
14. Hodges LM, Fyer AJ, Weissman MM, Logue MW, Haghighi F, Evgrafov O, et al. Evidence for linkage and association of GABRB3 and GABRA5 to panic disorder. Neuropsychopharmacology. 2014;39(10):2423–31.
15. Otowa T, Yoshida E, Sugaya N, Yasuda S, Nishimura Y, Inoue K, et al. Genome-wide association study of panic disorder in the Japanese population. J Hum Genet. 2009;54:122–6.
16. Otowa T, Tanii H, Sugaya N, Yoshida E, Inoue K, Yasuda S, et al. Replication of a genome-wide association study of panic disorder in a Japanese population. J Hum Genet. 2010;55(2):91–6.
17. Otowa T, Kawamura Y, Nishida N, Sugaya N, Koike A, Yoshida E, et al. Meta-analysis of genome-wide association studies for panic disorder in the Japanese population. Transl Psychiatry. 2012;2:e186.
18. Gregersen N, Dahl HA, Buttenschøn HN, Nyegaard M, Hedemand A, Als TD, et al. A genome-wide study of panic disorder suggests the amiloride-sensitive cátion channel 1 as a candidate gene. Eur J Hum Genet. 2012;20(1):84–90.
19. Kawamura Y, Otowa T, Koike A, Sugaya N, Yoshida E, Yasuda S, et al. A genome-wide CNV association study on panic disorder in a Japanese population. J Hum Genet. 2011;56(12):852–6.
20. Erhardt A, Spoormaker VI. Translational approaches to anxiety: focus on genetics, fear extinction and brain imaging. Curr Psychiatry Rep. 2013;15(12):417.
21. Cuellar-Barboza AB, Winham SJ, Colby C, Prieto M, Chauhan M, McElroy SL, et al. Genome-wide association study of bipolar disorder related phenotypes: rapid cycling, alcohol use disorders and panic attacks. Biol Psychiatry. 2014;75:1S–401.
22. Muiños-Gimeno M, Espinosa-Parrilla Y, Guidi M, Kagerbauer B, Sipilä T, Maron E, et al. Human microRNAs miR-22, miR-138-2, miR-148a, and miR-488 are associated with panic disorder and regulate several anxiety candidate genes and related-pathways. Biol Psychiatry. 2011;69:526–33.
23. Domschke K, Tidow N, Schrempf M, Schwarte K, Klauke B, Reif A, et al. Epigenetic signature of panic disorder: a role of glutamate decarboxylase 1 (GAD1)

DNA hypomethylation? Prog Neuropsychopharmacol Biol Psychiatry. 2013;46:189–96.

24. Shimada-Sugimoto M, Otowa T, Miyagawa T, Khor SS, Kashiwase K, Sugaya N, et al. Immune-related pathways including HLA-DRB1*13:02 are associated with panic disorder. Brain Behav Immun. 2015; 46:96–103.

25. Stein MB, Jang KL, Livesley WJ. Heritability of social anxiety-related concerns and personality characteristics: a twin study. J Nerv Ment Dis. 2002; 190(4):219–24.

26. Kendler KS, Karkowski LM, Prescott CA. Fears and phobias: reliability and heritability. Psychol Med. 1999;29(3):539–53.

27. Logue SF, Grauer SM, Paulsen J, Graf R, Taylor N, Sung MA, Zhang L, Hughes Z, Pulito VL, Liu F, Rosenzweig-Lipson S, Brandon NJ, Marquis KL, Bates B, Pausch M. The orphan GPCR, GPR88, modulates function of the striatal dopamine system: a possible therapeutic target for psychiatric disorders? Mol Cell Neurosci. 2009;42(4): 438–47.

Panic Disorder and Personality Disorder Comorbidity

10

Ricard Navinés, Elfi Egmond,
and Rocío Martín-Santos

Contents

Abstract

The present chapter systematically reviews the relationship between panic disorder, with and without co-occurring anxiety or depression, and current personality disorder. Data were collected with an advanced document protocol according to MOOSE (Meta-analysis of Observational Studies in Epidemiology) guidelines for observational studies. A comprehensive, computerized literature search was conducted in Medline, PsycINFO, and LILACS. Cohort, case-control and cross-sectional surveys studies evaluating the comorbidity between DSM panic disorder and personality disorders were included. Overall prevalence, comorbidity rates, and 95 % CI were calculated with a random effects model. From 97 initial selected papers, 24 entered in the review. Among patients with a current DSM-III/R/IV panic disorder, 44.3 % (34.6–54.2 %) had any personality disorder; 6.3 % (3.1–10.4 %) had cluster A; 17.9 % (12.2–24.2 %) cluster B, and 34.9 % (25.6–44.7 %) had cluster C. Among patients with a current panic disorder and co-occurring anxiety or

R. Navinés • R. Martín-Santos (✉)
Department of Psychiatry and Psychology, Hospital Clinic, Institut d'Investigació Biomèdica August Pi I Sunyer (IDIBAPS), Centro de Investigación Biomédica en Red en Salud Mental (CIBERSAM), G25, Universidad de Barcelona, Barcelona, Spain
e-mail: rnavines@clinic.ub.es; rmsantos@clinic.ub.es

E. Egmond
Department of Psychiatry and Psychology, Hospital Clinic, Barcelona, Spain

Department of Clinical and Health Psychology, Faculty of Psychology, Universidad Autònoma de Barcelona, Cerdanyola del Vallés, Barcelona, Spain
e-mail: egmond@clinic.ub.es

© Springer International Publishing Switzerland 2016
A.E. Nardi, R.C.R. Freire (eds.), *Panic Disorder*, DOI 10.1007/978-3-319-12538-1_10

depression, 61.8 % (44.6–77.7 %) had any personality disorder, 7.2 % (4.4–10.5 %) had cluster A; 24.0 % (17.6–30.9 %) cluster B, and 38.6 % (25.7–52.2 %) had cluster C. In conclusion, comorbidity between panic disorder and personality disorders is common. Cluster C was the most frequent personality disorder subtype related to panic disorder. Personality disorders were more prevalent among individuals with panic disorder and co-occurring anxiety or depression.

Keywords

Panic disorder • Personality disorder • Systematic review • Depression comorbidity • Cluster A • Cluster B • Cluster C • Treatment non-responders

10.1 Introduction

10.1.1 Epidemiological Data of Panic Disorder

Panic disorder is an anxiety disorder characterized by unexpected and repeated episodes of intense fear accompanied by physical symptoms that may include chest pain, heart palpitations, and shortness of breath, dizziness, or abdominal distress. Panic attacks usually produce a sense of unreality, a fear of impending doom, or a fear of losing control [1].

The estimated current prevalence rate for panic disorder is about 1–5 % of the adult population [2]. In the National Epidemiologic Survey on Alcohol and Related Conditions, the overall 12-month and lifetime prevalence rates for panic disorder (with or without agoraphobia) were 2.1 % and 5.1 %. The 12-month and lifetime prevalence rates for panic disorder with agoraphobia were 0.6 % and 1.1 %, while the corresponding rates for panic disorder without agoraphobia were 1.6 % and 4.0 %. Agoraphobia without panic disorder was uncommon (12-month prevalence 0.05 %; lifetime prevalence 0.17 %) [3].

Female gender, low socioeconomic status, and anxious childhood temperament are common risk factors for panic disorder. Panic disorder can produce marked distress and impairment, and is associated with significant suicide risk. Panic disorder appears to increase risk for all-cause mortality because it may increase risk for cardiovascular disease [2, 4].

10.1.2 Panic Disorder and Comorbidity

The diagnosis is frequently associated with other comorbid axis-I psychiatric disorders, especially with depressive and other anxiety disorders [5]. Moreover, panic disorder can also co-occur with comorbid axis-II psychiatric disorder, resulting in the diagnosis of personality disorder [6]. A personality disorder is a persistent and maladaptive pattern of internal experience and behavior, that have their beginning in the adolescence or first adult age, and that causes significant malaise or deterioration in the activity of the individual [DSM-IV]. Currently, personality disorders constitute an important medical and social pattern, shown by a high prevalence (10–15 % in general population and until 50 % in psychiatric patients), as well as the personal repercussions and partner-relatives who tolerate [7]. Although personality disorders have been defined categorically throughout the history of psychiatric nomenclatures, the DSM-5 Personality and Personality disorders Work Group proposed a substantial shift to a dimensional conceptualization and diagnosis of personality pathology [8]. The DSM-5 gives a categorical classification of personality disorders, grouped into three clusters (2013) [9].

The repercussion of personality disorders in panic disorder has been evaluated in different studies, suggesting that this comorbidity is associated with a greater severity of the symptoms of panic disorder [10], and also with a greater prevalence of comorbid agoraphobia [11]. Thus, the degree of fear and phobia is usually more pronounced in patients with axis II comorbidity, as is the level of general psychopathology [10, 12]. This is also reflected in health care cost analyses, showing that axis II personality disorder patients represent a markedly higher economic burden to the health care system than for example patients with depression and anxiety [13]. Moreover, having a personality disorder represents a strong vulnerability factor for developing other axis I disorders [14].

Moreover, treatment of comorbid personality disorder is normally more complex and has less favorable outcomes for panic disorder patients [15–17], higher drop-out rates [18], less positive patient expectations [19], and more challenges establishing a durable and flexible therapeutic alliance [19, 20]. Patients with cluster A or cluster B also appear to have a poorer treatment response than cluster C patients [21, 22]. Also, the degree of comorbid psychosocial impairment depends on the type of personality disorder. A higher degree of impairment seems to exist among schizotypal and borderline patients than among obsessive-compulsive or avoidant personality disorder patients [23].

10.1.3 What We Know About the Association Between Panic Disorder and Personality Disorders

Several studies of general associations between personality disorder and panic disorder have been published [6, 11, 14, 24]. The general conclusions from these reviews point out that the comorbid personality disorder among patients with a panic disorder diagnosis vary considerably, but the proportions of avoidant, dependent and compulsive (cluster C) personality disorders are the highest. These proportions are smaller in the schizoid, schizotypal and paranoid (cluster A) as well as the dramatic, borderline and anti-social personality disorders (cluster B).

Very few systematic reviews and meta-analyses on comorbidity between Axis II and anxiety disorders in general have been published. There is one previous systematic review and meta-analysis by Borenstein [25] that specifically focuses on examining comorbidity with dependent personality disorder. The ratio varied considerably between diagnostic groups of anxiety disorder, with panic disorder, obsessive-compulsive disorder, agoraphobia, and social phobia showing a higher prevalence of personality disorders, while generalized anxiety disorder and post-traumatic stress disorder showed no relationship between both. The study was also criticized [26] for using a personality disorder rather than an anxiety disorders as the main inclusion criteria. More recently, Friborg et al. [27] performed a systematic review and meta-analysis to identify the proportions of comorbid personality disorder across the major subtypes of anxiety disorders. The rate of any comorbidity in Axis II was high across all anxiety disorders, ranging from 35 % for post-traumatic stress disorder to 52 % for obsessive-compulsive disorder. Globally, cluster C occurred more than twice as often as cluster A or B. Within cluster C, the avoidant personality disorder occurred most frequently, followed by the obsessive-compulsive and the dependent subtype. Gender or duration of an anxiety disorder was not related to variation in personality disorder comorbidity.

In this chapter we undertook a systematic review, and meta-analysis when possible, of clinical observational studies, to summarize the relationship between personality disorders, and prevalence of panic disorder with and without anxiety or depression. The lack of previous systematic reviews or meta-analyses using panic disorder as the primary inclusion criteria was the main reason for the present study.

10.2   Methodology of the Review

10.2.1 Studies

For this review, we considered all relevant cohort, case-control and cross-sectional survey studies that evaluate the comorbidity between panic disorder and personality disorders.

10.2.2 Search Strategy

Data were collected with an advanced document protocol according to MOOSE (Meta-analysis of Observational Studies in Epidemiology) guidelines for observational studies [28]. A comprehensive, computerized literature search was conducted in Medline (1984–Jan 2012); EMBASE (1984–Dec 2014); PsychLIT (1984–Dec 2014); CINAHL (1984–Dec 2014); and LILACS (1984–Dec 2014), for studies in humans of the association between panic disorder and personality disorders. Our search terms included any combination of the key words "panic", "panic disorder", "personality", "personality disorder", "depression", and "comorbidity". We also reviewed reference lists of the identified studies and review articles to search for additional studies.

10.2.3 Data Extraction

The titles and abstracts were examined, and full-text articles of potentially relevant studies were obtained. Subsequently, inclusion and exclusion criteria were applied, and the selected articles were included in this systematic review.

Data was extracted from each study using a standardized spreadsheet. Information extracted included the following: title, author, year of publication, study design, sample size, age, sex, methods of interview, panic disorder diagnose, comorbid DSM (III, IV, IV-R) diagnoses, personality disorder diagnose, duration of panic disorder, and onset of panic disorder. We also extracted the N, % of any cluster personality disorder, and we calculated the 95 % confidence interval.

Two clinical researchers (RN, and EE), a psychiatrist and psychologist, performed each step in this literature research, study identification, study selection, and data extraction. Disagreements were resolved by discussion, and consensus was achieved in the selection of articles for analysis.

10.2.4 Inclusion and Exclusion Criteria

The inclusion criteria were clinical studies of subjects diagnosed with panic disorder using the Structured Clinical Interview for DSM-III, DSM-III-R or DSM-IV (SCID-I) [29–31] irrespective of gender, race, age, or nationality. All study participants also had to be diagnosed with categorical personality disorder using the Structured Clinical Interview for DSM-IV (SCID-II) [32].

The exclusion criteria were studies with patients with another axis I comorbidity, except other anxiety disorder or depressive disorders, and studies where panic disorder diagnose was secondary to medical or substance use pathology.

Studies were also excluded if they were not published as full reports, such as conference abstracts and letters to editors; or if N/% was not used in measuring the prevalence of personality disorders. If multiple published reports were available from the same study, we included only the one with the most detailed information on the relationship in question. Studies which evaluated only a subtype of personality disorder were also excluded.

10.2.5 Statistical Analysis

Cross-tabulations were used to calculate overall prevalence and comorbidity rates with a random effects model. Standard errors and 95 % confidence limits were estimated.

10.3 Results

Using keywords, 448 articles were identified and titles and abstracts were examined. At this stage, 351 articles were eliminated because they did not meet the selection criteria *a priori*. We obtained 99 potentially relevant papers, which were thoroughly examined. Twenty-four articles were rejected because they failed to meet inclusion criteria [11, 22, 33–54], 45 met exclusion criteria for comorbidity [22, 45, 46, 50, 55–68], no having a structured clinical diagnosis [33, 41, 44, 69–76], no measuring in N/% [45, 77–80], multiple published reports [56, 81–86], partial inclusion of personality disorders [45, 57, 61, 64, 82, 87], no specification of axis-I comorbidity [88], or because of language restriction [89]; were letters or reviews [43, 75, 90–95] (Fig. 10.1 shows the flow chart). We finally selected 24

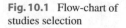

Fig. 10.1 Flow-chart of
studies selection

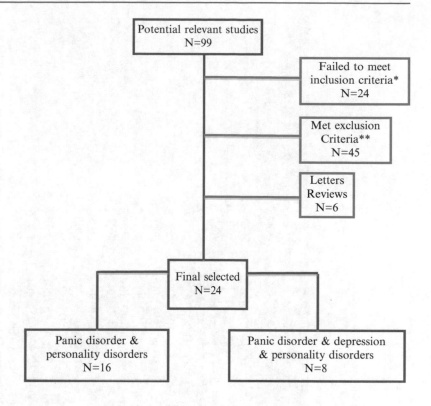

published studies, 16 evaluating the comorbidity between panic disorder and personality disorders [16, 17, 24, 86, 96–107], or eight between panic disorder and co-occurring major depression and personality disorders [10, 68, 97, 102, 103, 107–109], using the Structured Clinical Interview for DSM-III, DSM-III-R or DSM-IV (SCID-II).

Table 10.1 presents the characteristics of the selected studies of panic disorder patients and comorbid personality disorder [16, 17, 24, 86, 96–107], and Table 10.1 of panic disorder patients with co-occurring depression and comorbid personality disorder [10, 68, 97, 102, 103, 107–109].

Table 10.2 shows the overall prevalence and its 95% confidence interval of the selected studies of panic disorder patients and comorbid personality disorder, and Table 10.3 of panic disorder patients with co-occurring depression and comorbid personality disorder.

10.4   Discussion

10.4.1 Evidence-Based on the Review

10.4.1.1   Panic Disorder and Comorbid Personality Disorder

This systematic review confirms that the comorbidity between panic disorder with or without depression and personality disorder was common. Personality disorders were more prevalent among individuals with panic disorder and co-occurring depression. Furthermore, the associations between panic disorder alone or with co-occurring anxiety or depression and Axis II disorder were all high. The study confirms that between panic patients with a personality disorder, cluster C personality disorder was more frequently associated with panic disorder than with other personality disorders.

Table 10.1 Characteristics of the studies of panic disorder and comorbid personality disorders

Author	Year	N	Women (N, %)	Age (Mean ± SD)	Design	Panic disorder diagnose	Personality disorder diagnose	Duration of illness (years) (Mean ± SD)	Onset panic disorder (years) (Mean ± SD)
Mendoza et al.	2011	104	75 (71.1)	37.5 (8.8)	Cross-sectional	SCID-I (DSM-IV)	SCID-II	–	–
Telch et al.	2011	173	128 (73)	35.2 (–)	Cohort	SCID-I (DSM-IV)	SCID-II	9.1 (–)	–
Gutiérrez et al.	2008	157	108 (68.8)	34.9 (9.1)	Cross-sectional	SCID-I (DSM-IV)	SCID-II	–	–
Iketani et al.	2004	105	58 (55.2)	36.9 (12.2)	Cross-sectional	SCID-P (DSM-III-R)	SCID-II	–	33.3 (12.6)
Massion et al.	2002	386	260 (67.3)	39.8 (10.8)	Cohort	SCALUP(DSM-III-R)	IPDE	–	–
Barzega et al.	2001	184	112 (60.9)	31.8 (9.8)	Cross-sectional	SCID I (DSM-IV)	SCID-II	–	30.2 (10.3)
Dyck et al.	2001	230	153 (66.6)	41.0 (12.6)	Cohort	SCID I (DSM-III-R)	IPDE	–	–
Latas et al.	2000	60	45 (75)	33.6 (7.7)	Cross-sectional	SCID I (DSM-IV)	SCID-II	38.78 (44.05)	–
Ampollini et al.	1999	42	27 (64.3)	31.4 (9.6)	Case-Control	SCID I (DSM-III-R)	SIDP-R	–	–
Langs et al.	1998	49	27 (55.1)	34.8 (9.7)	Cross-sectional	SCID I (DSM-III-R)	SCID-II	4.4 (6.8)	30.4 (9.7)
Mauri et al.	1992	40	24 (60)	33.4 (9)	Cross-sectional	SCID I (DSM-III-R)	PDE	6.4 (7)	26.7 (9)
Mellman et al.	1992	23	17 (78)	37.7 (11.9)	Cohort	SADS	SID-P	12.7 (7.6)	25 (11.4)
Brooks et al.	1991	30	11 (37)	35 (12)	Cross-sectional	SCID-P (DSM-III-R)	SCID-II	–	–
Sciuto et al.	1991	48	32 (66.7)	34.9 (10.2)	Cross-sectional	SCID-I-DSM-III-R	SIDP-R	5.4 (5.7)	29.1 (10.3)
Alnaes et al.	1990	39	–	30 (19–51)	Cross-sectional	SCID-I (DSM-III)	SID-P	–	–
Reich et al.	1987	88	52 (59)	37.3 (1.9)	Cross-sectional	SCID I (DSM-III-R)	SIDP	–	–

IPDE (International Personality Disorder Examination)-DSM-III-R criteria. SCID-DSM-III (Structured Clinical Interview Diagnoses for DSM-III criteria. SCID-DSM-III-R (Structured Clinical Interview Diagnoses for DSM-III-R v.patient), SCID-DSM-IV (Structured Clinical Interview Diagnoses for DSM-IV v.patient). SID-P (Structured Interview for DSM-III Personality Disorders). SIDP-R (Revised Structured Interview for DSM-III-R Personality Disorders). PDE (Personality Disorder Examination-DSM-III-R and ICD-9)

Table 10.2 Characteristic of studies of panic disorder and co-occurring depression and comorbid personality disorders

Author	Year	N	Women (N, %)	Age (SD)	Design	Panic disorder diagnose	Personality disorder diagnoses	Duration of illness (years) (Mean ± SD)	Onset PD (years) (Mean ±SD)
Svanborg et al.	2008	15	9 (60)	30.5 (5.59)	Cohort	SCID-I (DSM-IV)	SCID-II	6.9 (4.9)	–
Gutiérrez et al.	2008	37	29 (78.4)	36.8 (9.4)	Cross-sectional	SCID-I (DSM-IV)	SCID-II	–	–
Marchesi et al.	2006	71	49 (69)	36.1 (10.7)	Cohort	SCID-I (DSM-IV)	SIDP	4.4 (5.5)	27.1 (10.1)
Ozkan et al.	2005	112	71 (63)	–	Cross-sectional	SCID-I (DSM-IV)	SCID-II	–	27.1 (7.5)
Ampollini et al.	1999	29	27 (93.1)	40.7 (11.5)	Case-control	SCID-I (DSM-III-R)	SIDP-R	–	–
Langs et al.	1998	35	27 (55.1)	35.7 (9.9)	Cross-sectional	SCID-I (DSM-III-R)	SCID-II	6.2 (8.1)	29.4 (8.9)
Alnaes et al.	1990	19	–	34 (19–59)	Cross-sectional	SCID-I (DSM-III)	SIDP	–	–
Pollach et al.	1990	100	63 (63)	40.4 (11.3)	Cohort	SCID-I (DSM-III-R)	PDQ-R	–	–

IPDE (International Personality Disorder Examination)–DSM-III-R criteria. SCID-DSM-III (Structured Clinical Interview Diagnoses for DSM-III-R,patient). SCID-DSM-III-R (Structured Clinical Interview Diagnoses for DSM-III-R v.patient). SCID-DSM-IV (Structured Clinical Interview Diagnoses for DSM-IV v.patient). SID-P (Structured Interview for DSM-III Personality Disorders). SIDP-R (Revised Structured Interview for DSM-III-R Personality Disorders). PDE (Personality Disorder Examination-DSM-III-R and ICD-9)

Table 10.3 Overall prevalence and 95 % confidence interval of comorbid panic disorder and personality disorders

Authors	Year	N	Any personality disorder (n; %, [95 % CI])	Any PD-A (n; %, 95 % CI)	Any PD-B (n; %, 95 % CI)	Any PD-C (n; %, 95 % CI)
Mendoza et al.	2011	104	56 (53.8, [43.8, 63.7])	2 (1.9, [0.2, 6.7])	18 (17.3, [10.6, 26.0])	36 (34.6, [25.6, 44.6])
Telch et al.	2011	173	54 (31.2, [24.4, 38.7])	12 (6.9, [3.6, 13.9])	15 (8.7, [4.9, 13.9])	42 (24.3, [18.1, 31.4])
Gutiérrez et al.	2008	157	42 (26.8, [20.0, 34.4])	1 (0.6, [0.0, 3.5])	9 (5.7, [2.6, 10.6])	16 (10.2, [5.9, 16.0])
Iketani et al.	2004	105	59 (56.2, [46.2, 65.9])	14 (13.3, [7.5, 21.4])	33 (31.4, [22.7, 41.2])	47 (44.8, [35.1, 54.8])
Massion et al.	2002	386	79 (20.5, [16.6, 24.8])	–	–	–
Barzega et al.	2001	184	125 (67.9, [60.7, 74.6])	12 (6.5, [3.4, 11.1])	46 (25.0, [18.9, 31.9])	84 (45.7, [38.3, 53.1])
Dyck et al.	2001	230	48 (20.9, [15.8, 26.7])	–	–	–
Latas et al.	2000	60	27 (45.0, [32.1, 58.4])	10 (16.7, [8.3, 28.5])	15 (25.0, [14.7, 37.9])	17 (28.3, [17.5, 41.4])
Ampollini et al.	1999	42	25 (59.5, [44.5, 73.0])	3 (7.1, [1.5, 19.5])	10 (23.8, [12.1, 39.5])	16 (38.1, [23.6, 54.4])
Langs et al.	1998	49	20 (40.8, [27.0, 55.8])	–	8 (16.3, [7.3, 29.7])	16 (32.7, [20.0, 47.5])
Mauri et al.	1992	40	18 (45.0, [29.3, 61.5])	–	–	–
Mellman et al.	1992	23	10 (43.5, [23.2, 65.5])	0 (0, [0.0, 14.8])	3 (13.0, [2.8, 33.6])	7 (30.4, [13.2, 52.9])
Ronald et al.	1991	30	16 (53.3, [34.3, 71.7])	6 (20.0, [7.7, 38.6])	8 (26.7, [12.3, 45.9])	22 (73.3, [54.1, 87.7])
Sciuto et al.	1991	48	13 (27.1, [15.3, 41.9])	–	–	–
Alnaes et al.	1990	39	28 (71.8, [55.1, 85.0])	–	–	–
Reich et al.	1987	88	50 (56.8, [45.8, 67.3])	5 (5.7, [1.9, 12.8])	13 (14.8, [8.1, 23.9])	31 (35.2, [25.3, 46.1])
Overall prevalence (Random effects model)			44.3 % [34.6, 54.2]	6.3 % [3.1,10.4][a]	17.9 % [12.2, 24.2]	34.9 % [25.6, 44.7]

[a] Assuming there had been 1 case instead of none in Mellman et al. [105]

Table 10.4 Overall prevalence and 95% confidence interval of comorbid panic disorder and co-occurring depression and personality disorders

Authors	Year	N	Any personality disorder (n; %, 95% CI)	Any PD-A (n; %, 95% CI)	Any PD-B (n; %, 95% CI)	Any PD-C (n; %, 95% CI)
Svanborg et al.	2008	15	9 (60.0, [32.3, 83.7])	0 (0.0, [0.0, 21.8])	3 (20.0, [4.3, 48.1])	12 (80.0, [51.9, 95.7])
Gutiérrez et al.	2008	37	15 (40.5, [24.8, 57.9])	1 (2.7, [0.1, 14.2])	4 (10.8, [3.0, 25.4])	7 (18.9, [8.0, 35.2])
Marchesi et al.	2006	71	53 (74.6, [62.9, 84.2])	7 (9.9, [4.1, 19.3])	14 (19.7, [11.2, 30.9])	26 (36.6, [25.5, 48.9])
Ozkan et al.	2005	122	38 (31.2, [23.1, 40.2])	8 (6.6, [2.9, 12.5])	26 (21.3, [14.4, 30.0])	28 (23.0, [15.8, 31.4])
Ampollini et al.	1999	29	24 (82.8, [64.2, 94.2])	4 (13.8, [3.9, 31.7])	9 (31, [15.3, 50.8])	19 (65.5, [45.7, 82.1])
Langs et al.	1998	35	24 (68.6, [50.7, 83.2])	–	15 (42.9, [26.3, 60.7])	15 (42.9, [26.3,60.7])
Pollack et al.	1990	100	40 (40.0, [30.3, 50.3])	10 (10.0, [4.9, 17.6])	28 (28.0, [19.5, 37.9])	24 (24.0, [16.0, 33.6])
Alnaes et al.	1990	19	18 (94.7, [74.0, 99.9])	–	–	–
Overall prevalence (Random effects model)			61.8 % [44.6, 77.7]	7.2 % [4.4, 10.5][a]	24.0 % [17.6, 30.9]	38.6 % [25.7, 52.2]

[a]Assuming there had been 1 case instead of none in Svanborg et al. [108]

This association may suggest some phenotypic similarities between personality traits and features of a panic disorder with or without co-occurring depression. However, since our study included clinical samples, it must be considered that a selection bias may operate to bring about a greater frequency of personality disorders among panic disorder patients who seek treatment (and, possibly, vice versa – a greater frequency on panic disorder among personality disordered patients who seek treatment). Nevertheless, epidemiological studies found similar results [3], showing that among individuals with a current mood or anxiety disorder, nearly half had at least one personality disorder. Moreover, the potential of anxiety-state confounding effect during the assessment of personality disorder trait measurement must be considered. Personality disorder traits do appear to lessen in response to treatment of panic disorder [11, 17], suggesting that state-trait confounding may occur in acute anxiety states.

10.4.1.2 Cluster C Personality Disorder

Cluster C personality disorder was the personality disorder most frequently associated with panic disorder. This could mean that avoidant, dependent and obsessive personalities may predispose to panic or agoraphobia in greater proportion than other types of personality disorders. Nevertheless, early panic symptoms may shape personality, and enhance avoidant, dependent or obsessive tendencies. In this sense, in a longitudinal study of Bienvenu et al. [11] using a cohort of a community sample, baseline *timidity* (avoidant, dependent, and related traits) predicted first-onset panic disorder or agoraphobia over the follow-up period beyond 10 years. These results suggest that avoidant and dependent personality traits are predisposing factors, or at least markers of risk, for panic disorder and agoraphobia, and not simply epiphenomena or consequences of panic attacks. Furthermore, an avoidant personality easily intensifies primary anxiety problems due to the loss of exposure to potentially corrective experiences constricting mobility and social participation of patients with a panic disorder.

A dependent personality disorder, if present, may develop an undue reliance of the patient on another person, hence representing a safety behavior strategy. The undue reliance also affects reciprocity in relationships with other people, which over time constrict the social network [27]. The result is also in concordance with epidemiological studies that found similar rates of comorbid cluster C personality disorder in patients with panic disorder [3, 110].

10.4.1.3 Cluster B Personality Disorder

Cluster B personality disorder was the second cluster more frequently associated with panic disorder in our review. This represents a slightly greater prevalence compared to that reported in epidemiological studies. For example, in the National Comorbidity Survey Replication (NCS-R) in the USA [111], the percentage of respondents with panic disorder diagnosis who met criteria for any cluster B personality disorder was 10.4 % [4.3], and the percent of respondents with a cluster B personality disorder who met criteria for a panic disorder diagnostic was 14.4 % [4.3], compared to the 17.9 % of the comorbidity observed in our review. In the national sample, any personality disorder, two or more personality disorders, borderline, and schizotypal personality disorder robustly predicted the persistence of panic disorder and other types of anxiety disorders over 3 years, consistent with the results of recent prospective clinical studies [112].

It is possible that patients with a cluster B personality disorder seek further treatment when they have panic attacks. In fact, it has been described that especially patients with a borderline cluster B personality disorder have more intense panic attacks, and also attract a lot more attention and concern from others. It is also regarded as the single personality disorder with the poorest prognosis for a successful treatment of a panic disorder [113]. In this review, since most of the included studies did not investigate the intensity of panic attacks, the relationship between the intensity of panic attacks and the type of personality disorder was not assessed.

10.4.1.4 Cluster A Personality Disorder

Higher associations were observed between Cluster A and several mood and anxiety disorders in epidemiological studies [110]. Prevalence of multiple imputed DSM-IV/IPDE personality disorders with a 12-month DSM-IV/CIDI panic disorder diagnostic in the Part II NCS-R was 15.7 % for cluster A. Specifically, paranoid and schizoid personality disorders were strongly related to panic disorder with agoraphobia. Panic disorder without agoraphobia was significantly associated with schizotypal personality disorder among men but not with women [114]. In our review, cluster A was the cluster with the lowest prevalence among patients with panic disorder, being 6.3 %. In a recent research on cluster A personality pathology in social anxiety disorder compared to panic disorder, patients with social anxiety disorder had more cluster A personality pathology than patients with a panic disorder, with the most solid indication for paranoid personality pathology [115]. The paranoid items were answered positively by 22–56 % of participants with social anxiety disorder, compared to 18–40 % of participants with panic disorder. As for the schizotypal dimension, more variation occurred, with numbers of 2–89 % in participants with social anxiety disorder, and 1–52 % in participants with panic disorder. Personality psychopathology should be assessed and addressed in treatment for all patients with a panic disorder.

10.4.2 Nature of the Relationship Between Panic Disorder and Personality Disorder

Taken together, these results suggest that panic disorder is generally comorbid with personality disorder. The nature of the relationship is not clear, and cannot be determined through a cross-sectional assessment. The studies we included in the review diagnosed personality disorder using a retrospective clinical interview. It is possible that personality disorders are risk factors for panic disorder alone or with comorbid conditions such as depression. Panic disorder and some subtypes of personality disorders may give different manifestations of the same underlying disease processes. It is also possible that personality disorders give secondary complications to having panic disorder beginning in early adulthood or before.

Assessing differences among individuals with panic disorder and/or co-occurring comorbidities, such as depression and the presentation of specific personality disorder, might shed greater light on environmental factors that may interact with genetic factors to determine the phenotypic expressions of personality disorder and panic disorder.

Although it has been assumed that the high prevalence between panic disorder and personality disorders is especially high in Cluster C, there are scarce studies that have evaluated this relationship in a systematic manner and through clinical interviews.

10.4.3 Impact of Personality on Panic Disorder

Another important question is the effect of personality traits on panic disorder outcomes. Personality psychopathology has been found to complicate the treatment of other psychiatric disorders [81, 116]. Several effective pharmacotherapeutic treatments exist for panic disorder; however, not all patients respond to treatment: between 20–40 % are non-responders. Predicting non-response to pharmacotherapy would be very helpful for both patients and clinicians since it takes several weeks before a clinical effect can be expected for most of these drugs. Personality disorders, or even personality traits, are possibly the most robust predictors of non response [117]. Moreover, data from the National Epidemiologic Survey on Alcohol and Related Conditions (NESARC) related to probability and predictors of first treatment contact for anxiety disorders in the United States, showed that several personality disorders decreased the probability of treatment contact [118].

There are currently three major psychotherapeutic approaches to the management of personality disorder: psychodynamic, cognitive-behavioral, and supportive. Though differing in basic conceptions and in methodology, all approaches aim at the amelioration of both the symptom-aspects that dominate the clinical picture at the outset, and the personality difficulties that remain apparent after the symptoms have been alleviated [119]. However, personality psychopathology was also found to exert a detrimental effect on the outcome of psychotherapies as the cognitive-behavioral treatment for panic disorder [120].

Interestingly, we observed that the overall prevalence of cases with comorbid personality disorders clearly increased (or was particularly high) in those cases with panic disorder and co-occurring depression, suggesting that a comorbid personality disorder place us in a more complex scenario of poorer treatment outcomes and higher prevalence of concurrent depression. Some studies clearly demonstrate that persons with anxiety disorders, who can be identified with certain comorbid personality disorders, are at risk of poorer outcomes [16]. The higher association between panic disorder and personality disorders found in the present review suggests the need of evaluating personality in patients with panic disorder. Moreover, in those patients with a comorbid personality disorders, psychotherapy could be particularly considered the treatment planning, in order to improve the rate of response to pharmacological treatment and course of panic disorder.

10.5   Conclusions

In conclusion, the comorbidity between panic disorder and personality disorders is common. Cluster C was the most frequent personality disorder subtype related to panic disorder. However, there are scarce studies that have evaluated this relationship in a systematic way and through clinical interviews. Personality disorders were more prevalent among individuals with panic disorder and co-occurring depression. Personality evaluations including both assessments categorical and

dimensional are needed. Further research is necessary to elucidate the impact of these concurrent conditions for panic disorder treatment and to define the most suitable therapeutic approach for each patient.

Acknowledgements This chapter was done in part with the support of Generalitat de Catalunya SGR2009/1435 and SGR2014/1411, and Centro de Investigación Biomédica en Red en Salud Mental (CIBERSAM), G25, Barcelona, Spain.

References

1. Oral E, Aydin N, Gulec M, Oral M. Panic disorder and subthreshold panic in the light of comorbidity: a follow-up study. Compr Psychiatry. 2012;53: 988–94.
2. Yates WR. Phenomenology and epidemiology of panic disorder. Ann Clin Psychiatry. 2009;21: 95–102.
3. Grant BF, Hasin DS, Stinson FS, Dawson DA, Goldstein RB, Smith S, et al. The epidemiology of DSM-IV panic disorder and agoraphobia in the United States: results from the National Epidemiologic Survey on Alcohol and Related Conditions. J Clin Psychiatry. 2006;67:363–74.
4. Byers AL, Yaffe K, Covinsky KE, Friedman MB, Bruce ML. High occurrence of mood and anxiety disorders among older adults: the National Comorbidity Survey Replication. Arch Gen Psychiatry. 2010; 67:489–96.
5. Westenberg HG, Liebowitz MR. Overview of panic and social anxiety disorders. J Clin Psychiatry. 2004;65 Suppl 14:22–6.
6. Bienvenu OJ, Stein MB. Personality and anxiety disorders: a review. J Pers Disord. 2003;17:139–51.
7. Svrakic DM, Draganic S, Hill K, Bayon C, Przybeck TR, Cloninger CR. Temperament, character, and personality disorders: etiologic, diagnostic, treatment issues. Acta Psychiatr Scand. 2002;106:189–95.
8. Samuel DB, Lynam DR, Widiger TA, Ball SA. An expert consensus approach to relating the proposed DSM-5 types and traits. Personal Disord. 2012; 3:1–16.
9. Diagnostic and statistical manual of mental disorders (DSM-5), 5th ed. American Psychiatric Association, Arlington, VA; 2013.
10. Ozkan M, Altindag A. Comorbid personality disorders in subjects with panic disorders: do personality disorders increase clinical severity? Compr Psychiatry. 2005;46:20–6.
11. Bienvenu OJ, Stein MB, Samuels JF, Onyike CU, Eaton WW, Nestadt G. Personality disorder traits as predictors of subsequent first-onset panic disorder or agoraphobia. Compr Psychiatry. 2009;50:209–14.

12. Dreessen L, Arntz A, Luttels C, Sallaerts S. Personality disorders do not influence the results of cognitive behavior therapies for anxiety disorders. Compr Psychiatry. 1994;35:265–74.

13. Soeteman DI, Verheul R, Busschbach JJ. The burden of disease in personality disorders: diagnosis-specific quality of life. J Pers Disord. 2008;22: 259–68.

14. Latas M, Milovanovic S. Personality disorders and anxiety disorders: what is the relationship? Curr Opin Psychiatry. 2014;27:57–61.

15. Reich J. The effect of Axis II disorders on the outcome of treatment of anxiety and unipolar depressive disorders: a review. J Pers Disord. 2003;17:387–405.

16. Massion AO, Dyck IR, Shea MT, Phillips KA, Warshaw MG, Keller MB. Personality disorders and time to remission in generalized anxiety disorder, social phobia, and panic disorder. Arch Gen Psychiatry. 2002;59:434–40.

17. Telch MJ, Kamphuis JH, Schmidt NB. The effects of comorbid personality disorders on cognitive behavioral treatment for panic disorder. J Psychiatr Res. 2011;45:469–74.

18. Sanderson C, Swenson C, Bohus M. A critique of the American psychiatric practice guideline for the treatment of patients with borderline personality disorder. J Pers Disord. 2002;16:122–9.

19. Martino F, Menchetti M, Pozzi E, Berardi D. Predictors of dropout among personality disorders in a specialist outpatients psychosocial treatment: a preliminary study. Psychiatry Clin Neurosci. 2012; 66:180–6.

20. Berger P, Sachs G, Amering M, Holzinger A, Bankier B, Katschnig H. Personality disorder and social anxiety predict delayed response in drug and behavioral treatment of panic disorder. J Affect Disord. 2004;80:75–8.

21. Hansen B, Vogel PA, Stiles TC, Götestam KG. Influence of co-morbid generalized anxiety disorder, panic disorder and personality disorders on the outcome of cognitive behavioural treatment of obsessive-compulsive disorder. Cogn Behav Ther. 2007;36:145–55.

22. Noyes Jr R, Reich J, Christiansen J, Suelzer M, Pfohl B, Coryell WA. Outcome of panic disorder. Relationship to diagnostic subtypes and comorbidity. Arch Gen Psychiatry. 1990;47:809–18.

23. Skodol AE, Gunderson JG, McGlashan TH, Dyck IR, Stout RL, Bender DS, et al. Functional impairment in patients with schizotypal, borderline, avoidant, or obsessive-compulsive personality disorder. Am J Psychiatry. 2002;159:276–83.

24. Brooks RB, Baltazar PL, McDowel DE, Munjack DJ, Bruns JR. Personality disorders co-occurring with panic disorder with agoraphobia. J Pers Disord. 1991;5:328–36.

25. Borenstein M, Hedges L, Higgins J, Rothstein H. Comprehensive meta-analysis: version 2.0. Englewood, NJ; Biostat, 2005.

26. Holmbeck GN, Durlak JA. Comorbidity of dependent personality disorders and anxiety disorders: conceptual and methodological issues. Clin Psychol Sci Pract. 2005;12:407–10.

27. Friborg O, Martinsen EW, Martinussen M, Kaiser S, Overgård KT, Rosenvinge JH. Comorbidity of personality disorders in mood disorders: a meta-analytic review of 122 studies from 1988 to 2010. J Affect Disord. 2014;152–154:1–11.

28. Stroup DF, Berlin JA, Morton SC, Olkin I, Williamson GD, Rennie D, et al. Meta-analysis of observational studies in epidemiology: a proposal for reporting. Meta-analysis of Observational Studies in Epidemiology (MOOSE) group. JAMA. 2000;283:2008–12.

29. First MB, Spitzer RL, Gibbon M, Williams JBW. Structured clinical interview for DSM-IV-TR axis I disorders, research version, patient edition (SCID-I/P). New York: Biometrics Research, New York State Psychiatric Institute; 2002.

30. First MB, Spitzer RL, Gibbon M, Williams JBW. Structured clinical interview for DSM-IV axis I disorders, research version, patient edition (SCID-I/P). New York: Biometrics Research, New York State Psychiatric Institute; 1990.

31. Spitzer RL, Williams JBW, Gibbon M, First MB. Structured clinical interview for DSM-III-R axis II disorders (SCID-II). Washington, DC: American Psychiatric Press, Inc.; 1990.

32. First MB, Gibbon M, Spitzer RL, Williams JBW, Benjamin LS. Structured clinical interview for DSM-IV axis II personality disorders (SCID-II). Washington, DC: American Psychiatric Press, Inc.; 1997.

33. Kotov R, Watson D, Robles JP, Schmidt NB. Personality traits and anxiety symptoms: the multilevel trait predictor model. Behav Res Ther. 2007:1485 503.

34. Goodwin RD, Faravelli C, Rosi S, Cosci F, Truglia E, de Graaf R, et al. The epidemiology of panic disorder and agoraphobia in Europe. Eur Neuropsychopharmacol. 2005;15:435–43.

35. Almeida Y, Nardi A. Psychological features in panic disorder. A comparison with major depression. Arq Neuropsiquiatr. 2002;60:553–7.

36. Kennedy BL, Schwab JJ, Hyde JA. Defense styles and Personality dimensions of research subjects with anxiety and depressive disorders. Psychiatry Q. 2001; 72:251–62.

37. Duijsens I, Spinhoven P, Goekoop J, Spermon T, Eurelings-Bontekoe E. The Dutch temperament and character inventory (TCI): dimensional structure, reliability and validity in a normal and psychiatric outpatient sample. Pers Indiv Differ. 2000;28: 487–99.

38. Dammen T, Ekeberg O, Arneses H, Friis S. Personality profiles in patients referred for chest pain: investigation with emphasis in panic disorder patients. Psychosomatics. 2000;413:269–76.

39. Battaglia M, Bertella S, Bajo S, Politi E, Bellodi L. An investigation of the co-occurrence of panic and somatization disorders through temperamental variables. Psychosom Med. 1998;60:726–9.

40. Hofmann SG, Shear MK, Barlow DH, Gorman JM, Hershberger G, Patterson M, et al. Effects of panic disorder treatment on personality disorders characteristics. Depress Anxiety. 1998;8:14–20.

41. Tyrer P, Johnson T. Establishing the severity of personality disorder. Am J Psychiatry. 1996;153: 1593–7.

42. Skodol AE, Oldham JM, Hyler SE, Stein DJ, Hollander E, Galaher PE, et al. Patterns of anxiety and personality disorders comorbidity. J Psychiat Res. 1995;29:361–74.

43. Clarck LA, Natson D, Mireka S. Temperament, personality and the mood and anxiety disorders. J Abnorm Psychol. 1994;103:103–16.

44. Jansen MA, Arntz A, Merckelbach H, Mersch PP. Personality disorders and features in social phobia and panic disorder. J Abnorm Psychol. 1994;103:391–5.

45. Reich J, Braginsky Y. Paranoid personality traits in a panic disorder population: a pilot study. Compr Psychiatry. 1994;35:260–4.

46. Reich J, Shera D, Dyck I, Vasile R, Goisman RM, Rodriguez-Villa F, et al. Comparison of personality disorders in different anxiety disorder diagnoses: panic, agoraphobia, generalized anxiety, and social phobia. Ann Clin Psychiatry. 1994;6:125–34.

47. Hoffart A, Martinsen EW. The effects of personality disorders and anxious-depressive comorbidity on outcome in patients with unipolar depression and with panic disorder and agoraphobia. J Pers Disord. 1993;7:304–11.

48. First MB, Vettorello N, Frances AJ, Pincus HA. Changes in mood, anxiety, and personality disorders. Hosp Community Psychiatry. 1993;44:1034–43.

49. Sanderson WC, Wetzler S, Beck BF. Prevalence of personality disorders among patients with anxiety disorders. Psychiatry Res. 1993;51:167–74.

50. Tyrer P, Seivewright N, Ferguson B, Tyrer J. The general neurotic syndrome: a coaxial diagnosis of anxiety, depression and personality disorder. Acta Psychiatr Scand. 1992;85:201–6.

51. Bagby RM, Cox BJ, Schuller DR, Levitt AJ, Swinson RP, Joffe RT. Diagnostic specificity of the dependent and self-critical personality dimensions in major depression. J Affect Disord. 1992;26:59–63.

52. Mavissakalian M, Hamann MS. DSM-II personality characteristics of panic disorder with agoraphobia patients in stable remission. Compr Psychiatry. 1992;33:305–9.

53. Saviotti FM, Grandi S, Savron G, Ermentini R, Bartolucci G, Conti S, et al. Characterological traits of recovered patients with panic disorder and agoraphobia. J Affect Disord. 1991;23:113–7.

54. Mavissakalian M, Hamann MS. Correlates of DSM-III personality disorder in panic disorder and agoraphobia. Compr Psychiatry. 1988;29:535–44.

55. Starcevic V, Latas M, Kolar D, Vucinic-Latas D, Bogojevic G, Milovanovic S. Co-occurrence of Axis I and Axis II disorders in female and male patients with panic disorder with agoraphobia. Compr Psychiatry. 2008;49:537–43.

56. Marchesi C, Cantoni A, Fontò S, Giannelli MR, Maggini C. The effect of temperament and character on response to selective serotonin reuptake inhibitors in panic disorder. Acta Psychiatr Scand. 2006; 114:203–10.

57. Albert U, Maina G, Forner F, Bogetto F. DSM-IV obsessive-compulsive disorder: prevalence in patients with anxiety disorders and in healthy comparison subjects. Compr Psychiatry. 2004;45: 325–32.

58. Marchesi C, Cantoni A, Fontò S, Giannelli MR, Maggini C. The effect of pharmacotherapy on personality disorders in panic disorder: a one year naturalistic study. J Affect Disord. 2005;89:189–94.

59. Iketani T, Kiriike N, Stein MB, Nagao K, Nagata T, Minamikawa N, et al. Personality disorder comorbidity in panic disorder patients with or without current major depression. Depress Anxiety. 2002;15: 176–82.

60. Starcevic V, Bogojevic G, Marinkovic J, Kelin K. Axis I and axis II comorbidity in panic/agoraphobic patients with and without suicidal ideation. Psychiatry Res. 1999;88:153–61.

61. Perugi G, Nassini S, Socci C, Lenzi M, Toni C, Simonini E, et al. Avoidant personality in social phobia and panic-agoraphobic disorder: a comparison. J Affect Disord. 1999;54:277–82.

62. Segui J, Márquez M, García L, Canet J, Salvador-Carulla L, Ortiz M. Diferential clinical features of early-onset panic disorder. J Affect Disord. 1999; 54:109–17.

63. Hoffart A. State and personality agoraphobic patients. J Pers Disord. 1994;8:333–41.

64. Flick SN, Roy-Byrne PP, Cowley DS, Shores MM, Dunner DL. DSM-III-R personality disorders in a mood and anxiety disorders clinic: prevalence, comorbidity, and clinical correlates. J Affect Disord. 1993;27:71–9.

65. Suarez A, Fernandez Vega F. Personality disorders in patients with panic disorder. Actas Luso Esp Neurol Psiquiatr Cienc Afines. 1992;20:241–5.

66. Pollack MH, Otto MW, Rosenbaum JF, Sachs GS. Personality disorders in patients with panic disorder: association with childhood anxiety disorders, early trauma, comorbidity, and chronicity. Compr Psychiatry. 1992;33:78–83.

67. Reich J. Avoidant and dependent personality traits in relatives of patients with panic disorder, patients with dependent personality disorder, and normal controls. Psychiatry Res. 1991;39:89–98.

68. Pollack MH, Otto MW, Rosenbaum JF, Sachs GS, O'Neil C, Asher R, et al. Longitudinal course of panic disorder: findings from the Massachusetts General Hospital Naturalistic Study. J Clin Psychiatry. 1990;51(Suppl A):12–6.

69. Newton-Howes G, Tyrer P, Anagnostakis K, Cooper S, Bowden-Jones O, Weaver T. COSMIC study team. The prevalence of personality disorder, its comorbidity with mental state disorders, and its clinical significance in community mental health teams. Soc Psychiatry Psychiatr Epidemiol. 2010;45:453–60.

70. Powers A, Westen D. Personality subtypes in patients with panic disorder. Compr Psychiatry. 2009;50:164–72.

71. Mula M, Pini S, Monteleone P, Iazzetta P, Preve M, Tortorella A, et al. Different temperament and character dimensions correlate with panic disorder comorbidity in bipolar disorder and unipolar depression. J Anxiety Disord. 2008;22:1421–6.

72. Wachleski C, Blaya C, Salum GA, Vargas V, Leistner-Segal S, Manfro GG. Lack of association between the serotonin transporter promoter polymorphism (5-HTTLPR) and personality traits in asymptomatic patients with panic disorder. Neurosci Lett. 2008;431:173–8.

73. Wachleski C, Salum GA, Blaya C, Kipper L, Paludo A, Salgado AP, et al. Harm avoidance and self-directedness as essential features of panic disorder patients. Compr Psychiatry. 2008;49:476–81.

74. Lana F, Fernández San Martín MI, Sánchez Gil C, Bonet E. Study of personality disorders and the use of services in the clinical population attended in the mental health network of a community area. Actas Esp Psiquiatr. 2008;36:331–6.

75. Tyrer P, Seivewright N, Ferguson B, Johnson T. Obsessional personality and outcome of panic disorder. Br J Psychiatry. 1998;172:187.

76. Wingerson D, Sullivan M, Dager S, Flick S, Dunner D, Roy-Byrne P. Personality traits and early discontinuation from clinical trials in anxious patients. J Clin Psychopharmacol. 1993;13:194–7.

77. Osma J, García-Palacios A, Botella C, Barrada JR. Personality disorders among patients with panic disorder and individuals with high anxiety sensitivity. Psicothema. 2014;26:159–65.

78. Kristensen AS, Mortensen EL, Mors O. The association between bodily anxiety symptom dimensions and the scales of the Revised NEO Personality Inventory and the Temperament and Character Inventory. Compr Psychiatry. 2009;50:38–47.

79. Blashfield R, Noyes R, Reich J, Woodman C, Cook BL, Garvey MJ. Personality disorder traits in generalized anxiety and panic disorder patients. Compr Psychiatry. 1994;35:329–34.

80. Reich J, Noyes Jr R, Troughton E. Dependent personality disorder associated with phobic avoidance in patients with panic disorder. Am J Psychiatry. 1987;144:323–6.

81. Marchesi C, Cantoni A, Fontò S, Giannelli MR, Maggini C. Predictors of symptom resolution in panic disorder after one year of pharmacological treatment: a naturalistic study. Pharmacopsychiatry. 2006;39:60–5.

82. Iketani T, Kiriike N, Stein MB, Nagao K, Nagata T, Minamikawa N, et al. Relationship between perfectionism, personality disorders and agoraphobia in patients with panic disorder. Acta Psychiatr Scand. 2002;106:171–8.

83. Ampollini P, Marchesi C, Signifredi R, Maggini C. Temperament and personality features in panic disorder with or without comorbid mood disorders. Acta Psychiatr Scand. 1997;95:420–3.

84. Alnaes R, Torgersen S. Clinical differentiation between major depression only, major depression with panic disorder and panic disorder only. Childhood, personality and personality disorder. Acta Psychiatr Scand. 1989;79:370–7.

85. Reich J, Troughton E. Frequency of DSM-III personality disorders in patients with panic disorder: comparison with psychiatric and normal control subjects. Psychiatry Res. 1988;26:89–100.

86. Reich J, Troughton E. Comparison of DSM-III personality disorders in recovered depressed and panic disorder patients. J Nerv Ment Dis. 1988;176:300–4.

87. Battaglia M, Bernardeschi L, Politi E, Bertella S, Bellodi L. Comorbidity of panic and somatization disorder: a genetic-epidemiological approach. Compr Psychiatry. 1995;36:41120.

88. Milrod BL, Leon AC, Barber JP, Markowitz JC, Graf E. Do comorbid personality disorders moderate panic-focused psychotherapy? An exploratory examination of the American Psychiatric Association practice guideline. J Clin Psychiatry. 2007;68:885–91.

89. Draganić-Rajić S, Lecić-Tosevski D, Paunović VR, Cvejić V, Svrakić D. Panic disorder-psychobiological aspects of personality dimensions. Srp Arh Celok Lek. 2005;133:129–33.

90. Moreno-Peral P, Conejo-Cerón S, Motrico E, Rodríguez-Morejón A, Fernández A, García-Campayo J, et al. Risk factors for the onset of panic and generalised anxiety disorders in the general adult population: a systematic review of cohort studies. J Affect Disord. 2014;168:337–48.

91. Bienvenu OJ. What is the meaning of associations between personality traits and anxiety and depressive disorders? Rev Bras Psiquiatr. 2007;29:3–4.

92. Brandes M, Bienvenu OJ. Personality and anxiety disorders. Curr Psychiatry Rep. 2006;8:263–9.

93. Marshall JR. Comorbidity and its effects on panic disorder. Bull Menninger Clin. 1996;60(2 Suppl A):A39–53.

94. Mavissakalian M. The relationship between panic disorder/agoraphobia and personality disorders. Psychiatr Clin North Am. 1990;13:661–84.

95. Reich J, Noyes R, Hirschfeld RP, Coryell W, O'Gorman TW. State and personality in depressed and panic patients. Am J Psychiatry. 1987;144:181–7.

96. Mendoza L, Navinés R, Crippa JA, Fagundo AB, Gutierrez F, Nardi AE, et al. Depersonalization and personality in panic disorder. Compr Psychiatry. 2011;52:413–9.

97. Gutiérrez F, Navinés R, Navarro P, García-Esteve L, Subirá S, Torrens M, et al. What do all personality disorders have in common? Ineffectiveness and

uncooperativeness. Compr Psychiatry. 2008;49: 570–8.

98. Iketani T, Kiriike N, Stein MB, Nagao K, Minamikawa N, Shidao A, et al. Patterns of axis II comorbidity in early-onset versus late-onset panic disorder in Japan. Compr Psychiatry. 2004;45:114–20.

99. Barzega G, Maina G, Venturello S, Bogetto F. Gender-related distribution of personality disorders in a sample of patients with panic disorder. Eur Psychiatry. 2001;16:173–9.

100. Dyck IR, Phillips KA, Warshaw MG, Dolan RT, Shea MT, Stout RL, et al. Patterns of personality pathology in patients with generalized anxiety disorder, panic disorder with and without agoraphobia, and social phobia. J Pers Disord. 2001;15:60–71.

101. Latas M, Starcevic V, Trajkovic G, Bogojevic G. Predictors of comorbid personality disorders in patients with panic disorder with agoraphobia. Compr Psychiatry. 2000;41:28–34.

102. Ampollini P, Marchesi C, Signifredi R, Ghinaglia E, Scardovi F, Codeluppi S, et al. Temperament and personality features in patients with major depression, panic disorder and mixed conditions. J Affect Disord. 1999;52:203–7.

103. Langs G, Quehenberger F, Fabisch K, Klug G, Fabisch H, Zapotoczky HG. Prevalence, patterns and role of personality disorders in panic disorder patients with and without comorbid (lifetime) major depression. Acta Psychiatr Scand. 1998;98:116–23.

104. Mauri M, Sarno N, Rossi VM, Armani A. Personality disorders associated with generalized anxiety, panic, and recurrent depressive disorders. J Pers Disord. 1992;6:162–7.

105. Mellman TA, Leverich GS, Hauser P, Kramlinger KL. Axis II pathology in panic and affective disorders: relationship to diagnosis, course of illness, and treatment response. J Pers Disord. 1992;6:53–63.

106. Sciuto G, Diaferia G, Battaglia M, Perna G, Gabriele A, Bellodi L. DSM-III-R personality disorders in panic and obsessive-compulsive disorder: a comparison study. Compr Psychiatry. 1991;32:450–7.

107. Alnaes K, Torgersen S. DSM-III personality disorders among patients with major depression, anxiety disorders and mixed conditions. J Nerv Ment Dis. 1990;178:693–8.

108. Svanborg C, Wistedt AA, Svanborg P. Long-term outcome of patients with dysthymia and panic disorder: a naturalistic 9-year follow-up study. Nord J Psychiatry. 2008;62:17–24.

109. Marchesi C, De Panfilis C, Cantoni A, Fontò S, Giannelli MR, Maggini C. Personality disorders and

response to medication treatment in panic disorder: a 1-year naturalistic study. Prog Neuropsychopharmacol Biol Psychiatry. 2006;30:1240–5.

110. Lenzenweger MF, Lane MC, Loranger AW, Kessler RC. DSM-IV personality disorders in the National Comorbidity Survey Replication. Biol Psychiatry. 2007;62:553–64.

111. Kessler RC, Merikangas KR. The National Comorbidity Survey Replication (NCS-R): background and aims. Int J Methods Psychiatr Res. 2004;13:60–8.

112. Skodol AE, Geier T, Grant BF, Hasin DS. Personality disorders and the persistence of anxiety disorders in a nationally representative sample. Depress Anxiety. 2014;31:721–8.

113. Nurnberg HG, Hurt SW, Feldman A, Suh R. Evaluation of diagnostic criteria for borderline personality disorder. Am J Psychiatry. 1988;145: 1280–4.

114. Pulay AJ, Stinson FS, Dawson DA, Goldstein RB, Chou SP, Huang B, et al. Prevalence, correlates, disability, and comorbidity of DSM-IV schizotypal personality disorder: results from the wave 2 national epidemiologic survey on alcohol and related conditions. Prim Care Companion J Clin Psychiatry. 2009;11:53–67.

115. O'Toole MS, Arendt M, Fentz HN, Hougaard E, Rosenberg NK. Cluster A personality pathology in social anxiety disorder: a comparison with panic disorder. Nord J Psychiatry. 2014;68:460–3.

116. Olatunji BO, Cisler JM, Tolin DF. A meta-analysis of the influence of comorbidity on treatment outcome in the anxiety disorders. Clin Psychol Rev. 2010;30:642–54.

117. Slaap BR, den Boer JA. The prediction of nonresponse to pharmacotherapy in panic disorder: a review. Depress Anxiety. 2001;14:112–22.

118. Iza M, Olfson M, Vermes D, Hoffer M, Wang S, Blanco C. Probability and predictors of first treatment contact for anxiety disorders in the United States: analysis of data from the National Epidemiologic Survey on Alcohol and Related Conditions (NESARC). J Clin Psychiatry. 2013; 74:1093–100.

119. Stone MH. Management of borderline personality disorder: a review of psychotherapeutic approaches. World Psychiatry. 2006;5:15–20.

120. Mennin DS, Heimberg RG. The impact of comorbid mood and personality disorders in the cognitive-behavioral treatment of panic disorder. Clin Psychol Rev. 2000;20:339–57.

Possible Mechanisms Linking Panic Disorder and Cardiac Syndromes

11

Sergio Machado, Eduardo Lattari, and Jeffrey P. Kahn

Contents

S. Machado (✉)
Laboratory of Panic and Respiration, Institute of Psychiatry, Federal University of Rio de Janeiro, Rio de Janeiro, Brazil

Physical Activity Neuroscience, Physical Activity Sciences Postgraduate Program, Salgado de Oliveira University, Niterói, Brazil
e-mail: secm80@gmail.com

E. Lattari
Laboratory of Panic and Respiration, Institute of Psychiatry, Federal University of Rio de Janeiro, Rio de Janeiro, Brazil
e-mail: eduardolattari@yahoo.com.br

J.P. Kahn
Department of Psychiatry, Weill-Cornell Medical College, Cornell University, New York, NY, USA
e-mail: JeffKahn@aol.com

Abstract

Since chest pain can indicate coronary artery disease, pulmonary embolus and other severe physical illness, a prompt and careful diagnosis is important. It can also be due to panic disorder, an anxiety disorder with serious morbidity and mortality consequences. The diagnosis of panic is often not obvious to all clinicians, and panic can also occur co-morbidly with physical heart disease. More specifically, panic anxiety often includes an abrupt feeling of fear accompanied by symptoms such as breathlessness, palpitations, chest pain, and thus patient fear of a heart attack. This concern may further confound physicians. The association between panic disorder and coronary artery disease has been extensively studied in recent years and, although some studies have shown anxiety disorders coexisting or increasing the risk of heart disease, no causal hypothesis has been well established. The aim of this chapter is to present the various ways in which the scientific community has been investigating the relations of panic disorder with cardiac syndromes.

Keywords

Anxiety disorders • Cardiac syndromes • Chest pain • Coronary artery disease • Panic disorder

11.1 Introduction

According to the diagnostic guidelines of the American Psychiatric Association, panic attack (PA) frequency can vary from a few each year, to several attacks over a single day [1]. The expression of this fear can includes additional cognitive symptoms such as derealization, depersonalization, as well as fears of going crazy, having a heart attack or stroke, or dying.

Among the 13 main symptoms of a PA, many are also frequent in heart disease, such as chest pain, hyperventilation, palpitations, nausea, feelings of shortness of breath, fear of dying or losing control, or unreality [1]. Typically, the individual with PA thinks that they have some serious physical illness, and often first seek care in the emergency room [2].

Chest pain, in particularly, is a PA symptom that seems to support the concern that a "heart attack" is in progress during the PA. Research suggests that PD is the primary diagnosis in some 30–60% of chest pain patients who seek care in emergency rooms [2, 3]. Thus, clinical evaluation of chest pain approach requires caution and careful differential diagnosis, and it may also be related to musculoskeletal disorders [4]. A primary initial focus should be the rapid identification of patients with a high probability of a true acute coronary syndrome (ACS), as well as diagnosis of other non-coronary heart disease, lung disease, musculoskeletal problems, and digestive tract disease [5]. Early treatment is essential.

Unfortunately, when the diagnostic possibilities of life-threatening events are removed, too little attention is given to the possibility that a mental disorder can be present or even being responsible for the patient's complaints, and far less attention is given to the effectiveness of, and need for treatment. In 130 patients who reported chest pain, 54.6% had no acute coronary syndrome or any other diagnosis for that complaint, and of these patients, 53.5% had anxiety and 25.3% had depression [6].

In most cases, the chest pain symptoms of PA are mild, without any corresponding physical disease. Nevertheless, it also possible that PA can occur in the presence of coronary disease,

sometimes not yet even diagnosed. In the study by Lynch and Galbraith [2] mentioned above, was observed in the group of PD patients, 25.5% of them also suffered from coronary disease. The authors called this combination of "lethal combination", given the documented increased risk of fatal events during long-term.

This association was first shown by Beitman et al. [7], who examined the prevalence of PD in 104 cardiac patients, including 30 with coronary artery disease (CAD). Sixteen patients with CAD fit diagnostic criteria for panic disorder (PD). Subsequently, Basha et al. [8] also examined the prevalence of PD in 49 patients with CAD who had typical and atypical chest pain, noting that 27% met criteria for DSM-III PD. Fleet et al. [9] also examined the prevalence of PD in patients with chest pain, where, 34% of with CAD patients also had PD. Despite the findings suggest that PD is associated with CAD, even without CAD, 43 patients were diagnosed for PD [7].

So the apparently prevalent of comorbidity between PD and cardiovascular disease raises an intriguing question: could there be a causal relationship between PD and CAD? Could the mechanisms involved in non-comorbid PA somehow be related to ischemia or to some other mechanism that explains the presence of non-CAD chest pain? For this, we investigate the relationship between anxiety and cardiovascular disease, with emphasis on the mechanisms involved in the pathophysiology of PD and mental stress. In summary, the association of PA with non-CAD chest pain raise questions of both cardiac and non-cardiac mechanisms.

11.2 Non-Cardiac Mechanisms

Some mechanisms have been bandied about as the cause of chest pain in PD, without direct involvement of the heart. Among them are a pain from musculoskeletal overload secondary to hyperventilation, changes in esophageal motility (including esophageal spasm), common in states of anxiety [10], and situations in which intense anxiety becomes interpreted as pain. Some patients tend to perceive their emotions in a predominantly somatic mode, experiencing physical pain instead of anxiety or sadness, for example. It was observed that the use of opioid analgesics in this population can enforce this distortion [11]. However, it has also been shown that opioids may offer partial treatment for the emotional pain of panic [12, 13].

11.3 Cardiac Mechanisms

Both autonomic activation and hyperventilation via respiratory alkalosis during the PA can lead to coronary spasm. Freeman et al. [14] observed that PA was studied with a hyperventilation protocol; patients had elevated levels of systolic and diastolic blood pressure (BP). In addition, PA can cause ischemic pain in patients with CAD, by increasing myocardial oxygen demand via increased heart rate (HR) and BP, as we shall see.

However, there are cases where acute non-CAD cardiac events can also occur, even without prior CAD. For example, the stress-induced or apical ballooning (i.e., Takotsubo) cardiomyopathy syndrome is typically triggered by acute mental stress. Takotsubo makes the heart to take the form of a pot ("Tsubo") for octopus ("Tako") fishing commonly used in Japan, where the syndrome was first described. With high prevalence in postmenopausal women, it is characterized by the presence of chest pain and ECG changes typical of acute myocardial infarction, but without evidence of arterial occlusive lesions in coronary angiography, as has been shown [15]. The exact pathophysiological mechanism correlating cases of acute stress with Takotsubo syndrome is still the subject of research, but some results converge to the correlation between acute stress and adrenergic hyperactivity. Wittstein et al. [16] clearly demonstrated the activation of the adrenal medulla and elevated plasma levels of adrenaline and noradrenaline from 7 until 9 days after the event of acute stress. It may well be that panic can contribute to this catecholaminergic cardiomyopathy [17].

Indeed, prolonged PAs often present as agitated depression in peri-menopausal women, and are associated with increased catecholamine levels [18]. Moreover, panic anxiety (associated with suppressed anger) is highly prevalent in cardiac transplant candidates with idiopathic dilated cardiomyopathy, but not in candidates with other end-stage cardiac disease [18]. There, too, the proposed mechanism is increased circulating catecholamines. Even in "healthy" panic patients, there is an increased prevalence of significant left ventricular enlargement [19].

These cardiomyopathy findings are also relevant to CAD. Cardiomyopathic hearts are large but inefficient, and thus result in reduced oxygen supply to the heart, even as the enlarged heart might require extra oxygen. This would aggravate existing CAD, and might even contribute to CAD development. Left ventricular hypertrophy can cause ischemia by increasing demand, rather than just by reducing supply, and is considered an independent risk factor for CAD events [20].

11.3.1 Myocardial Ischemia

Myocardial ischemia is the result of an imbalance between supply and demand: occurs when the coronary blood flow becomes inadequate to meet the needs of cellular oxygen (O_2) and metabolic substrates [21]. With that the heart starts to fail in its primary function, which is to act as a contractile hydraulic pump. The five main causes of imbalance and ischemia are (a) coronary artery thrombosis, (b) fixed non-thrombotic obstruction, (c) dynamic obstruction such as vasospasm, (d) inflammation (e) increased oxygen demand without obstruction, as in significant myocardial hypertrophy or in situations of myocardial hyperactivity and (f) reduced blood/oxygen supply to the heart as a result of cardiomyopathy or of other conditions [21].

One frequent cause of myocardial ischemia is the rupture of atheromatous plaque with release of thrombogenic substances and the consequent formation of a clot that prevents coronary blood flow [22]. In the presence of a state of persistent obstruction and ischemia, the process evolves to myocardial cell death or necrosis. In animal experiments myocardial cellular necrosis begins 20 min after coronary occlusion. In clinical practice, the extent of necrosis depends on the caliber of the occluded vessel, the level of myocardial oxygen demand and the presence of collateral circulation to the ischemic region. Acute coronary syndrome describes a spectrum of myocardial ischemia that has at one end stable angina, progressing to unstable angina and culminating in acute myocardial infarction and necrosis of cardiac muscle [23].

The most classic clinical manifestation of myocardial ischemia, regardless of the causal mechanism, is the thoracic pain of *angina pectoris*. Angina is a pain that is distributed diffusely in the retrosternal region, usually characterized as oppressive, burning or crushing. It may radiate to the throat, neck, ulnar site of arms, interscapular region, epigastric region, jaws and teeth. The intensity of pain can vary from a light weight or retrosternal discomfort, or tingling in one dermatome until excruciating pain. These findings are related to the cause of ischemia and are not completely specific for ischemia, which may be due to non-ischemic cardiac pain, as well as a non-cardiac cause. In a meta-analysis published in 2012, Haasenritter et al. [24] suggest that the accuracy of signs and symptoms for the diagnosis of myocardial ischemia varies between studies published according to the case definition of CAD.

Another significant finding in the diagnosis of chest pain of ischemic origin is the fact that it is triggered by physical exercise or emotional stress. Importantly many patients do not develop angina even after exhaustive physical activity, but do experience typical chest pain after a work meeting or event associated with strong emotional overload [25]. This aspect of chest pain, as we shall see, was responsible for lead to increasing research in recent years.

11.3.2 Mental Stress and Myocardial Ischemia

Provocative tests with laboratory studies indicate that myocardial ischemia can be induced in 50–70 % of patients with CAD, by mental stress [26]. Since a true PA is likely more emotionally potent than a mental stress test, it is quite possible that it can induce true ischemia in patients with existing CAD. During a PA intense activation of the cardiac sympathetic nervous system and increased secretion of adrenaline from two to six times the normal value occurs.

In the same way, sympathetic activation occurs during the release of neuropeptide Y (NPY) of the cardiac sympathetic nerve in the coronary sinus that may be related to coronary vasospasm [27]. Prospective study showed that mental stress-induced ischemia is associated with significantly higher rates of fatal independent and nonfatal cardiac events, age, baseline ejection fraction, and left ventricular myocardial infarction [28]. In this study mental stress-induced ischemia was a better predictor of future cardiac events than the exercise stress test, with a 60 month follow-up. Some caution is in order, though. PD might well act as a spontaneous "mental stressor" similar to provocative tests of mental stress. Though it has long been thought that the sympathetically-mediated increases in HR, BP and LV contractility present during a PA cause increased myocardial O_2 demand, and would be responsible for the appearance of myocardial ischemia [29]. However, it is known that the provocation tests for mental stress lead only to modest small increases in BP and HR, and only slightly affect myocardial O_2 demand. Along those lines, it should be noted that in one study of CAD patients with normal exercise stress tests, and stratified for those with and without PD, CO_2 induced panic was not associated with induced myocardial ischemia [30], even so, other mechanisms could be involved, such as vasospasm or a reduction in coronary flow [31].

New lines of research suggest that ischemia triggered by mental stress may be related to endothelial dysfunction in vessels with atherosclerotic disease, thus limiting coronary vasodilatation during mental stress. While arteries of patients without CAD dilate during mental stress, atherosclerotic arteries fail to dilate during mental stress, probably in response to changes in the activation of $\alpha2$ and $\beta2$ adrenoceptor and increased release of noradrenaline, which induces muscle contraction patients arteries [32, 33]. There is even angiographic evidence of coronary vasoconstriction at the site of atherosclerosis during arithmetic test performance [34]. Endothelial dysfunction has been described as potential contributors to the development of coronary spasm, by reducing the secretion of endothelial relaxing factor. In addition, "stress" causes ischemia and thus leads to damage and atherosclerosis, and also, "stress" causes ischemia which compounds the effects of already established atherosclerosis.

Thus, certain kinds of adrenergically active "stress" may provoke coronary vasoconstriction, consequent intimal damage and thus development of CAD. For example, the hostility and Type A behavior risk factors for CAD are quite highly correlated with directly measured adrenergic receptor densities on lymphocytes (vasodilative beta receptors) and platelets (vasoconstrictive alpha receptors). In particular, the alpha/beta receptor density ratio offers physiologic model for evidence of ongoing coronary vasoconstriction in healthy young adults [35].

11.3.3 Perfusion Defects in Panic Attack

In a groundbreaking study, Fleet et al. [36] sought to observe the possible myocardial ischemia triggered by an induced PA in patients with CAD. 65 selected patients, 35 with PD and 30 without PD. All were subjected to inhalation of gas mixture containing 65% oxygen and 35% carbon dioxide (CO_2), an established provocative test for PA. Upon CO_2 inhalation, infusion of technetium-99m was injected in all participants. The results showed that among those who had PA, 80.9% had myocardial perfusion defects, while only 46.4% of those who showed no attack showed perfusion defect. Even though all participants were using medications to control CAD, the study still shows that a PA can induce myocardial perfusion defects.

A remaining question is whether CO_2 induced PA in "healthy" PD patients without any CAD can also induce myocardial perfusion defects. *The cause-effect relationship* between these two variables is intriguing and requires further investigation.

For panic disorder, the CO_2 inhalation challenge test is highly sensitive (about 70%) and highly specific (about 100%). This approach offers some advantages such as ease of administration, good tolerance, and reliability in generating panic attacks similar to those experienced by patients outside the laboratory, self-limited and brief duration of panic provocation, without requiring the pharmacologic interventions of other panic provocations methods [37]. Studies could use gated myocardial perfusion scintigraphy, a test well supported by large studies as an effective method for detecting ischemia in the emergency room option [38]. In clinical practice cardiology myocardial perfusion imaging using radioisotope Technetium-99m sestamibi (sestamibi SPECT) is already incorporated into clinical practice in the study of myocardial ischemia.

The next topics present three additional perspectives: the first showing anxiety disorders as a predictor of heart disease, the second that anxiety disorders can be an aggravating factor for cardiovascular disease, and the third outlining a number of hypotheses of a possible causal relationship between anxiety and cardiovascular disease.

11.4   Anxiety Predictor of Heart Disease in Previously Healthy People

In the Myocardial Ischemia and Migraine Study, a longitudinal study involving 3369 community-dwelling healthy postmenopausal women assessing cardiovascular and cerebrovascular outcomes after 5.3 years, the authors observed a hazard ratio for all-cause mortality was 1.75 among those presenting a 6-month history of full-blown PA [39]. The Northwick Park Heart Study found that 1457 initially healthy men with high scores of phobic anxiety (PD is typically the underlying cause of most phobias), had a 3.77 relative risk of fatal CAD after 6.7 years of follow-up [40]. In a strong prospective long-term study, 49,321 young adults were followed for 37 years [41]. The authors found a multi-adjusted hazard ratios associated with anxiety of 2.17 for CAD and 2.51 for acute myocardial infarction (Table 11.1).

Table 11.1 Anxiety predictor of heart disease in previously healthy people

Authors	Study caracteristics	Population	Mean age	Objetives	Main results	Reference
Smoller et al.	Prospective cohort survey/5.3 years of follow-up	3369 community-dwelling, healthy postmenopausal women	51–83 years	To determine whether PA are associated with risk of cardiovascular morbidity and mortality in postmenopausal women	History of PA was associated with CAD and stroke The hazard ratio for all-cause mortality was 1.75	[39]
Haines et al.	Prospective study Northwick Park heart study	1457 white men	40–64 years	Study of the relation between PA and subsequent incidence of CAD	Consistent increase in risk of fatal CAD. The relative risk for score 5 and above on the phobic anxiety subscale was 3.77	[40]
Janszky et al.	Prospective Study/followed for CHD and for acute myocardial infarction for 37 years	49,321 young Swedish men	18–20 years	To investigate the long-term cardiac effects of depression and anxiety assessed at a young age	Multiadjusted hazard ratios associated with depression were 1.04 for AMI and 1.03 for CHD, respectively. The corresponding multiadjusted hazard ratios for anxiety were 2.17 and 2.51	[41]

CAD coronary artery disease, *CHD* coronary heart disease, *PA* panic attack

11.5   Anxiety Disorders Predict Poorer Prognosis in Established Cardiovascular Disease

In the same way, a meta-analysis that analyzed data from nearly 250,000 patients followed for more than 11 years showed that people with anxiety have both increased risk of CAD mortality and cardiac events, independent of demographic variables, biological risk factors, and health behaviors [42]. Walters et al. show data suggesting the presence of anxiety as an independent risk factor for CAD, both as a trigger of myocardial infarction and as a long-term precursor of cardiovascular diseases in younger people, under 50 years old [43]. Large-scale community-based studies, reported a significant relationship between anxiety disorders and cardiac death [40, 44, 45]. Although a relationship between anxiety levels and the occurrence of cardiac death has been detected in these studies, anxiety did not show any significant association with myocardial infarction. Moreover, mortality was mainly due to sudden cardiac death. A nearly twofold increased risk of nonfatal myocardial infarction or death in a an average follow-up of 3.4 years follow-up was detected in CAD population with anxiety [46]. CAD patients with anxiety symptoms during a 3-year follow-up showed a hazard ratio (HR) of 2.32 for a new ischemic event [47]. Indeed, epidemiological studies report a three to sixfold increase risk of myocardial ischemia and of sudden death [44, 45] (Table 11.2).

Some retrospective studies also attempted to demonstrate an association of PD with increased risk of mortality from cardiovascular causes in patients with CAD. Coryell et al. [48] examined mortality rates 35 years after psychiatric hospitalization and found that PD patients had mortality rates from cardiovascular causes twice as high as expected, compared with the same age, sex and length of hospital group. In a subsequent study, the same authors [49] observed high mortality rates in 155 patients, 12 years after admission, comparing with mortality rates of residents in the same age and gender.

11.6   Hypothesis of a Causal Relationship Between Anxiety and Cardiovascular Disease

Epidemiologic evidence suggests how anxiety may be a risk factor for the development of heart disease in initially healthy individuals and for complication in cardiac patients (Table 11.3). Kubzansky and Kawachi [50] examining papers published during the years 1980–1996 found some evidence suggesting that chronic anxiety could be a risk factor for the development of CAD by an influence on health behaviors, increased risk of hypertension and atherogenesis, and even triggering fatal coronary events, either through arrhythmia, plaque rupture, coronary vasospasm, or thrombosis. These hypotheses have gained strength in studies that showed associations between hypertension and anxiety disorders linked to an increased risk of a pro-inflammatory state and development of CAD [51] and acting as an independent risk factor for cerebrovascular accidents. In a follow-up period of 3-years, the hazard of stroke was estimated to be 2.37 times greater for patients with PD than for patients in the comparison cohort [52].

Some factors may contribute to CAD and stroke in PD patients. Studying hemostatic function in 96 subjects reporting frequent panic symptoms, von Kanel et al. [53] found higher levels of D-dimer, a common hypercoagulability marker, and lower fibrinogen levels than the 595 subjects reporting panic symptoms either 'not at all' or 'not very often'. The authors speculate that such a pro-coagulant state could contribute to increased coronary risk. Likewise, carotid-femoral pulse wave velocity (PWV), a well established way to evaluate arterial stiffness and to predict adverse cardiovascular outcomes, was measured in forty-two patients with PD, and 30 controls [54]. The authors found that PD was independently related to increase carotid-femoral PWV in a multivariate analysis (Table 11.3).

The striking presence of chest pain in some episodes of PA, sometimes with characteristics similar to an acute coronary syndrome, has led

Table 11.2 Studies demonstrating anxiety disorders leading to worsening prognosis in cardiovascular disease

Authors	Study caracteristics	Population	Mean age	Objetives	Main results	Reference
Roest et al.	A meta-analysis from 1980 to May 2009	249,846 persons	–	To assess the association between anxiety and risk of CAD	Anxious persons were at risk of CAD and death independent of risk factors and behaviors	[42]
Walters et al.	Cohort study	57,615 adults with PA/PD and 347,039 controls	–	To determine the risk of CAD, AMI, and CAD-related mortality in patients with PA/PD	PA/PD was associated to hazard of AMI in those under 50 years and CAD at all ages	[43]
Kawachi et al.	Prospective Study 2 years of follow-up	33,999 male health professionals	42–77 years	To examine association between phobic anxiety and risk of CAD	Relative risk of fatal CAD among anxiety men was 3.01	[44]
Kawachi et al.	Prospective Study 32 years of follow-up Cornell Medical Index	? Normative Aging Study	–	To examine increased risk of fatal CAD among PD patients and other anxiety disorders	Reports of two or more anxiety symptoms had elevated risks of fatal CAD and sudden death	[45]
Shibeshi et al.	Prospective cohort study	516 with CAD. 82% male	Mean age 68 years at entry	To examined the effect of anxiety on mortality and nonfatal AMI in patients with CAD	A high cumulative anxiety score was associated with an increased risk of nonfatal AMI or death	[46]
Rothenbacher et al.	Prospective cohort study	1052 patients with CAD	–	Observation of Fatal and non-fatal Cardiovascular disease events (CVDE)	Patients with anxiety had a HR of 2.32 for CVDE, Associated to depression raised HR to 3.31	[47]
Coryell et al.	Retrospective study. 35 years after admission	Data from 113 inpatients obtained by chart review	–	The author examined mortality rates among in-patients with PD	Patients with PD had excess mortality due to CAD	[48]
Coryell et al.	Prospective study 12-year follow-up	155 outpatients	–	To determine the Mortality rates linked to "anxiety neurosis"	Men were twice as likely to die due to CAD and suicide	[49]

CAD coronary artery disease, *PA* panic attack, *PD* panic disorder

Table 11.3 Studies demonstrating the hypothesis of a causal relationship between anxiety and cardiovascular disease

Authors	Study methodology	Population	Objectives	Main results	Reference
Kubzansky et al.	Review. Papers published during the years 1980–1996	–	To examine the association between anxiety and CAD	Chronic anxiety related to health behaviors, atherogenesis, coronary events, and arrhythmias	[50]
von Känel et al.	Self-rate panic screening and measurement of coagulation markers	691 employees 83 % men	Relation between panic and hypercoagulable state	Panic feelings related to higher D-dimer levels and lower fibrinogen levels	[53]
Cicek et al.	Case control study	42 patients and 30 control	Measurement of PWV as the surrogate of arterial stiffness in patients with panic disorder	Panic disorder independently related to PWV (βeta: 0.317, $p=0.011$)	[54]
Fleet et al.	Panic challenge test and myocardial scintigraphy	Sixty-five patients with CAD and positive nuclear exercise stress testing	Association between PA and ischemia in patients with CAD	80.9 % of patients with PA presented myocardial perfusion defect	[36]
Soares-Filho et al.	A case report. Panic challenge test and myocardial scintigraphy	Patient with no CAD	Association between PA and ischemia in patients without CAD	PA induced by inhalation of 35 % carbon dioxide triggering myocardial ischemia	[55]
Mansour et al.	Case report	Three patients with PD and documented cardiac ischemia	To study CP in PD. Patients without atherosclerosis	Presumed cases of coronary artery spasm	[59]
Esler et al.	Systematic review	–	Study PD and the range of cardiac complications	Sympathetic activation during PA leading adrenergic neural responses	[71]
Tanabe et al.	Case control study	Nine patients with panic disorder (7 men, 2 women) and 11 control subjects	To evaluate the cardiac sympathetic function in panic disorder using (123) I-metaiodobenzylguanidine	Myocardial scintigraphy showed impairment of cardiac sympathetic function in PD	[72]

(continued)

Table 11.3 (continued)

Authors	Study methodology	Population	Objectives	Main results	Reference
Tsuji et al.	Mortality risk in an elderly cohort	The Framingham Heart Study	The prognostic implications of alterations in heart rate variability	The estimation of heart rate variability offers prognostic information about traditional risk factors	[74]
Yeragani et al.	Comparison of postural changes in heart rate and the R-R interval variance	PD patients (n = 30), and normal controls (n = 20)	To investigate autonomic function in panic disorder patients	Supine R-R variance was significantly decreased in PD patients	[75]
McCraty et al.	Retrospective study of Holter records	38 PD and healthy, age- and gender-matched controls	Analysis of 24 h heart rate variability in patients with panic disorder	Low HRV in PD patients consistent with cardiovascular morbidity and mortality	[78]
Martinez et al.	Spectral analysis of HR and BP performed on PD and control subjects	30 PD and ten healthy subjects	To test if PD patients have a deregulated autonomic nervous system at rest and during orthostatic challenge	A consistent deregulation of autonomic arousal in PD patients	[80]
Yavuzkir et al.	PWD measured in PD patients. PAS and HDRS scored concomitantly	40 PD patients and 40 controls	To show an association between PWD and panic disorder	PWD was significantly greater in the PD group than in the controls	[82]
Atmaca et al.	Measurement of Q(max), Q(min), and QTd values. PAS and HDRS scored concomitantly	40 PD patients and 40 controls	To investigate whether QTd differs in PD patients compared to healthy controls	The mean corrected QTd was significantly greater in the patients than in the controls	[85]
Soares-Filho GL et al.	Case report and review	Report of a patient with SC with chest pain and an AMI-like ECG	To show an association between sympathetic hyperactivity and cardiovascular risk	Supports to the hypothesis that stress and sympathetic hyperactivity can lead to life risk conditions	[15]

BP blood pressure, *CAD* coronary artery disease, *ECG* electrocardiogram, *HR* heart rate, *PA* panic attack, *PD* panic disorder, *PWV* pulse wave velocity

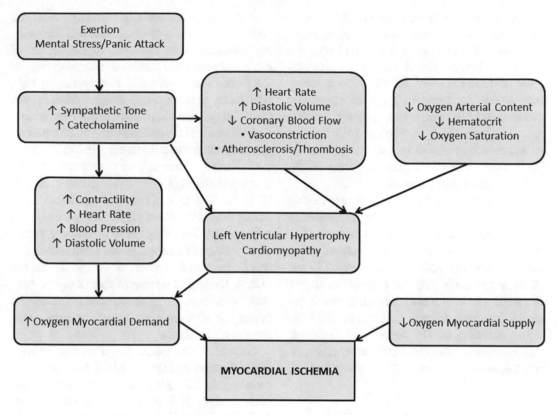

Fig. 11.1 Factors responsible for the myocardial work and oxygen demand

several authors to question whether some variant type of acute myocardial ischemia could be occurring. It is important to note again that an acute coronary syndrome is characterized by the imminent risk of life, so a rapid and appropriate assessment seeking early treatment is mandatory in all patients with symptoms suggesting an ACS (Fig. 11.1).

In two pioneering publications, myocardial perfusion was found to be clearly linked to PD. Fleet et al. [36], using carbon dioxide (CO_2) challenge test and assessing myocardial perfusion by single-photon emission computed tomography (SPECT), observed that PA could lead to myocardial perfusion defects in patients with CAD and PD, even maintaining treatment with cardiac drugs. Similarly, our group reported a clinical case of a patient with PD but no evidence of CAD and normal SPECT after treadmill exercise test that showed a deficit in myocardial

perfusion after CO_2 challenge test [55]. The patient described in this case reported very few symptoms and denied having a PA, but showed a raise in BP and in double product (HR × BP), a clinical estimation of myocardial oxygen demands. Thus, patients with stable CAD may present myocardial ischemia due to mechanisms present in mental stress response: indeed, an increase in coronary vasomotor tone (vasospasm) with decreased coronary blood flow or a sympathetic hyperactivity that determines an increase in HR, BP and myocardial contractility, may eventually lead to a rise in myocardial oxygen consumption [56–58].

Vasospasm or microcirculatory increased tone leading to chest pain (CP) when accompanied by normal coronary angiography is thought to be one possible cause of ischemic states. Mansour et al. [59] described the case of three patients with PD and CP, with documented cardiac ischemia, but no significant atherosclerosis at coronary angiography, and presumed the etiology to be due to coronary artery spasm. Vidovich et al. [60] also reported a female patient with depressive episodes associated with PA and CP. Since coronary arteries were angiographically normal, vasospasm was suggested as main etiologic factor. Reinforcing this hypothesis, Roy-Byrne [61] found a 40% prevalence of PD in patients with established micro-vascular angina and angiographically normal coronary arteries.

One potential mechanism proposed to explain coronary vasospasm in PA is hyperventilation. Indeed, hyperventilation test has been used to induce coronary spasm in clinical investigation, with a sensitivity and specificity of 62% and 100%, respectively [62]. Chelmowski et al. [63] described a case of a man presenting with myocardial infarction without CAD, apparently due to hyperventilation. The presence of autonomic nervous system alterations may play an important role in coronary vasospasm pathogenesis.

11.6.1 Sympathetic Hyperactivity

Acute stress situations in which a strong sympathetic activity is present have been reported as preceding episodes of ACS or sudden death that even nowadays shocks and surprises the medical community. Even more remarkable is how panic attacks are able to activate sympathetic discharges mediated by releasing high doses of catecholamines in the bloodstream, specifically epinephrine and norepinephrine [27]. For example, Biyik et al. [64] reported a patient presenting with myocardial infarction but no evidence of CAD at coronary angiography. Symptoms started after a fight, leaving him extremely agitated and afraid of being killed. Sympathetic hyperactivity by acute stress, comparable to a PA, was imputed as the most probable cause of his coronary event. According to Graeff [65] panic states correspond to the strong flight reaction evoked by a very close threat and appears to cause a major sympathetic activation, unlike chronic anxiety states, as seen in anticipatory anxiety, under a greater influence of hypothalamic-pituitary-adrenal (HPA) axis (Fig. 11.2).

Acute stress as a cause of coronary events and sudden death is widely documented in the medical literature. One example is the report of increased number of cardiac deaths on the day of the Northridge (California) Earthquake, in 1994 [66]. Many of these events affected patients with CAD, suggesting that stress caused by natural disaster can lead to acute rupture of atheromatous plaque, malignant arrhythmias, increased myocardial oxygen demand, or vasospasm. Similar cases were reported after an earthquake in Taiwan [67], airstrikes in Iraq [68] and emotional strain during football matches [69]. Those patients with pre-existing CAD may present coronary events after acute emotional stress due to an increase in shear stress and rupture of an atherosclerotic plaque, leading to release of thrombogenic substances and subsequent obstruction of the arterial lumen, or alternately to acute coronary vasospasm or arrhythmia.

Esler et al. [70] asserts that activation of sympathetic autonomic system of the heart related to catastrophes may act as a trigger for abnormal heart rhythm and sudden death, can activate platelets predisposing to thrombosis, reduce serum potassium concentrations and blood pressure surge can fissure coronary artery atherosclerotic plaques in CAD patients. He suggested that an epigenetic mechanism might sensitize patients with PD to cardiac symptoms, magnifying the sympathetic neural signal in the heart, underlying increased cardiac risk [71]. Muscle nerve neurograms have shown large sympathetic bursts, increases in cardiac norepinephrine spillover and peaks of epinephrine secretion during PA. Besides, neuropeptide Y released from cardiac sympathetics nerves can be involved in coronary artery spasm [27].

Some researchers have sought to document the sympathetic hyperactivity linked to PD. A study using I-123-metaiodobenzylguanidine (I-123-MIBG) myocardial scintigraphy was performed on nine patients with PD and showed impairment in cardiac sympathetic function [72]. Metaiodobenzylguanidine (MIBG) is a physiologic analog of norepinephrine and has been used for the evaluation of cardiac sympathetic function. In patients with PD has been demonstrated that myocardial uptake of I-123-MIBG was less

Fig. 11.2 Sympathetic
activity by situations of
acute stress

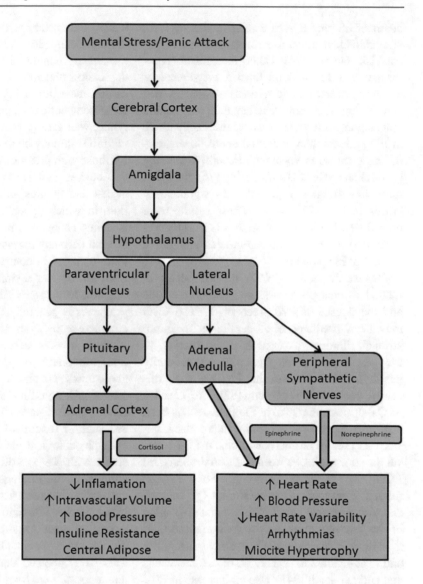

panic disorder patients than in healthy control values. Perhaps, the recurrent surges of sympathetic overactivity by panic attacks could lead to depletion or exhaustion of the sympathetic efferent system [72]. Patients with PD, evaluated by finger photoplethysmography, showed an increase in BP overshoot ("signal or function exceeds its target") compared with control groups [73]. Selective serotonin reuptake inhibitor treatment reduced significantly this BP overshoot towards normal values, confirming an increased sympathetic baroreflex function in PD.

Moreover, the autonomic nervous system plays a pivotal role in the triggering or sustaining of malignant ventricular arrhythmias. Sympathetic stimulation, opposed by vagal action, reduces the ventricular refractory period and the ventricular fibrillation threshold, and promotes triggered activity after potentials and enhances automaticity. HR variability measurements may assess cardiac autonomic function and predict cardiovascular events, especially cardiac death. In the Framingham Heart study, in a follow-up of 3.9 years, decreased HR variability predicted death of all causes with a relative risk of 1.70, after statistical adjustment for age, sex, and clinical risk factors [74]. Other researches hypothesize that PD patients have a heightened or deregulated autonomic nervous system at rest and during autonomic challenge. It was found that standing RR variance was significantly lower in PD patients than in normal controls, suggesting an increase in vagal withdrawal [75]. Using spectral analysis of HR variability [76], the absolute "low frequency power" was significantly diminished in PD patients. These results were reproduced in many other studies [77–81], demonstrating a consistently deregulated autonomic arousal in PD patients.

P-wave dispersion (PWD) has been also correlated to changes in systemic autonomic tone and the degree of PWD seen on 12-lead ECG may be a predictor of susceptibility to future atrial fibrillation. Yavuzkir et al. comparing 40 PD outpatients to 40 physically and mentally healthy controls showed that PWD was significantly greater in the PD group [82]. The presence of QT dispersion (QTd) in PD patients has been also investigated as a potential index of cardiovascular risk. In normal conditions, the QT interval duration varies between leads on the standard ECG [83]. QT dispersion is defined as the maximum QT interval minus minimum QT interval and was proposed to reflect the spatial dispersion of the ventricular recovery time. Increased QTd is a marker of myocardial electrical instability and is associated to higher risk of cardiac events and sudden death [84]. Once an increase in QTd indicates a cardiac autonomic imbalance, it has been investigated if its presence is associated with increased anxiety levels, thereby predisposing these patients to fatal heart disease. Atmaca et al. measured the QT interval in 40 PD patients and 40 healthy controls and noted values significantly higher in PD patients compared to control group [85]. Studying diurnal QTd measures in 32 normal adults and 22 PD patients using 24-h ECG monitoring, Yeragani et al. found that QTd measures at nighttime are significantly higher in PD patients compared with controls, indicating attenuated diurnal changes in ventricular repolar-

ization contrasting with a higher QT variability during sleep [86]. These findings were also present in population with PD associated to depression and may suggest a increase risk of cardiac mortality in this population [87]. Looking for a therapeutic tool, paroxetine, but not nortriptyline, was able to prevent QTd rise in patients with PD and may be a drug of choice, especially for those with previous cardiac disease [88].

Acute mental stress sometimes may trigger events that mimics an acute myocardial infarction, in which patients present typical CP and ischemic changes in ECG, but with no or insignificant coronary disease [89]. Case reports show these changes in contractility varying clinically from a mild and asymptomatic ventricular dysfunction to a severe circulatory failure, but the prognosis is generally favorable, with rapid and complete recovery in most cases. In a large series of cases, it was observed that during hospitalization, pulmonary congestion was present in 22 % and cardiogenic shock in 15 % of cases, but the rate of in-hospital mortality was 1 % and at discharge, 97 % of patients were in New York Heart Association functional class I [90]. The exact pathophysiological mechanism correlating acute stress with TS is still an object of research, although some findings have converged on the correlation between acute stress and adrenergic hyperactivity. Wittstein [16] clearly demonstrated activation of the adrenal medulla and elevated plasma levels of epinephrine and norepinephrine. This study showed that high plasma catecholamine levels were found in patients with ST until 7–9 days after the triggering event. Although some researches succeeded to show the occurrence of epicardial and microcirculation coronary spasm with typical ischemic ECG changes in response to provocative maneuvers during angiography in patients with TS the most accepted hypothesis is the neurogenically-stunned myocardium due to damage at cardiac muscle cells [91, 92]. Endomyocardial biopsy in TS patients showed mononuclear inflammatory infiltrates and extensive contraction-band necrosis caused by hyper-contraction of the sarcomeres, an indirect proof of elevated levels of catecholamines [16].

11.7   Conclusions

As we saw in this chapter, some epidemiological studies have established that anxiety is a risk factor for cardiovascular disease increasing the likelihood of ischemic heart disease, including myocardial infarction raising morbidity and mortality in this group [39, 40]. Recent efforts have started to propose pathophysiologic mechanisms anxiety could cause or potentiate cardiac disease.

Although chronic anxiety can be responsible for an influence on health behaviors like smoking, increasing risk of atherogenesis [48] there some evidence that anxiety may act as an independent risk for of a pro-inflammatory state [49] and hypercoagulability [52]. Fleet et al. [36] and Soares-Filho et al. [55] have shown PD as trigger of myocardial perfusion defect. Since anxiety and mental stress may be aspects of closely related mental processes, both may be associated with adrenal secretion of epinephrine, in the role of sympathetic hyperactivity. Both myocardial oxygen consumption [55–57] and coronary vasospasm [58–60] were suggested as causes of myocardial ischemia in these situations. Esler et al. suggests platelet activation and thrombosis, to coronary plaque fissures in CAD patients to explain ischemic events in CAD patients witnessing catastrophes [70]. The sympathetic hyperactivity linked to PD was demonstrated by myocardial scintigraphy [72], finger photoplethysmography [27], spectral analysis of HR variability [76–80], P-wave dispersion [81], QT dispersion [85–87]. Studying patients experiencing to stressful events and consequent Takotsubo cardiomyopathy [16].

More attention must be paid to anxiety patients complaining of cardiovascular symptoms, especially chest pain. Cardiologists should be able to recognize patients with psychiatric conditions and measure the cardiovascular consequences of anxiety. This review tried to shed light on important relationships between anxiety and heart disease, and opened the possibility of better clinical management and pathophysiological understanding.

References

1. Association AP. Diagnostic and statistical manual of mental disorders: DSM-5. ManMag; 2003.
2. Lynch P, Galbraith KM. Panic in the emergency room. Can J Psychiatry. 2003;48(6):361–6.
3. Fleet RP, Dupuis G, Marchand A, Burelle D, Beitman BD. Panic disorder, chest pain and coronary artery disease: literature review. Can J Cardiol. 1994;10(8):827–34.
4. Howell JM. Xiphodynia: a report of three cases. J Emerg Med. 1992;10(4):435–8.
5. Esporcatte R, Rangel F, Rocha R. Cardiologia Intensiva: bases práticas. Primeira edição Rio de. 2005.
6. Soares-Filho GL, Freire RC, Biancha K, Pacheco T, Volschan A, Valenca AM, et al. Use of the hospital anxiety and depression scale (HADS) in a cardiac emergency room: chest pain unit. Clinics. 2009;64(3):209–14.
7. Beitman BD, Basha I, Flaker G, DeRosear L, Mukerji V, Trombka L, et al. Atypical or nonanginal chest pain. Panic disorder or coronary artery disease? Arch Intern Med. 1987;147(9):1548–52.
8. Basha I, Mukerji V, Langevin P, Kushner M, Alpert M, Beitman BD. Atypical angina in patients with coronary artery disease suggests panic disorder. Int J Psychiatry Med. 1989;19(4):341–6.
9. Fleet RP, Dupuis G, Marchand A, Kaczorowski J, Burelle D, Arsenault A, et al. Panic disorder in coronary artery disease patients with noncardiac chest pain. J Psychosom Res. 1998;44(1):81–90.
10. Young LD, Richter JE, Anderson KO, Bradley LA, Katz PO, McElveen L, et al. The effects of psychological and environmental stressors on peristaltic esophageal contractions in healthy volunteers. Psychophysiology. 1987;24(2):132–41.
11. Fordyce WE. Behavioral methods for chronic pain and illness. St. Louis: CV Mosby; 1976.
12. Preter M, Klein DF. Panic, suffocation false alarms, separation anxiety and endogenous opioids. Prog Neuropsychopharmacol Biol Psychiatry. 2008;32(3):603–12. doi:10.1016/j.pnpbp.2007.07.029.
13. Preter M, Lee SH, Petkova E, Vannucci M, Kim S, Klein DF. Controlled cross-over study in normal subjects of naloxone-preceding-lactate infusions; respiratory and subjective responses: relationship to endogenous opioid system, suffocation false alarm theory and childhood parental loss. Psychol Med. 2011;41(2):385–93.doi:10.1017/S0033291710000838.
14. Freeman LJ, Nixon PG, Legg C, Timmons BH. Hyperventilation and angina pectoris. J R Coll Physicians Lond. 1987;21(1):46–50.
15. Soares-Filho GL, Felix RC, Azevedo JC, Mesquita CT, Mesquita ET, Valenca AM, et al. Broken heart or takotsubo syndrome: support for the neurohumoral hypothesis of stress cardiomyopathy. Prog Neuro-

psychopharmacol Biol Psychiatry. 2010;34(1):247–9. doi:10.1016/j.pnpbp.2009.10.013.

16. Wittstein IS, Thiemann DR, Lima JA, Baughman KL, Schulman SP, Gerstenblith G, et al. Neurohumoral features of myocardial stunning due to sudden emotional stress. N Engl J Med. 2005;352(6):539–48. doi:10.1056/NEJMoa043046.

17. Nguyen SB, Cevik C, Otahbachi M, Kumar A, Jenkins LA, Nugent K. Do comorbid psychiatric disorders contribute to the pathogenesis of tako-tsubo syndrome? A review of pathogenesis. Congest Heart Fail. 2009;15(1):31–4. doi:10.1111/j.1751-7133.2008. 00046.x.

18. Kahn JP, Stevenson E, Topol P, Klein DF. Agitated depression, alprazolam, and panic anxiety. Am J Psychiatry. 1986;143(9):1172–3.

19. Kahn JP, Gorman JM, King DL, Fyer AJ, Liebowitz MR, Klein DF. Cardiac left ventricular hypertrophy and chamber dilatation in panic disorder patients: implications for idiopathic dilated cardiomyopathy. Psychiatry Res. 1990;32(1):55–61.

20. Rautaharju PM, Soliman EZ. Electrocardiographic left ventricular hypertrophy and the risk of adverse cardiovascular events: a critical appraisal. J Electrocardiol. 2014;47(5):649–54. doi:10.1016/j.jelectrocard. 2014.06.002.

21. Braunwald E. Unstable angina: an etiologic approach to management. Circulation. 1998;98(21):2219–22.

22. Fuster V, Badimon L, Badimon JJ, Chesebro JH. The pathogenesis of coronary artery disease and the acute coronary syndromes (2). N Engl J Med. 1992; 326(5):310–8. doi:10.1056/NEJM199201303260506.

23. Aroney C, Boyden A, Jelinek M, Thompson P, Tonkin A, White H. Management of unstable angina: guidelines 2000. Med J Aust. 2000;173(8 Suppl):S65–88.

24. Haasenritter J, Stanze D, Widera G, Wilimzig C, Abu Hani M, Sonnichsen AC, et al. Does the patient with chest pain have a coronary heart disease? Diagnostic value of single symptoms and signs—a meta-analysis. Croat Med J. 2012;53(5):432–41.

25. Topol EJ, Califf RM. Textbook of cardiovascular medicine. Philadelphia: Lippincott Williams & Wilkins; 2007.

26. Rozanski A, Bairey CN, Krantz DS, Friedman J, Resser KJ, Morell M, et al. Mental stress and the induction of silent myocardial ischemia in patients with coronary artery disease. N Engl J Med. 1988;318(16):1005–12. doi:10.1056/NEJM198804213181601.

27. Esler M, Alvarenga M, Lambert G, Kaye D, Hastings J, Jennings G, et al. Cardiac sympathetic nerve biology and brain monoamine turnover in panic disorder. Ann N Y Acad Sci. 2004;1018:505–14. doi:10.1196/ annals.1296.062.

28. Jiang W, Babyak M, Krantz DS, Waugh RA, Coleman RE, Hanson MM, et al. Mental stress-induced myocardial ischemia and cardiac events. JAMA. 1996; 275(21):1651–6.

29. Hickam JB, Cargill WH, Golden A. Cardiovascular reactions to emotional stimuli. Effect on the cardiac output, arteriovenous oxygen difference, arterial

pressure, and peripheral resistance. J Clin Invest. 1948;27(2):290.

30. Fleet R, Foldes-Busque G, Gregoire J, Harel F, Laurin C, Burelle D, et al. A study of myocardial perfusion in patients with panic disorder and low risk coronary artery disease after 35 % CO_2 challenge. J Psychosom Res. 2014;76(1):41–5. doi:10.1016/j.jpsychores.2013. 08.003.

31. Loures DL, Sant Anna I, Baldotto CS, Sousa EB, Nobrega AC. Mental stress and cardiovascular system. Arq Bras Cardiol. 2002;78(5):525–30.

32. Dakak N, Quyyumi AA, Eisenhofer G, Goldstein DS, Cannon 3rd RO. Sympathetically mediated effects of mental stress on the cardiac microcirculation of patients with coronary artery disease. Am J Cardiol. 1995;76(3):125–30.

33. Krantz DS, Kop WJ, Santiago HT, Gottdiener JS. Mental stress as a trigger of myocardial ischemia and infarction. Cardiol Clin. 1996;14(2):271–87.

34. Yeung AC, Vekshtein VI, Krantz DS, Vita JA, Ryan Jr TJ, Ganz P, et al. The effect of atherosclerosis on the vasomotor response of coronary arteries to mental stress. N Engl J Med. 1991;325(22):1551–6. doi:10.1056/NEJM199111283252205.

35. Kahn JP, Perumal AS, Gully RJ, Smith TM, Cooper TB, Klein DF. Correlation of type A behaviour with adrenergic receptor density: implications for coronary artery disease pathogenesis. Lancet. 1987;2(8565):937–9.

36. Fleet R, Lesperance F, Arsenault A, Gregoire J, Lavoie K, Laurin C, et al. Myocardial perfusion study of panic attacks in patients with coronary artery disease. Am J Cardiol. 2005;96(8):1064–8. doi:10.1016/ j.amjcard.2005.06.035.

37. Nardi AE, Valenca AM, Lopes FL, Nascimento I, Veras AB, Freire RC, et al. Psychopathological profile of 35 % CO_2 challenge test-induced panic attacks: a comparison with spontaneous panic attacks. Compr Psychiatry. 2006;47(3):209–14. doi:10.1016/j. comppsych.2005.07.007.

38. Kontos MC. Evaluation of the Emergency Department chest pain patient. Cardiol Rev. 2001;9(5):266–75.

39. Smoller JW, Pollack MH, Wassertheil-Smoller S, Jackson RD, Oberman A, Wong ND, et al. Panic attacks and risk of incident cardiovascular events among postmenopausal women in the Women's Health Initiative Observational Study. Arch Gen Psychiatry. 2007;64(10):1153–60.

40. Haines AP, Imeson JD, Meade TW. Phobic anxiety and ischaemic heart disease. Br Med J. 1987; 295(6593):297–9.

41. Janszky I, Ahnve S, Lundberg I, Hemmingsson T. Early-onset depression, anxiety, and risk of subsequent coronary heart disease: 37-year follow-up of 49,321 young Swedish men. J Am Coll Cardiol. 2010;56(1):31–7. doi:10.1016/j.jacc.2010.03.033.

42. Roest AM, Martens EJ, de Jonge P, Denollet J. Anxiety and risk of incident coronary heart disease: a meta-analysis. J Am Coll Cardiol. 2010;56(1):38–46. doi:10.1016/j.jacc.2010.03.034.

43. Walters K, Rait G, Petersen I, Williams R, Nazareth I. Panic disorder and risk of new onset coronary heart disease, acute myocardial infarction, and cardiac mortality: cohort study using the general practice research database. Eur Heart J. 2008;29(24):2981–8.

44. Kawachi I, Colditz GA, Ascherio A, Rimm EB, Giovannucci E, Stampfer MJ, et al. Prospective study of phobic anxiety and risk of coronary heart disease in men. Circulation. 1994;89(5):1992–7.

45. Kawachi I, Sparrow D, Vokonas PS, Weiss ST. Symptoms of anxiety and risk of coronary heart disease. The Normative Aging Study. Circulation. 1994;90(5):2225–9.

46. Shibeshi WA, Young-Xu Y, Blatt CM. Anxiety worsens prognosis in patients with coronary artery disease. J Am Coll Cardiol. 2007;49(20):2021–7. doi:10.1016/j.jacc.2007.03.007.

47. Rothenbacher D, Hahmann H, Wusten B, Koenig W, Brenner H. Symptoms of anxiety and depression in patients with stable coronary heart disease: prognostic value and consideration of pathogenetic links. Eur J Cardiovasc Prev Rehabil. 2007;14(4):547–54. doi:10.1097/HJR.0b013e3280142a02.

48. Coryell W, Noyes R, Clancy J. Excess mortality in panic disorder. A comparison with primary unipolar depression. Arch Gen Psychiatry. 1982;39(6):701–3.

49. Coryell W, Noyes Jr R, House JD. Mortality among outpatients with anxiety disorders. Am J Psychiatry. 1986;143(4):508–10.

50. Kubzansky LD, Kawachi I, Weiss ST, Sparrow D. Anxiety and coronary heart disease: a synthesis of epidemiological, psychological, and experimental evidence. Ann Behav Med. 1998;20(2):47–58.

51. Player MS, Peterson LE. Anxiety disorders, hypertension, and cardiovascular risk: a review. Int J Psychiatry Med. 2011;41(4):365–77.

52. Chen YH, Hu CJ, Lee HC, Lin HC. An increased risk of stroke among panic disorder patients: a 3-year follow-up study. Can J Psychiatry. 2010;55(1):43–9.

53. von Kanel R, Kudielka BM, Schulze R, Gander ML, Fischer JE. Hypercoagulability in working men and women with high levels of panic-like anxiety. Psychother Psychosom. 2004;73(6):353–60. doi:10.1159/000080388.

54. Cicek Y, Durakoglugil ME, Kocaman SA, Guveli H, Cetin M, Erdogan T, et al. Increased pulse wave velocity in patients with panic disorder: independent vascular influence of panic disorder on arterial stiffness. J Psychosom Res. 2012;73(2):145–8. doi:10.1016/j.jpsychores.2012.05.012.

55. Soares-Filho GLF, Mesquita CT, Mesquita ET, Arias-Carrión O, Machado S, González MM, et al. Panic attack triggering myocardial ischemia documented by myocardial perfusion imaging study. A case report. Int Arch Med. 2012;5(1):24.

56. L'Abbate A, Simonetti I, Carpeggiani C, Michelassi C. Coronary dynamics and mental arithmetic stress in humans. Circulation. 1991;83(4 Suppl):II94–9.

57. Becker L, Pepine C, Bonsall R, Cohen J, Goldberg A, Coghlan C, et al. Left ventricular, peripheral vascular, and neurohumoral responses to mental stress in normal middle-aged men and women. Reference Group for the Psychophysiological Investigations of Myocardial Ischemia (PIMI) Study. Circulation. 1996;94(11):2768.

58. Sgoutas-Emch SA, Cacioppo JT, Uchino BN, Malarkey W, Pearl D, Kiecolt-Glaser JK, et al. The effects of an acute psychological stressor on cardiovascular, endocrine, and cellular immune response: a prospective study of individuals high and low in heart rate reactivity. Psychophysiology. 1994;31(3): 264–71.

59. Mansour VM, Wilkinson DJ, Jennings GL, Schwarz RG, Thompson JM, Esler MD. Panic disorder: coronary spasm as a basis for cardiac risk? Med J Aust. 1998;168(8):390–2.

60. Vidovich MI, Ahluwalia A, Manev R. Depression with panic episodes and coronary vasospasm. Cardiovasc Psychiatry Neurol. 2009;2009:453786. doi:10.1155/2009/453786.

61. Roy-Byrne PP, Schmidt P, Cannon RO, Diem H, Rubinow DR. Microvascular angina and panic disorder. Int J Psychiatry Med. 1989;19(4):315–25.

62. Nakao K, Ohgushi M, Yoshimura M, Morooka K, Okumura K, Ogawa H, et al. Hyperventilation as a specific test for diagnosis of coronary artery spasm. Am J Cardiol. 1997;80(5):545–9.

63. Chelmowski MK, Keelan Jr MH. Hyperventilation and myocardial infarction. Chest. 1988;93(5): 1095–6.

64. Biyik I, Yagtu V, Ergene O. Acute myocardial infarction triggered by acute intense stress in a patient with panic disorder. Turk Kardiyol Dern Ars. 2008; 36(2):111–5.

65. Graeff FG. Anxiety, panic and the hypothalamic-pituitary-adrenal axis. Rev Bras Psiquiatr. 2007;29 Suppl 1:S3–6.

66. Kloner RA, Leor J, Poole WK, Perritt R. Population-based analysis of the effect of the Northridge Earthquake on cardiac death in Los Angeles County, California. J Am Coll Cardiol. 1997;30(5):1174–80.

67. Tsai CH, Lung FW, Wang SY. The 1999 Ji-Ji (Taiwan) earthquake as a trigger for acute myocardial infarction. Psychosomatics. 2004;45(6):477–82. doi:10.1176/appi.psy.45.6.477.

68. Meisel SR, Kutz I, Dayan KI, Pauzner H, Chetboun I, Arbel Y, et al. Effect of Iraqi missile war on incidence of acute myocardial infarction and sudden death in Israeli civilians. Lancet. 1991;338(8768):660–1.

69. Baumhakel M, Kindermann M, Kindermann I, Bohm M. Soccer world championship: a challenge for the cardiologist. Eur Heart J. 2007;28(2):150–3. doi:10.1093/eurheartj/ehl313.

70. Esler M, Lambert E, Alvarenga M. Acute mental stress responses: neural mechanisms of adverse cardiac consequences. Stress Health. 2008;24(3): 196–202.

71. Esler M, Alvarenga M, Pier C, Richards J, El-Osta A, Barton D, et al. The neuronal noradrenaline transporter, anxiety and cardiovascular disease.

J   Psychopharmacol.   2006;20(4   Suppl):60–6. doi:10.1177/1359786806066055.

72. Tanabe Y, Harada H, Sugihara S, Ogawa T, Inoue Y. 123I-Metaiodobenzylguanidine myocardial scintigraphy in panic disorder. J Nucl Med. 2004;45(8): 1305–8.

73. Coupland NJ, Wilson SJ, Potokar JP, Bell C, Nutt DJ. Increased sympathetic response to standing in panic disorder. Psychiatry Res. 2003;118(1):69–79.

74. Tsuji H, Venditti Jr FJ, Manders ES, Evans JC, Larson MG, Feldman CL, et al. Reduced heart rate variability and mortality risk in an elderly cohort. The Framingham Heart Study. Circulation. 1994;90(2): 878–83.

75. Yeragani VK, Balon R, Pohl R, Ramesh C, Glitz D, Weinberg P, et al. Decreased R-R variance in panic disorder patients. Acta Psychiatr Scand. 1990;81(6): 554–9.

76. Yeragani VK, Pohl R, Berger R, Balon R, Ramesh C, Glitz D, et al. Decreased heart rate variability in panic disorder patients: a study of power-spectral analysis of heart rate. Psychiatry Res. 1993;46(1):89–103.

77. Klein E, Cnaani E, Harel T, Braun S, Ben-Haim SA. Altered heart rate variability in panic disorder patients.   Biol   Psychiatry.   1995;37(1):18–24. doi:10.1016/0006-3223(94)00130-U.

78. McCraty R, Atkinson M, Tomasino D, Stuppy WP. Analysis of twenty-four hour heart rate variability in patients with panic disorder. Biol Psychol. 2001;56(2):131–50.

79. Baumert M, Lambert GW, Dawood T, Lambert EA, Esler MD, McGrane M, et al. Short-term heart rate variability and cardiac norepinephrine spillover in patients with depression and panic disorder. Am J Physiol Heart Circ Physiol. 2009;297(2):H674–9. doi:10.1152/ajpheart.00236.2009.

80. Martinez JM, Garakani A, Kaufmann H, Aaronson CJ, Gorman JM. Heart rate and blood pressure changes during autonomic nervous system challenge in panic disorder patients. Psychosom Med. 2010; 72(5):442–9. doi:10.1097/PSY.0b013e3181d972c2.

81. Diveky T, Prasko J, Latalova K, Grambal A, Kamaradova D, Silhan P, et al. Heart rate variability spectral analysis in patients with panic disorder compared with healthy controls. Neuro Endocrinol Lett. 2012;33(2):156–66.

82. Yavuzkir M, Atmaca M, Dagli N, Balin M, Karaca I, Mermi O, et al. P-wave dispersion in panic disorder. Psychosom Med. 2007;69(4):344–7. doi:10.1097/ PSY.0b013e3180616900.

83. Cowan JC, Yusoff K, Moore M, Amos PA, Gold AE, Bourke JP, et al. Importance of lead selection in QT interval measurement. Am J Cardiol. 1988;61(1): 83–7.

84. Malik M, Batchvarov VN. Measurement, interpretation and clinical potential of QT dispersion. J Am Coll Cardiol. 2000;36(6):1749–66.

85. Atmaca M, Yavuzkir M, Izci F, Gurok MG, Adiyaman S. QT wave dispersion in patients with panic disorder. Neurosci Bull. 2012;28(3):247–52.

86. Yeragani VK, Pohl R, Balon R, Jampala VC, Jayaraman A. Twenty-four-hour QT interval variability: increased QT variability during sleep in patients with panic disorder. Neuropsychobiology. 2002; 46(1):1–6.

87. Yeragani VK, Pohl R, Jampala VC, Balon R, Ramesh C, Srinivasan K. Increased QT variability in patients with panic disorder and depression. Psychiatry Res. 2000;93(3):225–35.

88. Yeragani VK, Pohl R, Jampala VC, Balon R, Ramesh C, Srinivasan K. Effects of nortriptyline and paroxetine on QT variability in patients with panic disorder. Depress Anxiety. 2000;11(3):126–30.

89. Kodama K, Haze F, Hom M. Clinical aspects of myocardial injury: from ischaemia to heart failure. Tokyo: Kagahuhyouronsha; 1990.

90. Tsuchihashi K, Ueshima K, Uchida T, Oh-mura N, Kimura K, Owa M, et al. Transient left ventricular apical ballooning without coronary artery stenosis: a novel heart syndrome mimicking acute myocardial infarction.   J   Am   Coll   Cardiol.   2001;38(1): 11–8.

91. Kurisu S, Sato H, Kawagoe T, Ishihara M, Shimatani Y, Nishioka K, et al. Tako-tsubo-like left ventricular dysfunction with ST-segment elevation: a novel cardiac syndrome mimicking acute myocardial infarction. Am Heart J. 2002;143(3):448–55.

92. Elesber A, Lerman A, Bybee KA, Murphy JG, Barsness G, Singh M, et al. Myocardial perfusion in apical ballooning syndrome correlate of myocardial injury.   Am   Heart   J.   2006;152(3):469   e9–13. doi:10.1016/j.ahj.2006.06.007.

Panic Disorder and Cardiovascular Death: What Is Beneath?

12

Cristiano Tschiedel Belem da Silva
and Gisele Gus Manfro

Contents

C.T. Belem da Silva • G.G. Manfro (✉)
Anxiety Disorders Program, Hospital de Clínicas de
Porto Alegre (HCPA), Porto Alegre, Brazil

Department of Psychiatry, Federal University of Rio
Grande do Sul (UFRGS), Porto Alegre, Brazil
e-mail: cristianotbs@hotmail.com; gmanfro@gmail.com

Abstract

Considering extensive overlap among symptoms of panic disorder (PD) and cardiovascular diseases (CVD), the fact that so little attention has been drawn to studies addressing the relationship between both so far is somewhat intriguing. Therefore, this chapter will focus on the issue considering several perspectives. It starts delineating an historical background, showing that the interest on studying the relationship between anxiety symptoms and CVD began more than a century ago. Next, epidemiological research is reviewed, but considering lack of consistent findings, anxiety disorders and, more specifically, PD are deconstructed. Not only this, but biological mechanisms that could link anxiety symptoms and CVD, such as pleiotropy, heart rate variability, unhealthy lifestyle and atherosclerosis, are also explored. The chapter ends highlighting the importance of reversibility, that is, if PD and CVD are somewhat connected, intervention studies should prove the utility of prevention. As can be seen, the text is constructed using an epistemological perspective, since it constitutes a major area of indagation and research of the authors and is meant to raise the same sort of questioning in the readers.

Keywords

Anxiety disorders • Panic disorder • Neuroinflammation • Atherosclerosis • Vascular endothelium

12.1 Introduction

Common sense and stereotyping have been applied for a long time in order to attribute a causal role for anxiety in the development of CVD. Who never heard from a close relative: "if you keep up worrying so much about everything, you'll have a heart attack", or "John was so stressed, no wonder he ended up like that, dying suddenly". Trying to understand such association heuristically is not a surprise given the extensive overlap between anxiety and cardiovascular symptoms. As will be exposed, such overlap is present since the earliest descriptions of PD.

12.2 A Brief History

A syndrome resembling PD was first described by Jacob Mendes da Costa as "Irritable Heart Syndrome" [1]. He observed two-hundred soldiers as they left US army during Civil War (1861–1865) complaining of chest pain, palpitations, choking feelings and paresthesias. Despite a considerable number of such individuals, in fact, might have been suffering of some sort of cardiological disease – not only this, but also what we currently know as post-traumatic stress disorder (PTSD) – most of them had the shared feature of psychological distress. Let us not forget that this naturalistic study was carried on within a military hospital – so the fact that most patients were diagnosed as having a cardiological condition should cause no surprise.

Decades went by, and emotional dysregulation gained field as a possible cause of disease, especially after the psychoanalytical studies of Sigmund Freud (1856–1939) and his colleagues. Alternative nomenclatures, such as "neurasthenia", "effort syndrome" and "beta-adrenergic hyperactivity", all reflected attempts to describe etiologically what would be known as "panic disorder" only with the advent of DSM-III [2].

The second half of last century was marked by the interest of the scientific community over the possible association between psychiatric traits and symptoms and increased CVD [3]. The main concern was that a certain type of behavioral pattern, characterized by ambition, competitiveness, worries about time and money and aggressiveness, later called "type A" pattern, could raise the risk for CVD. Although several epidemiological studies confirmed such association [4, 5], subsequent investigations involving the files of tobacco companies unveiled dangerous pecuniary connections among reputed scientists and tobacco industry in order to replicate studies that could obtain favorable results and shift attention away from smoking as a cause of death [6]. Such tactic was effective for many years, until results from population prospective cohorts on the issue began to be published.

12.3 Epidemiological Studies

Results initially published from population cohorts in the early 2000s somewhat reflected what was mentioned above. Outcomes mainly consisted of negative affect measures, a dimension of symptoms that is common for both depression and anxiety, and present in "type A" behavioral pattern. Negative affect – characterized by anger, hostility, contempt, disgust, guilt and fear – was measured by psychometric scales such as Positive and Negative Affect Schedule (PANAS) [7] and Type D Scale [8]. A meta-analysis found a positive association between negative affect and major adverse cardiac events (OR = 3.16). All included studies were prospective and overall sample was relatively large ($n = 2903$) [9].

Increased prevalence of CVD in subjects with anxiety disorders, among which PD, began to intrigue scientists. Depending on the studied sample, comorbid CVD rates in patients with PD were as high as 15 times the ones found in otherwise healthy individuals [10].

Not long after, the first large-scale study reporting results on the association between panic attacks and the incidence of cardiovascular events and mortality was published. Smoller et al. [11, 12], analyzing data from 3369 women from the Women's Health Initiative (WHI), reported that those with panic attacks had 4,20 (95 % CI: 1.76–9.99) more risk for developing

MI and dying from cardiac causes than women with no panic attacks. As a strength of this study, we can mention that full statistical adjustment was performed. As a main limitation, we can consider the fact that panic attacks rather than PD were measured, which likely had little or no impact on the external validity of the study. Also, no male subjects were included.

Some critics of the above association claim that PD symptoms include features that can also be found in pre-clinical CVD. Indeed, it is quite plausible that patients are diagnosed as having panic attacks in their first contact with an emergency room due to inaccuracy of current diagnostic tools. In other words, subsyndromal CVD may be misdiagnosed. Indeed, such a possibility has already been raised in studies evaluating depression as a cause of increased CVD mortality [13].

12.4 Overcoming Existing Gaps: Going Above and Beyond Panic Disorder Research

Comorbidity rates between PD and depression may reach as much as 50% [14]. Considering the paucity of prospective studies on the association between PD and CVD, we shall add data from analogous conditions, such as depression and other anxiety disorders, on cardiovascular mortality, in order to enrich our discussion. Leung and colleagues [15], in a meta-analysis, found an increased risk of 1.79 (95% CI = 1.45–2.21) for all-cause mortality, cardiac mortality, rehospitalization, or major adverse cardiac events in patients with incident depression. Furthermore, they also showed that incident depression, after an acute coronary event, might also affect the same outcomes (effect size = 2.11; 95% CI = 1.66–2.68). Considering anxiety disorders, Phillips et al. [16] reported that subjects with generalized anxiety disorder (GAD) may have an increased risk of 2.89 (95% CI: 1.59–5.23) for incident CVD. After controlling for a number of confounders, this association was no longer significant (HR 1.84; 95% CI: 0.98–3.45).

However, a subsequent study by Roest et al. [17], after controlling for confounders, found an increased risk of 1.94 (95% CI: 1.14–3.30) for all-cause mortality and cardiovascular-related readmissions in patients with generalized anxiety disorder (GAD) after a myocardial infarct. Finally, Mykletun et al. [18], analysing data from the HUNT study, proposed that the relationship between GAD and mortality may be U-shaped, with highest mortality risks for low and severe anxiety loads. In this particular study, mortality risk for GAD was comparable to the risk associated with smoking. In accordance with Mykletun et al., a meta-analysis reported an increased risk of 1.83 (95% CI, 1.65–2.03) for all-cause mortality in smokers compared to non-smokers [19] (Fig. 12.1).

12.5 What is Beneath?

Since empirical data does not fully answer the question weather PD is a cause to CVD, biologically plausible pathophysiological mechanisms connecting both are going to be further discussed.

There are at least four mediators that could possibly be situated in the causal pathway between PD and CVD, none of which excluding the remaining mediators: (1) pleiotropy; (2) chronic changes in the heart rate variability (HRV); (3) unhealthy lifestyle and (4) atherosclerosis.

12.5.1 Pleiotropy

Conceptualized as genetic variations that can result in distinct phenotypes in different body systems, a classic example of pleiotropy is phenilketonuria (PKU). PKU is an autosomal recessive deficiency on the gene involved in the transcription of the enzyme phenylalanine hydroxylase, which is necessary for the metabolism of phenylalanine to tyrosine. The accumulated phenylalanine is then converted to a toxic substance (phenylpyruvate), which in turn can result in a range of complications in diverse body systems, such as skin lesions and mental retardation.

The idea of pleiotropy comes especially from studies on depression and CVD. For instance,

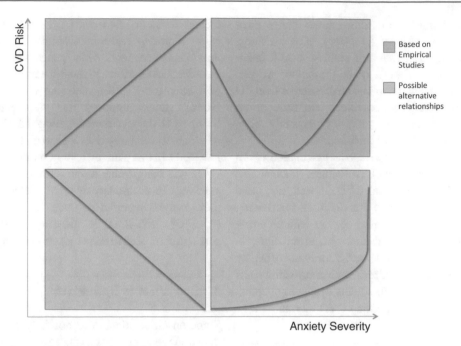

Fig. 12.1 Proposed models for the relationship between anxiety and CVD

Mannie et al. [20] found out that adolescents with no current or past mental illnesses whose mothers had at least one depressive episode had higher mean arterial blood pressure and increased insulin resistance than adolescents whose mothers had no history of depressive episodes. Considering high rates of comorbidity between PD and depression [14] and a recent transcriptome study by Gormanns et al. [21], in which significant changes in the enzymatic activation involved in glycolysis and vascular growth/modeling were detected in patients with anxiety disorders, the hypothesis that the same underlying genetic vulnerability could explain part of the association between PD and CVD is quite plausible. On the other hand, lack of specificity is a problem to be addressed by future research.

12.5.2 Chronic Changes in the Heart Rate Variability

Decreased vagal tone is considered a risk factor for a number of adverse outcomes, among which increased mortality is included. Decreased HRV is the most studied marker of impaired vagal tone, which is called "autonomic inflexibility" by some authors because it might indicate an impaired capacity of superior cortical structures (e.g., pre-frontal cortex) inhibit phylogenetically inferior regions (e.g., amygdala). Increased mortality for vagotomized patients or individuals with decreased HRV has been demonstrated in different settings, such as intensive-care units and emergency rooms, as compared to controls with no impairment in vagal activity [22, 23]. Furthermore, Pittig et al. showed that patients with PD had a decreased HRV compared to healthy controls [24] and Diveky et al. demonstrated that cognitive-behavioral therapy (CBT) had incremental effects on HRV in patients with PD allocated to intervention compared to controls [25]. Altogether, although preliminary, evidence point out that autonomic inflexibility could possibly mediate the relationship between PD and CVD.

12.5.3 Unhealthy Lifestyle

The prevalence of low levels of physical activity (PA) in individuals with panic attacks in the

WHI study was 45,8 % [11]. On the other hand, lower levels of physical activity practice are associated to increased anxiety levels [26]. Whether this relationship is causal or not is still unknown. Also, several studies showed benefit of different exercise protocols on anxiety disorders and symptoms, including PD [27]. The fact that avoidance of interoceptive threatening stimuli is a marked characteristic of PD, by itself, already justifies these findings on decreased PA practice, since physical exercise exacerbate physiological signs such as heart rate and breathing. Confirming such clinical observation, a recent study demonstrated that individuals with lifetime PD and higher somatic symptoms on Beck Anxiety Inventory [28] showed lower levels of PA. However, it is not clear to what extent the propensity to lower levels of PA during development is the consequence of a common trait that also leads to PD independently later in life. Although studying a sample with depressive disorder, such hypothesis was recently tested and confirmed in an elegant study by De Moor et al. [29].

The association between PD and an unhealthy lifestyle is also true for other factors, such as smoking [30] and higher body weight [31]. Indeed, much of the variance of the association among PD and anxiety symptoms and CVD can be attributed to such factors – when controlling for them in multivariate analysis, usually the effect sizes, although still significant, decrease in magnitude. Thus, the most plausible is that an unhealthy lifestyle partially – and not fully – mediate the relationship between PD and CVD.

12.5.4 Atherosclerosis

Atherosclerosis is essentially an inflammatory process involving vascular endothelium that can lead to the development of CVD. There are several ways of measuring atherosclerosis in research settings, the most frequent in studies involving patients with PD and anxiety symptoms are intima-media thickness (IMT), flow-mediated dilation (%FMD), pulse-wave velocity (PWV) and plethysmography. Two prospective

studies found an association between increased anxiety symptoms at baseline and poor vascular health. Paterniti et al. [32] reported that individuals with higher scores in the State-Trait Anxiety Inventory (STAI) [33] showed increased IMT 4 years later. Consistent with them, Belem da Silva et al. [34] described a positive association between anxiety symptoms assessed by the Hamilton Anxiety Questionnaire (HAM-A) [35] and a worse %FMD measured a median 8 years later in subjects with no reported history of previous CVD. Despite all subjects had a lifetime diagnosis of PD, no significant %FMD differences were found between patients with current PD *versus* remitted PD. On the other hand, after controlling for multiple confounders, Cicek et al. [36] found a positive cross-sectional association between PD and pathological PWV changes. In light of these studies, atherosclerosis may play a role in the development of CVD in patients with PD and no history of CVD. However, the way anxiety is assessed (e.g., cross-sectionally vs longitudinally; using symptom scales vs diagnostic scales) likely influence outcomes. For instance, weather psychometric scales with an important somatic component, such as HAM-A, might in fact indicate symptoms of subclinical CVD is still not clear. Interestingly, Stillman et al. [37] found that post-myocardial infarction (post-MI) patients with higher anxiety levels showed increased atherosclerosis compared to those with lower anxiety levels. Their findings support a possible role for the interaction between pre-existent CVD and anxiety symptoms in general. Supporting such an interaction, Fleet et al. [38], investigating a sample of patients with coronary artery disease, found that the group developing panic attacks after 35 % CO_2 inhalation presented higher levels of perfusion defects assessed by single-photon emission computed tomography (SPECT) than controls with no panic attacks. Interestingly, Soares-Filho et al. [39], reporting results of a case-series of patients with PD and no CVD, evidenced that the only case of perfusion defects on SPECT after 35 % CO_2 challenge occurred in a subject that reported no panic attacks. Both studies using SPECT suggest three possible pathophysiological mechanisms:

(1) perfusion defects induced by startle response, (2) a redistribution of blood flow in response to a CO_2-induced vasodilatation in certain coronary vessels and (3) a decreased tendency of PD-free individuals to refer PD-like symptoms. Most likely, a myriad of factors such as pre-existent atherosclerosis, adrenergic reactivity and altered interoception – or the capacity of sensing our own bodily physiological changes – interact for the genesis of CVD in predisposed individuals with PD.

12.6   To Prevent or Not to Prevent: The Question of Reversibility

Again, shall we apply lessons learned from the trials on depression and CVD to discuss implications and recommendations concerning PD and CVD: no matter which of the discussed relationships between the former and the later are valid, no conclusions should be drawn until consistent epidemiological data on successful PD treatment and reduction of mortality incidence by CVD are published. As mentioned previously, plenty of data exist on the association between depression and CVD. However, two large-scale trials, namely Sertraline Anti-Depressant Heart Attack Trial (SADHART) and Enhancing Recovery in Coronary Heart Disease (ENRICHD), failed to show benefit of antidepressant treatment and cognitive-behavioral therapy (CBT) on mortality reduction in patients with depression and CVD [40, 41]. To date, no epidemiological study has yet been designed to test such hypotheses in patients with PD; thus, the utility of recommending treatment of the underlying anxiety condition in order to prevent CVD and mortality by CVD remain elusive.

12.7   Conclusion

Although epidemiological studies cannot answer the question of weather PD and CVD are causally linked, the existence of plausible biological explanations for the association should fuel the design of longitudinal studies testing it as an *a priori* hypothesis. Once confirmed, the clinical utility of the association between PD and CVD

will only be proven when data on CVD prevention after PD treatment become available.

References

1. da Costa JM. On irritable heart: a clinical study of a form of functional cardiac disorder and its consequences. Am J Med Sci. 1871;61:17–52.
2. APA, American Psychiatry Association. Diagnostic and Statistical Manual of Mental Disorders (DSM-III). 3rd ed. Washington, DC: APA; 1980.
3. Jenkins CD. Psychologic and social precursors of coronary disease (first of two parts). N Engl J Med. 1971;284:244–55.
4. Rosenman RH, Brand RJ, Jenkins D, Friedman M, Straus R, Wurm M. Coronary heart disease in Western Collaborative Group Study. Final follow-up experience of 8 1/2 years. JAMA. 1975;233:872–7.
5. Haynes SG, Feinleib M, Kannel WB. The relationship of psychosocial factors to coronary heart disease in the Framingham Study: III. Eight-year incidence of coronary heart disease. Am J Epidemiol. 1980;111:37–58.
6. Petticrew MP, Lee K, McKee M. Type A behavior pattern and coronary heart disease: Philip Morris's "crown jewel". Am J Public Health. 2012;102:2018–25.
7. Watson D, Clark LA, Tellegan A. Development and validation of brief measures of positive and negative affect: the PANAS scales. J Pers Soc Psychol. 1988;54:1063–70.
8. Denollet J. DS14: standard assessment of negative affectivity, social inhibition, and Type D personality. Psychosom Med. 2005;67:89–97.
9. O'Dell KR, Masters KS, Spielmans GI, Maisto SA. Does type-D personality predict outcomes among patients with cardiovascular disease? A meta-analytic review. J Psychosom Res. 2011;71:199–206.
10. Chen YH, Lin HC. Patterns of psychiatric and physical comorbidities associated with panic disorder in a nationwide population-based study in Taiwan. Acta Psychiatr Scand. 2011;123:55–61.
11. Smoller JW, Pollack MH, Wassertheil-Smoller S, Jackson RD, Oberman A, Wong ND, et al. Panic attacks and risk of incident cardiovascular events among postmenopausal women in the Women's Health Initiative Observational Study. Arch Gen Psychiatry. 2007;64:1153–60.
12. Roest AM, Martens EJ, Jonge P, Denollet J. Anxiety and risk of incident coronary heart disease: a meta-analysis. J Am Coll Cardiol. 2010;56:38–46.
13. Roest AM, Carney RM, Freedland KE, Martens EJ, Denollet J, de Jonge P. Changes in cognitive versus somatic symptoms of depression and event-free survival following acute myocardial infarction in the Enhancing Recovery in Coronary Heart Disease (ENRICHD) study. J Affect Disord. 2013;149:335–41.
14. Kessler RC, Chiu WT, Jin R, Ruscio AM, Shear K, Walters EE. The epidemiology of panic attacks, panic

disorder, and agoraphobia in the National Comorbidity Survey Replication. Arch Gen Psychiatry. 2006;63:415–24.

15. Leung YW, Flora DB, Gravely S, Irvine J, Carney RM, Grace SL. The impact of premorbid and post-morbid depression onset on mortality and cardiac morbidity among patients with coronary heart disease: meta-analysis. Psychosom Med. 2012;74:786–801.

16. Phillips AC, Batty GD, Gale CR, Deary IJ, Osborn D, MacIntyre K, et al. Generalized anxiety disorder, major depressive disorder, and their comorbidity as predictors of all-cause and cardiovascular mortality: the Vietnam experience study. Psychosom Med. 2009;71:395–403.

17. Roest AM, Zuidersma M, de Jonge P. Myocardial infarction and generalized anxiety disorder: 10-year follow-up. Br J Psychiatry. 2012;200:324–9.

18. Mykletun A, Bjerkeset O, Overland S, Prince M, Dewey M, Stewart R. Levels of anxiety and depression as predictors of mortality: the HUNT study. Br J Psychiatry. 2009;195:118–25.

19. Gellert C, Schöttker B, Brenner H. Smoking and all-cause mortality in older people: systematic review and meta-analysis. Arch Intern Med. 2012;172:837–44.

20. Mannie ZN, Williams C, Diesch J, Steptoe A, Leeson P, Cowen PJ. Cardiovascular and metabolic risk profile in young people at familial risk of depression. Br J Psychiatry. 2013;203:18–23.

21. Gormanns P, Mueller NS, Ditzen C, Wolf S, Holsboer F, Turck CW. Phenome-transcriptome correlation unravels anxiety and depression related pathways. J Psychiatr Res. 2011;45:973–9.

22. Pontet J, Contreras P, Curbelo A, Medina J, Noveri S, Bentancourt S, et al. Heart rate variability as early marker of multiple organ dysfunction syndrome in septic patients. J Crit Care. 2003;18:156–63.

23. Peterson CY, Krzyzaniak M, Coimbra R, Chang DC. Vagus nerve and postinjury inflammatory response. Arch Surg. 2012;147:76–80.

24. Pittig A, Arch JJ, Lam CWR, Craske MG. Heart rate and heart rate variability in panic, social anxiety, obsessive–compulsive, and generalized anxiety disorders at baseline and in response to relaxation and hyperventilation. Int J Psychophysiol. 2013;87:19–27.

25. Diveky T, Prasko J, Kamaradova D, Grambal A, Latalova K, Silhan P, et al. Comparison of heart rate variability in patients with panic disorder during cognitive behavioral therapy program. Psychiatr Danub. 2013;25:62–7.

26. de Mello MT, Lemos VA, Antunes HK, Bittencourt L, Santos-Silva R, Tufik S. Relationship between physical activity and depression and anxiety symptoms: a population study. J Affect Disord. 2013;149:241–6.

27. Broocks A, Bandelow B, Pekrun G, George A, Meyer T, Bartmann U, et al. Comparison of aerobic exercise, clomipramine, and placebo in the treatment of panic disorder. Am J Psychiatry. 1998;155:603–9.

28. Belem da Silva CT, Schuch F, Costa M, Hirakata V, Manfro GG. Somatic, but not cognitive, symptoms of

anxiety predict lower levels of physical activity in panic disorder patients. J Affect Disord. 2014;164:63–8.

29. De Moor MH, Boomsma DI, Stubbe JH, Willemsen G, de Geus EJ. Testing causality in the association between regular exercise and symptoms of anxiety and depression. Arch Gen Psychiatry. 2008;65:897–905.

30. Mojtabai R, Crum RM. Cigarette smoking and onset of mood and anxiety disorders. Am J Public Health. 2013;103:1656–65.

31. Smith KJ, Béland M, Clyde M, Gariépy G, Pagé V, Badawi G, et al. Association of diabetes with anxiety: a systematic review and meta-analysis. J Psychosom Res. 2013;74:89–99.

32. Paterniti S, Zureik M, Ducimetière P, Touboul P, Fève J, Alpérovitch A. Sustained anxiety and 4-year progression of carotid atherosclerosis. Arterioscler Thromb Vasc Biol. 2001;21:136–41.

33. Spielberger CD. State-trait anxiety inventory: bibliography. 2nd ed. Palo Alto, CA: Consulting Psychologists Press; 1989.

34. Belem da Silva CT, Vargas da Silva AM, Costa M, Sant'Anna RT, Heldt E, Manfro GG. Increased anxiety levels predict a worse endothelial function in patients with lifetime panic disorder: results from a naturalistic follow-up study. Int J Cardiol. 2015;179:390–2.

35. Hamilton M. The assessment of anxiety states by rating. Br J Med Psychol. 1959;32:50–5.

36. Cicek Y, Durakoglugil ME, Kocaman SA, Guveli H, Cetin M, Erdogan T, et al. Increased pulse wave velocity in patients with panic disorder: independent vascular influence of panic disorder on arterial stiffness. J Psychosom Res. 2012;73:145–8.

37. Stillman AN, Moser DJ, Fiedorowicz J, Robinson HM, Haynes WG. Association of anxiety with resistance vessel dysfunction in human atherosclerosis. Psychosom Med. 2013;75:537–44.

38. Fleet R, Lespérance F, Arsenault A, Grégoire J, Lavoie K, Laurin C, et al. Myocardial perfusion study of panic attacks in patients with coronary artery disease. Am J Cardiol. 2005;96:1064–8.

39. Soares-Filho GL, Machado S, Arias-Carrión O, Santulli G, Mesquita CT, Cosci F, et al. Myocardial perfusion imaging study of CO_2-induced panic attack. Am J Cardiol. 2014;113:384–8.

40. O'Connor CM, Jiang W, Kuchibhatla M, Silva SG, Cuffe MS, Callwood DD, et al. Safety and efficacy of sertraline for depression in patients with heart failure: results of the SADHART-CHF (Sertraline Against Depression and Heart Disease in Chronic Heart Failure) trial. J Am Coll Cardiol. 2010;56:692–9.

41. Saab PG, Bang H, Williams RB, Powell LH, Schneiderman N, Thoresen C, et al. The impact of cognitive behavioral group training on event-free survival in patients with myocardial infarction: the ENRICHD experience. J Psychosom Res. 2009;67: 45–56.

Pulmonary Embolism in the Setting of Panic Attacks

13

Silvia Hoirisch-Clapauch,
Rafael Christophe R. Freire,
and Antonio Egidio Nardi

Contents

S. Hoirisch-Clapauch
Department of Hematology, Hospital Federal dos
Servidores do Estado, Ministry of Health,
Rio de Janeiro, Brazil
e-mail: sclapauch@gmail.com

R.C.R. Freire • A.E. Nardi (✉)
Laboratory of Panic and Respiration, Institute of
Psychiatry, Federal University of Rio de Janeiro,
Rio de Janeiro, Brazil
e-mail: rafaelcrfreire@gmail.com; antonioenardi@
gmail.com

Abstract

Panic attacks may present with conspicuous respiratory symptoms, which may also occur between attacks. Patients with panic attacks and respiratory symptoms have increased susceptibility to respiratory panicogenic challenges than patients without respiratory symptoms. Highly stressful situations, which may trigger a severe anxiety state such as panic attacks are characterized by a prothrombotic phenotype that increases the risk of thromboembolic events. Although episodes of pulmonary embolism might be accompanied by panic attacks, pulmonary embolism is seldom suspected when anxiety is the most likely alternative diagnosis. Given that recurrent thromboembolic disease may complicate with pulmonary artery hypertension and death, the diagnosis of pulmonary thromboembolism is fundamental. The diagnosis of pulmonary thromboembolism requires a high index of suspicion, because showers of microemboli are often asymptomatic or they may present with dyspnea, cough and wheezing, mimicking asthma. Notably, many patients with pulmonary thromboembolism have a relatively clear chest X-rays while severely hypoxemic. Confirmation of diagnosis usually depends on ventilation/perfusion lung scan or invasive imaging studies, such as computed tomographic pulmonary angiography. This paper discusses the characteristics of panic attacks that invite the suspicion of pulmonary embolism. It also suggests some diagnostic algorithms that help rule out thromboembolic disorders in the setting of panic attacks.

© Springer International Publishing Switzerland 2016
A.E. Nardi, R.C.R. Freire (eds.), *Panic Disorder*, DOI 10.1007/978-3-319-12538-1_13

Keywords
Asthma • Attacks • Coagulation • Panic • Pulmonary embolism

13.1 Introduction

Despite major advances in understanding the pathogenesis of panic attacks, some questions remain open. For example, is it possible that showers of microemboli could play a role in triggering or aggravating the attacks? Moreover, could psychiatrists help prevent pulmonary hypertension, a life-threatening disorder with great impact on exercise capacity and quality of life? Last, not least, is it possible that drugs commonly prescribed for panic disorder respiratory subtype could reduce thrombotic risk? Chances are high of an affirmative answer to all these questions. The objective of this chapter is to review the characteristics of panic attacks that might lead one to suspect pulmonary embolism.

13.2 Panic Attacks and Pulmonary Embolism

Some patients may present with prominent respiratory symptoms during and between panic attacks. These patients have a lower threshold to panicogenic challenges, such as inhalation of 35 % carbon dioxide, hyperventilation, sodium-lactate, breath-holding or caffeine [1, 2] than patients with panic attacks without respiratory symptoms. Patients with recurrent pulmonary embolism have increased ventilatory dead space, due to reduced blood flow to ventilated alveoli. An increased ventilatory drive and cardiovascular adjustments aiming at correcting hypoxia and hypercarbia [3] would probably lower the threshold of panicogenic challenges to induce panic attacks.

Highly stressful situations, such as panic attacks, are characterized by platelet activation and increased plasma levels of factors VII and VIII, fibrinogen and von Willebrand factor, all of which contribute to a prothrombotic state [4] that increase the risk of a thromboembolic event.

Although it is known that some patients with acute pulmonary thromboembolism may have a concurrent panic attack [5], the prevalence of pulmonary embolism in patients who present with panic attacks has not yet been estimated.

13.3 Suspecting of Pulmonary Embolism

The diagnosis of pulmonary embolism should be vigorously pursued, because recurrent pulmonary embolism may complicate with pulmonary hypertension, a devastating condition [6]. Bearing in mind that only 50–80 % of the patients with pulmonary hypertension related to chronic thromboembolism were aware of the thromboembolic events, there is good reason to assume that pulmonary embolism is dramatically underdiagnosed [7]. Indeed, pulmonary embolism is seldom suspected in psychiatric or in non-psychiatric emergency rooms when an anxiety disorder is the most likely alternative diagnosis [8].

Conditions that increase the risk of a thromboembolic event include: immobility, such as in heart failure or prolonged air travel, obesity and other inflammatory disorders, cancer, a previous episode of thromboembolism, varicose veins, pregnancy, the puerperium and estrogen use, as in oral contraception. Although moderate regular exercise decreases the risk of venous thromboembolism, both injury and dehydration increase the risk, which means that marathon athletes, for example, are not protected against thrombotic events [9].

Different from massive or sub-massive pulmonary thromboembolism, whose clinical presentation includes pleuritic chest pain, hemoptysis, hemodynamic instability and death, showers of microemboli are often asymptomatic or may present with dyspnea, cough, and wheezing, mimicking asthma [10, 11]. Windebank et al. [12] have shown that 4 % of 250 patients with acute pulmonary embolism demonstrated by pulmonary angiography had sufficient wheezing to be at first misdiagnosed with asthma.

The relationship between asthma and panic attacks is bidirectional. The frequency of panic spectrum symptoms in asthmatic patients has been reported to be from 6 % to 24 % [13, 14]. A large epidemiological study has demonstrated a more than threefold risk of pulmonary embolism for the asthmatic cohort, compared with the non-asthmatic cohort, after adjusting for sex, age, comorbidities and estrogen supplementation [15]. It has also been shown that patients with severe asthma and a comorbid mental disorder have a 12-fold greater risk for two or more asthma exacerbations and almost fivefold greater risk for two or more hospitalizations per year, compared to patients with severe asthma without mental disorders [16].

Suspicion of a thromboembolic event should be raised in patients with severe asthma [17], in those with frequent asthma exacerbation or respiratory symptoms severe enough to require hospitalization [15] and in asthmatic patients with concurrent panic attacks.

13.4 Confirming the Diagnosis of Pulmonary Embolism

Although some patients with pulmonary embolism may present with band atelectasis or elevation of a hemidiaphragm on chest X-ray film, many patients have a relatively clear chest X-rays while severely hypoxemic. Severity of hypoxemia is often intuitively interpreted with the amount of chest X-ray infiltrates, because pus, fluid, blood or proteinaceous material in the airspace may reduce alveolar ventilation and gas exchange. In case of acute pulmonary embolism, hypoxemia results from abnormal perfusion, while ventilation is not a problem [18].

It is recommended that the diagnosis of pulmonary embolism be confirmed with ventilation/perfusion lung scan or computed tomographic pulmonary angiography, an invasive procedure with recognized morbidity, such as contrast-induced nephropathy, and mortality. Different algorithms have been proposed to help rule out pulmonary embolism. For example, Wells et al. [19] have assigned scores to seven variables. Patients are considered low probability if the score is <2.0, moderate if the score is 2.0–6.0 and

Table 13.1 Probability of pulmonary embolism: Wells score

Clinical symptoms of deep venous thrombosis	3
No alternative diagnosis	3
Heart rate > 100	1.5
Immobilization or surgery in the previous 4 weeks	1.5
Previous deep venous thrombosis or pulmonary embolism	1.5
Hemoptysis	1
Malignancy	1

Probability of pulmonary embolism: sum of scores <2 = low; 2–6 = moderate; >6 = high

Table 13.2 Probability of pulmonary embolism: revised Geneva score

Age > 65 years	1
Previous deep venous thrombosis or pulmonary embolism	3
Surgery or fracture within 1 month	2
Active malignant condition	2
Unilateral lower limb pain	3
Hemoptysis	2
Heart rate of 75–94 beats per minute	3
Heart rate ≥ 95 beats per minute	5
Pain on lower-limb deep venous palpation and unilateral edema	4

Probability of pulmonary embolism: sum of scores <3 = low; 4–11 = intermediate; >11 = high

high if the score is >6.0 (Table 13.1). Measurement of plasma D-dimer levels by enzyme-linked immunosorbent antibody (ELISA) assay should be performed in all patients classified as low probability. D-dimer results from fibrin degradation and it is present in the blood after a blood clot is degraded by fibrinolytic enzymes. Pulmonary embolism can be ruled out if score is <2.0 and D-dimer levels are <500 μg/L. Wells score uses a subjective criterion: "is pulmonary embolism the most likely diagnosis?" which may lead to confusion when the disorder presents with a panic attack.

Another algorithm, known as the revised Geneva score [20], analyses nine variables. Probability of pulmonary embolism is low in patients with ≤3 points, intermediate in those with 4–10 points and high in patients with ≥11 points (Table 13.2). Again, if the patient is classified as low probability and D-dimer levels are <500 µg/L, the diagnosis of pulmonary embolism is highly improbable.

A computed tomography pulmonary angiography should be performed in all patients with elevated D-dimer levels or a prediction score indicating moderate or high risk [19, 20]. Given that pulmonary embolism is usually a complication of a deep venous thrombosis, a Doppler ultrasound of lower extremities should be also carried out. Compression sonography visualizing proximal deep vein thrombosis may be useful to confirm pulmonary embolism without the need for further imaging tests [21]. However, it is important to remember that after the thrombus detaches and moves through the bloodstream to the lung, it may not be detected on ultrasound exam. Therefore, the absence of a deep venous thrombosis does not rule out pulmonary embolism.

13.5 Therapeutic Aspects of Panic Attacks

Since a number of medications that are effective in preventing and treating panic disorder possess anticoagulant and/or profibrinolytic properties, in theory the therapeutic arsenal against panic attacks would also help prevent and treat thromboembolic events.

Similar to impending invasive procedures and other situations likely to provoke acute anxiety symptoms, panic attacks may hasten the clotting time of whole blood and evoke a hypercoagulable state [22]. Following this reasoning, medications that could reduce anxiety symptoms, such as benzodiazepines, would reduce hypercoagulability. In addition to their anxiolytic properties, clonazepam and diazepam have been shown to inhibit thrombin, adenosine diphosphate (ADP) and arachidonic acid-stimulated platelet aggregation *in vitro* [23]. Patients with panic disorder respiratory subtype have a fast and sustained response to clonazepam [24]. Of note, unlike many other benzodiazepines, clonazepam seems to interfere with serotonin metabolism [25].

Serotonin induces the expression in endothelial cells of both tissue factor and plasminogen activator inhibitor (PAI)-1 [26]. Tissue factor is the initial activator of the coagulation pathway, while PAI-1 is a major inhibitor of blood clot dissolution, a process known as fibrinolysis. Medications that interfere with serotonin metabolism, such as selective serotonin-reuptake inhibitors (SSRIs), may reduce thrombotic tendency and enhance fibrinolysis [27]. Resultant bleeding tendency is manifested by easy bruising, skin hematomas, gastrointestinal bleeding and increased bleeding following surgical procedures [27–29]. SSRIs such as sertraline, paroxetine, citalopram, escitalopram, fluoxetine and fluvoxamine are highly efficacious in the treatment of panic disorder [30].

Respiratory subtype of panic disorder patients may respond to imipramine [31], a tricyclic antidepressant that inhibits ADP-induced platelet aggregation [32, 33]. Respiratory panic disorder may also exhibit a good response to both nortriptyline and to clonidine, an alpha-2 adrenergic receptor agonist [34, 35]. Nortriptyline does not inhibit platelet activation [36], but it may increase norepinephrine levels. Norepinephrine stimulates plasminogen activator release, which increases fibrinolytic activity [37]. Although clonidine does not affect thrombotic tendency, it reduces both oxygen consumption and pulmonary artery pressure, without depressing the respiratory drive, which may benefit patients with pulmonary disorders such as pulmonary hypertension [38].

13.6 Conclusion

We strongly recommend that pulmonary thromboembolism be ruled out in the setting of panic attacks, especially when presenting with conspicuous respiratory symptoms.

References

1. Freire RC, Perna G, Nardi AE. Panic disorder respiratory subtype: psychopathology, laboratory challenge tests, and response to treatment. Harv Rev Psychiatry. 2010;18(4):220–9.
2. Parshall MB, Schwartzstein RM, Adams L, Banzett RB, Manning HL, Bourbeau J, et al. An official

American Thoracic Society statement: update on the mechanisms, assessment, and management of dyspnea. Am J Respir Crit Care Med. 2012;185(4):435–52.

3. Sudano I, Binggeli C, Spieker L, Lüscher TF, Ruschitzkam F, Noll G, et al. Cardiovascular effects of coffee: is it a risk factor? Prog Cardiovasc Nurs. 2005;20(2):65–9.

4. Austin AW, Wissmann T, von Känel R. Stress and hemostasis: an update. Semin Thromb Hemost. 2013;39(8):902–12.

5. Schlicht KF, Mann K, Jungmann F, Kaes J, Post F, Münzel T, et al. A 48-year-old woman with panic attacks. Lancet. 2014;384(9939):280.

6. Simonneau G, Robbins IM, Beghetti M, Channick RN, Delcroix M, Denton CP, et al. Updated clinical classification of pulmonary hypertension. J Am Coll Cardiol. 2009;54(1s1):S43–54.

7. Scheidl SJ, Englisch C, Kovacs G, Reichenberger F, Schulz R, Breithecker A, et al. Diagnosis of CTEPH versus IPAH using capillary to end-tidal carbon dioxide gradients. Eur Respir J. 2012;39(1):119–24.

8. Kabrhel C, McAfee AT, Goldhaber SZ. The probability of pulmonary embolism is a function of the diagnoses considered most likely before testing. Acad Emerg Med. 2006;13(4):471–4.

9. Hull CM, Harris JA. Venous thromboembolism and marathon athletes. Circulation. 2013;128(25):e469–71.

10. Tapson VF. Acute pulmonary embolism. N Engl J Med. 2008;358(10):1037–52.

11. Slaughter MC. Not quite asthma: differential diagnosis of dyspnea, cough, and wheezing. Allergy Asthma Proc. 2007;28(3):271–81.

12. Windebank WJ, Boyd G, Moran F. Pulmonary thromboembolism presenting as asthma. Br Med J. 1973;1(5845):90–4.

13. Zaubler TS, Katon W. Panic disorder in the general medical setting. J Psychosom Res. 1998;44(1):25–42.

14. Nascimento I, Nardi AE, Valença AM, Lopes FL, Mezzasalma MA, Nascentes R, et al. Psychiatric disorders in asthmatic outpatients. Psychiatry Res. 2003;110(1):73–80.

15. Chung WS, Lin CL, Ho FM, Li RY, Sung FC, Kao CH, et al. Asthma increases pulmonary thromboembolism risk: a nationwide population cohort study. Eur Respir J. 2014;43(3):801–7.

16. ten Brinke A, Ouwerkerk ME, Zwinderman AH, Spinhoven P, Bel EH. Psychopathology in patients with severe asthma is associated with increased health care utilization. Am J Respir Crit Care Med. 2001;163(5):1093–6.

17. Majoor CJ, Kamphuisen PW, Zwinderman AH, Ten Brinke A, Amelink M, Rijssenbeek-Nouwens L, et al. Risk of deep vein thrombosis and pulmonary embolism in asthma. Eur Respir J. 2013;42(3):655–61.

18. Tsang JY, Hogg JC. Gas exchange and pulmonary hypertension following acute pulmonary thromboembolism: has the emperor got some new clothes yet? Pulm Circ. 2014;4(2):220–36.

19. Wells PS, Anderson DR, Rodger M, Ginsberg JS, Kearon C, Gent M, et al. Derivation of a simple clinical model to categorize patients probability of pulmonary embolism: increasing the models utility with the SimpliRED D-dimer. Thromb Haemost. 2000;83(3):416–20.

20. Le Gal G, Righini M, Roy PM, Sanchez O, Aujesky D, Bounameaux H, et al. Prediction of pulmonary embolism in the emergency department: the revised Geneva score. Ann Intern Med. 2006;144(3):165–71.

21. Käberich A, Wärntges S, Konstantinides S. Risk-adapted management of acute pulmonary embolism: recent evidence, new guidelines. Rambam Maimonides Med J. 2014;5(4):e0040.

22. von Känel R, Kudielka BM, Schulze R, Gander ML, Fischer JE. Hypercoagulability in working men and women with high levels of panic-like anxiety. Psychother Psychosom. 2004;73(6):353–60.

23. Rajtar G, Zółkowska D, Kleinrok Z. Effect of diazepam and clonazepam on the function of isolated rat platelet and neutrophil. Med Sci Monit. 2002;8(4):I37–44.

24. Nardi AE, Valença AM, Nascimento I, Lopes LF, Mezzasalma MA, Freire RC, et al. A three-year follow-up study of patients with the respiratory subtype of panic disorder after treatment with clonazepam. Psychiatry Res. 2005;137(1–2):61–70.

25. Moroz G. High-potency benzodiazepines: recent clinical results. J Clin Psychiatry. 2004;65 Suppl 5:13–8.

26. Kawano H, Tsuji H, Nishimura H, Kimura S, Yano S, Ukimura N, et al. Serotonin induces the expression of tissue factor and plasminogen activator inhibitor-1 in cultured rat aortic endothelial cells. Blood. 2001;97(6):1697–702.

27. Hoirisch-Clapauch S, Nardi AE, Gris JC, Brenner B. Are the antiplatelet and profibrinolytic properties of selective serotonin-reuptake inhibitors relevant to their brain effects? Thromb Res. 2014;134(1):11–6.

28. de Abajo FJ, Rodríguez LAG, Montero D. Association between selective serotonin reuptake inhibitors and upper gastrointestinal bleeding: population based case-control study. BMJ. 1999;319(7217):1106–9.

29. Jeong BO, Kim SW, Kim SY, Kim JM, Shin IS, Yoon JS. Use of serotonergic antidepressants and bleeding risk in patients undergoing surgery. Psychosomatics. 2014;55(3):213–20.

30. Mochcovitch MD, Nardi AE. Selective serotonin-reuptake inhibitors in the treatment of panic disorder: a systematic review of placebo-controlled studies. Expert Rev Neurother. 2010;10(8):1285–93.

31. Briggs AC, Stretch DD, Brandon S. Subtyping of panic disorder by symptom profile. Br J Psychiatry. 1993;163:201–9.

32. Mohammad SF, Mason RG. Inhibition of human platelet-collagen adhesion reaction by amitriptyline and imipramine. Proc Soc Exp Biol Med. 1974;145(3):1106–13.

33. Gómez-Gil E, Gastó C, Carretero M, Díaz-Ricart M, Salamero M, Navinés R, et al. Decrease of the platelet 5-HT2A receptor function by long-term imipramine treatment in endogenous depression. Hum Psychopharmacol. 2004;19(4):251–8.

34. Nardi AE, Nascimento I, Valença AM, Lopes FL, Mezzasalma MA, Zin WA, et al. Respiratory panic disorder subtype: acute and long-term response to nortriptyline, a noradrenergic tricyclic antidepressant. Psychiatry Res. 2003;120(3):283–93.

35. Valença AM, Nardi AE, Mezzasalma MA, Nascimento I, Lopes FL, Zin WA, et al. Clonidine in respiratory panic disorder subtype. Arq Neuropsiquiatr. 2004;62(2B):396–8.

36. Pollock BG, Laghrissi-Thode F, Wagner WR. Evaluation of platelet activation in depressed patients with ischemic heart disease after paroxetine or nortriptyline treatment. J Clin Psychopharmacol. 2000;20(2):137–40.

37. Parmer RJ, Mahata M, Mahata S, Sebald MT, O'Connor DT, Miles LA. Tissue plasminogen activator (t-PA) is targeted to the regulated secretory pathway catecholamine storage vesicles as a reservoir for the rapid release of t-PA. J Biol Chem. 1992;272(2):1976–82.

38. Pichot C, Petitjeans F, Ghignone M, Quintin L. Is there a place for pressure-support ventilation and high positive end-expiratory pressure combined to alpha-2 agonists early in severe diffuse acute respiratory distress syndrome? Med Hypotheses. 2013;80(6):732–7.

Self-Consciousness and Panic

14

Thalita Gabínio, André B. Veras,
and Jeffrey P. Kahn

Contents

T. Gabínio • A.B. Veras (✉)
Dom Bosco Catholic University (UCDB),
Campo Grande, Brazil
e-mail: thalitagabinio@gmail.com;
barcielaveras@hotmail.com

J.P. Kahn
Department of Psychiatry, Weill Cornell Medical
College, Cornell University, New York, NY, USA
e-mail: jeffkahn@aol.com

Abstract

Authors present in this chapter some relationships between self-consciousness and panic disorder. Discussion is based on Karl Jaspers' phenomenological descriptions of the self-consciousness manifestations, aiming to explore the ideas of Jaspers about the relations between self-constitution and panic attacks. Some broad definitions of self-consciousness and some illustrated descriptions of their pathological manifestations are also presented, and the some psychological contributions for panic disorder. Panic disorder etiologies keep under investigation and stressful events in childhood and adulthood seem relevant. Another aspect addressed in this chapter is the association between brain regulation of self-consciousness, since neurological regulation of self-consciousness depends on brain structures responsible for programming and self-monitoring of our behavior. All common neurologic and clinical aspects between panic and schizophrenia seem to be connected by self-consciousness physiopathology and its psychopathological manifestations.

Keywords

Self-consciousness • Panic disorder • Psychosis • Psychopathological manifestations • Panic attacks

14.1 Self-Consciousness Definitions

We will present some relationships between self-consciousness and panic disorder. The reader will find four broad definitions of self-consciousness and some illustrated descriptions of their pathological manifestations. Although these afflictions of self-conscious are classically described in psychosis, other less severe effects are also commonly seen in panic attacks. When panic anxiety is combined with global self-consciousness deficits, there may even be a panic psychosis. The discussion is based on Karl Jaspers' phenomenological descriptions of the self-consciousness manifestations. Karl Jaspers (1883–1969) was a German psychiatrist and philosopher who decisively contributed to the development of Psychopathology through his book "General Psychopathology" [1].

Jaspers states that self-consciousness is the mind's ability to be conscious of itself. Self-consciousness differs from objective consciousness – which is the ability to be aware of perceptions of the outside world. Self-consciousness includes four different mental capacities: awareness of one's actions, unity of selfhood consciousness, self-identity consciousness and capacity to differentiate oneself from the environment and from other people. Through these characteristics and their containments, the self turns into a personality [1].

Each self-activity of the mind has its own subjective way of functioning, based on instinctive functions, which were considered by Jaspers as the inborn states of the self, and emotional feelings, considered as spontaneous self states. Personality is the individualization that occurs when a psychic life event, such as a bodily sensation or other perception, or the evocation of a memory, helps form a unique character-forming experience in each human being. It gives personal character to everyone's distinct mind. In addition, Jaspers recognized the importance of mind-body unity. That unity depends on unconscious manifestations of the brain, where our body functions and instincts come from [1]. When emotional feelings and automatic-unconscious manifestations match as expected,

they can be understood as a unique personal experience. When these events are experienced with a feeling of strangeness accompanied by an unexpected automatic response, then depersonalization as a symptom happens [1].

Patients who experience depersonalization usually report feelings of self-strangeness, as if they were acting mechanically or automatically. Living in a shadowy state, patients experience a feeling of vague existence, which is sometimes similar to a death feeling or a feeling of non-existence. These mind states described have been considered self-existence disorders. Another kind of manifestation of self-consciousness is self-activity disorder, it occurs when actions or psychic manifestations are felt by the patient as not belonging to him, but happening out of his control. A usual example is an obsessive patient who does not know the reason or the origin of his unwilled thoughts and, not being able to stop thinking about it, develops the sensation of non-self-generated thoughts, as if they were not produced by his own mind but by something from the outside world. Jaspers called these manifestations "influenced voluntary action".

Other examples are observed in psychosis, as when a patient believes that some thoughts were inserted in his head, started to control his actions and experiencing an inability to disobey [1]. In a less severe presentation, a patient cannot stop worrying about the possibility of suffering from a lethal disease and for this reason keeps looking for medical assistance. This patient, even though aware of the absurdity of his idea, feels that he lost control, and fears being crazy. Another common symptom in severe anxiety is the dissociative state. Although not the same as hallucinations or feelings of external control, dissociation is considered a kind of self-unity pathology. Jaspers describes dissociation as a "double personality" caused by the notion of an imbalanced consciousness, when self-split is experienced [1].

A third additional subtype of self-consciousness is self-identification. Self-identification is the ability of being aware of ourselves throughout existence. Self-identification distortions are usually seen in psychotic patients, as we hear their reports about how different they feel after psychosis, as if they have become a different person. Psychotic patients sometimes actually develop self-identity delusions, starting to introduce themselves and act as another person.

In patients with schizophrenia, a fourth self-consciousness manifestation is even more common and it occurs because of the inability of the patient to differentiate their inside world from the outside world. All is experienced as a mixture of what is felt or thought and what is perceived from the environment. Jaspers illustrated this kind of symptom with a patient belief that everyone could know what he was thinking, making it unnecessary for him to express himself. In another example, some patients report that they are not able to hide anything from others – convinced that all their thoughts are revealed [1]. This is similar to Eugen Bleuler's (Swiss psychiatrist; 1857–1939) view of schizophrenia as a disconnection between the real world and the unconscious emotional world.

Even in simpler psychotic manifestations, such as auditory hallucinations, this inability to differentiate self-phenomena from the world can be observed: voices are the patient's thoughts experienced as coming from the outside world. This inability to differentiate inner word and outer world is not exclusive to psychosis. It can be observed in a moderate to slight degree in anxiety disorders, and even among people without a mental disorder. For example, appropriate discrimination between an individual's feelings of exclusion and a true rejection by a group – which is the inner experience and which is the external reality – is difficult even for mature people [2], and is more pronounced in atypical depression. It could even be said that the complete and unequivocal distinction between internal and external is never fully acquired at any point in a lifetime.

14.2 Self-Consciousness Disorders in Panic Disorder

Here we aim to explore the ideas of Jaspers about the relations between self-constitution and panic attacks. His constitution concept is described as an entire notion of body life, where all bodily functions, including mental ones, are connected and integrated. There is also the idea of a psychic constitution, which is experienced in an way entirely connected with the body. This broad integration notion produces subjective characteristics to each person. It means that although psychopathological symptoms are classified in the same way to everyone, according to Jaspers each symptom has a different "color and meaning" in one's mind [1].

In panic disorder, an imbalance in this mind-body integrated notion can be seen. Patients experience symptoms in a confused way and may show some difficulty in naming them. It is rare that a patient presents to a psychiatrist with clearly self-described panic attacks, unless they have already been diagnosed elsewhere. For this reason, assisting these patients is a day by day multi-professional challenge. Patients are affected by bodily discomforts that are actually not from a physical illness, but clearly located in organs that after careful investigation do not show physical pathology sufficient to explain those symptoms.

Panic anxiety seems to translate as a breakdown in self-consciousness caused by a fear of self-disintegration and from the global anxiety provoked by a feeling of imminent death [2] or social exile. Patients afflicted with panic attacks have a real sensation of self-fragmentation, with a feeling of being about to loose control and of not being able to keep mind-body integrity. In this direction, self-consciousness would be affected: self-activity, represented by the ability of making decisions and acting by our own free will, is temporarily impaired by the panic episode's intense fear. Especially when panic symptoms are readily explained by real world situations, patients experience mere living as threat, since a panic attack can show up anytime and anywhere. Regarding self-unity, panic symptoms can seem like a stranger inside someone's mind, becoming the patient vulnerable to himself, having no control of symptoms and no idea of how to behave in a crisis.

Among other factors, an important psychological etiology of panic disorder is impaired psychological development during childhood.

Infants and young children who experienced physical or psychological traumas during important periods to personality development might be more susceptible to panic anxiety in later life. In cases of a more disorganized self-structure, who had an early failure in the protective network, emotional and self-consciousness development may be affected. Panic disorder etiologies keep under investigation and stressful events in childhood and adulthood seem relevant. Early parental loss is more common in panic disorder patients [3–5], but other stressful events just before the onset of panic manifestations happen in the majority of the cases [6, 7].

14.3 Brain Regulation of Self-Consciousness

Neurological regulation of self-consciousness depends on brain structures responsible for programming and self-monitoring of our behavior. Motor activity is firstly planned and programmed by frontal cortex areas. Then, when efferent activity is started, unconscious feedbacks from peripheral receptors, basal nucleus and cerebellum are simultaneously activated, continuously adjusting the motor behavior [8]. At the end, a matched sensory and programmed behavior brings the sensation of a self-generated one, while a mismatch between them leads to the impression of an environmental caused action [9]. Motor physiology usually seems easy to understand, but not every mental intention turns into action. At the same time, even non-motoric mental intentions have to be planned, and might be seen as hidden actions [10]. In that cases, different brain areas are activated as self-monitors, such as the anterior cingulate cortex (ACC) and speech areas (Broca and Wernick).

Recently, hidden actions self-monitoring brain structures have been better understood through Positron Emission Tomography (PET) and functional Magnetic Resonance Imaging (fMRI) studies. Examination of patients with severe psychological experiences of self-consciousness imbalance (such as the Schneiderian first-rank symptoms and auditory hallucinations), using functional imaging tools recently revealed important findings. Studies examining verbal production usually found a functional decrease in self-monitoring areas such as the ACC, superior parietal and temporal regions (speech and association areas), hippocampus (association) and cerebellum when patients were experiencing auditory hallucinations or producing speech, or even just imagining words [11, 12]. At the same time that this self-monitoring function decrease occurs, images of the brain show an increase in primary sensory areas such as the temporal auditory cortex [13–18]. The brain, not being able to integrate self experiences because of a disconnection between posterior-sensory and frontal-executive regions and because of a decreased function of self-monitoring areas, tends to work in a simpler mode using its sensory capacity to keep perceiving the world. However, with poor differentiation between internal and external worlds, many mental experiences become just one unconscious world.

Among all these neuroanatomical structures related with self-consciousness imbalances, some are also parts of the fear circuitry. The one which seems to deserve special attention is the ACC. The ACC, as part of the limbic system, participates in affective reactions [19, 20]. And, while the dorsal region of the ACC performs cognitive functions, its ventral portion is involved in the regulation of emotional responses [19]. The cognitive part of the ACC is connected with parietal and frontal areas, which play a role in spatial location and self-programming [19]. The emotional part of the ACC is directly linked to important structures of the fear circuit [21], particularly the amygdala, hypothalamus and periaqueductal gray matter [20]. Having these functions and connections, the ACC might be a pathway between two basic mental functions: self-consciousness and emotions.

Depersonalization (DP) and derealization are less severe psychopathological manifestations of self-consciousness, as previously described in this chapter. DP is a slight to moderate sensation of strangeness about our own body, movements, feeling and thoughts, where the patient keeps aware of this unusual mind state. Although DP may happen in many psychiatric disorders, such as schizophrenia, these manifestations are much more common in severe anxiety disorders. In panic disorder, DP reaches its highest prevalence, between 25 % and 80 % [22], and is included as one of the diagnostic symptoms of PD. In neurobiological studies, dysfunctional brain areas in DP patients overlap with the dysfunctional areas in patients with schizophrenia or first rank symptoms.

Sierra and Berrios [23] postulated a functional sensory–limbic disconnection. Sensory and prefrontal cortices are usually balanced with limbic areas through information interchanges through the cingulate cortex, mainly its anterior part – the ACC. According to this theory, a hyperactivity of the medial prefrontal cortex inhibits part of the amygdala and indirectly the ACC. At the same time, a hyperactivity of some amygdala circuits controlling both cholinergic and monoaminergic ascending circuits lead to the activation of attention in prefrontal cortex areas. The simultaneous activity of these two opposing mechanisms lead to a state of hyper-attention associated to a state of hypo-emotionality [22], characterizing the feelings of DP.

All the self-consciousness pathologies noted here seem to be related to a state of disconnection in the brain. Considering DP as one of the symptoms of a panic crisis, this self-consciousness symptom seems to be associated with – and perhaps caused by – a severe anxiety state. Depending on frequency and severity of the anxiety symptoms and the amount of hypofrontality a patient has, even more significant self-consciousness symptoms might happen, such as psychotic symptoms [2].

14.4 Panic Psychosis

While certain relationships between panic and schizophrenia have been observed for many years, recent reports have better established the findings and raised further questions for research and treatment. One observation is the common occurrence of panic anxiety in the prodromal phase of schizophrenia [24], a higher prevalence of panic in schizophrenia when comparing to

the general population [25], the existence of paroxysmal anxiety concomitant with auditory hallucinations and delusions in schizophrenia [26]. This latter finding supports clinical observation that voices may sometimes be audible panic attacks. Panic and schizophrenia also share some common etiological factors [27] such as heritage, biochemical factors, increased manifestations with marijuana use and overlapped brain areas. Indeed, alprazolam and clonazepam are antipanic benzodiazepines that can substantially improve both positive and negative schizophrenia symptoms when panic anxiety is also present [28]. This collection of findings relations even justifies the proposal of a novel diagnostic category, perhaps "panic psychosis" [26].

Briefly, panic psychosis may define a distinct subgroup of schizophrenic patients with paroxysmal panic attacks that are concurrent with abrupt onset of auditory hallucinations or delusions, as well as more frequent occurrences of positive symptoms [26, 29, 30]. They may differ from ordinary panic patients by hypofrontality leading to disruption of self-consciousness and consequent psychosis.

14.5 Conclusion

All these common neurologic and clinical aspects between panic and schizophrenia seem to be connected by self-consciousness physiopathology and its psychopathological manifestations, illustrating the intriguing importance of self-consciousness in panic to understand the neurobiology and etiology of this disorder.

References

1. Jaspers K. Psicopatologia Geral. São Paulo: Editora Ateneu; 2003.
2. Veras AB, Nardi AE, Kahn JP. Attachment and self-consciousness: a dynamic connection between schizophrenia and panic. Med Hypotheses. 2013;81(5):792–6.
3. Fergusson DM, Lynskey MT, Horwood LJ. Childhood sexual abuse and psychiatric disorder in young adulthood: I. Prevalence of sexual abuse and factors associated with sexual abuse. J Am Acad Child Adolesc Psychiatry. 1996;35(10):1355–64.

4. Kendler KS, Bulik CM, Silberg J, Hettema JM, Myers J, Prescott CA. Childhood sexual abuse and adult psychiatric and substance use disorders in women: an epidemiological and cotwin control analysis. Arch Gen Psychiatry. 2000;57(10):953–9.
5. Kendler KS, Neale MC, Kessler RC, Heath AC, Eaves LJ. Childhood parental loss and adult psychopathology in women. A twin study perspective. Arch Gen Psychiatry. 1992;49(2):109–16.
6. Faravelli C. Life events preceding the onset of panic disorder. J Affect Disord. 1985;9(1):103–5.
7. Salum GA, Blaya C, Manfro GG. Transtorno do pânico. Rev Psiquiatr RS. 2009;31(2):86–94.
8. Raveendran V, Kumari V. Clinical, cognitive and neural correlates of self-monitoring deficits in schizophrenia: an update. Acta Neuropsychiatr. 2007;19: 27–37.
9. Frith CD. The cognitive neuropsychology of schizophrenia. Erlbaum, UK: Taylor and Francis; 1992.
10. Jeannerod M, Pacherie E. Agency, simulation and self identification. Mind Lang. 2004;19:113–46.
11. Frith CD, Friston KJ, Herold S, Silbersweig D, Fletcher P, Cahill C,et al. Regional brain activity in chronic schizophrenic patients during the performance of a verbal fluency task. Br J Psychiatry. 1995;167:343–9.
12. Fu CH, Vythelingum GN, Andrew C, et al. Alien voices who said that? Neural correlates of impaired verbal selfmonitoring in schizophrenia. Neuroimage. 2001;13:S1052.
13. Hunter MD, Griffiths TD, Farrow TF, et al. A neural basis for the perception of voices in external auditory space. Brain. 2003;126:161–9.
14. McGuire PK, Shah GM, Murray RM. Increased blood flow in Broca's area during auditory hallucinations in schizophrenia. Lancet. 1993;342:703–6.
15. Shergill SS, Bullmore E, Simmons A, Murray R, McGuire P. Mapping auditory hallucinations in schizophrenia using functional magnetic resonance imaging. Arch Gen Psychiatry. 2000;59:468–89.
16. Silbersweig DA, Stern E, Frith C, et al. A functional neuroanatomy of hallucinations in schizophrenia. Nature. 1995;378:176–9.
17. Sokhi DS, Hunter MD, Wilkison ID, Woodruff PW. Male and female voices activate distinct regions in the male brain. Neuroimage. 2005;27:572–8.
18. Frith C. Neuropsychology of schizophrenia, what are the implications of intellectual and experiential abnormalities for the neurobiology of schizophrenia? Br Med Bull. 1996;52:618–26.
19. Bush G, Luu P, Posner MI. Cognitive and emotional influences in anterior cingulate cortex. Trends Cogn Sci. 2000;4(6):215–22.
20. Mobbs D, Marchant JL, Hassabis D, et al. From threat to fear: the neural organization of defensive fear systems in humans. J Neurosci. 2009;29(39): 12236–43.
21. Thomas H. A community survey of adverse effects of cannabis use. Drug Alcohol Depend. 1996;42:201–7.
22. Mula M, Pini S, Cassano GB. The neurobiology and clinical significance of depersonalization in mood and anxiety disorders: A critical reappraisal. J Affect Disord. 2007;99:91–9.
23. Sierra M, Berrios GE. Depersonalization: neurobiological perspectives. Biol Psychiatry. 1998;44: 898–908.
24. Craig T, Hwang MY, Bromet EJ. Obsessive-compulsive and panic symptoms in patients with first-admission psychosis. Am J Psychiatry. 2002;59: 592–800.
25. Achim AM, Maziade M, Raymond E, Olivier D, Merette C, Roy MA. How prevalent are anxiety disorders in schizophrenia? A meta-analysis and critical review on a significant association. Schizophr Bull. 2009;37:811–21.
26. Kahn JP, Meyers JR. Treatment of comorbid panic disorder and schizophrenia: evidence for a panic psychosis. Psychiatr Ann. 2000;30:29–33.
27. Hofmann SG. Relationship between panic and schizophrenia. Depress Anxiety. 1999;1995(9):101–6.
28. Kahn JP, Puertollano MA, Schane MD, Klein DF. Adjunctive alprazolam for schizophrenia with panic anxiety: clinical observation and pathogenetic implications. Am J Psychiatry. 1988;145:742–4.
29. Savitz AJ, Kahn TE, McGovern KE, Kahn JP. Carbon dioxide induction of panic anxiety in schizophrenia with auditory hallucinations. Psychiatry Res. 2011; 189:38–42.
30. Kahn JP. Consciousness lost and instinct run amok: schizophrenia and psychosis. In: Kahn JP, editor. Angst: origins of anxiety and depression. New York: Oxford University Press; 2013. Chapter 7.

Pharmacological Treatment with the Selective Serotonin Reuptake Inhibitors

15

Marina Dyskant Mochcovitch
and Tathiana Pires Baczynski

Contents

Abstract

The selective serotonin reuptake inhibitors (SSRI) are considered drugs of first choice for the treatment of panic disorder (PD) due to their clinical efficacy demonstrated by 11 randomized placebo-controlled trials and favorable side effects profile. SSRI's mechanism of action in PD treatment still not completely understood, but the role of serotonin in the pathophysiology of this disorder has been studied for the last few years by neuroimaging, neurochemical and challenge studies. In this chapter, efficacy studies, common side effects and drug interactions of the SSRI are described as well as general recommendations for their clinical use in PD.

Keywords

Selective serotonin reuptake inhibitors • Pharmacological treatment • Panic disorder • Serotonin • Mechanism of action • Randomized clinical trials

15.1 Introduction

The primary goal of pharmacological treatment in PD is to reduce the intensity and frequency of panic attacks. The anticipatory anxiety and agoraphobia arise as consequences of recurrent panic attacks, generating a vicious cycle, which leads to depression and demoralization syndrome.

M.D. Mochcovitch (✉) • T.P. Baczynski
Laboratory of Panic and Respiration, Institute of
Psychiatry, Federal University of Rio de Janeiro,
Rio de Janeiro, Brazil
e-mail: marimochco@yahoo.com.br;
tathipbac@gmail.com

© Springer International Publishing Switzerland 2016
A.E. Nardi, R.C.R. Freire (eds.), *Panic Disorder*, DOI 10.1007/978-3-319-12538-1_15

Systematic reviews demonstrate that a range of pharmacological [1, 2] psychological [3] and combination [4, 5] interventions are effective in the treatment of patients with PD. According to clinical guidelines, the antidepressants are considered drugs of choice for treating this disorder, reducing panic attacks and also acting in anticipatory anxiety and associated depression [6, 7].

Among the antidepressants, the monoamine oxidase inhibitors (IMAO) and tricyclic antidepressants were the classes initially used. Despite their proven efficacy for treating PD, these medications are not considered a primary choice anymore because of their unfavorable side effect and safety profiles, which often leads to treatment discontinuation. However, they are still used in clinical practice (especially tricyclic antidepressants) as second or third line drugs [2, 5, 6, 8].

The SSRI were developed as an effort to produce drugs that have similar therapeutic effect of tricyclic antidepressants, but act more selectively, producing therefore less troublesome side effects with a better safety profile. SSRI do not present four of the pharmacological properties that characterize the tricyclics class: norepinephrine reuptake inhibition, blockade of muscarinic, histaminic H1 and α1-adrenergic receptors. Thus, the inhibition of 5-hydroxytryptamine (5-HT) reuptake remained as the main pharmacological property. Therefore, although other classes of drugs can be used, the SSRI became the antidepressant class of choice for the treatment of PD.

15.2 Mechanism of Action

Different serotoninergic receptors show different physiological functions. The receptor most implicated in the antidepressant effect is the 5-HT-1 receptor, which also has an anxiolytic function and is responsible for raising the body temperature. On the other hand, the 5-HT-2 receptor appears to mediate anxiogenic effects, insomnia, reduction of body temperature and sexual dysfunction, among other aspects. Therefore, 5-HT-3 receptor would seem to be associated with reduced appetite and nausea and increased intestinal motility [9].

Two opposing hypotheses have been put forth to explain PD by serotonergic dysfunction: 5-HT excess or overactivity and 5-HT deficit or underactivity [10]. The 5-HT excess theory suggests that patients with PD show a hypersensitivity in postsynaptic 5-HT receptors – especially 5-HT-2c receptors, the 5-HT receptor subtype most associated with anxiogenic effect. The 5-HT deficit theory proposes that, in particular brain regions, such as the dorsal periaqueductal gray (PAG), 5-HT has a restraining effect on panic behavior and a 5-HT deficit may facilitate panic [10].

Considering the excess theory, it was suggested that the anxiolytic effect of SSRI occurs through desensibilization by downregulation of 5-HT-2C receptors [11, 12].

Besides that, to better understand the mechanism of action of the SSRI on anxiety, it is important to consider not only the different types of 5-HT receptors, but also the regions of the brain in which these receptors are found, or in other words, in which brain areas the SSRI act in the treatment of PD. The neuroanatomic model for PD, based on the "circuit of fear" suggests that the amygdala has a central function in the regulation of conditioned fear. Other brain areas that make up this circuit, with intense connections with the amygdala, are the hippocampus, the median prefrontal cortex and the median-dorsal nucleus of the thalamus [13].

Neuroimaging studies demonstrate functional and clinically relevant alterations in various elements of the 5-HT system affecting the neurocircuitry of panic [10]. A single-photon emission computed tomography study of the functional activity of 5-HT transporter (5-HTT) in PD showed that the patients with current PD had significantly lower 5-HTT binding in the midbrain raphe, in the temporal lobes, and in the thalamus than the healthy controls. On the other hand, the patients with PD in remission had normal 5-HTT-binding properties in the midbrain and in the temporal regions, but still a significantly lower thalamic 5-HTT binding. Another PET study has demonstrated that untreated PD patients showed reduced binding to 5-HT1A receptors in the raphe region as well as in the amygdala, and the orbitofrontal and temporal cortices [82]. PD patients who fully recovered after treatment with parox-etine in this study showed normalized density of postsynaptic receptors, but there remained a reduction in the density of 5-HT-1A receptors in the raphe and in the hippocampus [10].

Besides the functional alterations in panic neurocircuitry, more recent neuroimaging studies with structural and functional magnetic reso-nance with diffusion tensor have shown also structural alterations in PD patients after treat-ment with SSRI. Lai et al. [14, 15] have demon-strated improvements in white matter integrity of right uncinate fasciculus and left fronto-occipital fasciculus and increase in gray matter volume in the left superior frontal gyrus in first-episode PD patients after six weeks treatment with escitalo-pram (10–15 mg/day).

Challenge studies combined with the manipu-lation of the 5-HT system may also be informa-tive for clarifying the role of 5-HT in PD. One of the interventions that can be made for this purpose is the tryptophan depletion (TD), which provokes a decrease in 5HT levels, combined with the CO_2 challenge. TD increases the sensitivity to CO_2 in patients with PD. Miller et al. found that TD caused a greater panic and anxiogenic response and a higher rate of panic attacks after 5 % CO_2 inhalation in PD patients, but not in healthy sub-jects [16]. Treatment with SSRI significantly decreased the sensitivity of patients with PD to the panicogenic effects of CO_2 [17] and TD reversed the antipanic effect of chronic treatment with the SSRI paroxetine in PD patients [18].

Although these studies do not elucidate whether the 5-HT deficit is the primary dysfunc-tion in PD, they help to confirm that the SSRI exert their therapeutic effects in PD by increasing the synaptic availability of 5-HT, and that seroto-nergic enhancement by SSRI leads to antipanic effects [10, 18].

15.3   Efficacy Studies

The efficacy of SSRI in the treatment of PD has been widely studied. Several studies have dem-onstrated that SSRI are able not only to reduce the frequency and intensity of panic attacks, but

actually treat the whole panic syndrome. This includes reducing anticipatory anxiety and avoid-ant behavior, as well as treating comorbid depres-sion and improve the overall functioning of the patient [1, 2]. Fluoxetine was the first approved SSRI in USA, however, its most studies in PD treatment were open trials.

There are two open studies with fluoxetine, both showing symptoms improvements after 6–8 week follow-up [19, 20]. Two randomized con-trolled trial with this drug for PD treatment was performed by Michelson et al. [21, 22]. In the first study, patients received fluoxetine 10 and 20 mg/day and reduction of panic attacks were significantly different between the group taking 10 mg/day and placebo group (p=0.006), but not between the group taking 20 mg/day and placebo group (p=0.12). However, secondary endpoints anxiety symptoms assessed by the Hamilton Anxiety Scale (HAM-A) and overall improve-ment assessed by the CGI were higher for the group using 20 mg/day, compared to 10 mg/day, these results statistically significant [21]. In the second study, patients received fluoxetine at 20 mg/day for 12 weeks, but at the sixth week, patients who had failed to achieve a satisfactory response were eligible for dose escalation to a maximum of 60 mg of fluoxetine daily. In this study, fluoxetine 20 mg/day was associated with a statistically significantly greater proportion of panic-free patients compared with placebo after 6 weeks and at end-point [22].

Paroxetine was the most studied SSRI in PD, with four placebo-controlled studies. Ballenger [23] evaluated 278 subjects with fixed doses of 10, 20 and 40 mg/day for 10 weeks. There were greater percentages of patients free of panic attacks for all three active substance groups when compared to placebo, however, only the group receiving 40 mg/day showed statistical significance.

Lucubrier [24] compared paroxetine (doses varying from 20 to 60 mg/day) and clomipramine (doses varying from of and 50 to 150/day), for 12 weeks, in 367 patients. Between weeks 7 and 9, 50.9 % of patients receiving paroxetine were free of full PA, which occurred with 36.7 % of patients with clomipramine and 31.6 % of placebo patients. At this point, paroxetine effect was significantly different from placebo, which occurred between clomipramine and placebo just after the tenth week (between weeks 10 and 12). The authors concluded that although both drugs are effective for PD, the paroxetine effect could be initiated prior to that promoted by clomipramine. In addition, clomipramine group showed more gastrointestinal and central nervous system side effects, as agitation, emotional liability, anxiety and depression than the paroxetine group. Paroxetine was not significantly different from placebo concerning side effects.

Pollack [25] compared paroxetine, venlafaxine and placebo in 653 patients. Doses were fixed as 40 mg/day for paroxetine and 75 or 225 mg/day for venlafaxine and patients were followed for 12 weeks. At the end of the study, 58.3 % of patients taking paroxetine were free of PA, 64.7 % of patients taking venlafaxine 75 mg/day, 70 % of those taking venlafaxine 225 mg /day and 47.8 % of patients receiving placebo were panic-free. All active drugs showed statistically significant difference compared to placebo. Venlafaxine 225 mg/day was also significantly differentiated from paroxetine. These results suggest that higher doses of venlafaxine can lead to an additional efficacy. Both active drugs were well tolerated.

Sheehan [26] evaluated the efficacy of paroxetine CR versus placebo in 889 patients for 10 weeks. The dosage varied between 25 and 75 mg/day. In the study's tenth week, 73 % and 60 % of patients using placebo and paroxetine were respectively free of PA.

Three placebo-controlled studies with sertraline were performed. Londborg [27] used fixed doses of 50, 100 and 200 mg/day in 177 patients for 12 weeks. All three doses of the active drug were significantly superior to placebo, but not between them, when compared one to each other. Pollack [28] and Pohl [29] evaluated sertraline doses between 50 and 200 mg/day for 10 weeks, in 176 and 166 patients, respectively. Both studies showed significantly higher decrease in PA frequency than that observed in the placebo group.

Citalopram and escitalopram have also been tested and compared for use in PD. Stahl [30] compared the effect of escitalopram (10–20 mg/day) and citalopram (20–40 mg/day) versus placebo in 351 patients, for 10 weeks. The primary endpoint was the reduction in PA frequency (estimated by Panic and Agoraphobia Scale – PAS). Only escitalopram was statistically different from placebo. Side effects of both drugs were similar to placebo.

Fluvoxamine has demonstrated efficacy for the treatment of PD, as the others SSRI. Asnis [31] observed that fluvoxamine was superior to placebo. The used doses were 100–300 mg/day of fluvoxamine for 8 weeks in 188 patients. At the end of the study 69 % of subjects using the active drug was free of PA, which occurred in 45.7 % of those using placebo. The difference between placebo and active treatment was statistically significant. In this study, the need for large samples to demonstrate the superiority of active drug versus placebo in PD is discussed, since patients with this disorder usually show significant placebo response. Another clinical trial with fluvoxamine was performed by Nair et al. [32] and failed to show its efficacy when compared to placebo and imipramine. This result may have been due to the limited sample size in this study.

Some comparative studies without placebo group with PD patients have also been performed. Most of the comparative studies involving SSRI show similar efficacy for both drugs, which were: sertraline versus paroxetine [33], sertraline versus imipramine [34, 35], fluoxetine versus clomipramine [36], fluoxetine versus mirtazapine [37], paroxetine versus citalopram [38] and fluvoxamine versus inositol [39].

15.4 Side Effects

Patients with PD appear to be more sensitive to physical side effects of antidepressants [40]. They usually misinterpret side effects as anxiety symptoms, which can trigger vicious circle of escalating anxiety that can lead to panic attacks. For this reason, it is especially important that pharmacotherapy for panic disorder is well tolerated [40]. Many studies have showed SSRI are better tolerated than TCAs and serotonin and noradrenaline reuptake inhibitors (SNRI), due to a more favorable adverse effects profile, but also due to fewer drug interactions [40, 41].

The most common reported side effects of SSRI are dry mouth, constipation, diarrhea, anorexia, drowsiness, dizziness, lethargy, sleep disturbance, headache, tremor, anxiety, sweating, nausea and vomiting, asthenia, sexual side effects, heart rhythm disorders, abdominal pain and weight gain [40]. A recent article has investigated the influence of adverse effects on the dropout of SSRI treatment with data of 50,824 patients [42]. The adverse effects mentioned most frequently were: "discomfort" of the digestive system (10%), sleep disorders (8.6%), and heart rhythm disorders (4%), but they were of tolerable severity, as they did not significantly influence the dropout rate, whereas the occurrence of somnolence leads to discontinuation [42].

Comparing the incidences of side effects in short and long-terms studies, it has been suggested that certain adverse effects, as nausea and dry mouth, become less frequent with long-term treatment and the emergence of new side effects during long-term treatment is unusual [40]. The reduced incidence of collateral effects during long-term studies may be due to the remission of physical anxiety symptoms, which may be misinterpreted as side effects [40].

Concerning sexual adverse effects, its exact prevalence is unknown, ranging from very low percentages to more than 80% [43]. William et al. (2006) conducted a cross-sectional survey in 502 adults in France and United Kingdom and estimated a prevalence of 39.2% in the United Kingdom and 26.6% in France of SSRI/SNRI-induced sexual dysfunction [43]. Another interesting information is that most of studies have found no significant differences of overall sexual dysfunction rates between SSRI [43]. However, some trials have shown differences between the SSRI when consider analysis of phase-specific sexual functioning [43]. Erectile dysfunction, anorgasmia and vaginal lubrification seem to be more common in patients on paroxetine [43]. Montejo et al. [84] analyzed a population of 1022 patients and observed the following incidence-rates of sexual side effects: citalopram 72.7%, paroxetine 70.7%, sertraline 62.9%, fluvoxamine 62.3% and fluoxetine 57.7% (p<0.005). Sexual side effects also appear to differ between men and women [43]. Men experience sexual dys-function more often (specially impairment of desire and orgasm phase), whereas women experience sexual side effects of greater severity [43]. About tolerance, a study has shown 24.5% of patients showed a good tolerance of sexual dysfunction, 42.5% were discontent although he/she did not intend to discontinue the treatment for this reason, while 32.9% were very concerned about sexual side effects and considered to discontinue the treatment [43].

There is little data on long-term effects on SSRI on body weight [43]. During acute treatment of a randomized, double blind, placebo-controlled essay with fluoxetine, especially in the first 4 weeks, patients had a weight loss compared to placebo [43]. During continuation treatment, a significant weight gain was observed – more discrete than placebo in week 26, but without significant difference with placebo in weeks 39 and 50. This weight gain was attributed to poor appetite at study entry and to improvement in appetite after recovery [43]. Fava et al. [83] performed a randomized, double blind trial with fluoxetine, sertraline and paroxetine for 26–32 weeks [43]. The significant result refers to increase of weight in paroxetine group and a higher proportion of patients with >7% weight gained compared to fluoxetine and sertraline [43]. Similar results were seen in another long term, naturalistic and prospective study with five SSRI and clomipramine [43, 44]. Patients had gained a mean of 2.5% of their initial body mass (p<0.001) and, in 14% of them, weight increased by more than 7% [43, 44]. The percentages of patients with a >7% weight increase for each drug were as follows: 4.5% for sertraline, 8.7% for fluoxetine, 14.3% for citalopram 10.7% for fluvoxamine, 14.3% for paroxetine and 34.8% for clomipramine, although the difference was not statistically significant [43, 44].

About mania episodes induced by SSRI, many cases have been reported [45, 46]. Although, it is unclear if these patients have an "endogenous" bipolar disorder or if it is, in fact, a genuine SSRI side effect [45–47].

It cannot be assumed that adolescents have the same side effects of adults, since they have developmentally different pharmacodynamics and pharmacokinetics to adults [48]. The most common non-psychiatric side effects in adolescents are headache, nausea, vomiting, abdominal pain, dry mouth and discontinuation syndrome, while the most common of psychiatric adverse effects are insomnia, sedation, suicidal ideas and behaviors, hypomania or mania, akathisia, agitation, increased of anxiety, irritability, hypersensitivity, anger, worsening of depression, tremor and crying [48]. Despite of these side effects, discontinuation rates related to adverse effects range of 5–10 % [48]. There is a lack of research on reduction in expected growth and SSRI in adolescents [9]. Therefore, monitoring of growth in adolescents using SSRI should be encouraged [9]. An important issue about SSRI side effects in adolescents is associated to increase of suicidal ideas [48–50]. Differently from studies in adults, randomized and controlled trials indicate that adolescents taking SSRI are at small, but increased, risk of suicidal thoughts and behaviors, at least in short-term [48–50]. Paroxetine seems to be related to greater increased risk of suicide among SSRI, as well as venlafaxine, a SNRI [50]. These data warn for a closer monitoring of adolescents using SSRI, especially in the first weeks of use [49].

SSRI have been found to be well tolerated in the elderly compared with TCAs and SNRI [41, 51]. Amongst SSRI, fluvoxamine has been associated with highest discontinuation rate, followed by fluoxetine and sertraline [41]. The best tolerated appear to be escitalopram and citalopram [41]. Most common SSRI-induced side effects in elderly are very similar to younger adults and include nausea, dry mouth, constipation, diarrhea, anorexia, drowsiness, dizziness, lethargy, sleep disturbance, tremor and anxiety with their respective prevalence ranging from 1 % to 17 % [41, 51]. Nausea and vomiting were indicated as the commonest side effects in meta-analyses of 2004 with 17 % of prevalence rate [41]. Restlessness, sedation (especially during the first few days) and bradycardia are more prevalent in older patients, while gastrointestinal adverse effects, sweating and headache were less likely to be present in these patients [41, 51]. Some side effects seem to have greater impact in the elderly population, such as hyponatremia, bradycardia, gastrointestinal bleedings and falls, bone loss and fractures [41, 51]. Specifically about bradycardia, the US Food and Drug Administration (FDA)

recommends a maximum dose of 20 mg citalopram in elderly, although some studies state there is no clear evidence for this recommendation [41].

15.5　Drug Interactions

SSRI have been suggested as first choice for treatment of PD and are among the most frequently prescribed medications [52]. Besides, they are commonly prescribed in combination with other drugs used to treat co-morbid psychiatric or somatic disorders [53, 54]. For these reasons, drug interactions may occur and have become a challenge in clinical practice [52].

Drug interactions are classified as either pharmacodynamic (when target organs or receptor sites are involved) or pharmacokinetic (when absorption, distribution, metabolism or excretion is affected) [52, 54]. Because of a more selective mechanism of action and receptor profile, SSRI carry a relatively low risk for pharmacodynamic interactions when compared to MAOIs and tricyclic antidepressants (TCAs) [52–54]. However, SSRI are susceptible to pharmacokinetic drug interactions since they are metabolized in the liver by cytochrome P450 (CYP) isoenzymes, which are responsible for the oxidation of most drugs, environmental toxins and endogenous substrates [52–55]. The major CYP enzymes that play a role in drug metabolism are CYP1A2, CYP2B6, CYP2C9, CYP2C19, CYP2D6, and CYP3A4 [54, 55]. Drug interactions mediated by CYP enzyme system can produce enzyme inhibition or enzyme induction [52–55]. Despite this potential pharmacokinetic interaction of SSRI, clinically relevant consequences are not always observed because they also depend on other drug factors as therapeutic window, potency and concentration of the inhibitor/inducer, the contribution of the affected enzyme to overall drug elimination and presence of pharmacologically active metabolites, as well as on patient-related factors as genetic predisposition, ethnicity, age and co-morbidity [53–55].

SSRI exhibit drug interactions with impact on clinical practice through the ability to inhibit specific CYP enzymes [52–54]. Nevertheless, the inhibition profile is not the same between the

Table 15.1 Potential for interaction between Selective Serotonin reuptake inhibitors and Citochrome P450 isoenzymes

Drug	CYP enzyme inhibition				
	2D6	1A2	3A4	2C9	2C19
Citalopram	+	0	0	0	0
Escitalopram	+	0	0	0	0
Fluoxetine	+++	+	++	+++	++
Fluvoxamine	+	+++	++	++	+++
Paroxetine	+++	+	+	+	+
Sertraline	+	0	0	0	0

0=negligible; +=weak; ++=moderate; +++=strong
Adapted from Shellander and Donnerer [1]

different SSRI [52–55]. Fluoxetine and its metabolite norfluoxetine are potent inhibitors of CYP2D6, are moderate inhibitors of CYP2C9 and are mild inhibitors of CYP2C19 and CYP3A4 [52–54]. Fluvoxamine inhibits strongly CYP1A2 and CYP2C19, and it moderately inhibits CYP2C9 and CYP3A4 [52–54]. Paroxetine demonstrates a great capacity to inhibit CYP2D6, whereas sertraline inhibits this latter isoform in a dose-dependent manner [52–54]. Still, citalopram and escitalopram are weak inhibitors of CYP2D6 and have negligible effects on CYP1A2, CYP2C19, CYP2C9 and CYP3A4 [52–54]. Both citalopram and escitalopram have a very favorable drug interaction profile, but it has been suggested that escitalopram has an even more favorable profile than citalopram because of a weaker inhibition of CYP2D6 [1–3] (see Tables 15.1 and 15.2).

15.5.1 Serotonin Syndrome

The combination of SSRI with other serotonergic drugs as MAOIs, some TCAs, SNRI, buspirone, trazodone, *Hypericum* extracts, analgesics (tramadol, meperidine, fentanyl, oxycodone), drugs of abuse and linezolid may produce a potentially life-threatening pharmacodynamic interaction, the serotonin syndrome [52, 53]. The receptors involved in serotonin syndrome are $5HT_{1A}$ and $5HT_2$ based on animal models [52, 53, 56]. Serotonin syndrome symptoms include disturbed mental status, confusion, agitation, fever, hypertension, diaphoresis, diarrhea, tremor, hyperreflexia and myoclonus [52].

In 2006, the FDA released an alert to warn about the risks of combination of SSRI and triptans with serotonin syndrome based on 29 case reports gathered over a 5-year period [52, 53, 56]. Indeed, these drugs have serotoninergic effects and there are pharmacokinetic studies in healthy subjects that report around 20 % increase in triptan plasma concentration with concomitant administration of some SSRI [53]. These two classes are also frequently given simultaneously because of the high prevalence of depressive and anxiety disorders and migraine and *vice-versa* [56]. Gillman and Psych performed a review about the risk of triptans and SSRI cause serotonin syndrome [56]. They found that the validity of these case reports was questioned since most of them did not meet the diagnosis criteria for serotonin syndrome [53, 56]. Another issue raised was that triptans are agonists of $5HT_{1B}$, $5HT_{1D}$ and $5HT_{1F}$ receptors and serotonin syndrome seems to be mediated by $5HT_{1A}$ and $5HT_2$ receptors [53, 56]. Therefore, they concluded there is no clinical evidence or theoretical reason to maintain speculation about serious serotonin syndrome from triptans and SSRI [56]. Other experts have also suggested that there is no sufficient evidence to ensure that the use of triptans and SSRI together increase the risk of serotonin syndrome, but they advise caution [53, 57]. In 2010, the American Headache Society recommended that the FDA organize a review of the available data about triptans, SSRI and serotonin syndrome in order to reassess whether the warning should be maintained [53, 58].

15.5.2 Anticancer Drugs

Anticancer drugs are metabolized in the liver by CYP isoenzymes [55]. Studies *in vitro* have indicated that anticancer drugs interfere with CYP system not only as substrates, but also as inhibitors or inducers [55]. Unlike newer antidepressants, anticancer drugs usually have a narrow therapeutic index, what can result in consequences in clinical practice as excessive toxicity or reduced efficacy. Another important consideration is that patients with cancer can have underlying hepatic and renal impairments, which may lead into a reduced rate of drug elimination [55].

Table 15.2 Summary of SSRI-induced clinically relevant pharmacokinetic drug-drug interactions

SSRI	Drug with pharmacokinetic interaction with SSRI produces clinically relevant effects	Proposed mechanism
Citalopram	Desipramine	Inhibition of CYP2D6
Escitalopram	Desipramine	Inhibition of CYP2D6
Fluoxetine	TCAs Risperidone Clozapine Warfarin Propranolol and metoprolol Nifedipine and verapamil Tamoxifen	Inhibition of CYP2D6-mediated hydroxylation of TCAs; I Inhibition of CYP2D6 and, to a lesser extent, CYP3A4; Inhibition of CYP2D6, CYP2C19, CYP3A4; Inhibition of CYP2C9; Inhibition of CYP2D6; Inhibition of CYP3A4; Inhibition of CYP2D6
Fluvoxamine	TCAs (amitriptyline, imipramine, clomipramine) Clozapine Olanzapine Quetiapine Thoephyline Warfarin, Propranolol	Inhibition of CYP2C19 and, to a lesser extent, CYP1A2 and CYP3A4; Inhibition of CYP1A2 and, to a lesser extent, CYP2C19 and CYP3A4; Inhibition of CYP1A2; Inhibition of CYP3A4; Inhibition of CYP1A2; Inhibition of CYP2C9; Inhibition of CYP1A2 and CYP2C19
Paroxetine	Desipramine Perphenazine Clozapine Risperidone Atomoxetine Tamoxifen Tramadol	Inhibition of CYP2D6; Inhibition of CYP2D6; Inhibition of CYP2D6; Inhibition of CYP2D6; Inhibition of CYP2D6; Inhibition of CYP2D6; Inhibition of CYP2D6
Sertraline	Risperidone Lamotrigine	Inhibition of CYP2D6; Inhibition of glucuronidation

Adapted from Spina et al. [2]

SSRI are currently used in cancer patients to treat depressive and anxiety symptoms. SSRI can inhibit cytochrome P450 isoenzymes as mentioned above [52, 53, 55, 59]. A clinical relevant interaction between SSRI and anticancer drugs refers to tamoxifen, a selective estrogen receptor modulator used in the treatment and prophylaxis of breast cancer [55, 59]. Tamoxifen is converted into endoxifen by CYP2D6 in the liver [4, 8]. As Table 15.1 shows, fluoxetine and paroxetine are potent inhibitors of CYP2D6 and administration of them can reduce the benefits of tamoxifen and increases the risk of death [55, 59].

Borges et al. [85] published an article of a prospective trial of 158 patients with breast cancer taking tamoxifen with concomitant use of potent inhibitors of CYP2D6, such as fluoxetine and paroxetine, and demonstrated significantly decreased mean plasma concentrations of endoxifen. Simultaneous treatment with weak inhibitors of CYP2D6, such as sertraline (in low doses) and citalopram, resulted in a discrete reduction of plasma concentrations of endoxifen and co-administration of venlafaxine did not appear to alter endoxifen concentrations [55].

In 2009, Desmarais et al. reviewed seven clinical studies of women with breast cancer taking antidepressants and tamoxifen [60]. They concluded there is consistent evidence that paroxetine and fluoxetine can interfere on the metabolism of tamoxifen and should be avoided. Bupropion is also a strong inhibitor of CYP2D6 and Desmarais et al. stated there is indirect evidence that indicates bupropion may also have a large effect on the metabolism of tamoxifen (see Table 15.3) [55, 60, 61].

Table 15.3 SSRI and venlafaxine and their interaction with tamoxifen by CYP2D6

Drug	Effect on CYP2D6	Advice on the co-administration with tamoxifen
Citalopram	Mild	Consider use based on benefit-risk assessment (secondary choice)
Escitalopram	Mild	Consider use based on benefit-risk assessment (secondary choice)
Fluoxetine	Strong	Best to avoid
Fluvoxamine	Moderate	Consider use based on benefit-risk assessment
Paroxetine	Strong	Best to avoid
Sertraline	Moderate	Consider use based on benefit-risk assessment (best to avoid in high doses)
Venlafaxine	Minimal	Safest choice

Adapted from Harv Ment Health Lett [8], Desmarais and Looper [10]
Data also taken from Caraci et al. [4]

Kelly et al. conducted a population-based cohort with 2430 women treated with tamoxifen and a single SSRI and concluded that women with breast cancer who received simultaneously tamoxifen and paroxetine were at increase risk for death from breast cancer and death from any cause [55, 62]. Increases of 25 %, 50 %, and 75 % in the proportion of time on tamoxifen with overlapping use of paroxetine were associated with 24 %, 54 %, and 91 % increases in the risk of death from breast cancer, respectively [55, 62]. No relationship was found between the administration of another SSRI with tamoxifen and the increased breast cancer mortality [55, 62].

A recent review suggests caution in prescribing SSRI, especially paroxetine, and indicates the necessity of new, larger, randomized and controlled trials to better understand the use of SSRI with tamoxifen in women with breast cancer [63].

15.5.3 Risk of Bleeding

SSRI have been related to increase bleeding risk, especially of upper gastrointestinal bleeding, because they occasion serotonin blockade into platelets, reducing platelet aggregation and prolonging bleeding time. This risk seems to increase when other medications are taken simultaneously, such as nonsteroidal anti-inflammatory drugs (NSAIDs), oral anti-coagulants and anti-platelet drugs (including low-dose aspirin) as demonstrated in observational studies [52, 53, 64, 65].

With respect to the NSAIDs, studies demonstrated that the use combined of SSRI and NSAIDs increases 3- to 15-fold the risk of gastrointestinal bleeding, and one study showed no significant difference in risk of bleeding between the association of these two classes of medications and when they are not associated [53]. Recently, Anglin et al. conducted a review with case control and cohort studies about gastrointestinal bleeding and administration of SSRI [66]. They concluded SSRI are associated with a modest increase in the risk of upper gastrointestinal bleeding, which is lower than has previously been estimated, but this risk is significantly elevated when SSRI are used in combination with NSAIDs [66]. The authors suggest physicians should have caution to prescribe these drugs together and discuss this risk with patients [15].

Bak et al. [86], in a case control study, found a trend towards increased risk of hemorrhagic stroke when SSRI and NSAIDs were associated. However, in general, studies on intracerebral bleeding and SSRI-NSAID interaction are not yet conclusive [53].

Regarding the co-administration of SSRI and warfarin, the interaction between these two classes of drugs have been suggested to be related to the increased risk of bleeding due to prolongation of bleeding time (especially for agents with the highest degree of inhibition of serotonin reuptake as sertraline, fluoxetine and paroxetine), but also due to pharmacokinetic interaction through inhibition of CYP2C9 [53]. In the latter case, drugs like fluoxetine and fluvoxamine could interact more, as they are stronger inhibitors of CYP2C9 [53].

Some studies, mainly the earlier ones, failed to demonstrate a relationship between the association of warfarin and SSRI with increased bleeding, as Kharofa et al. noted no increase in hemorrhagic stroke risk with concomitant use of SSRI and warfarin [52, 67, 68]. However, other observational studies have shown this relationship. This is the case of Wallerstedt et al. that observed increased risk of non-gastrointestinal bleeding when warfarin and SSRI were used simultaneously [69], and Cochran et al. that verified relationship between the co-administration of these drugs with increased risk of any bleeding event [70]. More recently, Löppönen et al. concluded concurrent use of warfarin and the SSRI, relative to warfarin alone, seemed to increase the case fatality rate for primary intracerebral hemorrhage [67].

In 2012, an epidemiological meta-analysis suggested SSRI increase the risk of intracranial hemorrhage. According to the authors, the risk of bleeding rises when oral anticoagulants are combined with SSRI as compared to the risk when oral anticoagulants are used alone. Nevertheless, the authors believe that the absolute risk is low because it is a rare event [71].

Thus, physicians should be alert to the association of SSRI with nonsteroidal anti-inflammatory drugs (NSAIDs), oral anti-coagulants and anti-platelet drugs, especially in elderly patients. They should seek strategies as replacing SSRI, prescribing proton pump inhibitors for gastroprotection, using NSAIDs with the lowest gastrointestinal toxicity (for example: ibuprofen and cyclooxygenase 2 inhibitors), adjusting dosage of warfarin/aspirin and testing of blood clotting [52].

15.6   Clinical Management

SSRI may have a transient stimulating effect and, therefore, worsen anxiety, tremor and restlessness early in treatment, leading sometimes to an increased number of panic attacks. For this reason, the initial dose of SSRI in PD patients should be lower than the usual therapeutic dose. The association with benzodiazepines in this phase of treatment may be necessary [72]. Usual initial and maintenance doses are described in Table 15.4.

It should be emphasized to the patient that treatment response is not immediate and that prolonged courses are needed to maintain an initial treatment response [6].

Table 15.4 Usual initial and maintenance dosages for the treatment of PD

Medication	Initial dose (mg/day)	Maintenance dose (mg/day)
Fluoxetine	5–10	20–60
Paroxetine	10	20–60
Sertraline	25–50	75–150
Citalopram	10	20–60
Escitalopram	5–10	10–20
Fluvoxamine	50	50–300

The selection of a specific drug within the SSRI's class should be determined by the adverse effects and drug interaction profiles and by whether the patient has previous experience of treatment with that compound [6, 7]. For example, choose fluoxetine if the patient needs to loose wait, paroxetine if the he presents insomnia or lack of appetite and citalopram or escitalopram if the patient is polymedicated.

It is recommended to wait at least 8 weeks after the treatment initiation to evaluate its efficacy since a complete response may not happen before 8–12 weeks of treatment [6, 7, 72]. Consider increasing the dose if there is insufficient response, but the clinician must have in mind that the evidence for a dose response relationship with SSRI is inconsistent [6]. If initial treatment fails, consider switching to another evidence-based treatment such as other SSRI, an SNRI or a benzodiazepine. If it is still not efficient, combine evidence-based drugs or pharmacological and psychological treatments [6, 7].

The discontinuation syndrome may occur with abrupt discontinuation of SSRI, especially the ones with shorter half-life (paroxetine, sertraline and fluvoxamine) [73, 74]. Commonly, this syndrome presents with flu-like symptoms such as malaise, nausea and headache for 2–7 days after drug interruption. Dizziness, paraesthesia, agitation and mood changes are also reported [74]. Among the SSRI, fluoxetine is the one with a longer half-life and therefore show a lower risk of discontinuation syndrome [74]. However, to interrupt a short half-life SSRI intake, it is advisable to conduct a gradual reduction over several weeks, as 25 % each week [73]. In the presence of withdrawal symptoms, drug dose should be increased again to the usual dosage and then restart the discontinuation in a slower way. It is important to differentiate discontinuation symptoms from disorder relapse symptoms [6, 73]. For general recommendations for the treatment with SSRI for PD patients, see Box 15.1.

15.7 Treatment Duration

The discontinuation of pharmacotherapy in PD has been reported as a challenge, given many patients present relapse with drug withdrawal [2]. This information can indicate medication should be maintained for a longer period [2]. At the same time, studies suggest that more than half patients with PD discontinue treatment within several months to years [2].

Therefore, the optimal duration of treatment would be the one that allow patients to discontinue pharmacotherapy relatively safely and allow not taking medication longer than necessary [2]. Unfortunately, research on optimal duration of drug therapy, including SSRI, is scarce and the limited data on this issue do

---

Box 15.1. General Recommendations for the Treatment with SSRI in PD Patients

Initial side effects can be minimized by slowly increasing the dose or by adding a benzodiazepine for a few weeks.

Advise the patient that treatment periods of up to 12 weeks may be needed to assess efficacy.

Drug choice is determined by their side effects and drug interaction profiles, as well as patient's previous experience.

Wait at least 8 weeks before increasing the dose or changing the drug.

Discontinuation syndrome may be minimized by gradual drug discontinuation (25 % a weak).

---

not indicate a safe period to withdraw medication [2]. The American Psychiatric Association refrains from recommendations [2, 72]. Most guidelines refer to expert consensus and suggest maintaining medication for at least a year [2, 72, 75, 76], but other guidelines have also suggested a shorter period [2, 6, 77]. Long-term followed-up studies with citalopram, fluoxetine, sertraline, paroxetine, venlafaxine XR and moclobemide have shown maintained benefits and continued improvements over 6–12 months of ongoing treatment [6, 77]. A naturalistic study by Nardi et al. [78] showed that the efficacy of paroxetine and clonazepam for PD was maintained for 34 months of follow-up.

However, when medication is being discontinued, it is suggested focusing on the use of a slow taper over weeks to months in order to avoid withdrawal symptoms and relapse [2, 79]. Another strategy to enhance long-term outcome of treatment is providing psychotherapy to PD patients [2, 80]. The literature indicates that either cognitive behavioral therapy or brief psychodynamic psychotherapy may reduce relapse rates in panic disorder [2, 81].

An important field of research relates to the predictor factors for relapse. From the moment it is possible to identify those at highest risk of relapse, it would be easier to propose maintenance treatment with better adherence, given the high costs of recurrence of PD [2].

15.8 Conclusion

The SSRI are well established as first choice drugs for the treatment of PD reducing PA and improving anticipatory anxiety, as well as depressive symptoms. They are overall well tolerated but unpleasant long-term side effects as sexual dysfunction and weight gain are commonly reported and still not widely studied. Comparing to older antidepressants, the SSRI are safer and better tolerated, showing also less drug interactions. Patient should be informed about the latency period and initial side effects and drug discontinuation should be gradual. More long-term studies are necessary to determinate the optimal treatment duration.

References

1. Andrisano C, Chiesa A, Serretti A. Newer antidepressants and panic disorder: a meta-analysis. Int Clin Psychopharmacol. 2013;28:33–45.
2. Batelaan NM, Van Balkom AJLM, Stein DJ. Evidence-based pharmacotherapy of panic disorder: an update. Int J Neuropsychopharmacol. 2012;15:403–15.
3. Schmidt NB, Keogh ME. Treatment of panic. Annu Rev Clin Psychol. 2010;6:241–56.
4. Furukawa TA, Watanabe N, Churchill R. Combined psychotherapy plus antidepressants for panic disorder with or without agoraphobia. Cochrane Database Syst Rev. 2007;1:CD004364.
5. Watanabe N, Churchill R, Furukawa TA. Combination of psychotherapy and benzodiazepines versus either therapy alone for panic disorder: a systematic review. BMC Psychiatry. 2007;7:18.
6. Baldwin DS, Anderson IM, Nutt DJ, Allgulander C, Bandelow B, den Boer JA, et al. Evidence-based pharmacological treatment of anxiety disorders, post-traumatic stress disorder and obsessive-compulsive disorder: a revision of the 2005 guidelines from the British Association for Psychopharmacology. J Psychopharmacol. 2014;28(5):403–39.
7. Katzman MA, Bleau P, Blier P, Chokka P, Kjernisted K, Van Ameringen M, et al. Canadian clinical practice guidelines for the management of anxiety, posttraumatic stress and obsessive-compulsive disorders. BMC Psychiatry. 2014;14 Suppl 1:S1.
8. Bakker A, van Balkom AJ, Spinhoven P. SSRI vs. TCAs in the treatment of panic disorder: a meta-analysis. Acta Psychiatr Scand. 2002;106(3):163–7.
9. Stahl SM. Mechanism of action of serotonin selective reuptake inhibitors: serotonin receptors and pathways mediate therapeutic effects and side effects. J Affect Disord. 1998;51:215–35.
10. Maron E, Shlik J. Serotonin function in panic disorder: important, but why? Neuropsychopharmacology. 2006;31:1–11.
11. Kennett GA, Lightowler S, de Biasi V, et al. Effect of chronic administration of selective 5-hydroxytryptamine and noradrenaline uptake inhibitors on a putative index of 5-HT2C/2B receptor function. Neuropharmacology. 1994;33(12):1581–8.
12. Yamauchi M, Tatebayashi T, Nagase K, Kojima M, Imanishi T. Chronic treatment with fluvoxamine desensitizes 5-HT2C receptor-mediated hypolocomotion in rats. Pharmacol Biochem Behav. 2004;78(4):683–9.
13. Gorman JM, Kent JM, Sullivan GM, Coplan JM. Neuroanatomical hypothesis of panic disorder, revised. Am J Psychiatry. 2000;157(4):493–505.
14. Lai CH, Wu YT, Yu PL, Yuan W. Improvements in white matter micro-structural integrity of right uncinate fasciculus and left fronto-occipital fasciculus of remitted first-episode medication-naïve panic disorder patients. J Affect Disord. 2013;150(2):330–6.
15. Lai CH, Wu YT. Changes in gray matter volume of remitted first-episode, drug-naïve, panic disorder patients after 6-week antidepressant therapy. J Psychiatr Res. 2013;47(1):122–7.
16. Miller HE, Deakin JF, Anderson IM. Effect of acute tryptophan depletion on CO2-induced anxiety in patients with panic disorder and normal volunteers. Br J Psychiatry. 2000;176:182–8.
17. Bertani A, Perna G, Arancio C, Caldirola D, Bellodi L. Pharmacologic effect of imipramine, paroxetine, and sertraline on 35% carbon dioxide hypersensitivity in panic patients: a double-blind, random, placebo-controlled study. J Clin Psychopharmacol. 1997;17:97–101.
18. Bell C, Forshall S, Adrover M, Nash J, Hood S, Argyropoulos S, et al. Does 5-HT restrain panic? A tryptophan depletion study in panic disorder patients recovered on paroxetine. J Psychopharmacol. 2002;16:5–14.
19. Gorman JM, Liebowitz MR, Fyer AJ, Goetz D, Campeas RB, Fyer MR, et al. An open trial of fluoxetine in the treatment of panic attacks. J Clin Psychopharmacol. 1987;7(5):329–32.
20. Schneier FR, Liebowitz MR, Davies SO, Fairbanks J, Hollander E, Campeas R, et al. Fluoxetine in panic disorder. J Clin Psychopharmacol. 1991;10(2):119–21.
21. Michelson D, Lydiard RB, Pollack MH, et al. Outcome assessment and clinical improvement in panic disorder: evidence from a randomized controlled trial of fluoxetine and placebo. Am J Psychiatry. 1998;155:1570–7.
22. Michelson D, Allgulander C, Dantendorfer K, Knezevic A, Maierhofer D, Micev V, et al. Efficacy of usual antidepressant dosing regimens of fluoxetine in panic disorder: randomised, placebo-controlled trial. Br J Psychiatry. 2001;179:514–8.
23. Ballenger JC, Wheadon DE, Steiner M, Bushnell W, Gergel IP. Double-blind, fixed-dose, placebo-controlled study of paroxetine in the treatment of panic disorder. Am J Psychiatry. 1998;155:36–42.
24. Lecrubier Y, Bakker A, Dunbar G, Judge R; the Collaborative Paroxetine Panic Study Investigators. A comparison of paroxetine, clomipramine and placebo in the treatment of panic disorder. Acta Psychiatr Scand. 1997; 95: 145–52.
25. Pollack M, Mangano R, Entsuah R, Tzanis E, Simon NM. A randomized controlled trial of venlafaxine ER and paroxetine in the treatment of outpatients with panic disorder. Psychopharmacology (Berl). 2007; 194:233–42.
26. Sheehan DV, Burnham DB, Iyengar MK, Perera P. Efficacy and tolerability of controlled-release paroxetine in the treatment of panic disorder. J Clin Psychiatry. 2005;66:34–40.
27. Londborg PD, Wolkow R, Smith WT, et al. Sertraline in the treatment of panic disorder: a multi-site, double-blind, placebo-controlled, fixed-dose investigation. Br J Psychiatry. 1998;173(7):54–60.
28. Pollack MH, Otto MW, Worthington JJ, Manfro GG, Wolkow R. Sertraline in the treatment of panic disorder: a flexible-dose multicenter trial. Arch Gen Psychiatry. 1998;55:1010–6.

29. Pohl RB, Wolkow RM, Clary CM. Sertraline in the treatment of panic disorder: a double-blind multicenter trial. Am J Psychiatry. 1998;155:1189–95.

30. Stahl SM, Gergel I, Li D. Escitalopram in the treatment of panic disorder: a double blind, randomized, placebo-controlled trial. J Clin Psychiatry. 2003;64:1322–7.

31. Nair NPV, Bakish D, Saxena B, Amin M, Schwartz G, West TEG. Comparison of fluvoxamine, imipramine and placebo in the treatment of outpatients with panic disorder. Anxiety. 1996;2:192–8.

32. Asnis GM, Hameedia FA, Goddardb AW, Potkin SG, Black D, Jameel M, et al. Fluvoxamine in the treatment of panic disorder: a multi-center, double-blind, placebo-controlled study in outpatients. Psychiatry Res. 2001;103:1–14.

33. Bandelow B, Behnke K, Lenoir S, Hendriks GJ, Alkin T, Goebel C, et al. Sertraline versus paroxetine in the treatment of panic disorder: an acute, double-blind noninferiority comparison. J Clin Psychiatry. 2004;65:405–13.

34. Lepola U, Arato M, Zhu Y, Austin C. Sertraline versus imipramine treatment of comorbid panic disorder and major depressive disorder. J Clin Psychiatry. 2003;64:654–62.

35. Mavissakalian MR. Imipramine vs. sertraline in panic disorder: 24-week treatment completers. Ann Clin Psychiatry. 2003;15:171–80.

36. Cavaljuga S, Licanin I, Kapic E, Potkonjak D. Clomipramine and fluoxetine effects in the treatment of panic disorder. Bosn J Basic Med Sci. 2003;3:27–31.

37. Ribeiro L, Busnello JV, Kauer-Sant'Anna M, et al. Mirtazapine versus fluoxetine in the treatment of panic disorder. Braz J Med Biol Res. 2001;34:1303–7.

38. Perna G, Bertani A, Caldirola D, et al. A comparison of citalopram and paroxetine in the treatment of panic disorder: a randomized, single-blind study. Pharmacopsychiatry. 2001;34:85 90.

39. Palatnik A, Frolov K, Fux M, Benjamin J. Double-blind, controlled, crossover trial of inositol versus fluvoxamine for the treatment of panic disorder. J Clin Psychopharmacol. 2001;21:335–9.

40. Baldwin DS, Birtwistle J. The side effect burden associated with drug treatment of panic disorder. J Clin Psychiatry. 1998;59 Suppl 8:39–44. discussion 45–6.

41. Topiwala A, Chouliaras L, Ebmeier KP. Prescribing selective serotonin reuptake inhibitors in older age. Maturitas. 2014;77(2):118–23.

42. Kostev K, Rex J, Eith T, Heilmaier C. Which adverse effects influence the dropout rate in selective serotonin reuptake inhibitor (SSRI) treatment? Results for 50,824 patients. Ger Med Sci. 2014;12:Doc15.

43. Demyttenaere K, Jaspers L. Review: Bupropion and SSRI-induced side effects. J Psychopharmacol. 2008;22(7):792–804.

44. Maina G, Albert U, Salvi V, Bogetto F. Weight gain during long-term treatment of obsessive-compulsive disorder: a prospective comparison between serotonin reuptake inhibitors. J Clin Psychiatry. 2004;65(10):1365–71.

45. Kimmel RJ, Seibert J. Is antidepressant-associated mania always an evidence of a bipolar spectrum disorder? A case report and review of the literature. Gen Hosp Psychiatry. 2013;35(5):577.

46. Mendhekar DN, Gupta D, Girotra V. Sertraline-induced hypomania: a genuine side-effect. Acta Psychiatr Scand. 2003;108(1):70–4.

47. Akiskal HS, Hantouche EG, Allilaire JF, Sechter D, Bourgeois ML, Azorin JM, et al. Validating antidepressant-associated hypo- mania (bipolar III): a systematic comparison with spontaneous hypomania (bipolar II). J Affect Disord. 2003;73(1–2):65–74.

48. Gordon M, Melvin G. Selective serotonin re-uptake inhibitors—a review of the side effects in adolescents. Aust Fam Physician. 2013;42(9):620–3.

49. Mahendran R. The risk of suicidality with selective serotonin reuptake inhibitors. Ann Acad Med Singapore. 2006;35(2):96–9.

50. Barbui C, Esposito E, Cipriani A. Selective serotonin reuptake inhibitors and risk of suicide: a systematic review of observational studies. CMAJ. 2009;180(3):291–7.

51. Chemali Z, Chahine LM, Fricchione G. The use of selective serotonin reuptake inhibitors in elderly patients. Harv Rev Psychiatry. 2009;17(4):242–53.

52. Schellander R, Donnerer J. Antidepressants: clinically relevant drug interactions to be considered. Pharmacology. 2010;86(4):203–15.

53. Spina E, Trifirò G, Caraci F. Clinically significant drug interactions with newer antidepressants. CNS Drugs. 2012;26(1):39–67.

54. Spina E, Santoro V, D'Arrigo C. Clinically relevant pharmacokinetic drug interactions with second-generation antidepressants: an update. Clin Ther. 2008;30(7):1206–27.

55. Caraci F, Crupi R, Drago F, Spina E. Metabolic drug interactions between antidepressants and anticancer drugs: focus on selective serotonin reuptake inhibitors and hypericum extract. Curr Drug Metab. 2011;12(6):570–7.

56. Gillman PK. Triptans, serotonin agonists, and serotonin syndrome (serotonin toxicity): a review. Headache. 2010;50(2):264–72.

57. Shapiro RE, Tepper SJ. The serotonin syndrome, triptans, and the potential for drug-drug interactions. Headache. 2007;47(2):266–9.

58. Evans RW, Tepper SJ, Shapiro RE, Sun-Edelstein C, Tietjen GE, Evans RW, et al. The FDA alert on serotonin syndrome with use of triptans combined with selective serotonin reuptake inhibitors or selective serotonin-norepinephrine reuptake inhibitors: American Headache Society position paper. Headache. 2010;50(6):1089–99.

59. Harvard Medical School. Antidepressants and tamoxifen. Drug interactions may increase risk of cancer recurrence or death. Harv Ment Health Lett. 2010;26(12):6–7.

60. Desmarais JE, Looper KJ. Interactions between tamoxifen and antidepressants via cytochrome P450 2D6. J Clin Psychiatry. 2009;70(12):1688–97.

61. Desmarais JE, Looper KJ. Managing menopausal symptoms and depression in tamoxifen users: implications of drug and medicinal interactions. Maturitas. 2010;67(4):296–308.

62. Kelly CM, Juurlink DN, Gomes T, Duong-Hua M, Pritchard KI, Austin PC, et al. Selective serotonin reuptake inhibitors and breast cancer mortality in women receiving tamoxifen: a population based cohort study. BMJ. 2010;340(8):c693.

63. Carvalho AF, Hyphantis T, Sales PM, Soeiro-de-Souza MG, Macêdo DS, Cha DS, et al. Major depressive disorder in breast cancer: a critical systematic review of pharmacological and psychotherapeutic clinical trials. Cancer Treat Rev. 2014; 40(3):349–55.

64. Hersh EV, Pinto A, Moore PA. Adverse drug interactions involving common prescription and over-the-counter analgesic agents. Clin Ther. 2007;29(Suppl):2477–97.

65. Serebruany VL. Selective serotonin reuptake inhibitors and increased bleeding risk: are we missing something? Am J Med. 2006;119(2):113–6.

66. Anglin R, Yuan Y, Moayyedi P, Tse F, Armstrong D, Leontiadis GI. Risk of upper gastrointestinal bleeding with selective serotonin reuptake inhibitors with or without concurrent nonsteroidal anti-inflammatory use: a systematic review and meta-analysis. Am J Gastroenterol. 2014;109(6):811–9.

67. Löppönen P, Tetri S, Juvela S, Huhtakangas J, Saloheimo P, Bode MK, et al. Association between warfarin combined with serotonin-modulating antidepressants and increased case fatality in primary intracerebral hemorrhage: a population-based study. J Neurosurg. 2014;120(6):1358–63.

68. Kharofa J, Sekar P, Haverbusch M, Moomaw C, Flaherty M, Kissela B, et al. Selective serotonin reuptake inhibitors and risk of hemorrhagic stroke. Stroke. 2007;38(11):3049–51.

69. Wallerstedt SM, Gleerup H, Sundström A, Stigendal L, Ny L. Risk of clinically relevant bleeding in warfarin-treated patients—influence of SSRI treatment. Pharmacoepidemiol Drug Saf. 2009;18(5):412–6.

70. Cochran KA, Cavallari LH, Shapiro NL, Bishop JR. Bleeding incidence with concomitant use of antidepressants and warfarin. Ther Drug Monit. 2011; 33(4):433–8.

71. Hackam DG, Mrkobrada M. Selective serotonin reuptake inhibitors and brain hemorrhage: a meta-analysis. Neurology. 2012;79:1862–65.

72. American Psychiatric Association. Practice guidelines for the treatment of patients with panic disorder. 2nd ed. Washington, DC: American Psychiatric Association; 2009.

73. Lejoyeux M, Adès J. Antidepressant discontinuation: a review of the literature. J Clin Psychiatry. 1997;58 Suppl 7:11–5. discussion 16.

74. Judge R, Parry MG, Quail D, Jacobson JG. Discontinuation symptoms: comparison of brief interruption in fluoxetine and paroxetine treatment. Int Clin Psychopharmacol. 2002;17(5):217–25.

75. Andrews G. Australian and New Zealand clinical practice guidelines for the treatment of panic disorder and agoraphobia. Aust N Z J Psychiatry. 2003;37: 641–56.

76. Bandelow B, Zohar J, Hollander E, Kasper S, Möller HJ; WFSBP Task Force on Treatment Guidelines for Anxiety,Obsessive-Compulsive and Post-Traumatic Stress Disoders, et al. World Federation of Societies of Biological psychiatry (WFSBP) Guidelines for the pharmacological treatment of anxiety, obsessive-compulsive and post-traumatic stress disorders— first revision. World J Biol Psychiatry. 2008;9: 248–312.

77. Canadian Psychiatric Association (CPA). Clinical practice guidelines management of anxiety disorders. Can J Psychiatry. 2006;51(Suppl):9–91.

78. Nardi AE, Freire RC, Mochcovitch MD, Amrein R, Levitan MN, King AL, et al. A randomized, naturalistic, parallel-group study for the long-term treatment of panic disorder with clonazepam or paroxetine. J Clin Psychopharmacol. 2012;32(1):120–6.

79. Ballenger JC. Long-term pharmacologic treatment of panic disorder. J Clin Psychiatry. 1991;52(Suppl):18–23. discussion 24–5.

80. Wright J, Clum GA, Roodman A, Febbraro GA. A bibliotherapy approach to relapse prevention in individuals with panic attacks. J Anxiety Disord. 2000; 14:483–99.

81. Wiborg IM, Dahl AA. Does brief dynamic psychotherapy reduce the relapse rate of panic disorder? Arch Gen Psychiatry. 1996;53:689–94.

82. Nash MS, Willets JM, Billups B, John Challiss RA, Nahorski SR. Synaptic activity augments muscarinic acetylcholine receptor-stimulated inositol 1,4,5-trisphosphate production to facilitate Ca2+ release in hippocampal neurons. J Biol Chem. 2004;279(47): 49036–44.

83. Fava M, Judge R, Hoog SL, Nilsson ME, Koke SC. Fluoxetine versus sertraline and paroxetine in major depressive disorder: changes in weight with long-term treatment. J Clin Psychiatry. 2000;61(11):863–7.

84. Montejo A, Majadas S, Rizvi SJ, Kennedy SH. The effects of agomelatine on sexual function in depressed patients and healthy volunteers. Hum Psychopharmacol. 2011;26(8):537–42.

85. Borges S, Desta Z, Li L, Skaar TC, Ward BA, Nguyen A, et al. Quantitative effect of CYP2D6 genotype and inhibitors on tamoxifen metabolism: implication for optimization of breast cancer treatment. Clin Pharmacol Ther. 2006;80(1):61–74.

86. Bak S, Tsiropoulos I, Kjaersgaard JO, Andersen M, Mellerup E, Hallas J, et al. Selective serotonin reuptake inhibitors and the risk of stroke: a population-based case-control study. Stroke. 2002;33(6): 1465–73.

Benzodiazepines in Panic Disorder

16

Roman Amrein, Michelle Levitan,
Rafael Christophe R. Freire, and Antonio E. Nardi

Contents

R. Amrein
Private Practice, Basel, Switzerland

Laboratory of Panic and Respiration, Institute of
Psychiatry, Federal University of Rio de Janeiro,
Rio de Janeiro, Brazil
e-mail: r.amrein@bluewin.ch

M. Levitan • R.C.R. Freire (✉)
• A.E. Nardi
Laboratory of Panic and Respiration, Institute of
Psychiatry, Federal University of Rio de Janeiro,
Rio de Janeiro, Brazil
e-mail: milevitan@gmail.com;
rafaelcrfreire@gmail.com; antonioenardi@gmail.com

Abstract

Benzodiazepines are efficacious and well tolerated in clinical use. Besides the anxiolytic effects,
they present sedative, muscle relaxant and anticonvulsive properties. Concerns regarding their
use rely on drug dependence after prolonged used
and difficult to manage withdrawal symptoms
during drug discontinuation. Studies show that
treatment duration up to 12 weeks is insufficient
to reach maximal possible therapeutic effect and
relapse during or shortly after drug discontinuation is frequent. Rebound and intolerable adverse
events during drug discontinuation can be avoided
in most cases if the dose is slowly down titrated in
small decrements over a prolonged period of time.
Most recent treatment guidelines recommend
selective serotonin reuptake inhibitors (SSRI) as
the first choice for the treatment of PD and the
benzodiazepines have the reputation to cause
dependence, especially if they are taken for long-
term and in high doses. However these recommendations are mainly based on expert opinions
mainly supported by a large number of clinical
trials with SSRI since evidence coming from
direct drug comparisons in PD is sparse. The
SSRI have side effects and need a gradual tapering out too. There is no doubt that every substance

© Springer International Publishing Switzerland 2016
A.E. Nardi, R.C.R. Freire (eds.), *Panic Disorder*, DOI 10.1007/978-3-319-12538-1_16

must be prescribed with clinical concerns and indications. The benzodiazepines are efficacious in PD and the physician can manage the concern about dependence and withdrawal. It is very important to include this class of drugs in the armamentarium for treating PD.

Keywords

Pharmacology • Anti-anxiety agents • Drug-related side effects and adverse reactions • Substance-related disorders • GABA-A receptor agonists

16.1 History

Benzodiazepines are psychoactive drugs, characterized as central nervous system depressants, with anxiolytic, sedative hypnotic, anticonvulsant, and muscle relaxant properties. They act as positive allosteric modulators on the gamma-aminobutyric acid-A (GABA-A) receptor, producing a inhibitory effect in the brain.

The first two benzodiazepines, chlordiazepoxide and diazepam were introduced in 1960/63. In comparison to previous drugs for the treatment of neuroses they were more effective, more potent, much less toxic and less prone to produce drug dependence and quickly became the drugs of choice in such treatments. Diazepam became within few years the most frequently prescribed drug worldwide and was prescribed uncritically resulting in the fact that "misuse and abuse of diazepam became an increasingly common medical problem" [1].

Chouinard published in 1982 the first controlled study on the use of alprazolam in PD. Patients were diagnosed according the Research Diagnostic Criteria [2] on which the DSM-III category "Panic disorder" was based [3]. PD patients taking alprazolam at an average final dose of 2.25 mg/day (N=14) improved more than placebo patients (N=6) as shown in HAM-A and Physician's Global Impression Scale. Drowsiness was the most prevalent side effect. Beaudry et al. hypothesized that clonazepam with its high potency combined with its intermediate to long half-life, and its serotonergic properties would

appear to be an advantageous anti panic agent, minimizing withdrawal symptoms or rebound. Based on an open study Beaudry et al. proposed "clonazepam as a new alternative treatment for the severe anxiety seen in patients suffering from PD and agoraphobia with panic attacks" [4].

16.2 Short-Term Controlled Trials

16.2.1 Alprazolam Versus Placebo

Alprazolam CT (compressed tablet) is the classical formulation with quick breakup. Dunner compared the effect of alprazolam, diazepam, and placebo on anxiety and panic attacks in PD [5]. An initial week of single-blind placebo administration was followed by 6 weeks of active treatment with doses titrated upward as tolerated (alprazolam 4 mg/day, diazepam 44 mg/day). At endpoint ANOVA did not show significant difference on the frequency of panic attacks between the treatment and placebo groups. Pre-post comparisons showed on the other hand significant reduction of panic attacks in the active treatment groups but not in the placebo group.

The efficacy of alprazolam, imipramine, and placebo [6] was evaluated in a total of 1168 patients in 12 centers randomly assigned to an 8 weeks double blind treatment of their panic attacks. The study was with unusual high intensity and effort planned, followed and monitored with the obvious intention to become the biggest, most powerful CNS study ever done, realizing the latest concepts of DSM III and methodology. During the first month of treatment there was a clear difference in efficacy between alprazolam and placebo (82 %/43 % better or moderately improved) but afterwards the study was handicapped by an unacceptable high rate of dropouts in the placebo group, in which a total of 43.7 % placebo patients did not complete the study (alprazolam 17.4 %). Endpoint analysis revealed that alprazolam and imipramine were significantly more effective than placebo in reducing the frequency of panic attacks and the intensity on measures of anticipatory anxiety, depressive and phobic symptoms but completer analysis failed to demonstrate such effects.

Lydiard et al. [7] investigated fixed-dose study of alprazolam 2 mg, alprazolam 6 mg, and placebo in PD in a total of 91 patients over 6 weeks. A significant dose effect was present: 6 mg alprazolam superior over alprazolam 2 mg and superior over placebo. At week 6 both doses were superior over placebo but there was no statistically significant difference between them. The authors conclude that many patients may derive substantial improvement in several important clinical measures while receiving a relatively low dose (2 mg/day) of alprazolam.

16.2.2 Alprazolam XR Versus Placebo

Alprazolam is rapidly absorbed and has a half-life of about 11 h. After high single doses it produces also in panic patients excessive sedation and drowsiness, at lower doses it is short acting. To avoid inter-dose rebound anxiety, so called clock watching [8, 9], "a four times-a-day dosing schedule has been found to be most effective, which is not only an inconvenience but a possible contributing factor to medication noncompliance" [10].

That short half-life is problematic for treating PD, as shown by Herman when he switched PD patients from alprazolam to placebo [9]. He observed in this open trial a reduction of early morning anxiety and of emerging anxiety between dosing although the frequency of administration was reduced with clonazepam. The extended release formulation alprazolam XR was brought to the market to compensate for the disadvantages of short elimination half-life. Alprazolam XR prolongs the absorption time, what results in smoother plasma-concentration curves during several hours after drug intake while elimination half-life is unchanged. Significant improvements indicated by a decrease in numbers of panic attacks and by clinical global improvement were observed with both doses of alprazolam-XR within the first week of dose up-titration and after 4 weeks of fixed dosing but not at treatment end. At this time in all three-treatment group every second patient was free of panic attacks s and in three out of four panic frequency was reduced to half. This study was handicapped not only by a high drop out and placebo

response rate but also by group differences at baseline and by study-center bias.

Pecknold et al. [11] compared in a double-blind, placebo-controlled, flexible-dose, multicenter, 6-week study regular alprazolam (compressed tablet, CT), given four times per day with alprazolam-XR, given once in the morning. In contrast to the previously reported studies patients were mainly recruited by referrals and the study was performed in only three centers. Starting dose was 2 mg/day. Each week the dose was increased by 1 mg until the patient was panic free or persistent dose-limiting side effects occurred. The daily dose was in the two drug treated groups over time similar and reached at study end 4 mg/day for alprazolam CT and 4.4 mg/day for Alprazolam-XR. Dropout rate for CT, XR and placebo was 19%, 20% and 36%. After 1 week of treatment reduction of panic attacks was with CT and XR greater than with placebo as was the rate of panic free patients. During the following weeks panic attack frequency was further reduced in all groups but the difference between active treatments and placebo was not maintained. At treatment end, both alprazolam treatment groups differed significantly from placebo in CGI-S, CGI-I, overall phobia score, HAM-A and the Sheehan Patient rated anxiety scale. There were some marginal positive difference of Ct over XR but altogether this study demonstrated that XR given in the morning can replace CT given four times a day.

Schweizer et al. [10] compared the XR formulation with placebo during 6 weeks in a total of 194 patients with a diagnosis of agoraphobia with panic attacks or PD with limited phobic avoidance. Treatment was initiated at 1 mg/day, administered in the morning, and increased every 3–4 days as tolerated to a maximum permissible dose of 10 mg/day (or ten tablets of placebo). The mean dose was during week 2.5 mg and increased afterwards up to 4.7 mg/day. As in most previous studies the dropout rate was significantly higher with placebo (34%) than with alprazolam (20%). In the last observation carried forward analysis alprazolam was superior over placebo in reducing panic attacks at all controls (week 1–6) whereas a difference for patients still in study could be demonstrated only for week 1 and 2. After 6 weeks alprazolam XR was in the LOCF analysis superior over placebo in all efficacy parameters and completer analysis showed superiority for HAM-A, CGI-S, CGI-I and Sheehan Patient-Rated Anxiety Scale.

16.2.3 Clonazepam Versus Placebo

In a multicenter dose-finding study published by Rosenbaum et al., 413 patients suffering from panic attacks with or without agoraphobia were randomly assigned to receive placebo or one of five fixed daily doses (0.5 mg, 1.0 mg, 2.0 mg, 3.0 mg, or 4.0 mg) of clonazepam [12]. Doses of 1 mg and above were all more efficacious than placebo or 0.5 mg of clonazepam. Within the dose range of 1–4 mg, no clear-cut dose effect for efficacy was apparent but the 3.0 mg and 4.0 mg dose groups had the highest AE rates. In the four higher dose groups, 69 % of patients on average were free of panic attacks at the end of the study.

Moroz and Rosenbaum reported the treatment of 438 PD patients over 6 weeks with either placebo or clonazepam at individually adjusted doses [13]. The mean optimized clonazepam dose was 2.3 mg/day. Clonazepam was superior to placebo in reducing the number of panic attacks, extent of fear and avoidance, and duration of anticipatory anxiety, and in improving the different items on the Clinical Global Impression (CGI) scales. At endpoint, 62 % of patients receiving clonazepam were free of full panic attacks, compared to 37 % of placebo-treated patients. The gradual tapering of clonazepam was not associated with symptoms suggestive of withdrawal syndrome.

Valença et al. [14] tested a fixed dose of clonazepam (2 mg/day) versus placebo in 24 PD patients with agoraphobia. After 6 weeks, the response rate (reducing anticipatory anxiety, scores of phobia, and CGI) was 61 % with clonazepam and 1 % with placebo. Three years later, Valença et al. [15] compared 34 PD patients with agoraphobia regarding the clinical efficacy of clonazepam in a fixed dosage (2 mg/day) versus placebo. The patients were divided in respiratory and non-respiratory subtypes. After 6 weeks, there was a statistically significant clinical superiority of clonazepam over placebo concerning remission of panic attacks (p<0.001) and decrease in anxiety (p=0.024). Response was similar in patients with respiratory and non-respiratory subtypes of PD, without significant differences between the two groups.

16.3 Studies Comparing Alprazolam and Clonazepam

Tesar et al. [16] compared clonazepam with alprazolam in a double blind, placebo controlled study over 6 weeks in 72 subjects with PD. He hypothesized that this high potency benzodiazepine with its relatively long half-life (39 ± 8.3 h, [39]) and a more potent influence on central serotonergic function than other benzodiazepines will be at least as effective as alprazolam in reducing the frequency of panic attacks and associated symptoms. 92 % (24) patients randomized to clonazepam, 83 % (20) to alprazolam but only 8 (36 %) placebo recipients completed the study. The daily dose was fractioned in four daily intakes and was up titrated during the first 3 weeks of treatment up to reach maximal therapeutic benefit or undesirable side effects and reached at treatment end 5.39 ± 2.89 mg/day for alprazolam and 2.50 ± 0.94 mg/day for clonazepam. Comparison of the two active treatments revealed no significant differences, but in some variables a trend for patients receiving clonazepam to experience greater benefit than those receiving alprazolam (see Fig. 16.1).

16.4 Studies Comparing High Potency Benzodiazepines with Antidepressants

Imipramine was the first drug with proven efficacy in PD. In 1961 an experimental placebo controlled study with imipramine and chlorpromazineprocyclidine was initiated at Hillside Hospital in 311 psychiatric inpatients investigating further possible indications of the two drugs. Patients could enter in study regardless of symptomatology or diagnosis. Against expectations a group of patients characterized by "sudden onset of subjectively inexplicable panic attacks, accompanied by hot and cold flashes, rapid breathing, palpitations, weakness, unsteadiness and a feeling of impend-

Fig. 16.1 Clinical improvement with alprazolam, clonazepam and placebo. *CGI* Clinical Global Impression. Tesar et al. [16]

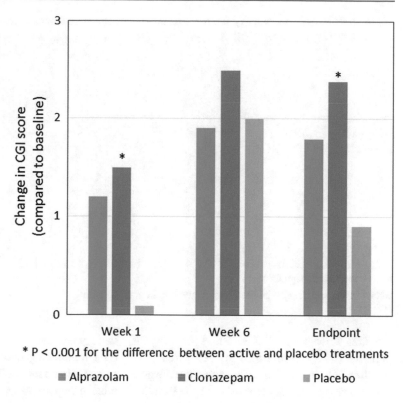

ing death" improved much with imipramine but not with chlorpromazine [17, 18].

Studies comparing approved benzodiazepines with approved SSRI or SNRI in PD patients are ideal to demonstrate the advantages and disadvantages of the two treatment strategies. Unfortunately they are sparse. A report describing a comparison of alprazolam with sertraline in 40 PD patients [19] gives insufficient and contradictory information on the design of the study and on efficacy and does not present at all adverse events. The presented graphs seem to reveal that both drugs were similarly effective. This study is not helpful to interpret the relative value of alprazolam and SSRI in treating PD.

Nardi et al. [20] compared the efficacy and safety of clonazepam and paroxetine in patients with PD referred to or consulted spontaneously the Panic and Respiration Laboratory at the Federal University of Rio de Janeiro during acute and long-term treatment and during drug discontinuation. Patients were afterwards systematically followed over additional 6 years. This is to our best knowledge, the only study that compared

valid a high potency benzodiazepine with an SSRI in PD under realistic therapeutic conditions and demonstrated long-term outcome. A total of 120 patients with PD were randomized to receive clonazepam (N=63) or paroxetine (N=57). Treatment was started with 1 mg clonazepam or 10 mg of paroxetine and up-titrated to maintenance doses of 2 mg clonazepam and 40 mg paroxetine that most reached in most instances at the end of the second treatment week.

Panic attacks decreased in both groups already during the first treatment week with marginal advantage for clonazepam. Patients had at the end of the first month in average 0.1 PA/week under clonazepam and 0.5 PA/week under paroxetine ($p < 0.5$). After 8 weeks of treatment the number of PA/week was nearly identical in both groups (0.16/0.15) and 90 % of the clonazepam patients were free from PA (paroxetine: 82 %). CGI-S improved in both treatment groups similarly over time but improvement in HAMA was in both groups only small during the 8 weeks of treatment (Clonazepam from 11.9±2.4 to10.5±3, paroxetine from 11.5±2.7 over a pejoration during

Table 16.1 Adverse events during short-term treatment with clonazepam or paroxetine (Nardi et al. [20])

	Before treatment (N=120)	Clonazepam (N=63)	Paroxetine (N=57)	Statistic
Frequent adverse events ≥ 10%	%	%	%	P*
Drowsiness/fatigue	0	57	81	<0.01
Sexual dysfunction	13	11	70	<0.01
Nausea/vomiting	16	0	61	<0.01
Appetite/weight change	1	2	54	<0.01
Dry mouth	3	0	47	<0.01
Excessive sweating	2	0	32	<0.01
Diarrhea/constipation	0	0	26	<0.01
Shaking/trembling	44	0	25	<0.01
Memory/concentration difficulties	27	24	40	ns
Weakness	27	10	12	ns
Anxiety/irritability	58	0	2	ns
Insomnia/nigthmare	45	0	0	ns

ns non-statistically significant

*Exact Fisher with Benjamini-Hochberg correction for multiple testing

the first 2 weeks to 10.4 ± 3.7). A majority of patients (64%) complained of AE already at baseline. The following table shows the adverse events present during the month before baseline and during study (see Table 16.1).

All participants were invited to continue treatment during the long-term study [21]. Patients with a good primary response (≤1 PA 7 month, CGI-S <3) to monotherapy continued with the same drug and dose. Partial responders or no responders were invited to switch to combination therapy with clonazepam and paroxetine. Target doses for all patients were 2 mg/day clonazepam and 40 mg/day paroxetine (both taken at bedtime). The therapeutic success reached after 8 weeks of treatment was maintained during long-term treatment and patients improved further somewhat: 87% of clonazepam-treated patients were free of PAs and 55% of both PAs and AEs during the entire long-term period. Patients switched to combination therapy improved also but remained always in a somewhat less favorable conditions compared to the two other groups.

At the end of 3-years treatment resulted for both monotherapy groups in an excellent treatment success (PA/month, CGI-S, CGI-I, 0.1 HAM-A for clonazepam 0.1 ± 0.3, 1.1 ± 0.3, 1.2 ± 0.4, 6.4 ± 2.8 and for paroxetine $0.2 \pm 0.5, 1.1 \pm 0.3, 1.1 \pm 0.3$, 7.5 ± 3.1. Altogether

there were only small differences between the three treatment groups concerning efficacy: some more patients left in the combination and paroxetine group treatment early and some efficacy parameters were with clonazepam slightly better. All three treatments were in general well tolerated but with AE profiles favoring clonazepam (Table 16.2).

Discontinuation of drug treatment aimed to reach drug free status within two months. The dosage of clonazepam was decreased by increments of 0.5 mg clonazepam (paroxetine: 10 mg) in 2-week intervals until reaching 1 mg/day (paroxetine: 20 mg/day) Subsequently, dose reduction was by 0.25 mg clonazepam per week (5 mg paroxetine). If the patient did not tolerate the tapering because of anxiety or withdrawal symptoms the intervals of dose reduction were prolonged or another medication, such as mirtazapine or carbamazepine was temporally added. Withdrawal was altogether at least partially successful in 73% of patients (clonazepam 86%, paroxetine 60%, combination 63%) with statistical advantage for clonazepam (p<0.005).

After drug discontinuation, nearly all patients relapsed within the following 6 years, but renewed treatment with clonazepam or paroxetine for at least 1 year was successful in most instances. Nearly every third patient needed

Table 16.2 Adverse events during long-term treatment with clonazepam or paroxetine (Nardi et al. [21])

Frequent adverse events during long-term treatment (≥10% in a group)	Clonazepam	Paroxetine	Statistic	Combination
	(N=47)	(N=37)		(N=21)
	%	%	P*	%
Appetite/weight change	0	97.3	<0.001	86
Excessive sweating	0	78.4	<0.001	67
Nausea/vomiting	6.4	78.4	<0.001	62
Diarrhea/constipation	21.3	81	<0.001	71
Dry mouth	25.5	83.8	<0.001	81
Sexual dysfunction	46.8	97.3	<0.001	90
Shaking/trembling/tremor	0	40.5	<0.001	67
Headache	6.4	35.1	<0.05	33
Drowsiness	76.6	100	<0.05	95
Memory/concentration problems	59.6	86.5	ns	71
Insomnia/nightmares	4.3	21.6	ns	10
Paresthesias	4.3	21.6	ns	14
Weakness	29.8	40.5	ns	24
Depersonalization	17	24.3	ns	24
Metallic taste	21.3	18.9	ns	29

ns non-statistically significant

*Exact Fisher with Benjamini-Hochberg correction for multiple testing

drug treatment during each follow up year. Altogether was the procedure of drug discontinuation and restart of treatment successful: 90% of patients were in average in remission during the follow up period (36% full remission [<2 PA/mth], 54% partial remission [<3 PA/mth]) during the 6 years after drug withdrawal. 73% of patients were in average free of panic attacks during the 6 follow up years, 91.1% had a GCI-S score of 1, and 38.8% had a HAMA between 5 and 10 points.

The study has shown that clonazepam and paroxetine were highly effective for treating PD during acute and long-term treatment (with minor advantages in favor of clonazepam). All patients tolerated stepwise drug discontinuation after 3-years of treatment without rebound and with few AE and only slight worsening of the clinical status. Clonazepam was better tolerated during treatment and drug discontinuation. The large majority of patients (94%) relapsed during the follow-up period of 6 years. Retreatment after relapse with either clonazepam or paroxetine was again successful in most instances but again better tolerated by patients taking clonazepam.

16.5   Long-Term Experience

In the naturalistic study of Worthington [22] a total of 93 patients receiving clonazepam alone or clonazepam plus another medication was followed over 2 years. Patients on monotherapy took between month 6 and 24 in average 1.6 mg/day clonazepam. Daily clonazepam dose of patients on combined therapy was 2.3 mg in average at month 6 and 2 mg between month 12 and 24. Treatment with clonazepam alone and in combination was associated with achievement and maintenance of therapeutic benefit without the development of tolerance as manifested by dose escalation or worsening of clinical status.

Nardi et al. [23] studied 35 patients with the respiratory subtype of PD and 32 patients with the non-respiratory subtype in an open prospective long-term study lasting for 3 years. Patients with at least 4 PA/month, a HAMA ≥ 18 and $HAMD_{21} < 17$, negative screens for benzodiazepines and barbiturates were admissible. The dose was 1 mg/day during the first week and, in the absence of dose-limiting adverse reactions, was increased to 2 mg/day in the second week and could afterwards be increased up to a maximum of 4 mg/day. The average dose at endpoint was 2.8 mg/day, whereby two third of patients took 2 mg/day. A significant improvement was shown in number of panic attacks, CGI, anticipatory anxiety, HAM-A, quality of life after 6 month and 3 years of treatment. The most common adverse events recorded were (%) somnolence (76), fatigue (46), memory complaints (39), dry mouth (40), decreased libido (34), ataxia (22), constipation (21), and lightheadedness (23).

16.6 Combined Treatment with SSRI

Co-therapy with SSRI or augmentation therapy SSRI are effective in the treatment of PD and are regarded by many clinicians as first-choice treatment [23]. This is also reflected in the APA guidelines for the treatment of PD. SSRI have for the treatment of PD a number of handicaps including a slow onset of action (several weeks) and unpleasant side effects, such as worsening of anxiety in the initial treatment phase, sexual and cognitive disturbances, insomnia, agitation, and weight gain [24]. Adding clonazepam for the initial 6–8 weeks is effective in bridging the time until the desired SSRI effect is achieved. This helps to prevent the occurrence of anxiety states, insomnia, and agitation during the initial phase of treatment.

In a double-blind trial, Goddard et al. [25, 26] treated 50 PD patients with open-label sertraline for 12 weeks. In addition, patients were randomly assigned to either clonazepam (1.5 mg/day) or placebo for 4 weeks. The clonazepam dose was then tapered during 3 weeks and finally discontinued. There were significantly more responders in the sertraline/clonazepam group than the sertraline/placebo group at the end of week 1 of the trial (41% vs. 4%, p=0.003). Moreover, there was a significant between-group difference in the percentage of responders at the end of week 3 (63% of the sertraline/clonazepam group vs. 32% of the sertraline/placebo group, p=0.05). The authors concluded that stabilization of panic symptoms can be safely achieved with a sertraline/clonazepam combination, demonstrating the clinical value of the combination for facilitating early improvement of panic symptoms relative to sertraline alone.

In a randomized study in 60 PD patients, Pollack et al. [27] compared the efficacy and safety of paroxetine and placebo vs. paroxetine co-administered with clonazepam. The initial treatment phase was followed by a tapered benzodiazepine discontinuation phase or ongoing combination treatment. All treatment groups demonstrated marked improvement. There was a significant advantage for combined treatment early on but subsequently, the outcome in all three groups was similar. The authors concluded that combined treatment with paroxetine and clonazepam resulted in a more rapid initial response than SSRI treatment alone, but there was no difference beyond the initial few weeks of therapy.

Initiating combined treatment, followed by benzodiazepine taper after a few weeks, may provide fast benefit while avoiding the potential adverse consequences of long-term combination therapy. Published studies suggest that between 30% and 60% of patients receiving antidepressants including SSRI may experience some form of treatment-induced sexual dysfunction [28]. In an opinion survey among 439 psychiatrists, most psychiatrists appeared to favor switching to clonazepam monotherapy if major AEs occurred with SSRI, rather than continuing with combination therapy [29].

Initial co-therapy of high potency benzodiazepines with SSRI and serotonin norepinephrine reuptake inhibitors (SNRI) has also clear disadvantages: treatment schedule is becoming very complex including a drug discontinuation period for benzodiazepines after some weeks of treatment and the high probability that adverse events (AE) of both drugs will combine. Cotherapy should therefore only be considered if there are definite reasons not to initiate monotherapy with high potency benzodiazepines (concomitant depression, history of alcohol or drug abuse) and the patients is not willing to accept this initial disadvantages of SSRI/SNRI treatment.

16.7 Adverse Events During Short-Intermediate Term Placebo Controlled Studies

Benzodiazepines, including clonazepam and alprazolam, are well tolerated in clinical use. Benzodiazepines have addition to its anxiolytic effects three other basic pharmacological effects; they are in addition sedative, muscle relaxant and anticonvulsive. In the big placebo controlled studies complaints were systematically recorded at baseline and at each control during drug administration. This allowed identifying untoward signs and symptoms that appeared or worsened during treatment. AE have the same frequency with the active treatment and placebo are most probably not drug related. In certain situations patients certain adverse events may be more frequent with placebo than with active treatment. This indicates that symptoms occurring during PD but not considered to be core symptoms are controlled by the active treatment but not by placebo.

Detailed information on AE were for alprazolam reported by Noyes et al. [30], Tesar et al. [16], the Cross-National Collaborative Panic Study [6], Cassano et al. [31] and Leon et al. [32]. The above-mentioned studies reported AE on patients treated with alprazolam and patients treated with placebo. Sedation, fatigue, ataxia, slurred speech, amnesia were the most frequent AE of alprazolam during the acute treatment of PD as can be seen from the following table. They are well known pharmacological effects of benzodiazepines that are unwanted in the specific situation. They were dose related and many of them decreased or disappeared with ongoing treatment. On the other hand had alprazolam a positive effect on other symptoms that are associated with PD, as can be seen from Table 16.3.

Significant differences between alprazolam-XR and placebo were found for sedation, somnolence, coordination abnormal, dysarthria, memory impairment, disturbance in attention, balance impaired, libido decreased, ataxia, constipation, fatigue. Most of the reported drug related AE are expressions of unwanted pharmacological effects. The placebo-controlled studies of clonazepam in PD [12–14, 33] comprise in total 626 cases treated with clonazepam and 341 placebo cases. The common drug related adverse events are nearly identical with those of alprazolam as can be seen from Table 16.4.

16.8 Adverse Events During Long-Term Treatment

Most patients with PD need long-term treatment to prevent recurrence of panic attacks. Drug related adverse events are most often reason for abandoning prematurely chronic treatment. Good tolerability is therefore pivotal for long-term success. Unfortunately information on AE's during long-term treatment of PD with benzodiazepines is very sparse. Schweizer et al. [34] gives detailed information on AE's present during the 32-week treatment with alprazolam (N = 30) and placebo (N = 10). The frequency of AE decreased during long-term treatment compared to the initial 6-week treatment and there was no significant difference between alprazolam and placebo in the frequency of any AE, but sedation, dry mouth, impaired coordination and decreased libido were numerically more frequent with alprazolam and nervousness was numerically more frequent with placebo. Nardi et al. compared the adverse events present before and during long-term treatment with clonazepam or paroxetine [21]. Comparison of the frequencies of AEs experienced at baseline and during long-term treatment may enable distinguishing between drug-related AEs and those associated with PD. Insomnia/nightmares, anxiety, shaking/trembling/tremor, paresthesia, and headache were present at study baseline, and these were significantly reduced during treatment with either clonazepam or paroxetine, whereas nausea/vomiting, weakness, and memory/concentration difficulties were reduced only by clonazepam. Clonazepam treatment led to a significant increase in drowsiness during the entire long-term treatment (2 % vs 23 %, P G 0.01). During long-term paroxetine treatment, sexual dysfunction (14 % vs 49 %, P G 0.01), excessive sweating (0 % vs 20 %, P G 0.05), drowsiness (0 % vs

Table 16.3 Adverse effects of alprazolam and placebo in panic disorder treatment

	Alprazolam	Placebo	Statistic
Number of subjects	760	733	P*
AE more frequent with alprazolam			
Sedation	58.7%	23.6%	<0.001
Fatigue/weakness	20.8%	15.3%	<0.05
Ataxia	19.6%	9.1%	<0.001
Increased appetite	18.2%	8.6%	<0.001
Constipation	11.3%	6.7%	<0.05
Slurred speech	11.2%	4.4%	<0.001
Amnesia	5.2%	3.0%	ns
Weight change	4.4%	0.6%	<0.001
AE more frequent with placebo			
Dry mouth	14.9%	20.7%	<0.05
Headache	11.3%	17.5%	<0.05
Nausea/vomiting	7.0%	15.3%	<0.001
Dizziness/faintness	10.9%	15.0%	<0.05
Insomnia	5.8%	13.2%	<0.001
Nervousness	4.9%	11.5%	<0.001
Tremor	6.3%	11.5%	<0.05
Depression, depressive feelings	3.0%	9.0%	<0.001
Abdominal discomfort	4.6%	8.5%	<0.05
Tachycardia	4.9%	8.3%	<0.05
Chest pain	4.2%	7.8%	<0.05
Excessive sweating	4.8%	7.8%	<0.05
Hyperventilation	3.6%	6.4%	<0.05
Palpitations	2.4%	5.9%	<0.05
Other sleep disturbance	2.0%	5.9%	<0.001
Malaise	3.0%	5.9%	<0.05
Faintness	3.0%	5.2%	ns

AE adverse events, *ns* non-statistically significant
*Pearson's Chi-square test with Benjamini-Hochberg correction for multiple testing

41%, P G 0.001), diarrhea/ constipation (0% vs 23%, P G 0.01), and appetite/weight change (5% vs 51%, P G 0.001) were increased.

16.9 Drug Discontinuation

Withdrawal, concern on drug dependence after prolonged use of benzodiazepines and difficult to manage withdrawal symptoms during drug discontinuation were mentioned in nearly all publications on therapeutic use of benzodiazepines that appeared in the nineties of the last and at begin of this century. There is no doubt that after abrupt drug discontinuation severe withdrawal symptoms, including seizures, were observed also in panic patients treated with benzodiazepines [35, 36].

According to Greenblatt et al. the abrupt discontinuation of short-acting agents, following extended use at high doses, would be expected to produce a more severe withdrawal syndrome [37]. Drug discontinuation from drugs with short half-life of elimination is more difficult and demanding than discontinuation from drugs with intermediate or long half-life: Withdrawal reactions for fluoxetine (half-life 4–6 days) are not reported but for venlafaxine (half-life 4 h) severe withdrawal symptoms were observed few hours after omitting a dose [28, 37, 38]. Extended release formulations (XR, CR) prolong the absorption phase, reduce C_{max} and smooth there-

Table 16.4 Adverse events of alprazolam XR in panic disorder treatment

	Clonazepam	Placebo	Statistic
	N = 626	N = 342	P*
Somnolence/sedation	37.1 %	1.8 %	<0.001
Dizziness/drowsiness	9.3 %	0.6 %	<0.001
Fatigue	6.4 %	0.0 %	<0.001
Memory/attention problems	6.7 %	0.3 %	<0.001
Ataxia	5.6 %	0.0 %	<0.001
Coordination abnormal	5.4 %	0.9 %	<0.05
Depression	6.7 %	3.5 %	<0.05

*Pearson's Chi-square test with Benjamini-Hochberg correction for multiple testing [45]

fore the plasma-concentration-profile between doses, but this has no essential effect on the elimination constants and does therefore not reduce the withdrawal risk associated with half-life of elimination.

Withdrawal reactions are not limited to treatment with benzodiazepines. It is now widely accepted that PD is related to abnormalities in the function of a variety of neurotransmitters including serotonin, noradrenaline, GABA, dopamine, cholecystokinin, and endogenous opioids [39, 40]. Treatment with antipanic agents results in a new homeostasis that would be disrupted by sudden stop or quick discontinuation of antipanic treatment. A similar situation is present after long-lasting treatment with corticosteroids; there the drug discontinuation takes weeks or even months, although these substances have half-lives of few hours. Abrupt discontinuation of long-lasting treatment with corticosteroids is medical malpractice and this is in our opinion also the case for abrupt discontinuation of long-lasting treatment with antipanic drugs, especially if they do not have ultra-long half-life.

It was already in the early short-term studies shown that withdrawal reactions can be avoided by slow down-titration. Fyer et al. documented [41] tapered alprazolam discontinuation after an average treatment duration of 6 months in 18 out of 30 patients that had participated in a drug trial [42]. Patients took 2.5–8.5 mg/day alprazolam before drug discontinuation. Alprazolam was decreased every 3 days by 0.5 or 1 mg, at as close to a rate of 10 % of the original dose. Only four of the 18 subjects completed withdrawal from alprazolam by following the fixed protocol, 15 had a

recurrence or increase in panic and all except one suffered from withdrawal symptoms. The authors discuss critically the problematic outcome of this study and mention as possible reasons for withdrawal symptoms and relapse: the length of drug treatment, the rate of drug decrease, and drug half-life.

Abelson et al. [43] gives an example of modern, patient centered drug discontinuation after long-term treatment of panic attacks. Nineteen patients had participated in a clinical study [44] and were afterwards switched to long-term treatment. After 1 year of treatment in average and 1–3 months of stable remission, alprazolam dose reductions were initiated to reach maintenance of a non-impaired, minimal symptom state and drug discontinuation if feasible. The pace of reductions was at the discretion of the patients, but they were cautioned to go no faster than 0.25 mg every 2 weeks. Drug tapers lasted in average 7.7 months. Alprazolam was discontinued in 78 % of patients. Relapse occurred in 36 % of these, an average of 6.4 months after drug discontinuation. The total group remained at follow up, 21 months after initiation of alprazolam treatment, significantly improved, compared to pre-treatment, despite the presence of five relapsers.

Nardi et al. [45] report the discontinuation of clonazepam in PD patients treated continuously for at least 3 years. In total, 73 Patients free of PD symptoms for at least 1 year and wishing to discontinue clonazepam therapy participated in the systematic discontinuation with a planned dose decrease by increments of 0.5 mg in 2-week intervals until reaching 1 mg/day. Subsequently, dose reduction was by 0.25 mg/week. If the patient did not tolerate the tapering because of anxiety or withdrawal symptoms, tapering intervals could be prolonged or another medication could be added. Patients received before tapering a clonazepam dose of 2.7 ± 1.2 mg/day. It was aimed to reach drug free state without reappearance of panic attacks latest 4 month after begin of the drug discontinuation. This goal was reached in 70% of cases; in additional 19% of patients it took between 4 and 6 month to be drug free and without panic attacks. As many as 89% of patients completely stopped intake of any PA medication and were free of PAs within 6 months at most. In the remaining 11% of patients the clonazepam dose was reduced to 0.5 mg/day and later on replaced by mirtazapine. During the follow-up period of 8 month no patient had to go back to the original treatment and 82% of patients remained free of panic attacks and all others had one PA/month at most.

All the presented studies show unequivocally the same: Treatment duration up to 12 (24) weeks is insufficient to reach maximal possible therapeutic effect and relapse during or shortly after drug discontinuation is frequent. Rebound and intolerable adverse events during drug discontinuation can be avoided in most cases if the dose is slowly down titrated in small decrements over a prolonged period of time.

We follow since many years and with excellent success the following schema for drug discontinuation after long-term treatment with antipanic agents: first the dose is decreased in 2-week intervals by decrements corresponding to 20–25% of daily dose at treatment end. Afterwards the dose is decreased each week but the dose diminution is reduced to have, corresponding to approximately 10–15% of the daily dose at treatment end. If the patients does not well tolerate the tapering than the intervals have to be prolonged. This scheme worked well and most of the patients became drug free within 6–8 weeks and the temporary use of another medication such as carbamazepine or mirtazapine was necessary only very rare cases. In our 3-year-long-term study comparing clonazepam with paroxetine we have for instance used the following schema: The dosage of clonazepam was decreased by increments of 0.5 mg in 2-week intervals until reaching 1 mg/day. Subsequently, dose reduction was by 0.25 mg/week. The dosage of paroxetine was decreased by increments of 10 mg in 2-week intervals until reaching 20 mg/day. Subsequently; dose reduction was by 5 mg/week [46].

16.10 Pharmacokinetics

Panic attacks manifest often out-of-the blue and this can happen at any time of the day. Benzodiazepines effects are dose- and concentration dependent. The patient is not protected against panic attacks if the concentration is too low. On the other hand excessive high concentrations should be avoided to avoid oversedation lack of attention, fatigue, weakness and ataxia leading to unstable gait and slurred speech. Alprazolam has a short half-life of elimination, accounting for approximately 12 h. In the early clinical studies alprazolam intake was therefore fractioned in multiple doses. But even dose fractioning could not prevent daytime symptom recurrence, so called "clock watching" and early morning "rebound" [8]. These symptoms were no longer observed when patients were switched to the long acting clonazepam [9]. To avoid these problems an additional galenical formulation of alprazolam was introduced to the market: Alprazolam-XR, an extended-release formulation prolonging the absorption period over several hours. Glue [47] investigated the single dose pharmacokinetics of alprazolam XR in 12 adolescent and 12 adult subjects. Blood samples were obtained predose and at 1, 2, 3, 4, 6, 8, 10, 12, 24, 36, and 48 h after the intake of 1 and 3 mg. The mean concentration–time profiles after 1 and 3 mg doses were similar in both populations, and were characterized by a fast plasma concentration increase during the first 2 h after intake, followed by a quasi-plateau phase until 12 h, and the final elimination with a half-life of 15.6 h. PD needs chronic treatment over months or even years. The usual alprazolam XR dose in PD is 3 mg/day to be taken in the morning and the usual clonazepam dose is 2 mg/day to be taken before going to bed. The graph shows a simulation of steady state plasma levels of 3 mg/day alprazolam XR given in the morning and of 2 mg/day clonazepam given in the evening. The therapeutic dose differences mirror clonazepam's higher affinity to the benzodiazepine receptors and therefore higher potency. The simulation is based on single dose data provided by Glue for alprazolam XR [47] and by Crevoisier for clonazepam [48]. Plasma level fluctuation over 24 h is for alprazolam XR and clonazepam similar: $C_{max}/C_{min} = 1.68$ for alprazolam and 1.64 for clonazepam. Plasma level fluctuation during day is with alprazolam more important: $C_{morning}/C_{evening} = 1.68$ for alprazolam and 1.16 for clonazepam. Plasma levels of alprazolam and clonazepam decrease during night by 50 % resp. increase by 60 %. Clonazepam may therefore be given preference for patients that suffer from comorbid sleep disorders. After alprazolam XR given in the morning induces maximal CP around noon and may therefore have a negative impact on alertness and performance during this period. The maximal plasma concentration of clonazepam during nigh could eventually increase in elderly patients the increased risk of falls during night. From a pharmacokinetic standpoint of view single evening doses of clonazepam offer some advantage over single alprazolam XR morning doses (see Fig. 16.2).

16.11 Substance Use Disorder

Most recent treatment guidelines recommend SSRI as the first choice for the treatment of PD. Benzodiazepines have the reputation to cause dependence, especially if they are taken for long-term and in high doses. These recommendations are mainly based on expert opinions mainly supported by a large number of clinical trials and panels with SSRI since evidence coming from direct drug comparisons in PD is sparse.

The study reported by Fisecovic et al. [19] lacks validity, as reported above. The direct comparison of clonazepam with paroxetine [49, 50] showed that both drugs were similarly highly effective during acute treatment but clonazepam was better tolerated. All patients agreed to discontinue drug treatment after 3 years of treatment and a dose reduction was possible in all patients. The risk for benzodiazepine dependence and abuse was in the past by several authors considered as clear argument against any long-term use of benzodiazepines. Benzodiazepine dependence and abuse was mainly observed when benzodiaz-

Fig. 16.2 Pharmacokinetics of alprazolam XR and clonazepam

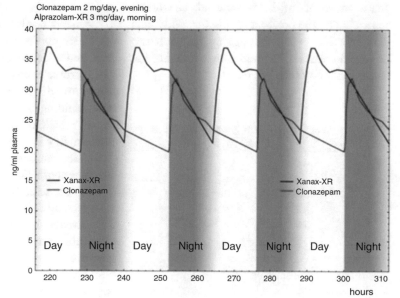

epines were taken without precise indication, for recreational purpose or by patients with a history of alcohol or substance abuse. There is no doubt that every substance must be prescribed with clinical concerns and indications.

Fujii et al. [51] published recently the results of a cross-sectional study on the dependence on benzodiazepines in patients with PD. Among outpatients with this disorder, 60.8 % had a total score higher than 4 in the SDS. This was interpreted as a 60.8 % incidence of psychological dependence. The proportion of patients with dependence was significantly lower in remitted patients (i.e. a total score lower than 5 in the PDSS) (44.1 %, n = 15/34) than those who were not remitted (94.1 %, n = 16/17). We found that the study has limitations that make the interpretation difficult: Severity of Dependence Scale (SDS) (Box 16.1) was developed to measure the degree of psychological dependence experienced by users of different types of illicit drugs (heroin, amphetamine, cocaine and others). De las Cuevas et al. [52] has adapted and validated the SDS with the help of CIDI for benzodiazepine dependence in regular benzodiazepine users. A cut-off of 7 resulted in a sensitivity of 97.9 % and in a specificity of 94.2 %. Fujii et al. [51] gave no explanation for the low cut-off point at 5 points. From a clinical standpoint of view it seems normal that a PD patient having experienced the positive therapeutic effects of drug treatment would be worried somewhat by the prospect of missing a dose and would not like to stop the ongoing treatment also he would wish that he could stop it once he is in remission for a prolonged time.

Box 16.1. Severity of Dependence Scale [52]

Did you think your use of tranquillizers was out of control? Did the prospect of missing a dose make you anxious or worried? Did you worry about your use of tranq-uillizers? Did you wish you could stop? How difficult would you find it to stop or go without your tranquillizers?

We appraise that the publication from Fujii et al. [51] overestimates the problem. We have therefore asked ten psychiatrists treating predominately anxiety disorders at our institute to give an educated guess on problematic use of benzodiazepines and SSRI/SNRI's during the treatment of PD patients. There was consensus that impaired control (items 1 and 2, seldom 3) would be the main characteristics for substance use disorder during the treatment of PA with benzodiazepines and SSRI/SNRI. There was consensus that the therapeutic use both substance classes may lead to mild/moderate substance abuse disorder in ~one out of four patients. Dose increase over the prescribed dose is more often observed with benzodiazepines but treatment for longer than intended, and unsuccessful efforts to stop or cut down use is a more frequent problem with SSRI/SNRI.

The risk for substance abuse disorder subsequent treatment of PD is not negligible but of similar importance for benzodiazepines and SSRI/SNRI. Since no hard data exist a group of psychiatrists treating mainly anxiety disorders made an educated guess on problematic drug use during and after PD treatment with high potency benzodiazepines or SSRI/SNRI (Table 16.5).

16.12 Conclusions

The benzodiazepines are efficacious in PD treatment. Despite the importance relying on the concern about dependence and withdrawal, other psychotropic drugs as antidepressants, are also related to withdrawal syndrome even few hours after omitting a dose. Studies show that withdrawal reactions can be avoided by slow downtitration, in which the physician can manage.

Another important concern regards benzodiazepines abuse. Despite the few studies available, a research with psychiatrists working mainly with anxiety disorders reached a consensus that the increase over the prescribed dose is more often observed with benzodiazepines but treatment for longer than intended, and unsuccessful efforts to stop or cut down use is a problem more frequent with SSRI/SNRI. Also, the risk for

Table 16.5 Educated guess on problematic drug use during and after PD treatment with high potency benzodiazepines or SSRI/SNRI

Educated guess from ten psychiatrists – Laboratory of Panic & Respiration – Institute of Psychiatry – Rio de Janeiro, Brazil Currently ~3500 patients are treated for PD with benzodiazepines and ~2800 with SSRI/SNRI	High potency benzodiazepines	SSRI, SNRI
Patients not being alcohol- or drug abuser before start of PD treatment develop dose increase that could be considered abuse	<2 %[a]	<1 %[b]
Patient increased himself the daily dose and reached uncommon high level	~5 %	~2 %
After an appropriate length of treatment with positive outcome (patient was free from PA since at least 1 year): Patient does not agree to start drug discontinuation Drug discontinuation failed	~20 % of good responders ~25 % of good responders, mainly due to increase in basal anxiety levels. Not necessary associated with panic attacks	~60 % of good responders ~50 % of good responders, mainly due to increase in basal anxiety levels. Not necessary associated with panic attacks
Drug treatment was not successful. Nevertheless patient is not prepared to initiate treatment with another class of drugs	Very rare. Almost always very easy to take slowly out the benzodiazepine when there is no therapeutic response	Very rare. Almost always very easy to take slowly out the SSRI when there is no therapeutic response
Patients being treated for PD with benzodiazepine/SSRI, SNRI and suffering from mild/moderate substance use disorder according to DSM-5	25 %	25 %

[a]Easiest group to taper out
[b]Easy to change and correct the treatment

substance abuse disorder subsequent treatment of PD is not negligible but of similar importance for benzodiazepines and SSRI/SNRI.

Despite the reputation to cause dependence based on expert opinions mainly supported by a large number of clinical trials and panels with SSRI, benzodiazepines are very important in PD treatment and it is essential to have this group in the armamentarium for treating this disorder.

References

1. Woody GE, O'Brien CP, Greenstein R. Misuse and abuse of diazepam: an increasingly common medical problem. Int J Addict. 1975;10(5):843–8.
2. Spitzer R, Endicott J, Robins E. Research diagnostic criteria. 3rd ed. Biometrics research. New York; New York State Psychiatric Institute: 1977.
3. Williams JB, Spitzer RL. Research diagnostic criteria and DSM-III: an annotated comparison. Arch Gen Psychiatry. 1982;39(11):1283–9.
4. Beaudry P, Fontaine R, Chouinard G, Annable L. An open clinical trial of clonazepam in the treatment of patients with

recurrent panic attacks. Prog Neuropsychopharmacol Biol Psychiatry. 1985;9(5–6):589–92.
5. Dunner DL, Ishiki D, Avery DH, Wilson LG, Hyde TS. Effect of alprazolam and diazepam on anxiety and panic attacks in panic disorder: a controlled study. Clin Psychiatry. 1986;47(9):458–60.
6. Study C-NCP. Drug treatment of panic disorder. Comparative efficacy of alprazolam, imipramine, and placebo. Cross-National Collaborative Panic Study, Second Phase Investigators. Br J Psychiatry. 1992;160:191–202; discussion 205.
7. Lydiard RB, Lesser IM, Ballenger JC, Rubin RT, Laraia M, DuPont R. A fixed-dose study of alprazolam 2 mg, alprazolam 6 mg, and placebo in panic disorder. J Clin Psychopharmacol. 1992;12(2):96–103.
8. Herman JB, Brotman AW, Rosenbaum JF. Rebound anxiety in panic disorder patients treated with shorter-acting benzodiazepines. J Clin Psychiatry. 1987;48(Suppl):22–8.
9. Herman JB, Rosenbaum JF, Brotman AW. The alprazolam to clonazepam switch for the treatment of panic disorder. J Clin Psychopharmacol. 1987;7(3):175–8.
10. Schweizer E, Patterson W, Rickels K, Rosenthal M. Double-blind, placebo-controlled study of a once-a-day, sustained- release preparation of alprazolam for the treatment of panic disorder. Am J Psychiatry. 1993;150(8):1210–5.

11. Pecknold J, Luthe L, Munjack D, Alexander P. A double-blind, placebo-controlled, multicenter study with alprazolam and extended-release alprazolam in the treatment of panic disorder. J Clin Psychopharmacol. 1994;14(5):314–21.

12. Rosenbaum JF, Moroz G, Bowden CL. Clonazepam in the treatment of panic disorder with or without ago-raphobia: a dose-response study of efficacy, safety, and discontinuance. Clonazepam Panic Disorder Dose-Response Study Group. J Clin Psychopharmacol. 1997;17(5):390–400.

13. Moroz G, Rosenbaum JF. Efficacy, safety, and gradual discontinuation of clonazepam in panic disorder: a placebo-controlled, multicenter study using optimized dosages. J Clin Psychiatry. 1999;60(9):604–12.

14. Valenca AM, Nardi AE, Nascimento I, et al. Double-blind clonazepam vs placebo in panic disorder treatment. Aeq Neuro Psiquiatr. 2000;58(4):1025–9.

15. Valença AM, Nardi AE, Mezzasalma MA, Nascimento I, Zin WA, Lopes FL, et al. Therapeutic response to benzodiazepine in panic disorder subtypes. Sao Paulo Med J. 2003:12 1(2):77–80.

16. Tesar GE, Rosenbaum JF, Pollack MH, Otto MW, Sachs GS, Herman JB, et al. Double-blind, placebo-controlled comparison of clonazepam and alprazolam for panic disorder. J Clin Psychiatry. 1991; 52(2):69–76.

17. Klein D. Importance of psychiatric diagnosis in pre-diction of clinical drug effects. Arch Gen Psychiatry. 1967;16:118–26.

18. Klein DF. Delineation of two drug-responsive anxiety syndromes. Psychopharmacologia. 1964;17:397–408.

19. Fisekovic S, Loga-Zec S. Sertraline and alprazolam in the treatment of panic disorder. Bosn J Basic Med Sci. 2005;5(2):78–81.

20. Nardi AE, Valenca AM, Freire RC, Amrein R, Sardinha A, Levitan MN, et al. Randomized, open naturalistic, acute treatment of panic disorder with clonazepam or paroxetine. J Clin Psychopharmacol. 2011;31(2):259–61.

21. Nardi AE, Freire RC, Mochcovitch MD, et al. A ran-domized, naturalistic, parallel-group study for the long-term treatment of panic disorder with clonaze-pam or paroxetine. J Clin Psychopharmacol. 2012;32(1):120–6.

22. Worthington IJJ, Pollack MH, Otto MW, McLean RYS, Moroz G, Rosenbaum JF. Long-term experience with clonazepam in patients with a primary diagnosis of panic disorder. Psychopharmacol Bull. 1998;34(2):199–205.

23. Nardi AE, Valenca AM, Nascimento I, et al. A three-year follow-up study of patients with the respiratory subtype of panic disorder after treatment with clonaz-epam. Psychiatry Res. 2005;137(1–2):61–70.

24. Noyes Jr R, Clancy J, Coryell WH, et al. A with-drawal syndrome after abrupt discontinuation of alprazolam. Am J Psychiatry. 1985;142(1):114–6.

25. Klein E, Colin V, Stolk J, et al. Alprazolam with-drawal in patients with panic disorder and generalized anxiety disorder: vulnerability and effect of carbam-azepine. Am J Psychiatry. 1994;151(12):1760–6.

26. Goddard AW, Brouette T, Almai A, et al. Early coad-ministration of clonazepam with sertraline for panic disorder. Arch Gen Psychiatry. 2001;58(7):681–6.

27. Pollack MH, Simon NM, Worthington JJ, et al. Combined paroxetine, and clonazepam treatment strategies compared to paroxetine monotherapy for panic disorder. J Psychopharmacol. 2003;3:276–82.

28. Sabljic V, Ruzic K, Rakun R. Venlafaxine withdrawal syndrome. Psychiatr Danub. 2011;23(1):117–9.

29. Dording CM, Mischoulon D, Petersen TJ, et al. The pharmacologic management of SSRI-induced side effects: a survey of psychiatrists. Ann Clin Psychiatry. 2002;14(3):143–7.

30. Noyes Jr R, DuPont Jr RL, Pecknold JC, et al. Alprazolam in panic disorder and agoraphobia: results from a multicenter trial: II. Patient acceptance, side effects, and safety. Arch Gen Psychiatry. 1988;45(5): 423–8.

31. Cassano GB, Toni C, Petracca A, et al. Adverse effects associated with the short-term treatment of panic disorder with imipramine, alprazolam or pla-cebo. Eur Neuropsychopharmacol. 1994;4:47–53.

32. Leon CA, De Arango MV, Arevalo W, et al. Comparison of the effect of alprazolam, imipramine and placebo in the treatment of panic disorders in Cali, Colombia. Acta Psiquiatr Psicol Am Lat. 1990;36(1–2):59–72.

33. Beauclair L, Fontaine R, Annable L, et al. Clonazepam in the treatment of panic disorder: a double-blind, placebo-controlled trial investigating the correlation between clonazepam concentrations in plasma and clinical response. J Clin Psychopharmacol. 1994;14(2):111–8.

34. Schweizer E, Rickels K, Weiss S, Zavodnick S. Maintenance drug treatment of panic disorder: I. Results of a prospective, placebo-controlled com-parison of alprazolam and imipramine. Arch Gen Psychiatry. 1993;50(1):51–60.

35. Browne JL, Hauge KJ. A review of alprazolam with-drawal. Drug Intell Clin Pharm. 1986;20(11):837–41.

36. O'Sullivan GH, Swinson R, Kuch K, et al. Alprazolam withdrawal symptoms in agoraphobia with panic dis-order: observations from a controlled Anglo-Canadian study. J Psychopharmacol. 1996;10(2):101–9.

37. Greenblatt DJ, Shader RI, Abernethy DR. Clinical use of benzodiazepines. N Engl J Med. 1983;309(7):410–6.

38. Dallal A, Chouinard G. Withdrawal and rebound symptoms associated with abrupt discontinuation of venlafaxine. J Clin Psychopharmacol. 1998;18(4): 343–4.

39. Johnson MR, Lydiard RB, Ballenger JC. Panic disor-der. Pathophysiology and drug treatment. Drugs. 1995;49(3):328–44.

40. Graeff FG. New perspective on the pathophysiology of panic: merging serotonin and opioids in the periaq-ueductal gray. Braz J Med Biol Res. 2012;45(4): 366–75.

41. Fyer AJ, Liebowitz MR, Gorman JM, et al. Discontinuation of alprazolam treatment in panic patients. Am J Psychiatry. 1987;144(3):303–8.

42. Liebowitz MR, Fyer AJ, Gorman JM, et al. Alprazolam in the treatment of panic disorders. J Clin Psychopharmacol. 1986;6(1):13–20.
43. Abelson JL, Curtis GC. Discontinuation of alprazolam after successful treatment of panic disorder: a naturalistic follow-up study. J Anxiety Disord. 1993;7(2):107–17.
44. Coryell W, Noyes R. HPA axis disturbance and treatment outcome in panic disorder. Biol Psychiatry. 1988;24(7):762–6.
45. Nardi AE, Freire RC, Valença AM, et al. Tapering clonazepam in patients with panic disorder after at least 3 years of treatment. J Clin Psychopharmacol. 2010;30(3):290–3.
46. Smith RB, Kroboth PD, Vanderlugt JT, Phillips JP, Juhl RP. Pharmacokinetics and pharmacodynamics of alprazolam after oral and IV administration. Psychopharmacology (Berl). 1984;84(4):452–6.
47. Glue P, Fang A, Gandelman K, Klee B. Pharmacokinetics of an extended release formulation of alprazolam (Xanax XR) in healthy normal adolescent and adult volunteers. Am J Ther. 2006;13(5):418–22.
48. Crevoisier C, Delisle MC, Joseph I, et al. Comparative single-dose pharmacokinetics of clonazepam following intravenous, intramuscular and oral administration to healthy volunteers. Eur Neurol. 2003;49(3):173–7.
49. Nardi A, Freire R, Machado S, Mochcovitch M, Silva A, Amrein R. Ultra long-term follow-up of panic disorder patients: randomized 3 years treatment with clonazepam, paroxetine, or their combination and follow-up for additional 6 years. Poster Nr. 131 presented at anxiety and depression (ADAA) annual conference Miami Florida, 2015.
50. Nardi A, Valença A, Freire R, et al. Psychopharmacotherapy of panic disorder: 8-week randomized trial with clonazepam and paroxetine. Braz J Med Biol Res. 2011;44:366–72.
51. Fujii K, Uchida H, Suzuki T, et al. Dependence on benzodiazepines in patients with panic disorder: a cross-sectional study. Psychiatry Clin Neurosci. 2015;69(2):93–9.
52. Delascuevas C, Sanz EJ, Delafuente JA, et al. The Severity of Dependence Scale (SDS) as screening test for benzodiazepine dependence: SDS validation study. Addiction. 2000;95(2):245–50.

Repetitive Transcranial Magnetic Stimulation in Panic Disorder

17

Sergio Machado, Flávia Paes,
and Oscar Arias-Carrión

Contents

S. Machado (✉)
Laboratory of Panic and Respiration, Institute of
Psychiatry, Federal University of Rio de Janeiro, Rio
de Janeiro, Brazil

Physical Activity Neuroscience, Physical Activity
Sciences Postgraduate Program, Salgado de Oliveira
University, Niterói, Brazil
e-mail: secm80@gmail.com

F. Paes
Laboratory of Panic and Respiration, Institute of
Psychiatry of Federal University of Rio de Janeiro,
Rio de Janeiro, Brazil
e-mail: flavia.paes.psi@gmail.com

O. Arias-Carrión
Unidad de Trastornos del Movimiento y Sueño
(TMS), Hospital General Dr. Manuel Gea González,
México, DF, Mexico

Unidad de Trastornos del Movimiento y Sueño
(TMS), Hospital General Ajusco Medio,
México, DF, Mexico
e-mail: arias@email.ifc.unam.mx

Abstract

The available treatment methods for panic
disorder (PD; pharmacotherapy and cognitive
behavioral therapy) are well documented as
safe and effective. However, few patients
remain free of panic attacks or in complete
remission. With the advancement in the understanding of the neurobiological mechanisms
involved in PD, new treatments have been proposed. One such method is transcranial magnetic stimulation (TMS), a non-invasive method
of focal brain stimulation. TMS is based on the
Faraday's law of electromagnetic induction,
where an electric current is influenced by the
magnetic field into the brain, inducing an electric current that depolarizes or hyperpolarizes
neurons. Unlike for depression, only few studies are available today investigating the therapeutic effects of rTMS for PD. Thus, this
chapter aimed to provide information on the
current research approaches and main findings
regarding the therapeutic use of rTMS in the
context of PD. So far, there is no conclusive evidence of the efficacy of rTMS as a treatment for
PD. While positive results were found in most
of studies, various treatment parameters, such
as location, frequency, intensity and duration

have been used unsystematically, making difficult the interpretation of results and providing little guidance about which treatment parameters (i.e., the stimulus location and frequency) may be more useful for the PD treatment. Therefore, further studies are needed to clearly determine the role of rTMS in PD treatment.

Keywords

Anxiety • Dorsolateral prefrontal cortex • Panic attack • Panic disorder • Repetitive transcranial magnetic stimulation • rTMS

17.1   Introduction

The available treatment methods for panic disorder (PD; pharmacotherapy and cognitive behavioral therapy) are well documented as safe and effective [1–3]. However, few patients remain free of panic attacks or in complete remission [4]. With the advancement in the understanding of the neurobiological mechanisms involved in PD, new treatments have been proposed. One such method is transcranial magnetic stimulation (TMS), originally introduced in 1985 as a non-invasive method of focal brain stimulation [5]. TMS is based on the Faraday's law of electromagnetic induction, where an electric current is influenced by the magnetic field into the brain, inducing an electric current that depolarizes or hyperpolarizes neurons [6].

Despite being increasingly used in the treatment of some psychiatric disorders, the effects of repetitive transcranial magnetic stimulation (rTMS) on PD treatment are poorly understood. The use of rTMS is considered a treatment based on brain neuromodulation due to the focus on directly reach the neural circuitry of psychiatric disorders. rTMS operates by changing or modulating the function of the neural circuitry into the brain which is believed to be disorganized in these disorders [7].

This chapter aims to provide current concepts and findings on the therapeutic effects of rTMS for PD, in order to determine whether it may become feasible as a clinical application in the future. The historical aspects, basic principles, efficacy, main experimental findings and safety of rTMS will be addressed.

17.2   Historical Aspects of Transcranial Magnetic Stimulation

Since the first experiments in 1831, the English physicist and chemist Michael Faraday noted that it was possible to convert electrical energy into magnetic fields and magnetic fields into electricity. Faraday demonstrated in his experiments that electrical currents are produced by an alternating magnetic field, rather than static. Thus, if this concept is applied to a living organism, the body tissues would be considered the secondary circuit. The primary circuit would be composed of a stimulating coil and a stimulator, which promotes the discharge of electric pulses creating a magnetic field. Consequently, the magnetic field generates an electric flow without direct contact with the tissue [8].

The French physician and physicist D'Arsonval performed the first clinical observation of electromagnetic stimulation in the brain, in 1896. In Paris, at the meeting of the Society of Biology, he reported the perception of light flashes (phosphenes) "seen" or perceived visually by subject undergoing to a magnetic field applied to the head. Some years later, in 1902, Beer also reported similar findings of D'Arsonval, who seem to be due to direct stimulation of the retina rather than to a central stimulation, since the retina is the most sensitive structure of the human body to induced currents [8].

In 1910, Silvanus Thompson confirmed that visual sensations can be induced by alternating magnetic field. He also noted that after an exposure lasting 2 min tactile sensations were also produced. Other researchers [9] subsequently reported these findings.

After five decades of neglect the use of TMS in research was resumed. In 1965, Bilckford and Fremming conducted an experiment where they contracted the muscles of animals and humans.

Researchers stimulated target muscles with a sinusoidal magnetic field of 500 Hz with a peak amplitude of 4 T, reduced to zero in approximately 40 milliseconds (ms). Bickford and Fremming supposed that was possible to induce currents into the brain with magnitude enough to non-invasively stimulate cortical structures [10].

In 1974, Anthony Barker, an English biomedical engineer, began to investigate the use of short-pulsed magnetic fields. The first TMS device with practical utility was developed in the next year in an attempt to produce nerve stimulation with selective speed, leading them to investigate the possibility of creating a new unit for clinical purposes [9]. The equipment used by Barker reached a peak current of 2300 amperes (A), and rise time of approximately 120 microseconds (µs), and that resulted in a magnetic field with a peak in the coil center (transfer efficiency ~ 20 %) of 2 Tesla (T). Thus, by applying a magnetic stimulus at a given joint, for example, the elbow, a clear feeling could be felt and light muscle contractions were observed [5, 9].

However, the development and implementation of a biomedical device capable of modulating brain circuits through the electromagnetic field have only been developed and published in 1985 by Anthony Barker and colleagues [5]. The equipment reached a peak current of 6800 A and a magnetic field of 2.2 T, being more efficient in energy transfer from the capacitor to the coil (transfer efficiency of ~80 %) than the apparatus previously developed. Since then, the group managed to Barker stimulated for first time the motor cortex. A coil of 100-mm diameter was placed on the vertex of a healthy volunteer. They observed hand muscular movements and motor evoked potentials recorded from abductor muscle of the little finger by using superficial electromyography [5, 9].

The use of TMS emerged to be a diagnostic tool to study motor conduction in patients with neurological diseases, such as multiple sclerosis [6, 9]. Within this context, a single pulse was fired and the motor conduction time, i.e. from motor cortex to the peripheral response, was measured. In 1990, with the introduction of repeated pulses, also called rTMS, it became a promising tool for therapeutic use [9].

17.3 Basic Principles of Transcranial Magnetic Stimulation

rTMS has several relevant foundations that must be highlighted in order to provide a better understanding about its modes of action. In this section, we will present the main terminology of rTMS that has so far been studied.

rTMS is a non-invasive technique able to modulate human cortical excitability not just at the site of stimulation, but also at remote areas. rTMS consists of essentially two pieces of hardware, the main unit – a capacitive high voltage, high current discharge system – and a stimulating coil [11]. rTMS relies on Faraday's principle of electromagnetic induction. During stimulation the main unit discharges a strong, brief current through the stimulation coil. This in turn induces a relatively brief (~100 µs), focal and rapidly changing magnetic field, perpendicular to the plane of the coil. When the coil is held against the scalp, the magnetic field passes unimpeded through the scalp and skull [2]. The time-varying magnetic field induces a weak and short-lived current, flowing in loops parallel to the orientation of the coil, at the site of stimulation that results in neuronal depolarization or spiking. The induced current is dependent on rate of change of the current discharged through the coil [11].

In addition to being able to apply a single pulse of TMS, pulses can be applied in pairs or in trains, respectively referred to as *paired-pulse TMS* and *rTMS* [1, 2]. The protocol (i.e. the pattern and frequency of the pulse trains) of rTMS can be classified as being *conventional* or *patterned*. Conventional rTMS can be further subdivided into low frequency (<1 Hz) and high frequency rates (>1 Hz) of stimulation, and also whether it is applied during or immediately after a subject performs a task (referred to as *online* TMS) or whether the stimulation and task are separated in time (*offline* TMS) [12]. The success of online TMS relies on the pulse train altering neural activity during processing while the success of offline TMS relies on the stimulation effects outlasting the stimulation period. Patterned TMS refers to protocols where short trains of pulses are separated by periods of no stimulation, and is only employed in an offline approach. The behavioral impact of rTMS depends on the stimulation parameters used, though the neurophysiology of these differences remains unclear [2]. Despite of the fact that the effects of conventional rTMS in primary motor cortex can be measured objectively by recording motor evoked potentials [4], the effect of rTMS on the majority of brain areas has no visible outcome thus must be indexed by either changes in accuracy or response times (or both) in an appropriate task. If the stimulated area is causally involved during the processing of a task [13, 14], and the rTMS is administered at an appropriate time [11] then rTMS temporarily affects task performance (e.g. decreases in accuracy or increases in reaction time – RTs). Consequently, online TMS is sometimes referred to as using TMS in 'virtual lesion' mode [13, 15]. Under certain circumstances TMS can improve task performance, for example if the target area is not required for the task or the TMS pulse is administered at an inappropriate time, then the TMS either has no effect or can facilitate task performance, the latter effect is consistent with intersensory facilitation [16]. Unlike offline rTMS (conventional or patterned), the effects of online rTMS appear to be short lived. This allows the investigation of when an area is causally involved in a cognitive function and is often referred to as using TMS in neurochronometric mode [15].

Recently, a novel pattern of rTMS called theta-burst stimulation (TBS) was developed to produce changes in the human cerebral cortex excitability [17]. The main advantage of the TBS paradigm compared with conventional rTMS protocols is that a shorter period (between 20 and 190 s) of subthreshold stimulation causes changes in cortical excitability that outlast the time of stimulation for at least 15–20 min. Huang et al. [18] proposed a TBS protocol consisting of bursts of three pulses given at 50 Hz repeated every 200 ms (5 Hz), thus, mimicking the coupling of theta and gamma rhythms in the brain [16]. Two main modalities of TBS have been tested: intermittent TBS induces facilitation of motor cortical excitability; continuous TBS leads to inhibition for 15–30 min after application [17, 19].

Given the fact that rTMS induces a current in the brain, one might reasonably ask *where* is this current induced? Where is the locus of excitation? Considering that magnetic field strength decreases rapidly with distance, it seems reasonable to expect that elements in the cortical mantle are preferentially affected, with minimal impact on subadjacent white matter. Indeed, this was backed up by evidence from simulations of the effects of rTMS on increasingly realistic whole heads [20, 21] and somewhat less realistic phantom heads (containers filled with physiological saline solution) [22]. Within the gray matter, mathematical modeling and empirical evidence suggest that excitation is most likely to occur at axonal bends, terminals and hillocks (where the cell body joins the axon), that is locations where the spatial derivative of the induced voltage exceeds a particular negative value [23]. Neurons with lower thresholds activate first and can propagate the excitation along the axons and therefore to connected regions [24].

As the magnetic field intensity decays rapidly with distance [25, 26], successful stimulation of a target tissue depends on choosing appropriate stimulation intensity [27, 28]. However, at present there is no consensus on the optimum way to do this. Some researchers use a set intensity for all subjects in a study [29]. However, given that there are likely to be intersubject differences in factors such as the depth of the target [30] and the target area's local connectivity the minimum stimulation threshold is likely to vary across individuals and thus the choice of an arbitrary fixed stimulation intensity may lead to subjects being under- or over-stimulated. If the targeted area is under-stimulated, the probability that the rTMS-induced current will affect processing is reduced. Over-stimulation increases the area of cortex being stimulated [11] and increases transsynaptic current spread [31], potentially confounding interpretation of results. Furthermore, overstimulation increases the likelihood of adverse reactions to rTMS, such as peripheral nerve stimulation and importantly increasing the possibility of seizure [32].

The alternative is to attempt to adjust the stimulation intensity for each participant in an experiment, most commonly by calibrating the stimulation intensity to a percentage of each participant's motor threshold (MT) [11, 32]. When rTMS is applied over primary motor cortex (M1), the induced currents depolarize neurons in the corticospinal tract. If the intensity of stimulation is high enough, this results in a motor evoked potential or a visible twitch in the muscle corresponding to the stimulated cortical area [1, 2]. The MT is often defined as the minimum rTMS intensity that elicits a response (either motor evoked potential or twitch) in the contralateral thumb (abductor pollicis brevis) or index finger (first dorsal interosseous muscle) in at least 50% of the trials. MT is lowered by voluntarily contracting the muscle (active motor threshold) relative to the resting motor threshold [1, 2, 11, 32]. Initial studies found no relationship between a subject's MT and their phosphene threshold (the minimum intensity required to elicit phosphenes on 50% of trials of visual cortex stimulation), calling into the question the suitability of using MT as a means of calibrating intensity for areas outside M1 [11, 32]. However, more recent work has shown that thresholds for motor and visual cortices are correlated when the thresholding procedure is the same across sites [33], providing a validation of the use of MT-calibrated stimulation intensities in non-motor areas. An alternative approach, where the MT is scaled according to the difference in depth between M1 and the cortical site of interest, is also promising [26] though awaits empirical testing.

The focality of the induced current is affected by the shape of the stimulating coil, and while a circular coil generates the strongest magnetic field, the most commonly used coil is the figure-of-8 design [1, 2, 11]. In the latter, the induced electric field under the intersection of the 8, sometimes referred to as the 'hot spot', is double the magnitude as that induced at the wings. For certain target sites and coil orientations, the wings may be sufficiently distant from the scalp that the magnetic field produced by them may be discounted. However, this may not always be the case and the field produced by the wings may be sufficient to induce unwanted current in peripheral muscles, nerves or possibly neural tissue [11]. Approximations of the actual extent of induced neural current can be obtained from mathematical three dimensional head models [20] and phantom head simulations however, these estimates do not provide information regarding the physiologically effective spatial resolution, i.e. the distance between two points where stimulation produces different responses. For example, though the area that experiences at least 90% of the maximum induced current is thought to be greater than 1 cm^2, using single pulse TMS it is possible to resolve sites less than 1 cm apart in motor and premotor cortex [34]. It is likely that the effective spatial resolution will be affected by the stimulation intensity and protocol, with higher intensities and increasing number of pulses decreasing the resolving power. Nonetheless, sites as close as 2 cm have been effectively resolved using short pulses of high frequency rTMS [29]. The effective spatial resolution is also affected by properties of the stimulated tissue, including the orientation of neurons, the areas of local and distant connectivity, as well as the pre-existing activation state of the network, though to what extent is unknown.

It is important to distinguish between the time that the neural effects persist from the time that the behavioral effects persist. As discussed above, invasive animal studies suggest that changes in neural activity induced by single, paired or repetitive TMS may persist for greater than 500 ms, however, the effect on behavior in humans is considerably shorter than this. For example, by using single and paired pulse TMS it is possible to show disruption in a task (for example, increased RTs) during a particular time window much smaller than half a second. For example, a single pulse TMS significantly impairs semantic processing when it is delivered at 250 ms post-stimulus onset, but not 200 ms or before nor 300 ms and onwards [35]. Paired pulse TMS to the frontal eye fields impairs stimulus discriminability when the pair of pulses occurs at 40 and 80 ms, but not at 0 and 40 or 80 and 120 ms onwards [36]. These studies suggest that the effective temporal resolution of TMS is approximately 40–50 ms, though this is likely to be dependent on the stimulation parameters and properties of the stimulated tissue.

17.4   Effects of Sham-rTMS
and Stimulation Parameters

An important issue in the rTMS research regarding the design of randomized, sham-controlled clinical trials is the use of appropriate control conditions that provide a reliable blinding of patients and investigators [37]. Within this context, different control conditions can be used to try to ensure that changes in performance be ascribed to rTMS effects upon a specific brain area. One of the most common strategies is the use of sham stimulation (sham-rTMS) [38]. rTMS is indeed associated with a number of sensory perceptions that can nonspecifically interfere with task performance. For instance, the discharging coil produces a click sound that may induce arousal, thereby modulating task performance, irrespective of the experimental demands (i.e., *via* intersensory facilitation) [39]. An alternative way that is routinely used in the cognitive rTMS literature is vertex stimulation because the auditory and somatosensory activations caused by vertex rTMS can be equivalent to those of real stimulation. Of course, the underlying assumption is that vertex rTMS does not affect the cognitive network active during task execution [40, 41].

In general, sham-rTMS has been applied by tilting the coil away from the scalp [42], so that both sound and scalp contact are roughly similar to those experienced during active stimulation, whereas the magnetic field does not reach cortical neurons or cutaneous receptors or superficial muscles. Although sham coils produce an analogous sound artifact, they do not induce the same scalp sensations or muscle twitches, so that they can rest tangential to the scalp surface, exactly as they are during active stimulation [43, 44]. Another important consideration that must be taking into account in order to determine the specific efficacy of rTMS in clinical trials and to create a credible placebo (i.e., sham-rTMS) condition, is that patients in randomized trials should be naive to rTMS, in other words, rTMS studies should not have a crossover design. With respect to this issue, the ideal sham condition should not have a real stimulation effect, and it should not

be recognized as sham by patients, particularly when considering that real stimulation conditions come along with rTMS specific side effects.

In line with that, Herwig et al. [45] investigating the antidepressant effects of rTMS, asked for patients to give their impression whether they received the sham or the real treatment, and if they would recommend the treatment to others. From 15 patients with real stimulation, 11 suggested that they obtained true stimulation, and four to have obtained sham. From 14 sham stimulated subjects, nine suggested that they obtained the real condition and five to have been sham stimulated. There was no significant difference between these and in addition, the majority of patients in both stimulation conditions would recommend rTMS to others. In both conditions, the majority of subjects believed they had received the real condition. This implies suitability of the sham condition used since subjects appeared not to be able to accurately identify or differentiate this condition from sham. The results imply the feasibility of a valid sham condition with a "real" coil.

However, there is evidence that some types of sham manipulations used in clinical trials actually do exert some effects on the brain [46, 47]. The tilting does reduce any discomfort from scalp stimulation associated with active rTMS and, thus, may have the potential to interfere to some degree with the adequacy of study blinding. Studies guard against this by recruiting only rTMS-naïve patients, so that subjects are not cued to discriminate between active and sham conditions based on scalp sensation. Even if a form of coil-tilt sham that does not exert measurable brain effects is used, studies rarely report data on the integrity of the blind on the part of the patients and raters. It is reasonable to assume that crossover trials with coil-tilt sham conditions are likely to be unblended because active and sham rTMS do not feel the same [48, 49]. Other option include the one used in a recent experiment consisting of a sensor strip between the electromagnet and the scalp, which can counter-stimulate during pulse delivery so as to reduce the scalp sensation perceived from active rTMS [50].

The matter of placebo effects is especially important in some conditions, such as studies investigating the efficacy of treatments [38]. For such purposes alternative methods of brain stimulation to provide suitable control conditions have been proposed. For instance, Rossi et al. [51] developed a new method of sham stimulation, known as real electromagnetic placebo, in which a fake coil (made of wood) with the same shape as a real coil is attached to the real coil. This fake coil has two functions: to block the magnetic field from the real coil, and to house a bipolar electrical stimulator in contact with the scalp. This device is more likely to be judged as real stimulation by naive rTMS subjects. The difficulty in blinding makes the comparison of rTMS with a gold standard treatment (e.g., psychopharmacology) complex. In the case of pharmacologic agents, it would be possible to use a "double-dummy" design in which some patients would receive sham rTMS plus active medication, whereas other patients would receive active rTMS and a placebo pill. An additional challenge in the design of clinical trials with rTMS pertains to the standardization of the dosage. Just as it is critical to control the dosage of medication administered during drug trials, it is likewise essential to control the amount of rTMS administered and the location of the brain region stimulated [52].

Other important considerations to be taken into account are the parameters of stimulation, e.g., pulse width, number of stimulation sessions, frequency, intensity and site of stimulation [53]. A protocol composed of repeated sessions may be superior to a single session, due to its cumulative effect related to amount of stimulation required to induce a sustained effect. Indeed, although some studies have shown a relatively long-lasting effect (i.e., of 2 weeks), this period is short if the goal is to induce a clinically meaningful result. Maintenance treatments or other patterns of stimulation that might induce longer-lasting modulation of cortical excitability should be explored. One possibility is to increase the total number of sessions, as in a recent study of major depression, in which up to 30 sessions of rTMS were administered [54]. Novel patterns

of stimulation, for example primed 1 Hz stimulation [55] or theta burst stimulation [18], might offer advantages, as they seem to induce longer-lasting long-term-depression-like phenomena. Careful consideration of cortical targets seems to be critical, and this might need to be individualized for each patient and underlying pathology.

In summary, a number of parameters need to be taken into account in order to optimize the clinical effects of rTMS. Predictions with regard to the efficacy of clinical effects of rTMS are hampered due to the relative paucity of parametric studies performed on these variables. Moreover, individualizing stimulation parameters, taking into account the underlying pathophysiology and the stimulation settings by online physiological and neuroimaging measures, seems to be a crucial procedure to adopt [37, 38].

17.5   Factors Influencing the Individual Response to rTMS

During the last years, genetic diversity in human population has been a crucial topic in clinical research. It has been hypothesized that common genetic variants may contribute to genetic risk for some diseases and that they might influence the subject's response to rTMS [56, 57]. One could speculate that a profound knowledge on genetic variants might help to predict whether participants will respond or not to magnetic stimulation and in which direction the modulation will take place.

The Brain Derived Neurotrophic Factor (BDNF) gene has been associated to the individual response to rTMS. This gene has 13 exons and it encodes a precursor peptide (pro-BDNF) which in turn is cleaved to form the mature protein. A single nucleotide polymorphism (SNP) located at nucleotide 196 (guanine (G)/adenosine (A)) has been identified. The result is an amino acid substitution Valine (Val)-to-Methionine (Met) at codon 66, and it has been hypothesized that this SNP though located in the pro-BDNF alters intracellular processing and secretion of

BDNF [58]. In healthy subjects it has been associated with mild memory impairments, reduction in hippocampal and frontal cortical areas and some personality traits [58]. This Val66Met polymorphism could be also associated to psychiatric disorders such as depression and risk of schizophrenia, as well as to the pathogenesis of some neurodegenerative diseases, i.e., Alzheimer's disease, Parkinson's disease and amyotrophic lateral sclerosis [58].

The strong evidence, on the one hand, of a functional role for this BDNF common polymorphism, and on the other hand, the implication of this gene in LTP process yielded to analyze whether a BDNF genotype influences the response to rTMS delivered over M1. Little is known regarding this topic. The first investigation demonstrated that the facilitation following the performance of fine-motor tasks, reflected as an increase in the amplitude of cMAPs, was more pronounced in Val/Val polymorphism carriers as compared to Val/Met or Met/Met carriers [57]. A second study explored the inhibitory effect of the cTBS protocol in healthy carriers of different polymorphisms of the BDNF gene. The findings suggested that Val/Met or Met/Met (Non-Val/Val) carriers have a reduced response to cTBS as compared to those subjects with Val66Val polymorphism [56].

Beside genetic variations a second factor influences the individual response to rTMS: the physiological state of neurons at the time of stimulation. Synaptic plasticity can be modulated by prior synaptic activity. The direction and the degree of modulation seem to depend on the previous state of the network. This kind of plasticity is called metaplasticity [59, 60]. For example, external stimulation that activates the resting network could decrease the same network if it was not at rest at the moment of stimulation. In animal models, it has been related to the NMDA receptor activation, Caþ2 influx, CaM, CaMKII and to modifications of inhibition of GABA release [61].

The phenomenon of metaplasticity has been demonstrated applying rTMS at cortical regions that have previously been modulated by means of cathodal or anodal transcranial direct current

stimulation [62]. One-minute of muscular contraction of the abductor pollicis brevis (APB) during TBS over M1 suppressed the effect of the cTBS and iTBS effect on the cMAPS amplitude. When the contraction was hold immediately after TBS, it enhanced the facilitatory effect of iTBS and reversed the usual inhibitory effect of cTBS into facilitation. In a second study, the application of 300 pulses of cTBS facilitated cMAPs amplitude, whereas the same train of stimulation preceded by voluntary contraction of 5 min or 600 pulses of cTBS with the muscle at rest decreased it. The results suggest that 300 pulses of cTBS may have a similar mechanism than iTBS and may prime neuronal elements to undergo inhibition by the late cTBS with 600 pulses. Similarly, the change in the TBS effects before or after a muscular contraction provides evidence for metaplasticity of corticospinal excitability in the human M1. These findings must be considered when applying TBS in clinical trials.

17.6   Repetitive Transcranial Magnetic Stimulation and Panic Disorder

Anxiety is a normal adaptive response to stress that allows coping with adverse situations. However, when anxiety becomes excessive or disproportional in relation to the situation that evokes it or when there is not any special object directed at it, such as an irrational dread of routine stimuli, it becomes a disabling disorder and is considered to be pathological [63, 64]. The term "anxiety disorders" subsumes a wide variety of conditions of abnormal and pathological fear and anxiety, such as PD [65, 66]. The anxiety disorders comprise the most frequent psychiatric disorders and can range from relatively beginning feelings of nervousness to extreme expressions of terror and fear.

Based on the idea of an interhemispheric imbalance and/or deficit in the limbic-cortico control, Ressler and Mayberg [67] proposed a model for human anxiety based on the theory so called "valence-hypothesis", which has been formerly proposed for [68]. According to this model, withdrawal-related emotions such as anxiety are located to the right hemisphere, whereas approach related emotions such as joy or happiness are biased to the left hemisphere. In line with this hypothesis, Keller et al. [69] examined and found an increased right hemispheric activity in anxiety disorders, reinforcing an association between increased right hemispheric activity and anxiety. The first evidence of this model was observed by the use of 1 Hz-rTMS on the right prefrontal cortex (PFC) has demonstrated effects in some studies involving healthy individuals [65].

However, Pallanti and Bernardi also argued that rTMS over the left dorsolateral prefrontal cortex (DLPFC), especially above 5 Hz-rTMS, reduces the symptoms of anxiety in PTSD and panic disorders [66]. Therefore, to further elucidate the putative anxiolytic action of rTMS in anxiety patients future studies have to be conducted.

Other studies set out to investigate the hypothesis of high-rTMS efficacy in anxiety disorders treatment [66]. Specifically, the cerebral hyperexcitability and behavioral or cognitive activation observed in neuropsychiatric disorders support this hypothesis [70]. The studies demonstrated that the activity of fronto-subcortical circuits can arguably be diminished by increasing the activity in the indirect pathway by stimulating the left DLPFC by high-rTMS [66, 71]. In this section, we will discuss the mechanisms and circuitries involved in panic disorders and the therapeutic effects of rTMS.

In the current literature, few studies have been conducted in PD. Sakkas et al. [72] described the improvement of panic symptoms comorbid to depression in a patient who presented symptoms of myocardial infarction using a protocol of 20 Hz-rTMS administered to 110 % LM in the left PFC, five double sessions per week during 3 weeks (1600 pulses/day). In 2007, Montavani [73] described a study of six patients with PD and comorbid depression that after 2 weeks (five sessions per week) of application of 1 Hz-rTMS over the right DLPFC at 110 % LM (1200 pulses/day) with an improvement of anxiety and decrease in symptoms of depression since the

first week, which were sustained for up to 6 months. In 2009, Dresler et al. reported the case of a PD patient with comorbid depression who received one session of high-frequency rTMS over the left prefrontal cortex per day five times per week during 3 weeks. Near-infrared spectroscopy (NIRS) was used to measure the brain activity in response to an emotional Stroop task before and after the rTMS treatment. Authors found that high-frequency rTMS modulated panic-related prefrontal brain dysfunctions [74]. More recently, Machado and colleagues (2014) reported a PD patient with comorbid depression and resistant to cognitive behavior therapy (CBT) and pharmacotherapy. Patient was treated with a combined protocol of rTMS with a sequential stimulation of 1 Hz-rTMS over right DLPFC and 10-Hz-rTMS over left DLPFC, three times per week during 4 weeks, with 1 month follow-up. The authors found improvements in mood and panic symptoms that were sustained for 1 month [75].

In 2007, Prasko et al. [76] conducted the first randomized placebo-controlled clinical study comparing the effects of active rTMS and sham rTMS-like strategy of therapeutic potentiation in PD patients non responders to SSRIs. Antidepressants used were paroxetine 20 mg, citalopram 20 mg, fluoxetine 20 mg, sertraline 50 mg, and venlafaxine 75. A protocol of low-frequency stimulation (1 Hz) was administered for 30 min at 110 % MT on the right DLPFC, in a total of 1800 pulses per section, i.e. five sessions per week during 2 weeks. The same design was applied to control patients. Symptoms were assessed just after the first session, after ten sessions (second week) and after 4 weeks. The scales used were the Clinical Global Impression-Severity of Illness (CGI), the Panic Disorder Severity Scale (PDSS), the Hamilton Rating Scale for Anxiety (HAM-A) and the Beck Anxiety Inventory (BAI). Results showed that low-frequency rTMS (1 Hz) was not significantly higher than those of sham-rTMS in improving symptoms of PD patients, the score was slightly better than placebo results.

After the study of Prasko et al. [76], important questions were raised: (1) Was the average duration of PD before treatment with rTMS was very long (9 years)? (2) What would be the rTMS efficacy for PD patients with short duration of disease? (3) Are the area and frequency used in this treatment ideal for PD? (4) Could PD respond to low-frequency rTMS applied to contralateral regions, such as epilepsy and depression respond (indirect propagation)?

Despite the efforts of RCT studies later, it is still possible to answer the many questions that arose with the study of Prasko et al. [76]. In this sense, two recent studies tried to answer some questions [77, 78]. Mantovani et al. [77] conducted a randomized double-blind study where 1Hz-rTMS at 110 % MT (1800 stimuli/day) was applied to the right DLPFC during 4 weeks of 25 PD patients. Authors found a significant improvement in PD symptoms with active compared to sham-rTMS. At the end of the protocol, response rate for PD was 50 % with active rTMS and 8 % with sham-rTMS. Non-responder patients could choose for more 4 weeks of treatment, and after this period of active rTMS, response rate was 67 % for PD and 50 % for depressive symptoms, with clinical improvements maintained for 6 months follow-up. One year later, Deppermann et al. [78] investigated the effects of iTBS above the DLPFC on cognition in 44 PD patients. Patients were divided into two groups, one receiving 15 sessions of real-iTBS sessions above the left DLPFC plus psychoeducation and the other receiving 15 sessions of sham-iTBS plus psychoeducation. Before first and after last iTBS-treatment, cortical activity during a verbal fluency task was assessed via functional near-infrared spectroscopy (fNIRS) and compared to healthy control individuals. At baseline, a hypofrontality (i.e., including DLPFC) was observed in PD patients significantly different from activation of controls; however when real-iTBS was applied there was no increase in prefrontal activity in PD patients. Interestingly, after application of sham-iTBS, a significant increase in the left inferior frontal gyrus (IFG) was found during the phonological task.

The limitation of these findings are the lack of RCTs, the most of studies were case reports and open studies with a statistically small sample and without a standardized protocol that difficult

making generalizations. The randomized double blind studies showed inconclusive findings related to therapeutic effectiveness and thus opened more questions and avenues to be discovered than possible conclusions. Among them, we can highlight the disease duration, the appropriate area of stimulation, as well as the type of frequency used. So many questions raise the need for better understanding the PD pathophysiology in order to reach therapeutic effectiveness.

17.7 Safety and rTMS

Despite the safety of rTMS application, the assessment of clinical conditions and side effects should get special attention. The use of hearing protection and hearing evaluation in patients with complaints of hearing loss or losses observed, as well as the electroencephalography record before, during and after rTMS, allow some precautions during the procedure. Among the adverse effects known, the seizure induction, it is quite rare standing in the range of 1.4 % in epileptic patients and less than 1 % in normal patients. Other potential adverse effects arising from a consensus conference held in Siena in 2008 [32], are summarized in Table 17.1.

In addition to adverse effects, there are contraindications to rTMS, and the presence of metals implanted in the body (aneurysm clip), cochlear prosthesis, or any condition that may increase the risk of seizure development. Other relevant factor deals with the treatment protocols, not exceeding the recommended limits with respect to frequency. Cancer, infections, metabolic brain injuries even without a history of seizures, alcoholism, drug use and pregnancy are still other factors that should be considered. Considering the contraindications to rTMS, a security questionnaire composed of 15 questions was developed in a consensus conference [32]:

1. Do you have epilepsy or have you ever had a convulsion or a seizure?
2. Have you ever had a fainting spell or syncope? If yes, please describe in which occasion(s).

Table 17.1 Possible adverse effects related to rTMS use

Adverse effects	Single pulse TMS	Paired TMS	Low-frequency rTMS	High-frequency rTMS	Theta burst
Induction of seizures	Rare	Not reported	Rare (protective effect)	Possible (1.4 % risk in epileptic and less than 1 % in normal persons)	Possible (seizure in a normal individual with cTBS)
Momentary induction of acute hypomania	No	No	Rare	Possible by stimulating the left prefrontal cortex	Not reported
Syncope	Possible as epiphenomenon	Possible as epiphenomenon	Possible as epiphenomenon	Possible as epiphenomenon	Possible
Momentary headache, local pain, neck pain, teeth pain, paresthesia	Possible	Possible, but not ported	Frequent	Frequent	Possible
Momentary changes in hearing	Possible	Possible, but not ported	Possible	Possible	Not reported
Momentary cognitive changes	Not reported	Not reported	Usually negligible	Usually negligible	Momentary loss in working memory
Induced currents in electric circuits	Theoretically possible	Theoretically possible	Theoretically possible	Theoretically possible	Theoretically possible
Structural brain changes	Not reported	Not reported	Inconsistent	Inconsistent	Not reported
Histotoxicity	No	No	Inconsistent	Inconsistent	Not reported
Other momentary biological effects	Not reported	Not reported	Not reported	Momentary changes in hormone serum levels	Not reported

3. Have you ever had severe (i.e., followed by loss of consciousness) head trauma?
4. Do you have any hearing problems or ringing in your ears?
5. Are you pregnant or is there any chance that you might be?
6. Do you have metal in the brain/skull (except titanium)? (e.g., splinters, fragments, clips, etc.).
7. Do you have cochlear implants?
8. Do you have an implanted neurostimulator? (e.g., DBS, epidural/subdural, VNS).
9. Do you have a cardiac pacemaker or intra-cardiac lines or metal in your body?
10. Do you have a medication infusion device?
11. Are you taking any medications? (Please list).
12. Did you ever have a surgical procedure to your spinal cord?
13. Do you have spinal or ventricular derivations?
14. Did you ever undergo rTMS in the past?
15. Did you ever undergo MRI in the past?

Positive answers to one or more questions taking into account the numbers 1–13 do not represent absolute contraindications to rTMS, but the risk/benefit ratio should be carefully balanced by the main researcher responsible for the treatment.

17.8 Conclusion

So far, there is no conclusive evidence of the efficacy of rTMS as a treatment for PD. While positive results were found in most of studies, various treatment parameters, such as location, frequency, intensity and duration have been used unsystematically, making difficult the interpretation of results and providing little guidance about which treatment parameters (or that is, the stimulus location and frequency) may be more useful for the PD treatment [1, 2, 37, 38]. A possible explanation with regard to the efficacy of rTMS in the PD treatment is limited by the focal nature of stimulation, since only the superficial layers of the cerebral cortex can be directly affected.

Effects on subcortical areas are idealized to occur indirectly through synaptic cross-connections [2, 37]. Therefore, further studies are needed to clearly determine the role of rTMS in PD treatment.

References

1. Paes F, Machado S, Arias-Carrión O, Velasques B, Teixeira S, Budde H, et al. The value of repetitive transcranial magnetic stimulation (rTMS) for the treatment of anxiety disorders: an integrative review. CNS Neurol Disord Drug Targets. 2011; 10:610–20.
2. Machado S, Paes F, Velasques B, Teixeira S, Piedade R, Ribeiro P, et al. Is rTMS an effective therapeutic strategy that can be used to treat anxiety disorders? Neuropharmacology. 2012;62(1):125–34.
3. Nardi AE, Valença AM, editors. Transtorno de pânico: diagnóstico e tratamento. Rio de Janeiro: Guanabara Koogan; 2005.
4. Katschnig H, Amering M, Stolk JM, Klerman GL, Ballenger JC, Briggs A, et al. Long-term follow-up after a drug trial for panic disorder. Br J Psychiatry. 1995;167(4):487–94.
5. Barker AT, Jalinous R, Freeston IL. Non-invasive magnetic stimulation of human motor cortex. Lancet. 1985;1:1106–7.
6. Tyc F, Boyadjian A. Cortical plasticity and motor activity studied with transcranial magnetic stimulation. Rev Neurosci. 2006;17:469–95.
7. Nahas Z, Lomarev M, Roberts DR, Shastri A, Lorberbaum JP, Teneback C, et al. Unilateral left prefrontal transcranial magnetic stimulation (TMS) produces intensity-dependent bilateral effects as measured by interleaved BOLD fMRI. Biol Psychiatry. 2001;50:712–20.
8. Mills KR. Magnetic stimulation of the human nervous system. New York: Oxford University Press; 1999.
9. Cadwell J. Principles of magnetoelectric stimulation. In: Chokroverty S, editor. Magnetic stimulation in clinical neurophysiology. Boston: Butterworths; 1990.
10. Barker AT. The history and basic principles of magnetic nerve stimulation. Electroencephalogr Clin Neurophysiol Suppl. 1999;51:3–21.
11. Hallett M. Transcranial magnetic stimulation: a primer. Neuron. 2007;55:187–99.
12. Machado S, Bittencourt J, Minc D, Portella CE, Velasques B, Cunha M, et al. Therapeutic applications of repetitive transcranial magnetic stimulation in clinical neurorehabilitation. Funct Neurol. 2008;23(3):113–22.
13. Walsh V, Rushworth M. A primer of magnetic stimulation as a tool for neuropsychology. Neuropsychologia. 1999;37(2):125–35.
14. Pascual-Leone A, Bartres-Faz D, Keenan JP. Transcranial magnetic stimulation: studying the

brain-behaviour relationship by induction of 'virtual lesions'. Philos Trans R Soc Lond B Biol Sci. 1999;354(1387):1229–38.

15. Pascual-Leone A, Walsh V, Rothwell J. Transcranial magnetic stimulation in cognitive neuroscience-virtual lesion, chronometry, and functional connectivity. Curr Opin Neurobiol. 2000;10(2):232–7.

16. Terao Y, Ugawa Y, Suzuki M, Sakai K, Hanajima R, Gemba-Shimizu K, et al. Shortening of simple reaction time by peripheral electrical and submotor-threshold magnetic cortical stimulation. Exp Brain Res. 1997;115(3):541–5.

17. Hoffman RE, Hawkins KA, Gueorguieva R, Boutros NN, Rachis F, Carroll K, et al. Transcranial magnetic stimulation of left temporoparietal cortex and medication-resistant auditory hallucinations. Arch Gen Psychiatry. 2003;60:49–56.

18. Huang YZ, Edwards MJ, Rounis E, Bhatia KP, Rothwell JC. Theta burst stimulation of the human motor cortex. Neuron. 2005;45(2):201–6.

19. Eichammer P, Johann M, Kharraz A, Binder H, Pittrow D, Wodarz N, et al. High-frequency repetitive transcranial magnetic stimulation decreases cigarette smoking. J Clin Psychiatry. 2003;64(8):951–3.

20. Wagner T, Gangitano M, Romero R, Théoret H, Kobayashi M, Anschel D, et al. Intracranial measurement of current densities induced by transcranial magnetic stimulation in the human brain. Neurosci Lett. 2004;354(2):91–4.

21. Salinas FS, Lancaster JL, Fox PT. 3D modeling of the total electric field induced by transcranial magnetic stimulation using the boundary element method. Phys Med Biol. 2009;54(12):3631–47.

22. Roth Y, Amir A, Levkovitz Y, Zangen A. Three-dimensional distribution of the electric field induced in the brain by transcranial magnetic stimulation using figure-8 and deep H-coils. J Clin Neurophysiol. 2007;24(1):31–8.

23. Nagarajan SS, Durand DM. A generalized cable equation for magnetic stimulation of axons. IEEE Trans Biomed Eng. 1996;43(3):304–12.

24. Rotem A, Moses E. Magnetic stimulation of one-dimensional neuronal cultures. Biophys J. 2008; 94(12):5065–78.

25. Bohning DE, Pecheny AP, Epstein CM, Speer AM, Vincent DJ, Dannels W, et al. Mapping transcranial magnetic stimulation (TMS) fields in vivo with MRI. Neuroreport. 1997;8(11):2535–8.

26. Stokes MG, Chambers CD, Gould IC, English T, McNaught E, McDonald O, et al. Distance-adjusted motor threshold for transcranial magnetic stimulation. Clin Neurophysiol. 2007;118(7):1617–25.

27. Cukic M, Kalauzi A, Ilic T, Miskovic M, Ljubisavljevic M. The influence of coil-skull distance on transcranial magnetic stimulation motor-evoked responses. Exp Brain Res. 2009;192(1):53–60.

28. Zangen A, Roth Y, Voller B, Hallett M. Transcranial magnetic stimulation of deep brain regions: evidence

for efficacy of the H-coil. Clin Neurophysiol. 2005;116(4):775–9.

29. Pitcher D, Walsh V, Yovel G, Duchaine B. RTMS evidence for the involvement of the right occipital face area in early face processing. Curr Biol. 2007; 17(18):1568–73.

30. McConnell KA, Nahas Z, Shastri A, Lorberbaum JP, Kozel FA, Bohning DE, et al. The transcranial magnetic stimulation motor threshold depends on the distance from coil to underlying cortex: a replication in healthy adults comparing two methods of assessing the distance to cortex. Biol Psychiatry. 2001; 49(5):454–9.

31. Paus T, Jech R, Thompson CJ, Comeau R, Peters T, Evans AC. Transcranial magnetic stimulation during positron emission tomography: a new method for studying connectivity of the human cerebral cortex. J Neurosci. 1997;17(9):3178–84.

32. Rossi S, Hallett M, Rossini PM, Pascual-Leone A. Safety of RTMS Consensus Group. Safety, ethical considerations, and application guidelines for the use of transcranial magnetic stimulation in clinical practice and research. Clin Neurophysiol. 2009;120(12): 2008–39.

33. Deblieck C, Thompson B, Iacoboni M, Wu AD. Correlation between motor and phosphene thresholds: a transcranial magnetic stimulation study. Hum Brain Mapp. 2008;29(6):662–70.

34. Schluter ND, Rushworth MF, Mills KR, Passingham RE. Signal-, set-, and movement-related activity in the human premotor cortex. Neuropsychologia. 1999; 37(2):233–43.

35. Devlin JT, Matthews PM, Rushworth MF. Semantic processing in the left inferior prefrontal cortex: a combined functional magnetic resonance imaging and transcranial magnetic stimulation study. J Cogn Neurosci. 2003;15(1):71–84.

36. O'Shea J, Muggleton NG, Cowey A, Walsh V. Timing of target discrimination in human frontal eye fields. J Cogn Neurosci. 2004;16(6):1060–7.

37. de Graaf TA, Sack AT. Null results in RTMS: from absence of evidence to evidence of absence. Neurosci Biobehav Rev. 2011;35(3):871–7.

38. Sandrini M, Umiltà C, Rusconi E. The use of transcranial magnetic stimulation in cognitive neuroscience: a new synthesis of methodological issues. Neurosci Biobehav Rev. 2011;35(3):516–36.

39. Marzi CA, Miniussi C, Maravita A, Bertolasi L, Zanette G, Rothwell JC, et al. Transcranial magnetic stimulation selectively impairs interhemi- spheric transfer of visuo-motor information in humans. Exp Brain Res. 1998;118(3):435–8.

40. Dormal V, Andres M, Pesenti M. Dissociation of numerosity and duration processing in the left intraparietal sulcus: a transcranial magnetic stimulation study. Cortex. 2008;44(4):462–9.

41. Knops A, Nuerk HC, Sparing R, Foltys H, Willmes K. On the functional role of human parietal cortex in

number processing: how gender mediates the impact of a 'virtual lesion' induced by rTMS. Neuropsychologia. 2006;44(12):2270–83.

42. Sandrini M, Rossini PM, Miniussi C. The differential involvement of inferior parietal lobule in number comparison: an rTMS study. Neuropsychologia. 2004;42(14):1902–9.

43. Cappelletti M, Barth H, Fregni F, Spelke ES, Pascual-Leone A. rTMS over the intraparietal sulcus disrupts numerosity processing. Exp Brain Res. 2007;179(4): 631–42.

44. Cohen Kadosh R, Cohen Kadosh K, Schuhmann T, Kaas A, Goebel R, Henik A, et al. Virtual dyscalculia induced by parietal-lobe TMS impairs automatic magnitude processing. Curr Biol. 2007;17(8): 689–93.

45. Herwig U, Cardenas-Morales L, Connemann BJ, Kammer T, Schönfeldt-Lecuona C. Sham or real-post hoc estimation of stimulation condition in a randomized transcranial magnetic stimulation trial. Neurosci Lett. 2010;471(1):30–3.

46. Lisanby SH, Gutman D, Luber B, Schroeder C, Sackeim HA. Sham RTMS: intracerebral measurement of the induced electrical field and the induction of motor-evoked potentials. Biol Psychiatry. 2001; 49(5):460–3.

47. Loo CK, Taylor JL, Gandevia SC, McDarmont BN, Mitchell PB, Sachdev PS. Transcranial magnetic stimulation (TMS) in controlled treatment studies: are some "sham" forms active? Biol Psychiatry. 2000; 47(4):325–31.

48. Tsubokawa T, Katayama Y, Yamamoto T, Hirayama T, Koyama S. Chronic motor cortex stimulation in patients with thalamic pain. J Neurosurg. 1993;78: 393–401.

49. Shah DB, Weaver L, O'Reardon JP. Transcranial magnetic stimulation: a device intended for the psychiatrist's office, but what is its future clinical role? Expert Rev Med Devices. 2008;5(5):559–66.

50. O'Reardon JP, Solvason HB, Janicak PG, Sampson S, Isenberg KE, Nahas Z, et al. Efficacy and safety of transcranial magnetic stimulation in the acute treatment of major depression: a multi-site randomized controlled trial. Biol Psychiatry. 2007;62(11): 1208–16.

51. Rossi S, Ferro M, Cincotta M, Ulivelli M, Bartalini S, Miniussi C, et al. A real electro-magnetic placebo (REMP) device for sham transcranial magnetic stimulation (TMS). Clin Neurophysiol. 2007;118(3): 709–16.

52. Lisanby SH, Kinnunen LH, Crupain MJ. Applications of TMS to therapy in psychiatry. J Clin Neurophysiol. 2002;19(4):344–60.

53. Dileone M, Profice P, Pilato F, Ranieri F, Capone F, Musumeci G, et al. Repetitive transcranial magnetic stimulation for ALS. CNS Neurol Disord Drug Targets. 2010;9(3):331–4.

54. Fitzgerald PB, Benitez J, de Castella A, Daskalakis ZJ, Brown TL, Kulkarni J. A randomized, controlled trial of sequential bilateral repetitive transcranial

magnetic stimulation for treatment-resistant depression. Am J Psychiatry. 2006;163:88–94.

55. Iyer MB, Schelper N, Wassermann EM. Priming stimulation enhances the depressant effect of low-frequency repetitive transcranial magnetic stimulation. J Neurosci. 2003;23:10867–72.

56. Cheeran B, Talelli P, Mori F, Koch G, Suppa A, Edwards M, et al. A common polymorphism in the brain-derived neurotrophic factor gene (BDNF) modulates human cortical plasticity and the response to rTMS. J Physiol. 2008;586:5717–25.

57. Kleim JA, Chan S, Pringle E, Schallert K, Procaccio V, Jimenez R, et al. BDNF val66met polymorphism is associated with modified experience dependent plasticity in human motor cortex. Nat Neurosci. 2006; 9:735–7.

58. Egan MF, Kojima M, Callicott JH, Goldberg TE, Kolachana BS, Bertolino A, et al. The BDNF val66met polymorphism affects activity-dependent secretion of BDNF and human memory and hippocampal function. Cell. 2003;112:257–69.

59. Abraham WC, Bear MF. Metaplasticity: the plasticity of synaptic plasticity. Trends Neurosci. 1996;19: 126–30.

60. Turrigiano GG, Leslie KR, Desai NS, Rutherford LC, Nelson SB. Activity dependent scaling of quantal amplitude in neocortical neurons. Nature. 1998;391: 892–6.

61. Davies CH, Starkey SJ, Pozza MF, Collingridge GL. GABA autoreceptors regulate the induction of LTP. Nature. 1991;349:609–11.

62. Siebner HR, Lang N, Rizzo V, Nitsche MA, Paulus W, Lemon RN, et al. Preconditioning of low-frequency repetitive transcranial magnetic stimulation with transcranial direct current stimulation: evidence for homeostatic plasticity in the human motor cortex. J Neurosci. 2004;24:3379–85.

63. Coutinho FC, Dias GP, do Nascimento Bevilaqua MC, Gardino PF, Pimentel Range B, Nardi AE. Current concept of anxiety: implications from Darwin to the DSM-V for the diagnosis of generalized anxiety disorder. Expert Rev Neurother. 2010;10:1307–20.

64. Tallman JF, Paul SM, Skolnick P, Gallager DW. Receptors for the age of anxiety: pharmacology of the benzodiazepines. Science. 1980;207:274–81.

65. Zwanzger P, Fallgatter AJ, Zavorotnyy M, Padberg F. Anxiolytic effects of transcranial magnetic stimulation e an alternative treatment option in anxiety disorders? J Neural Transm. 2009;116:767–75.

66. Pallanti S, Bernardi S. Neurobiology of repeated transcranial magnetic stimulation in the treatment of anxiety: a critical review. Int Clin Psychopharmacol. 2009;24:163–73.

67. Ressler KJ, Mayberg HS. Targeting abnormal neural circuits in mood and anxiety disorders: from the laboratory to the clinic. Nat Neurosci. 2007;10:1116–24.

68. Heller W, Nitschke JB, Etienne MA, Miller GA. Patterns of regional brain activity differentiate types of anxiety. J Abnorm Psychol. 1997;106: 376–85.

69. Keller J, Nitschke JB, Bhargava T, Deldin PJ, Gergen JA, Miller GA, et al. Neuropsycological differentiation of depression and anxiety. J Abnorm Psychol. 2000;109:3–10.
70. Hoffman RE, Cavus I. Slow transcranial magnetic stimulation, long-term depotentiation, and brain hyperexcitability disorders. Am J Psychiatry. 2002; 159:1093–102.
71. George MS, Wassermann EM, Post RM. Transcranial magnetic stimulation: a neuropsychiatric tool for the 21st century. J Neuropsychiatry Clin Neurosci. 1996;8:373–82.
72. Sakkas P, Psarros C, Papadimitriou GN, Theleritis CG, Soldatos CR. Repetitive transcranial magnetic stimulation (rTMS) in a patient suffering from comorbid depression and panic disorder following a myocardial infarction. Prog Neuropsychopharmacol Biol Psychiatry. 2006;30(5):960–2.
73. Mantovani A, Lisanby SH, Pieraccini F, Ulivelli M, Castrogiovanni P, Rossi S. Repetitive Transcranial Magnetic Stimulation (rTMS) in the treatment of panic disorder (PD) with comorbid major depression. J Affect Disord. 2007;102(1–3):277–80.
74. Dresler T, Ehlis AC, Plichta MM, Richter MM, Jabs B, Lesch KP, et al. Panic disorder and a possible treatment approach by means of high-frequency rTMS: a case report. World J Biol Psychiatry. 2009;10(4 Pt 3): 991–7.
75. Machado S, Santos V, Paes F, Arias-Carrión O, Carta MG, Silva AC, et al. Repetitive transcranial magnetic stimulation (rTMS) to treat refractory panic disorder patient: a case report. CNS Neurol Disord Drug Targets. 2014;13(6):1075–8.
76. Prasko J, Zálesk R, Bares M, Horácek J, Kopecek M, Novák T, et al. The effect of repetitive transcranial magnetic stimulation (rTMS) add on serotonin reuptake inhibitors in patients with panic disorder: a randomized, double blind sham controlled study. Neuro Endocrinol Lett. 2007;28(1):33–8.
77. Mantovani A, Aly M, Dagan Y, Allart A, Lisanby SH. Randomized sham controlled trial of repetitive transcranial magnetic stimulation to the dorsolateral prefrontal cortex for the treatment of panic disorder with comorbid major depression. J Affect Disord. 2013;144(1–2):153–9.
78. Deppermann S, Vennewald N, Diemer J, Sickinger S, Haeussinger FB, Notzon S, et al. Does rTMS alter neurocognitive functioning in patients with panic disorder/agoraphobia? An fNIRS-based investigation of prefrontal activation during a cognitive task and its modulation via sham-controlled rTMS. Biomed Res Int. 2014;2014:542526.

Exercise in Panic Disorder: Implications for Disorder Maintenance, Treatment and Physical Health

18

Aline Sardinha and Claudio Gil Soares de Araújo

Contents

A. Sardinha (✉)
Laboratory of Panic and Respiration, Institute of Psychiatry, Federal University of Rio de Janeiro, Rio de Janeiro, Brazil

Cognitive Therapy Association of Rio de Janeiro (ATC-Rio), Rio de Janeiro, Brazil
e-mail: contato@alinesardinha.com

C.G.S. de Araújo
Heart Institute Edson Saad, Federal University of Rio de Janeiro, Rio de Janeiro, Brazil

Exercise Medicine Clinic (CLINIMEX), Rio de Janeiro, Brazil
e-mail: cgaraujo@iis.com.br

Abstract

According to cognitive behavioral models (CBT), panic attacks (PA) arise from distorted and catastrophic interpretations of bodily symptoms. As exercising involves exposure to physiological stimuli similar to those experienced during PAs, patients often experience anxiety and avoid exercising. Exercise avoidance and low levels of everyday physical activity turn out to promote a sedentary lifestyle as an indirect effect of Panic Disorder (PD), with deleterious health impacts. Interoceptive exposure techniques – the voluntary exposure of the patient to autonomic manifestations – have a therapeutic effect in the treatment of panic by promoting habituation to physiological cues contributing to break the hypervigilance-anxiety-panic-avoidance cycle. Preliminary results indicating a potential positive role of aerobic exercise interventions designed to promote interoceptive habituation in the treatment of PD have been reported. The inclusion of an exercise protocol in the context of CBT interventions, may enhance motivation to participate and endure the discomfort and anxiety provoked by exercise using exposure therapy rationale. This chapter reviews and discusses the main results available in the literature of including exercise in the treatment of PD, as well as provides an unique cognitive behavioral perspective to the understanding of the relationship between panic disorder and exercise. We also

© Springer International Publishing Switzerland 2016
A.E. Nardi, R.C.R. Freire (eds.), *Panic Disorder*, DOI 10.1007/978-3-319-12538-1_18

provide specific recommendations for exercise testing, adherence promotion and particular issues that need to be addresses to successfully include PD patients in exercise protocols, based on authors experience.

Keywords

Panic disorder • Anxiety • Cognitive behavioral therapy • Cognitive therapy • Exercise • Physical activity • Anxiety sensitivity

18.1 Cognitive Behavioral Approach to Panic Disorder

Panic disorder (PD) patients present anxiety regarding the occurrence of benign body symptoms and autonomic arousal, so that situations that could elicit the feared somatic manifestations are commonly avoided. These patients also present significant global functioning deficits and psychological problems, a greater incidence of psychiatric disorders, suicidal ideation, psychological stress, activity restrictions and chronic physical diseases, and poorer indices of physical and mental health [1].

According to cognitive behavioral models, panic attacks (PA) arise from distorted and catastrophic interpretations of bodily symptoms. Such interpretations increase arousal and intensify bodily sensations, generating more catastrophic interpretations and anxiety in a rapid spiral that leads to panic. Repetition of attacks make individuals increasingly more sensitive to internal stimuli and to the situations in which they occur, as well as to heighten surveillance of any physical sensation. Combined with that is anticipatory anxiety and catastrophic interpretations of symptoms. The fear-conditioned behavior leads the individual to avoid somatic symptoms or places associated with previous attacks [2].

Cognitive behavioral therapy (CBT) is considered, along with pharmacological treatment with some antidepressants, the gold-standard treatment for PD [3]. The main goals of CBT are correcting catastrophic interpretations, providing

anxiety coping strategies and extinguishing conditioned fears of body sensations, through interoceptive exposure techniques, and avoidances [2]. The rationale for this intervention is that patients can learn how to manage anxiety states and symptoms, as well as reinterpret the environmental and interoceptive cues that trigger a PA in a more adaptive fashion, in order not to fear somatic arousal. In this sense, the cognitive behavioral approach is the one that provides the best rationale for the proposal of introducing exercise in the treatment of PD. Exercise can be used, in the context of CBT, as a trigger for somatic arousal to which the patient would be exposed in order to promote habituation.

One of the main clinical strategies in the cognitive behavioral treatment of PD is interoceptive exposure. Interoceptive exposure techniques – the voluntary exposure of the patient to autonomic manifestations – have a therapeutic effect in the treatment of PD by promoting habituation to physiological cues and consequent anxiety reduction, contributing therefore to break the hypervigilance-anxiety-panic-avoidance cycle [4]. This technique consists in having patients repeatedly induce and experience their feared physical sensations (e.g., shortness of breath, heart palpitations, dizziness) as a mean of reducing their fear of those sensations through habituation and cognitive restructuring [5]. The exercises should be associated to symptoms that cause discomfort or fear in each individual. Although literature shows that interoceptive exposure techniques were successfully in producing fear habituation [6, 7], there is still a discussion regarding its efficacy as a treatment alone or in the context of CBT. The overall idea, however, is that it should be included in the treatment package paired with cognitive restructuring.

More recently, aerobic exercise training has been associated with improvement of symptoms in PD patients and acute antipanic effect in patients [8–10] and control subjects [11]. The data on the use of exercise in patients with PD allow us to consider that the sensitivity to the symptoms of anxiety could be treated within the context of CBT, using the supervised practice of physical exercises as a desensitization tool for interoceptive exposure. Despite the scarce investigation on this field until recently, it is now starting to be considered as a useful tool in the treatment of anxiety and PD [12].

18.2   The Role of Anxiety in Promoting Sedentary Behavior

Anxiety sensitivity (AS), or the belief that anxiety-related sensations can be threatening or induce negative consequences, plays an important role in the etiology and maintenance of PD [13]. The single presence of AS, regardless meeting full criteria for PD, often predicts panic symptoms in response to biological challenges that provoke feared bodily sensations, consisting of a risk factor for panic attacks (PA) and perhaps PD [14]. It has been hypothesized, additionally, that high AS might be the missing piece to explain agoraphobic avoidance in the absence of PA [15]. Individuals with PD or those sensitive to anxiety – who fear somatic arousal – would be more sensitive to interoceptive stimuli, leading autonomic alterations, whether natural or provoked, to transform into factors that trigger anxiety and PAs [16]. See Fig. 18.1.

As aerobic exercise involves exposure to physiological stimuli similar to those experienced during anxiety reactions, PD patients often experience anxiety and avoid exercising. In this sense, it is possible that the autonomic alterations triggered by physical exercise become the factor responsible for the phobic avoidance of this activity [16, 17]. See Fig. 18.2.

In a study designed to evaluate the acceptance of an exercise program in patients with anxiety disorders, only 45 % of participants presented the recommended levels of physical activity for health at baseline [18]. Consistent with that, in a cardiovascular fitness assessment, PD patients exhibited lower maximum oxygen uptake (VO_2 max) and decreased exercise tolerance when compared to healthy subjects [19]. These findings are consistent in literature and confirmed by direct interviews indicating low physical activity habits [20].

A common characteristic in PD patients is the presence of health-related anxiety. This specific type of anxiety that leads to increased worrying about one's health and the belief that normal bodily symptoms are threatening, harmful and medically serious, despite evidence of the contrary. More specifically related to avoidance of aerobic exercises might be a particular subtype of health-related anxiety called cardiac anxiety, which is a condition characterized by abrupt and recurrent sensations of thoracic pain, without a physical disease to explain it, that are associated with significant concern over its potential consequences. This symptom is also associated with safety behaviors, such as verification of the heart rate, hypervigilance of symptoms and repeated medical consultations [21].

The cognitive behavioral model proposed by Zvolensky et al. to explain cardiac anxiety highlights the role of selective attention focused on cardiovascular symptoms and of interoceptive conditioning in the origin of PA with limited symptoms (only cardio-respiratory symptoms and acute thoracic pain) [22]. The presence of such characteristics could predispose susceptible individuals to the development of the syndrome.

There is considerable evidence implicating heartbeat perception accuracy and AS in the development of panic in adults and children, as increased panic/somatic symptoms are associated with an enhanced ability to perceive internal physiological cues and fear of such sensations [23, 24]. Ehlers et al. found that patients who seek medical help for benign palpitations can be distinguished from those with clinically significant arrhythmias [25]. Patients with arrhythmias rarely reported palpitations and were more likely to perceive their heartbeats accurately than control subjects with sinus rhythm . Individuals with awareness of sinus rhythm could be distinguished from those with arrhythmia by several variables: female sex, higher prevalence of PD, higher heart rate, lower levels of physical activity, fear of bodily sensations and depression. In fact, there is evidence of differential effects of anxiety sensitivity and heart-focused anxiety as a function of gender, with higher prevalence of heart-focused anxiety in women. Cardiac anxiety could be,

Fig. 18.1 Anxiety sensitivity cycle

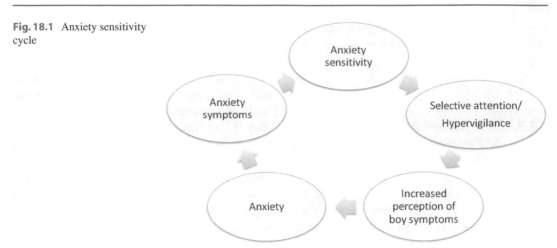

Fig. 18.2 Exercise as a trigger for panic attacks

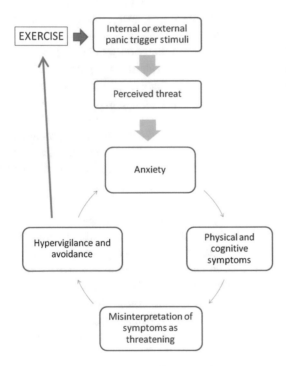

patients and that it plays a special role in exercise avoidance. It may be possible, for example, that some individuals believe that exercise is harmful and avoid it, because it produces some short-term discomfort and cardiac sensations. In the Cardiac Anxiety Questionnaire, physical activities appear as avoided situations in both formal exercise sessions and everyday activities that involve physical effort. The instrument items such as "I avoid physical exertion", "I avoid exercise or other physical work", "I avoid activities that make my heart beat faster" and "I avoid activities that make me sweat" are clear examples of how cardiac anxiety can promote sedentary behavior in PD patients [26].

The rationale of the positive impact of physical activity on anxiety sensitivity relies on the assumption that exercising could be seen as an indirect desensitization process, since autonomic alterations triggered by exercising are similar to anxiety symptoms, but elicited in a safe context [23, 30], using the same underlying assumption of interoceptive exposure in CBT.

therefore, a potentially distinct subtype of panic disorder, with specific characteristics and treatment strategies comparable with those that occur with respiratory subtype, although this hypothesis still needs to be verified [12].

A third component of cardiac anxiety is avoidant behavior related to situations that could potentially trigger anxiety. It is probable that cardiac anxiety is related to the AS observed in these

18.3 Sedentary Behavior in the Panic Cycle Maintenance

It is possible that reduced aerobic fitness might even contribute to the pathophysiology of PD [8]. Once sedentary individuals tend to present increased sympathetic response to physical effort during daily activities [27], everyday situations might trigger more anxiety responses and consequent avoidance. Smits and Zvolensky [28] found that current physical inactivity was significantly associated with greater levels of AS and severity of PD in a clinical sample. Ehlers et al. [25] reported that patients that seek medical help for benign palpitations could be distinguished from those with clinically relevant arrhythmias, among other variables, by presenting lower levels of physical activity and higher levels of AS. In addition, a case-control study performed with sedentary and exercising cardiac patients found that individuals regularly attending to a supervised exercise program presented significantly less PD and cardiac anxiety [29].

18.4 Panic Disorder and Physical Health

Once we understand how human psychophysiology functions as an integrated system, both the physiological and the psychological variables involved could be manipulated to improve the organism's homeostasis. As science evolves and more complex mechanisms linking psychological and physiological functioning are disclosed, it becomes mandatory to address health in a holistic manner.

Anxiety has recently received special attention as a potential risk and aggravation factor for cardiovascular diseases, notably in terms of the interface between the autonomic changes caused by chronic anxiety and PAs, and the autonomic mechanisms of cardiac regulation [12, 30]. Anxiety triggers activation of the human stress system through behavioral and physiological changes that improve the ability of the organism to adjust homeostasis and increase its chances for survival. These processes appear to adversely affect autonomic and hormonal regulation, resulting in metabolic abnormalities, inflammation, insulin resistance, and endothelial dysfunction [31].

The association of anxiety, particularly PD, with coronary artery disease may even increase cardiac mortality in this population [32].

Moreover, patients who present an association between panic and physical diseases experience more deficits in a wide range of modalities than those who suffer from only one of these conditions [33].

In the long run, exercise avoidance and low levels of everyday physical activity turn out to promote a sedentary lifestyle as an indirect effect of PD [34]. Paradoxically, elevated health-related anxiety, AS and PD seem to contribute to an unhealthy lifestyle instead of motivating the patient towards more healthy habits, such as exercising. As indicated by Meyer et al. [35], low fitness observed in PD patients seem to be a byproduct of physical activity avoidance, and can be directly impacted by interventions aiming to achieve non-clinical control levels. As reduced fitness capacity is directly associated with a higher all-cause mortality [36] and increased cardiovascular risk [37] in healthy subjects, the inclusion of an exercise protocol in the treatment of PD can, per se, be useful as a cardio protective strategy for those patients.

Moreover, cohort studies found that PD is a significant independent risk for new onset coronary heart disease, acute myocardial infarction, and cardiac mortality [12, 38, 39]. One of hypothesis is that a sedentary lifestyle and it's metabolic consequences might mediate this association [12, 40]. On the other hand, a recent study showed that, despite presenting a higher risk than normal subjects, PD patients without cardiovascular artery disorder, when submitted to CO_2 challenges, presented a low risk of cardiovascular ischemia [41].

Additionally, patients with anxiety disorders normally present more cortisol in the urine than individuals without psychiatric disorders, while there seems to be no difference in the excretion of catecholamines and serotonin [42]. There is considerable evidence from clinical, cellular and molecular studies that elevated cortisol, particularly when combined with secondary inhibition of sex steroids and growth hormone secretions, causes accumulation of fat in visceral adipose tissues as well as metabolic abnormalities [43].

Glucocorticoid exposure is also followed by stress-induced overeating behavior with increased food intake and leptin-resistant obesity, perhaps disrupting the balance between leptin and neuropeptide Y [43]. Consistent with that, a study using a rodent model of social stress found that the consumption of a high-fat diet during social stress enhances the effect of chronic stress on body composition, adding to the body of knowledge about the mechanisms responsible for the development of obesity, diabetes and, ultimately, metabolic syndrome [44]. The most accepted underlying mechanism relies on the hypothesis that increased activation of the HPA axis could be pathophysiologically involved in the concomitant occurrence of the typical metabolic syndrome risk factors and stress.

Metabolic syndrome is a construct that defines a closely related cluster of factors that increase the risk of coronary artery disease and diabetes mellitus type 2. Although increasing amounts of research efforts have been allocated lately in the attempt to propose a universally accepted pathogenic mechanism and clearly define the diagnostic criteria, there is still a lot of debate concerning these issues [45]. The currently most widely accepted definition was proposed by Alberti et al., in the Joint Interim Statement. This document proposes the presence of three of the following criteria: (1) elevated waist circumference – according to population and country-specific definitions; (2) triglycerides 150 mg/dl or greater; (3) HDL-cholesterol lower than 40 mg/dl in men and 50 mg/dl in women; (4) blood pressure of 130/85 mmHg or greater and (5) fasting glucose 100 mg/dl or greater. This definition highlights that there should be no obligatory component but rather all individual components should be considered on cardiovascular risk prediction [31].

In a previous review, Sardinha et al. [40] found evidence for the hypothesis of an indirect relationship between anxiety and metabolic syndrome as a byproduct of a sedentary lifestyle. Combined with the already established weight gain side-effect of psychotropic medication used to treat anxiety disorders, particularly antidepressants, sedentary behavior might also play a role in raising the risk of obesity and its consequences among anxious patients [46]. Furthermore, a higher prevalence of sedentary habits, obesity and metabolic syndrome associated with anxiety disorders might be possible underlying mechanisms that contribute to increasing cardiovascular risk in this population [12, 34, 40]. See Fig. 18.3.

Once both metabolic and psychological disorders play a role in cardiovascular outcomes, these issues should be considered when addressing cardiac health and designing treatment and prevention interventions. In this sense, mental health professionals should also take responsibility for addressing detectable signs of metabolic disorders, such as visceral obesity, and unhealthy behaviors, such as tabagism, sedentarism and high-fat diet. In this sense, it is possible that exercise plays a double role in the treatment of PD: psychological symptoms and cardiovascular risk reduction.

18.5 Exercise Testing in PD Patients

Exercise testing has been considerably underutilized in the clinical assessment of PD patients, mainly due to misconceptions or false beliefs that these patients will not be able to tolerate or collaborate with the procedure [47]. In this context, it is not surprising to note that there is scarce literature regarding this topic. There is no doubt that assessing aerobic fitness is often a difficult and challenging task in PD patients [12, 47]. Nevertheless, recent clinical experience and limited research evidence suggest that if some characteristics of the PD are taken in account, it is possible in about 2/3 of the PD patients to achieve a truly maximal exercise effort. Even in those patients not willing or amenable to do so, relevant clinical information can be obtained by assessing submaximal exercise responses [47].

In general, PD patients tend to be sedentary, having rarely participated in competitive sports or regularly exercising very vigorously through lifetime [48, 49] and are poorly motivated to perform a true maximal exercise test [47]. Moreover, often PD patients fear and present lower tolerance to exercise-induced distress and tend to associate movement, exercise or sports to

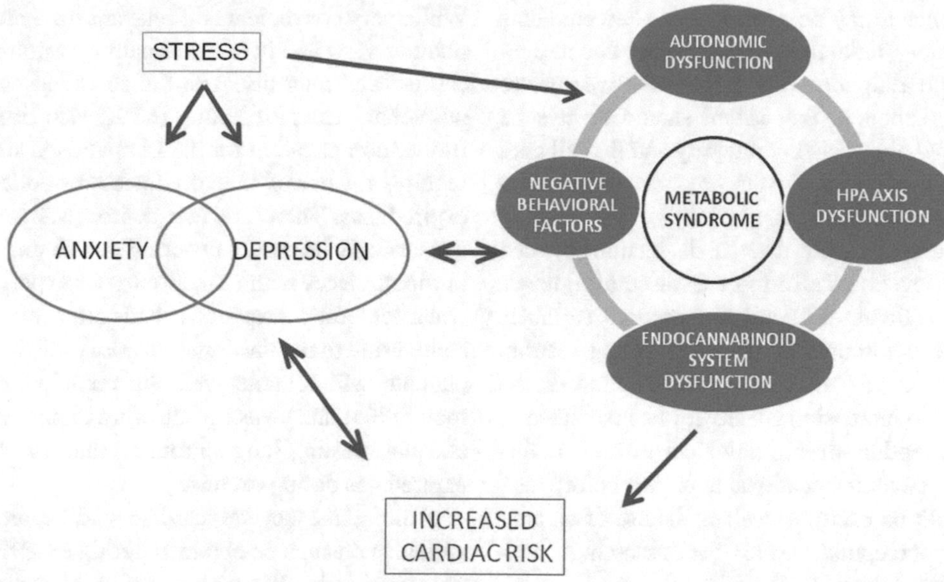

Fig. 18.3 Psychological variables associated with metabolic syndrome

unpleasant situations and even with potentially more relevant clinical symptoms of dyspnea, palpitations or chest pain/discomfort [50, 51]. While by asking the correct questions regarding exercise habits during clinical interview is possible to estimate that aerobic fitness is below the age-predicted level, frequently there is a need for obtaining more precise measurements of exercise cardiopulmonary performance. This is particularly relevant for those PD patients that refer exercise intolerance or other exercise-related clinical abnormalities, situations in which an indication for exercise testing seems to be surely warranted [47].

Analyzing in a broader context, there are several epidemiological studies that have shown that exercise capacity and/or aerobic fitness are strongly related to all-cause mortality in both healthy and unhealthy adults [52–54]. More interestingly, recent data indicated that among adults with no known cardiac diseases and already classified as having low cardiorespiratory fitness there is a clear and significant trend for a poorer prognosis and higher mortality among those in the lower extreme of the distribution [55]. Moreover, it is worthwhile to comment that low exercise capacity and aerobic fitness are

potentially treatable by regular physical exercise and that there are some evidences showing that, if adequate psychological advice and support is provided, PD patients can safely engage and benefit of supervised exercise program [56], including interval training strategies and resistance exercises [57]. So that, even for PD patients with no abnormal cardiorespiratory complaints at rest or exercise, it is reasonable to suggest that, at least, exercise testing data would be useful to more precisely quantity aerobic fitness and to provide relevant information for exercise prescription [47, 57].

In order to carry out a good quality exercise testing in PD patients, several issues should be adequately addressed. First, there is a clear need to establish a good and trustful relationship between the supervising health professional and the patient before the testing. Face-to-face conversations in which the objectives of testing are clearly discussed and the protocol is fully explained are very important steps for a successful outcome. Appropriate information about the testing and the expected body reactions should be provided aiming to lower anticipatory anxiety. Also, the presence of a physician and reassurance about the safety of the test and the symptoms experienced may be considered as key elements for anxiety reduction. While verbal encouragement and a supportive but rather assertive posture are fundamental, the patient should always be reassured about his/her integrity and that all care is being taken to prevent any relevant clinical complication during testing [47].

In our experience, a verbal distinction between anxiety levels and effort-related discomfort levels helped patients differentiate these two variables that are commonly confounded during testing. This could prevent an early interruption of the test due to increasing anxiety levels associated to the perception of physiological effort. In this sense, a psychoeducational approach before testing could be useful as well as the use of an analogue perceptual scale for assessing both variables at the same time.

There are several ways and possibilities to perform a maximal exercise testing. Considering the characteristics of this clinical entity, in our experience after testing over 100 patients with PD, a ramp protocol – small progressive and constant increments of exercise intensity at very short time intervals – carefully chosen in order to achieve exhaustion in about 8–12 min is very convenient. In terms of equipment choice, we strongly favor use cycling in a stationary cycle ergometer rather walking/running in a treadmill. The main reason for this cycling preference is due to the fact that is possible to explain to the patient that in any case or undue discomfort, he/she will be easily able to stop the test by simply ceasing to pedal with no risk of fall or other kind of injury. Stopping the exercise would not be so easy or practical to in a treadmill. Additionally, it is not only possible by using cycling protocols to start at a very low intensity level and to slowly progress towards the maximum but to obtain also a higher quality in electrocardiogram recordings and in blood pressure measurements, further reassuring the physician about the significance of the results.

Since it is common for PD patients to complain of respiratory symptoms [50], we consider that the combination of expired gas analysis is useful for a more comprehensive evaluation of cardiorespiratory responses during exercise.

While very convenient and relevant for a clinical point of view and in terms of pathophysiology of cardiac and lung diseases, the so called cardiopulmonary exercise testing (CPX) also brings a further complication for the PD patients, since it requires the use of an extra device for collecting expired gas. There are two approaches for this question: facial masks covering mouth and nose or mouthpieces with noseclip. In our experience, when carefully explained and after allowing some trial time, the large majority of the PD patients will tolerate well the combination of mouthpiece and noseclip, allowing to perform an adequate resting lung spirometry and to collect expired gas during exercise.

Among the several cardiac and respiratory variables that can be obtained during a CPX [58–61], considering PD patients, there are some that deserve special comments, as briefly described below:

- Heart rate: measured at rest, during submaximal and maximal effort and along the first minutes of post-exercise period. Evaluates the normalcy of cardiac chronotropic response, provides information about autonomic integrity – sympathetic and parasympathetic branches – and offer useful information for exercise prescription by establishing target zones.
- Blood pressure: recorded before, during (ideally at 1-min intervals) and along the first 5-min after exercise. Evaluates normalcy of some components of hemodynamic response at rest and during exercise and provides objective clues regarding abnormal findings or symptoms such as dizziness and vertigo.
- Electrocardiogram: recorded at rest and during/immediately after the end of exercise. Provides relevant information regarding the presence or absence of cardiac arrhythmias, offering chance to relate these findings to reported symptoms of chest discomfort, palpitations, dizziness or pre-syncopal episodes.
- Expired ventilation (VE): represents the amount of air that enters and leaves the lungs at each minute and it is measured in L/min. This value is substantially increased during

exercise and tends to disproportionally increase once the anaerobic threshold exercise intensity is reached. Maximal possible ventilation can be predicted from results obtained during resting spirometry allowing interpreting if exercise performance is hindered by an abnormal ventilatory reserve.

- Aerobic fitness (VO_2 max): this is reflected by the maximal oxygen uptake obtained in a given minute of the exercise testing, almost always achieved at last minute of exercise testing. Represents the main criteria of cardiorespiratory fitness and is strongly and directly related to health and long-term survival in adults [62, 63].
- Anaerobic threshold (% of VO_2 max): corresponds to the maximal exercise intensity, more often expressed as percent of VO_2 max or a given exercise intensity, that can be tolerated before significant metabolic acidosis (as reflected by blood lactate levels and an exponential increase in pulmonary ventilation as related to exercise intensity) ensues. Approximately represents the maximal tolerable intensity of exercise that can be sustained during several minutes before inducing volitional fatigue [63].
- Cardiorespiratory optimal point (COP): represents the minimal value of VE/VO_2 obtained in a given minute of an incremental and graded maximal exercise testing and reflects the best interaction between respiration and circulation. Normal COP values indicate that gas exchange and alveolar dead space are also normal during submaximal exercise [47, 64, 65].

A careful understanding of the cognitive processing of PD patients by the health professionals supervising CPX is advisable in order to adjust the testing to the specific needs of these patients. This should not mean that all PD patients would react to the testing in the same way, but rather that a cognitive-behavioral comprehension of the patient could help the health professional pose the right questions. Depending on each patient's specific beliefs, small adaptations or changes could be introduced in order to minimize the interference of anxiety in mimicking test results. For example, verbal or equipment (e.g. making heart rate and other physiological measurements visible to the patient) feedback can be anxiety-relieving or anxiety-provoking depending on how the patient will handle or interpret these stimuli. Also, equipment noises and signals and the presence of CPR equipment might be interpreted as safe or dangerous stimuli. In this sense, simply asking whether the patient prefer to receive feedback as well as provide explanation about the expected use of any equipment could be of great help and easily feasible.

In summary, starting with appropriate explanation and working to minimize the anticipatory anxiety of the PD patient, exercise testing, preferably including collection and analysis of expired gases (CPX), is well-tolerated and able to provide several clinical relevant results. It also helps to establish an etiology for the cardiorespiratory symptoms and to allow the quantification of the aerobic fitness (VO_2 max), anaerobic threshold and COP, variables that can not only contribute to the understanding of the etiology of effort intolerance but also collaborate to develop a more individually-tailored exercise prescription and to estimate survival.

18.6 Exercise as a Therapeutic Tool in the Treatment of Anxiety and PD

Physical activity is known to positively impact the prognoses of numerous chronic diseases, particularly cardiovascular problems [66]. The prescription of exercise is particularly useful for preventing premature death from all causes, ischemic heart disease, stroke, hypertension, colon and breast cancer, type 2 diabetes, metabolic syndrome, obesity, osteoporosis, sarcopenia, functional dependence and falls in the elderly, cognitive impairment, anxiety and depression. This benefit is observed in both sexes and increases with the volume or intensity of exercise [67]. Regular exercising also seems to play an important role in mental health maintenance [68]. As we have proposed elsewhere, it is plausible that regular exercise could be a useful and simple strategy to address psychological disorders and metabolic alterations simultaneously in both high-risk patients and the general population [12].

Abounding evidence demonstrate the role of exercise in lowering the prevalence of anxiety, raising the question whether exercise may be used in the prevention of psychiatric disorders [68], and particularly in the treatment of PD [69]. It is plausible that the exercise intervention might be more effective in achieving even more symptom improvement if it was conducted as a systematic interoceptive exposure tool in the context of CBT, tailored to involve all of the patients' feared cues systematically rather than just incidental to the exercise schedule. In this sense, if the experience of exercise were accompanied by cognitive restructuring, it could be as effective as regular interoceptive exposure therapy but with other potential benefits, specially cardiac risk reduction [70]. Therefore, it would be possible to reduce both sensitivity to anxiety and cardiac anxiety to favor lifestyle changes, reducing behavioral risk factors and facilitating the adhesion to cardiac rehabilitation programs [71]. See Fig. 18.4.

Preliminary results indicating a potential positive role of aerobic exercise interventions designed to promote interoceptive habituation in the treatment of PD have been reported. Although the evidence for positive effects of exercise on anxiety is growing, its clinical use as an adjunct tool to established treatment approaches like CBT or pharmacotherapy, is still insipient [68].

A systematic review of exercise in the treatment of anxiety disorders hypothesized that the induction of bodily sensations mimic those homeostatic changes associated with anxiety, and thus may alleviate anxiety sensitivity [72]. A study by Faulkner and colleagues [73] suggested that task-unfamiliarity with those aerobic and resistance exercises that are recommended in heart failure patients was associated with higher anticipatory anxiety levels. Such anticipatory anxiety is important to identify and remedy considering that exercise avoidance is common in panic disorder generally and not restricted to the respiratory subtype [68]. As previously proposed while exercise avoidance seem to limit adherence to treatment, increase sedentary habits and worsen cardiovascular prognosis in CHD patients, cognitive behavioral techniques could enhance physical exercise participation as an interoceptive exposure intervention, contributing, therefore, to reduce panic symptoms along with increasing patients fitness and reducing cardiovascular risk [12].

Although exercise has been tested with positive results in the treatment of PD patients [72], the sole presence of anxiety sensitivity, cardiac anxiety or panic disorder negatively impacts adherence to this intervention [67, 68]. In our experience, the use of an exercise protocol in the context of CBT interventions, may enhance motivation to participate and endure the discomfort and anxiety provoked by exercise using the same rationale of an exposure procedure. We have been testing both continuous aerobic exercise protocols and interval training with promising results in reducing exercise avoidance, cardiac anxiety levels and panic disorder symptoms intensity, as well as improving fitness. For further details of continuous and interval aerobic exercise protocols, please see [56, 57].

An acute immediate anti panic effect of aerobic exercise had been demonstrated in laboratory

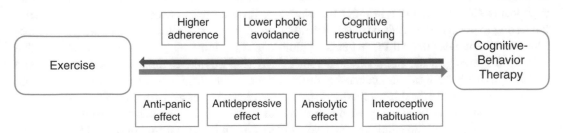

Fig. 18.4 Exercise and cognitive behavioral therapy in the treatment of panic disorder

panic induction protocols using cholecystokinin tetrapeptide (CCK-4) [11] and carbon dioxide (CO_2) inhalation [74]. This effect has been demonstrated both in healthy individuals [75] and PD patients [76], with some evidence for a dose-response relationship favoring moderate to intense exercise in the reduction of panic symptoms. Strohle et al. [77] advocate that this effect is probably due to the observed brain derived neurotrophic factor (BDNF) serum concentration increase yield in PD patients, approaching healthy controls reported levels, after a 30-min exercise session. Another study by the same group also found that the anxiolytic activity of exercise was correlated with the increase in plasma atrial natriuretic peptide (ANP) concentrations [78].

Comparative studies have shown a therapeutic effect of aerobic exercise in the treatment of PD [50]. Broocks et al. [8] compared a 10-week running program with clomipramine and placebo pills and found that regular aerobic exercise alone, versus placebo, was associated with significant clinical improvement, but less effective than clomipramine. It was hypothesized that this effect was mediated by exercise impact in the serotoninergic system, based on the findings that regular physical activity is associated with decreasing somatization, probably through adaptive mechanisms of serotonergic circuits implicated in anxiety and nociception regulation [79]. This was supported further by studies that showed reduced 5-HT1A responsivity in patients with PD [80] and a downregulation of central 5-HT2C receptors in healthy volunteers after 10-weeks of jogging three times a week [81].

Despite the positive effects of exercise alone, a randomized controlled trial that compared the addition of a 10-week aerobic exercise protocol and relaxation training to paroxetine found no differences between interventions [10]. Nocturnal cortisol excretion analysis of those patients yield no relationship between hypothalamo-pituitary-adrenomedullary (HPA) activity and treatment response nor with exercise [82].

Another study that added a home-based walking program to CBT in patients with anxiety disorders found significant decrease in stress, depression and anxiety levels in comparison to the group that received regular CBT [9]. When compared versus CBT, however, a 12-week exercise protocol showed poorer results in the Mobility Inventory (MI), the Agoraphobic Cognitions Questionnaire (ACQ) and Body Sensations Questionnaire (BSQ) than the established treatment [83]. See Table 18.1.

Interestingly, while AS predicts exercise avoidance and low adherence, aerobic exercise-based interventions seem to reduce it. Six 20-min exercise sessions were sufficient to demonstrate a decrease in AS levels [13]. Considering exercise intensity, results by Broman-Fulks et al. [84] indicated that both high- and low-intensity exercise reduced AS. However, high-intensity exercise accounted for more expressive outcomes and was the only intensity found to reduce fear of anxiety-related bodily sensations. Another recent study [85] found that a single session of psycho-education combined with interoceptive exposure was effective to reduce AS and outcomes were maintained in a 6-month follow-up. Yet, while in PD patients exercise accounts for poorer results than CBT, when it comes to reducing AS levels in high-AS score individuals, no additional effect of a cognitive restructuring intervention was observed in a 2-week trial, being both conditions more effective than waiting list control.

It seems that changes in AS mediate the beneficial effects of exercise on anxious and depressed mood [86]. It is plausible that, as aerobic exercise involves exposure to physiological cues similar to those experienced during anxiety reactions, the participation in exercise programs might have a similar role in the treatment of AS and panic to the already established interoceptive exposure techniques. Thus, if exercise is administered in gradual increase intensity, it may promote a habituation effect on the interoceptive conditioning underlying AS and panic [12].

It is also possible that other constructs that participate in the psychopathology of PD, such as cardiac anxiety, agoraphobic cognitions, fear of body symptoms, hypervigilance and avoidance might be differentially influenced by exercise

Table 18.1 Available evidence on the use of exercise in the treatment of PD

Reference	Sample	Anxiety manifestation	Exercise intervention	Duration	Outcome
Broocks et al. [8]	n=46 (exercise group, n=16 and control n=30)	PD with and without agoraphobia	Walking or running a 4-mile route, three times a week	10 weeks	Exercise was associated with significant clinical improvement in PD patients but less effective than treatment with clomipramine
Wedekind et al. [10]	n=75	PD with and without agoraphobia	Running	10 weeks	While paroxetine was superior to placebo, aerobic exercise did not differ from relaxation training in most efficacy measures
Merom et al. [9]	n=74	PD, Generalized anxiety disorder or Social phobia	30-min sessions of moderate-intensity walking (150 min per week)	10 weeks	Exercise+CBT was more effective that CBT alone in reducing scores in self-report depression, anxiety, and stress scales
Esquivel et al. [76]	n=18	PD patients submitted to a panic induction maneuver with 35 % CO_2	Moderate/hard exercise or very-light exercise	Acute intervention	Panic reactions to CO_2 were smaller in patients that performed moderate/hard exercise in contrast to those that performed very-light exercise
Strohle et al. [77]	n=24	PD patients × healthy controls	Aerobic treadmill exercise (30 min at an intensity of 70 % of the maximal oxygen uptake)	Acute intervention	Compared to healthy control subjects, patients with panic disorder had significantly reduced BDNF concentrations at baseline and 30 min of exercise significantly increased BDNF concentrations only in these patients
Strohle et al. [68]	n=24	PD patients × healthy controls submitted to a panic induction maneuver with CCK-4	Aerobic treadmill exercise (30 min at an intensity of 70 % of the maximal oxygen uptake)	Acute intervention	Patients with PD showed increased somatic but not anxiety symptoms after exercise. Exercise reduced the severity of CCK-4-induced panic and anxiety
Esquivel et al. [75]	n=20	Healthy subjects submitted to a panic induction maneuver with 35 % CO_2	Aerobic exercise in a bicycle ergometer reaching >6 mm of blood lactate	Acute intervention	Subjects under the exercise condition reported less panic symptoms than controls after a CO_2 challenge
Strohle et al. [11]	n=15	Healthy subjects submitted to a panic induction maneuver with CCK-4	Aerobic treadmill exercise (30 min at an intensity of 70 % of the maximal oxygen uptake)	Acute intervention	Panic attacks occurred in 12 subjects after rest but in only six subjects after exercise

(continued)

Table 18.1 (continued)

Reference	Sample	Anxiety manifestation	Exercise intervention	Duration	Outcome
Hovland et al. [83]	n=36	PD patients treated with CBT×physical exercise	Three weekly sessions of physical exercise	12 weeks	Group CBT was more effective than group physical exercise for PD
Gomes et al. [56]	n=4	PD patients in an exercise+pharmacotherapy protocol. No control group	Two weekly sessions of treadmill walking at controlled intensity (75% VO$_2$ max)	12 weeks	Exercise was shown to be safe and potentially useful tool as adjunct non-pharmacologic treatment of PD
Sardinha et al. [57]	n=1	Case report of CBT+pharmacotherapy +exercise	12 sessions of interval training treadmill walking	6 weeks	Addition of exercise protocol contributed to reduce anxiety and cardiac anxiety

interventions. It is consistent with recent data from Belem da Silva et al. in which patients with high somatic anxiety showed a significantly higher prevalence of low level of physical activity as compared to those with low somatic anxiety. In this study, somatic symptoms of anxiety remained the only important predictors of low level of physical activity (odds ratio=2.81) [48].

Likewise, even though addition of exercise interventions to regular therapy seem promising, adherence is often mentioned as a challenge in this population, with dropout rates yielding 30%, approximately [8]. A pilot study of acceptability and adherence to CBT+exercise showed a significant drop in exercise session participation along time [18], which is consistent with the high exercise avoidance in PD patients mentioned previously. To the moment, there is a gap in the knowledge on how to best deal with anxiety-related symptoms which hinder patients to participate and benefit from exercise protocols [49].

Recently, an interesting clinical strategy was proposed by Gomes et al., in a pilot-study of aerobic exercise in the treatment of PD. These authors proposed the inclusion of familiarization sessions before initiating the exercise protocol, which consisted of taking one or two sessions to have the patients walk in the treadmill until anticipatory anxiety have decreased and they reported feeling safe to engage in the exercise program [56].

18.7 Discussion

To the moment, aerobic exercise demonstrated no significant effect for the treatment of anxiety disorders alone. Exercise appears to reduce anxiety symptoms but it is less effective than antidepressant medication. Exercise combined with CBT and/or antidepressant medication seem to improve treatment outcomes [87]. The studies involving exercise presented a wide variety of control situations and the effect size of aerobic exercise interventions seem to be highly influenced by the type of control condition. Trials utilizing waitlist/placebo controls and trials that did not control for exercise time reported large effects of aerobic exercise while other trials report no effect of aerobic exercise [88]. In this sense, current evidence does not support the use of aerobic exercise as an effective treatment for anxiety disorders as compared to the control conditions. The present outcomes point to an incremental role of exercise and suggests its usefulness as an adjunct therapy to be added to already established interventions such as pharmacotherapy and CBT in the treatment of PD [9, 12]. Specific investigations on its clinical effects, interaction with standard treatment approaches and details on the optimal type, intensity, frequency and duration that might further support the clinical administration in PD patients are still insipient, with no established guideline in terms of the exercise protocol.

Also, there is scarce information about the specific effect of exercise in the psychological constructs implicated in the psychopathology of panic and whether the observed gains are generalized or mediated by any of these variables. Along with that, although laboratory research on exercise and anxiety present a more rigorous control of physiological variables and exercise intensity [11, 60, 62], longitudinal studies have failed to present elucidative data on the cardiorespiratory changes and fitness condition before and after participation and on the interactions between those variables and anxiety symptoms improvement. Besides, exercise protocols reported so far include general and unsupervised exercise prescriptions, with poor control of exercise intensity, cardiorespiratory and psychological parameters during sessions [8, 9, 69], with an only recent preoccupation in detailing the exercise protocols in the literature [56, 59].

A research design aiming to provide interesting information on the behavior of several variables implicated in the psychopathology of PD could shed light on the specific effect of exercise in each of these. We could hypothesize that the variables that include cognitive information processing, like restructuring of health-related worries and fears and agoraphobic cognitions could suffer an indirect impact of exercise, possibly through naturalistic evidence of safety provided by the previous reduction in panic manifestations and fear of body symptoms.

It is also possible that the reported neurophysiological changes due to exercise [75, 78], promote an immediate anti panic modulation, as patients can experience autonomic arousal induced by exercise in the absence of panic or in the presence of more attenuated anxiety reactions [11, 75, 76, 89]. In this sense, exercise could have an additional advantage over regular interoceptive exposure in terms of the intensity of symptoms that can be elicited without panic reactions, which possibly potentiates habituation.

18.8 Conclusion

The positive role of exercise training in symptom reduction in addition to traditional pharmacotherapy and psychotherapy treatment to PD in authors experience is consistent with previous reports in the literature, that point to a superior effect of the combination of regular therapy with exercise. To the moment, aerobic exercise has demonstrated no significant effect for the treatment of anxiety disorders when administered alone. As exercise is often feared and avoided by PD patients, adherence implies the use of the exposure therapy rationale, in the context of a broader cognitive behavioral intervention. Also, a cognitive behavioral comprehensive approach is helpful to adequately perform exercise testing in this population. The specific therapeutic effects of exercise interventions in each of the variables implicated in the psychopathology of PD, however, deserves further understanding in order to better adapt the inclusion of exercise interventions to individual needs.

References

1. Kinley DJ, Cox BJ, Clara I, Goodwin RD, Sareen J. Panic attacks and their relation to psychological and physical functioning in Canadians: results from a nationally representative sample. Can J Psychiatry. 2009;54(2):113–22.
2. Manfro GG, Heldt E, Cordioli AV, Otto MW. Cognitive-behavioral therapy in panic disorder. Rev Bras Psiquiatr. 2008;30 Suppl 2:s81–7.
3. Mitte K. A meta-analysis of the efficacy of psycho- and pharmacotherapy in panic disorder with and without agoraphobia. J Affect Disord. 2005;88(1):27–45.
4. Domschke K, Stevens S, Pfleiderer B, Gerlach AL. Interoceptive sensitivity in anxiety and anxiety disorders: an overview and integration of neurobiological findings. Clin Psychol Rev. 2010;30(1):1–11.
5. Craske MG, Barlow DH, Meadows E. Mastery of your anxiety and panic: therapist guide for anxiety, panic and agoraphobia (MAP-3). San Antonio, TX: Graywind/Psychological Corporation; 2000.
6. Antony MM, Ledley DR, Liss A, Swinson RP. Responses to symptom induction exercises in panic disorder. Behav Res Ther. 2006;44(1):85–98.
7. Lee K, Noda Y, Nakano Y, Ogawa S, Kinoshita Y, Funayama T, et al. Interoceptive hypersensitivity and interoceptive exposure in patients with panic disorder:

specificity and effectiveness. BMC Psychiatry. 2006;6:32.

8. Broocks A, Bandelow B, Pekrun G, George A, Meyer T, Bartmann U, et al. Comparison of aerobic exercise, clomipramine, and placebo in the treatment of panic disorder. Am J Psychiatry. 1998;155(5):603–9.

9. Merom D, Phongsavan P, Wagner R, Chey T, Marnane C, Steel Z, et al. Promoting walking as an adjunct intervention to group cognitive behavioral therapy for anxiety disorders—a pilot group randomized trial. J Anxiety Disord. 2008;22(6):959–68.

10. Wedekind D, Broocks A, Weiss N, Engel K, Neubert K, Bandelow B. A randomized, controlled trial of aerobic exercise in combination with paroxetine in the treatment of panic disorder. World J Biol Psychiatry. 2010;11(7):904–13.

11. Strohle A, Feller C, Onken M, Godemann F, Heinz A, Dimeo F. The acute antipanic activity of aerobic exercise. Am J Psychiatry. 2005;162(12):2376–8.

12. Sardinha A, Araujo CGS, Soares-Filho GL, Nardi AE. Anxiety, panic disorder and coronary artery disease: issues concerning physical exercise and cognitive behavioral therapy. Expert Rev Cardiovasc Ther. 2011;9(2):165–75.

13. Broman-Fulks JJ, Storey KM. Evaluation of a brief aerobic exercise intervention for high anxiety sensitivity. Anxiety Stress Coping. 2008;21(2):117–28.

14. McNally RJ. Anxiety sensitivity and panic disorder. Biol Psychiatry. 2002;52(10):938–46.

15. Hayward C, Wilson KA. Anxiety sensitivity: a missing piece to the agoraphobia-without-panic puzzle. Behav Modif. 2007;31(2):162–73.

16. Story TJ, Craske MG. Responses to false physiological feedback in individuals with panic attacks and elevated anxiety sensitivity. Behav Res Ther. 2008;46(9):1001–8.

17. Sardinha A, Nardi AE, Zin WA. Are panic attacks really harmless? The cardiovascular impact of panic disorder. Rev Bras Psiquiatr. 2009;31(1):57–62.

18. Phongsavan P, Merom D, Wagner R, Chey T, von Hofe B, Silove D, et al. Process evaluation in an intervention designed to promote physical activity among adults with anxiety disorders: evidence of acceptability and adherence. Health Promot J Austr. 2008;19(2):137–43.

19. Schmidt NB, Lerew DR, Santiago H, Trakowski JH, Staab JP. Effects of heart-rate feedback on estimated cardiovascular fitness in patients with panic disorder. Depress Anxiety. 2000;12(2):59–66.

20. Broocks A, Meyer TF, Bandelow B, George A, Bartmann U, Ruther E, et al. Exercise avoidance and impaired endurance capacity in patients with panic disorder. Neuropsychobiology. 1997;36(4):182–7.

21. Sardinha A, Nardi AE, Araujo CGS, Ferreira MC, Eifert GH. Brazilian Portuguese validated version of the cardiac anxiety questionnaire. Arq Bras Cardiol. 2013;101(6):554–61.

22. Zvolensky MJ, Feldner MT, Eifert GH, Vujanovic AA, Solomon SE. Cardiophobia: a critical analysis. Transcult Psychiatry. 2008;45(2):230–52.

23. Barsky AJ. Palpitations, arrhythmias, and awareness of cardiac activity. Ann Intern Med. 2001;134(9 Pt 2):832–7.

24. Eley TC, Stirling L, Ehlers A, Gregory AM, Clark DM. Heart-beat perception, panic/somatic symptoms and anxiety sensitivity in children. Behav Res Ther. 2004;42(4):439–48.

25. Ehlers A, Mayou RA, Sprigings DC, Birkhead J. Psychological and perceptual factors associated with arrhythmias and benign palpitations. Psychosom Med. 2000;62(5):693–702.

26. Eifert GH, Thompson RN, Zvolensky MJ, Edwards K, Frazer NL, Haddad JW, et al. The cardiac anxiety questionnaire: development and preliminary validity. Behav Res Ther. 2000;38(10):1039–53.

27. Mueller PJ. Exercise training and sympathetic nervous system activity: evidence for physical activity dependent neural plasticity. Clin Exp Pharmacol Physiol. 2007;34(4):377–84.

28. Smits JA, Zvolensky MJ. Emotional vulnerability as a function of physical activity among individuals with panic disorder. Depress Anxiety. 2006;23(2):102–6.

29. Sardinha A, Araujo CGS, Nardi AE. Psychiatric disorders and cardiac anxiety in exercising and sedentary coronary artery disease patients: a case-control study. Braz J Med Biol Res. 2012;45(12):1320–6.

30. Frasure-Smith N, Lesperance F. Depression and anxiety as predictors of 2-year cardiac events in patients with stable coronary artery disease. Arch Gen Psychiatry. 2008;65(1):62–71.

31. Alberti KG, Zimmet P, Shaw J. The metabolic syndrome—a new worldwide definition. Lancet. 2005;366(9491):1059–62.

32. Fleet R, Lesperance F, Arsenault A, Gregoire J, Lavoie K, Laurin C, et al. Myocardial perfusion study of panic attacks in patients with coronary artery disease. Am J Cardiol. 2005;96(8):1064–8.

33. Marshall EC, Zvolensky MJ, Sachs-Ericsson N, Schmidt NB, Bernstein A. Panic attacks and physical health problems in a representative sample: singular and interactive associations with psychological problems, and interpersonal and physical disability. J Anxiety Disord. 2008;22(1):78–87.

34. de Wit LM, Fokkema M, van Straten A, Lamers F, Cuijpers P, Penninx BW. Depressive and anxiety disorders and the association with obesity, physical, and social activities. Depress Anxiety. 2010;27(11):1057–65.

35. Meyer T, Broocks A, Bandelow B, Hillmer-Vogel U, Ruther E. Endurance training in panic patients: spiroergometric and clinical effects. Int J Sports Med. 1998;19(7):496–502.

36. Lollgen H, Bockenhoff A, Knapp G. Physical activity and all-cause mortality: an updated meta-analysis with different intensity categories. Int J Sports Med. 2009;30(3):213–24.

37. Kodama S, Saito K, Tanaka S, Maki M, Yachi Y, Asumi M, et al. Cardiorespiratory fitness as a quantitative predictor of all-cause mortality and cardiovascular events in healthy men and women: a meta-analysis. JAMA. 2009;301(19):2024–35.

38. Chen YH, Tsai SY, Lee HC, Lin HC. Increased risk of acute myocardial infarction for patients with panic disorder: a nationwide population-based study. Psychosom Med. 2009;71(7):798–804.
39. Walters K, Rait G, Petersen I, Williams R, Nazareth I. Panic disorder and risk of new onset coronary heart disease, acute myocardial infarction, and cardiac mortality: cohort study using the general practice research database. Eur Heart J. 2008;29(24):2981–8.
40. Sardinha A, Nardi AE. The role of anxiety in metabolic syndrome. Expert Rev Endocrinol Metab. 2012;7(1):63–71.
41. Fleet R, Foldes-Busque G, Gregoire J, Harel F, Laurin C, Burelle D, et al. A study of myocardial perfusion in patients with panic disorder and low risk coronary artery disease after 35 % CO_2 challenge. J Psychosom Res. 2014;76(1):41–5.
42. Epel ES. Psychological and metabolic stress: a recipe for accelerated cellular aging? Hormones (Athens). 2009;8(1):7–22.
43. Bjorntorp P. Do stress reactions cause abdominal obesity and comorbidities? Obes Rev. 2001;2(2):73–86.
44. Tamashiro KL, Hegeman MA, Sakai RR. Chronic social stress in a changing dietary environment. Physiol Behav. 2006;89(4):536–42.
45. Kassi E, Pervanidou P, Kaltsas G, Chrousos G. Metabolic syndrome: definitions and controversies. BMC Med. 2011;9:48.
46. Smits JA, Rosenfield D, Mather AA, Tart CD, Henriksen C, Sareen J. Psychotropic medication use mediates the relationship between mood and anxiety disorders and obesity: findings from a nationally representative sample. J Psychiatr Res. 2010;44(15):1010–6.
47. Ramos PS, Sardinha A, Nardi AE, Araujo CGS. Cardiorespiratory optimal point: a submaximal exercise variable to assess panic disorder patients. PLoS One. 2014;9(8):e104932.
48. Belem da Silva CT, Schuch F, Costa M, Hirakata V, Manfro GG. Somatic, but not cognitive, symptoms of anxiety predict lower levels of physical activity in panic disorder patients. J Affect Disord. 2014;164:63–8.
49. Muotri RW, Bernik MA. Panic disorder and exercise avoidance. Rev Bras Psiquiatr. 2014;36:68–75.
50. Pollard CA. Respiratory distress during panic attacks associated with agoraphobia. Psychol Rep. 1986;58(1):61–2.
51. Soares-Filho GL, Arias-Carrion O, Santulli G, Silva AC, Machado S, Valenca AM, et al. Chest pain, panic disorder and coronary artery disease: a systematic review. CNS Neurol Disord Drugs Targets. 2014;13(6):992–1001.
52. Hung RK, Al-Mallah MH, McEvoy JW, Whelton SP, Blumenthal RS, Nasir K, et al. Prognostic value of exercise capacity in patients with coronary artery disease: the FIT (Henry Ford ExercIse Testing) project. Mayo Clinic Proc. 2014;89(12):1644–54.
53. Kokkinos P, Myers J, Faselis C, Panagiotakos DB, Doumas M, Pittaras A, et al. Exercise capacity and mortality in older men: a 20-year follow-up study. Circulation. 2010;122(8):790–7.
54. Myers J, Prakash M, Froelicher V, Do D, Partington S, Atwood JE. Exercise capacity and mortality among men referred for exercise testing. N Engl J Med. 2002;346(11):793–801.
55. Farrell SW, Finley CE, Haskell WL, Grundy SM. Is there a gradient of mortality risk among men with low cardiorespiratory fitness? Med Sci Sports Exerc. 2014 (online first).
56. Gomes RM, Sardinha A, Araújo CGS, Nardi AE, Deslandes AC. Aerobic training intervention in panic disorder: a case-series study. Med Express. 2014;1(4):195–201.
57. Sardinha A, Araújo CGS, Nardi AE. Interval aerobic training as a tool in the cognitive-behavioral treatment of panic disorder. J Bras Psiquiatr. 2011;60(3):227–30.
58. American Thoracic S, American College of Chest P. ATS/ACCP Statement on cardiopulmonary exercise testing. Am J Respir Crit Care Med. 2003;167(2):211–77.
59. Araújo CGS. Teste cardiopulmonar de exercício: breves considerações sobre passado, presente e futuro. Rev DERC. 2012;18(4):104.
60. Balady GJ, Arena R, Sietsema K, Myers J, Coke L, Fletcher GF, et al. Clinician's guide to cardiopulmonary exercise testing in adults: a scientific statement from the American Heart Association. Circulation. 2010;122(2):191–225.
61. Guazzi M, Adams V, Conraads V, Halle M, Mezzani A, Vanhees L, et al. EACPR/AHA Joint Scientific Statement. Clinical recommendations for cardiopulmonary exercise testing data assessment in specific patient populations. Eur Heart J. 2012;33(23):2917–27.
62. Araújo CGS, Herdy AH, Stein R. Maximum oxygen consumption measurement: valuable biological marker in health and in sickness. Arq Bras Cardiol. 2013;100(4):e51–3.
63. Ricardo DR, De Almeida MB, Franklin BA, Araújo CGS. Initial and final exercise heart rate transients: Influence of gender, aerobic fitness, and clinical status. Chest. 2005;127(1):318–27.
64. Ramos PS, Araújo CG. Análise da estabilidade de variável submáxima em teste cardiopulmonar de exercício: ponto ótimo cardiorrespiratório. Rev Bras Ativ Fis Saúde. 2013;18(5):585–93.
65. Ramos PS, Ricardo DR, Araujo CG. Cardiorespiratory optimal point: a submaximal variable of the cardiopulmonary exercise testing. Arq Bras Cardiol. 2012;99(5):988–96.
66. Kruk J. Physical activity in the prevention of the most frequent chronic diseases: an analysis of the recent evidence. Asian Pac J Cancer Prev. 2007;8(3):325–38.

67. Subirats Bayego E, Subirats Vila G, Soteras Martinez I. Exercise prescription: indications, dosage and side effects. Med Clin (Barc). 2012;138(1):18–24.

68. Strohle A. Physical activity, exercise, depression and anxiety disorders. J Neural Transm. 2009;116(6):777–84.

69. Zschucke E, Gaudlitz K, Strohle A. Exercise and physical activity in mental disorders: clinical and experimental evidence. J Prev Med Public Health. 2013;46 Suppl 1:S12–21.

70. Teachman BA, Marker CD, Clerkin EM. Catastrophic misinterpretations as a predictor of symptom change during treatment for panic disorder. J Consult Clin Psychol. 2010;78(6):964–73.

71. Gary RA, Dunbar SB, Higgins MK, Musselman DL, Smith AL. Combined exercise and cognitive behavioral therapy improves outcomes in patients with heart failure. J Psychosom Res. 2010;69(2):119–31.

72. Asmundson GJ, Fetzner MG, Deboer LB, Powers MB, Otto MW, Smits JA. Let's get physical: a contemporary review of the anxiolytic effects of exercise for anxiety and its disorders. Depress Anxiety. 2013;30(4):362–73.

73. Faulkner J, Westrupp N, Rousseau J, Lark S. A randomized controlled trial to assess the effect of self-paced walking on task-specific anxiety in cardiac rehabilitation patients. J Cardiopulm Rehabil Prev. 2013;33(5):292–6.

74. Smits JA, Meuret AE, Zvolensky MJ, Rosenfield D, Seidel A. The effects of acute exercise on CO(2) challenge reactivity. J Psychiatr Res. 2009;43(4):446–54.

75. Esquivel G, Schruers K, Kuipers H, Griez E. The effects of acute exercise and high lactate levels on 35% CO_2 challenge in healthy volunteers. Acta Psychiatr Scand. 2002;106(5):394–7.

76. Esquivel G, Diaz-Galvis J, Schruers K, Berlanga C, Lara-Munoz C, Griez E. Acute exercise reduces the effects of a 35% CO_2 challenge in patients with panic disorder. J Affect Disord. 2008;107(1–3):217–20.

77. Strohle A, Stoy M, Graetz B, Scheel M, Wittmann A, Gallinat J, et al. Acute exercise ameliorates reduced brain-derived neurotrophic factor in patients with panic disorder. Psychoneuroendocrinology. 2010;35(3):364–8.

78. Strohle A, Feller C, Strasburger CJ, Heinz A, Dimeo F. Anxiety modulation by the heart? Aerobic exercise and atrial natriuretic peptide. Psychoneuroendocrinology. 2006;31(9):1127–30.

79. Kornreich C. Panic, somatization and exercise. Rev Med Brux. 2006;27(2):78–82.

80. Broocks A, Meyer T, Opitz M, Bartmann U, Hillmer-Vogel U, George A, et al. 5-HT1A responsivity in patients with panic disorder before and after treatment with aerobic exercise, clomipramine or placebo. Eur Neuropsychopharmacol. 2003;13(3):153–64.

81. Broocks A, Meyer T, Gleiter CH, Hillmer-Vogel U, George A, Bartmann U, et al. Effect of aerobic exercise on behavioral and neuroendocrine responses to meta-chlorophenylpiperazine and to ipsapirone in untrained healthy subjects. Psychopharmacology (Berl). 2001;155(3):234–41.

82. Wedekind D, Sprute A, Broocks A, Huther G, Engel K, Falkai P, et al. Nocturnal urinary cortisol excretion over a randomized controlled trial with paroxetine vs. placebo combined with relaxation training or aerobic exercise in panic disorder. Curr Pharm Des. 2008;14(33):3518–24.

83. Hovland A, Nordhus IH, Sjobo T, Gjestad BA, Birknes B, Martinsen EW, et al. Comparing physical exercise in groups to group cognitive behaviour therapy for the treatment of panic disorder in a randomized controlled trial. Behav Cogn Psychother. 2013;41(4):408–32.

84. Broman-Fulks JJ, Berman ME, Rabian BA, Webster MJ. Effects of aerobic exercise on anxiety sensitivity. Behav Res Ther. 2004;42(2):125–36.

85. Keough ME, Schmidt NB. Refinement of a brief anxiety sensitivity reduction intervention. J Consult Clin Psychol. 2012;80(5):766–72.

86. Smits JA, Berry AC, Rosenfield D, Powers MB, Behar E, Otto MW. Reducing anxiety sensitivity with exercise. Depress Anxiety. 2008;25(8):689–99.

87. Jayakody K, Gunadasa S, Hosker C. Exercise for anxiety disorders: systematic review. Br J Sports Med. 2014;48(3):187–96.

88. Bartley CA, Hay M, Bloch MH. Meta-analysis: aerobic exercise for the treatment of anxiety disorders. Prog Neuropsychopharmacol Biol Psychiatry. 2013;45:34–9.

89. Taylor S, Cox BJ. An expanded anxiety sensitivity index: evidence for a hierarchic structure in a clinical sample. J Anxiety Disord. 1998;12(5):463–83.

Pharmacological Treatment of Panic Disorder with Non-Selective Drugs

19

Patricia Cirillo and Rafael Christophe R. Freire

Contents

P. Cirillo • R.C.R. Freire (✉)
Laboratory of Panic and Respiration, Institute of
Psychiatry, Federal University of Rio de Janeiro,
Rio de Janeiro, Brazil
e-mail: pat_cirillo@hotmail.com; rafaelcrfreire@
gmail.com

Abstract

Until now, little information exists about pharmacological strategies to follow in the treatment of panic disorder (PD) patients with unsatisfactory response to first line medications. Furthermore, there is no consensus about the definition of concept and management of treatment-resistant PD. The physiopathology of panic disorder is related to serotoninergic, noradrenergic and GABAergic systems. Based on this knowledge, antidepressants, anxiolytics, atypical antipsychotics and anticonvulsants have been studied in PD treatment. Besides selective serotonin reuptake inhibitors, tricyclics, benzodiazepines and venlafaxine, there is limited scientific evidence on the effectiveness of other medications. More studies are needed for a better elucidation of the effectiveness and tolerability of medications in PD, especially atypical antipsychotics, anticonvulsants and other new drugs. For the moment, it is important to use existing scientific evidences regarding effectiveness, tolerability and safety, combined with clinical experience, to plan a rational treatment sequence for PD patients.

Keywords

Panic disorder • Treatment-resistant • Antidepressive agents • Antipsychotic agents • Anticonvulsants • Drug therapy

19.1 Introduction

In clinical trials with panic disorder (PD) patients, response and remission rates vary between 40–70% and 20–47%, respectively [1]. After 2 years, 21.4% of patients with remission experience recurrence of panic attacks [2]. Currently, little information exists about pharmacological strategies to follow in the treatment of PD patients with unsatisfactory response to first line medications. Moreover, there is no consensus about the definition of treatment resistant PD. Therefore, each study uses its own criteria, making it difficult to compare results.

Currently, there are many pharmacological treatment options for PD including different classes of antidepressants, anxiolytics, atypical antipsychotics and anticonvulsants, among others. This is why selecting the best option of treatment for each patient becomes difficult. According to current guidelines, usually the first line pharmachological treatments for PD are selective serotonin reuptake inhibitors (SSRI) or serotonin and norepinephrine reuptake inhibitor (SNRI) venlafaxine extended release (XR) [3–6]. Up to now there is no evidence to support greater effectiveness of one antidepressant class over another nor combination over monotherapy treatment [4]. SSRI, benzodiazepines, venlafaxine XR and tricyclic antidepressants (TCA) have demonstrated similar efficacy. In clinical practice, the choice of pharmacological treatment is based on a series of factors such as tolerability, previous treatments and outcomes, clinical or psychiatric comorbidity, treatment cost and patient preference [4].

SSRI have easy dosage regimen and usually the therapeutic dose is rapidly achieved. This class of antidepressants have fewer side effects and less dropouts than TCA and monoamine oxidase inhibitors (MAOI) as well as less toxicity in overdose and no dietary restrictions as MAOI. SSRI are usually presented as a safer option when compared with benzodiazepines (BZD) due to risk of dependency and should be the choice to treat patients with substance abuse. This preference for SSRI is also due to likelihood of BZD might cause mild cognitive impairment.

Although, BZD can be associated at the beginning of treatment for a quicker symptom control due to its fast action. Also regarding patients with comorbid psychiatric disorders (i.e. depression), which is common in PD, SSRI treat both disorders. Venlafaxine XR is considered first option along with SSRI because both have similar efficacy compared to TCA but with fewer adverse events (AE), less dropouts and more comfortable posology [4]. The most common side effects of SSRI and venlafaxine are somnolence, dry mouth and insomnia [7]. Sexual dysfunction is also very common in both of these treatments but in a larger scale with SSRI [8].

The objectives of the treatment of PD are to reduce the frequency and intensity of panic attacks, anticipatory anxiety and agoraphobic avoidance. Accordingly, treatment failure occurs when the patient does not have satisfactory improvement of the mentioned symptoms or if there is recurrence of symptoms [4]. Although several medications have shown good results in PD treatment, a significant percentage of patients remain symptomatic after an adequate trial. Patients who do not respond or who dropout due to AE reach 30% [9]. In this group of patients that does not respond, first it is important to exclude causes such as inadequate dose and/or duration, noncompliance, poor tolerability, psychosocial stressors and clinical or psychiatric comorbidity. If the patient has inadequate response despite those factors, there is a need to choose new strategies and have further therapeutic options. At that point, it is important to consider switching or augmentation [10].

If some improvement is obtained, augmentation is indicated. Although, if current treatment have not provided satisfactory response, switching should be considered. Primarily, it is recommended to choose treatments with a higher level of evidence. Therefore, regardless of adding or switching, another first-line medication should be chosen. It is also a common strategy to associate BZD for residual symptoms [10]. Thus, this chapter analyzes existing evidence of new pharmacological treatment options for patients who showed unsatisfactory response to the treatment of PD with SSRI, venlafaxine and BZD.

Panic attacks are acute responses to misinterpreted stimuli in people with hypersensitivity in fear networks. The physiopathology of PD is also related to genetic susceptibility as well as serotoninergic, noradrenergic and GABAergic systems. This abnormal functioning may exacerbate response to anxiety, including autonomic symptoms [11]. Based on this knowledge, many medications have been studied in treatment-resistant PD.

19.2 Treatments

19.2.1 Antidepressants

19.2.1.1 Venlafaxine

Venlafaxine is considered as an early choice for the treatment of PD. Actually, among newer antidepressants is the best-studied for pharmacological treatment of PD. Venlafaxine was effective to control panic symptoms in short and long term follow-up studies and has shown to be safe and well tolerated [12]. This serotonin-norepinephrine reuptake inhibitor demonstrated significantly higher response and remission rates than placebo and was as effective as paroxetine. Until now, there are six published controlled trials, most of them randomized and double-blind, and two open label studies lasting from 8 to 26 weeks that evaluated antipanic properties of venlafaxine. Symptoms like frequency of panic attacks, anticipatory anxiety, fear and avoidance were evaluated as well as response and remission rates. Considering all studies, more than 1900 patients were evaluated and venlafaxine was compared to placebo and paroxetine [13–19].

Two multicenter, double-blind, randomized studies of 12 weeks, compared venlafaxine XR at doses of 75 mg/day, 150 mg/day and paroxetine 40 mg/day [13, 14]. In addition, one study also included a venlafaxine 225 mg/day group and a placebo group [13]. In total, more than 1200 patients were included in both studies. All groups were superior to placebo and both venlafaxine groups at doses of 75 and 150 mg/day were equally effective to paroxetine 40 mg/day group. However, venlafaxine 225 mg/day group was significantly superior to paroxetine group in improving panic symptoms and remission rates [13, 14].

In a long-term follow up study, venlafaxine prevented relapse of panic symptoms and improved quality of life and disability when compared to placebo [16]. Patients who were responders in a venlafaxine XR open label study of 12 weeks were randomized to continue receiving venlafaxine XR or switch to placebo. Patients in venlafaxine XR group have shown 22.5% relapse while in placebo group were 50% (p<0.001). In general, AE were mild to moderate and similar to paroxetine group and to those studies with venlafaxine for depression and other anxiety disorders [17].

In one randomized, double-blind, multicenter study of 10 weeks, comparing flexible-dose venlafaxine XR (75–225 mg/day, N=155) to placebo (N=155), dropout rates were 4% for placebo group and 7% for venlafaxine ER-treated group [20]. In this study, AE were the primary or secondary cause of withdrawal. On the other hand, a study of 8 weeks with a small sample (N=25), found more dropouts in placebo group (66%) than in venlafaxine-treated group (15%) [15]. This probably occurred because of lack of therapeutic effect.

19.2.1.2 Duloxetine

A small open label study, with 15 patients, evaluated duloxetine in the treatment of PD with or without agoraphobia [21]. Authors hypothesized that duloxetine, a SNRI, with a similar mechanism of action of venlafaxine, would be effective for the treatment of panic disorder. Patients received daily doses ranging from 60 to 120 mg/day for 8 weeks. The majority of patients (53%) had a prior failed pharmacological treatment. At the endpoint, patients had significant improvement of panic and depressive symptoms, overall anxiety and quality of life. Duloxetine was more effective in patients without psychiatric comorbidity than in patients with comorbidity. The most common AE were nausea, sedation and sexual dysfunction [21]. However, this is a restricted result and randomized controlled studies with larger samples are needed.

19.2.1.3 Trazodone

There are few trials with trazodone in PD and the results are controversial [22, 23]. A study compared the efficacy of imipramine, alprazolam and trazodone in 74 patients with PD for 8 weeks [22]. Imipramine and alprazolam reduced the symptoms of anxiety, frequency of panic attacks and phobic avoidance. Alprazolam was effective on the first week of treatment as imipramine only from the fourth week. However, trazodone has not been effective and was poorly tolerated. Trazodone group had only two patients with good improvement or remission and only 17 patients completed least 4 weeks of treatment [22].

A single-blind trial with patients with PD and agoraphobia evaluated trazodone 300 mg/day for 8 weeks [23]. Only 11 patients completed the study, and they had significant improvements of anxiety, depression and phobias, including agoraphobic avoidance [23]. Therefore, in this study, trazodone was effective but studies with larger samples are needed.

19.2.1.4 Reboxetine

Reboxetine was compared to SSRI in two single-blind studies and SSRI were superior especially if there is comorbid depression [24, 25]. However, in two other studies, reboxetine showed efficacy and superiority to placebo [26, 27]. In addition, reboxetine demonstrated efficacy in patients whose previous treatment with SSRI failed [9, 25].

A single-blind, randomized trial compared reboxetine and paroxetine in 68 patients with PD for 3 months [24]. Paroxetine was more effective than reboxetine for panic attacks but there was no difference regarding anticipatory anxiety and avoidance. Furthermore, reboxetine group showed less weight gain and sexual dysfunction [24].

Another study evaluated the efficacy of reboxetine in 29 outpatients with PD whose previous treatment with SSRI failed [9]. These 6 weeks open label study used reboxetine up to 8 mg/day. The 24 (82.76%) patients who completed the study showed significant improvement in frequency of panic attacks, overall anxiety, global functioning and depressive symptoms. Five patients (17.24%) discontinued because of AE [9].

Another single-blind randomized trial compared reboxetine and citalopram in 19 patients with a crossover design [25]. After the first 8 weeks with one of those AD and a 2 weeks washout, patients were switched to the other AD. Seven of 13 (54%) patients of reboxetine-treated group and 9 of 11 (82%) of citalopram-treated group showed improvement. There were no difference in efficacy between both AD related to panic attacks, but there was a superiority of citalopram for depressive symptoms. One patient that remained symptomatic with citalopram responded to reboxetine and three patients that remained symptomatic with reboxetine responded to citalopram. Based on these data, citalopram is a better option especially for PD with comorbid depression [25].

A placebo-controlled, randomized, double-blind, parallel-group study with 82 patients with PD with or without agoraphobia evaluated reboxetine (6–8 mg/day) for 8 weeks [26]. Seventy-five patients finished the trial. In the reboxetine group, there was a significant reduction in mean number of panic attacks and phobic symptoms, as well as depressive symptoms and functioning. Reboxetine was well tolerated and common AE were dry mouth, constipation and insomnia [26]. A meta-analyses evaluated dropout rates of several antidepressants in the treatment of PD and reboxetine showed a high rate in comparison to placebo [27].

19.2.1.5 Mirtazapine

Several small, uncontrolled studies, with 8–12 weeks duration, indicated efficacy of mirtazapine in reducing significantly the number and the intensity of panic attacks and anticipatory anxiety [28–31]. In the only double-blind, randomized trial of mirtazapine published until today, this antidepressant was as effective as fluoxetine reducing panic attacks and mirtazapine-treated group had better improvement of phobic anxiety during the 8 weeks of the study [32]. In addition, patients self-evaluation of their improvement was better in mirtazapine group. Regarding AE, the most common in mirtazapine group was weight gain while in fluoxetine group were nausea and paresthesia [32].

Mirtazapine was also compared to another SSRI but in this case to paroxetine and in an open label study of 8 weeks [30]. The improvement of PD was similar in both groups. Indeed mirtazapine significantly reduced the number of panic attacks since the third week suggesting a faster response than with paroxetine. After a follow-up 6 months later, 95 % of the patients that responded at week 8 remained asymptomatic. Patients with PD and comorbid depression improved in both groups [30]. Overall, mirtazapine was well tolerated in all studies and withdrawal was low (6.3 %) [28–32]. Despite the positive results, a meta-analysis of antidepressants for the treatment of PD including 53 studies and 5236 patients, found that mirtazapine was superior to placebo for overall anxiety symptoms but not for panic symptoms. This meta-analysis confirmed the low dropout rate [27].

19.2.1.6 Phenelzine

MAOI have been largely studied for depressive disorders, though for PD, few studies exist. Irreversible MAOI, like phenelzine and tranylcypromine, require a low tyramine diet and have unfavorable side effects like hypertensive crisis. Therefore, in clinical practice, MAOI are reserved for patients who have tried several standard treatments either as monotherapies or as adjunctive therapies that not showed satisfactory response.

Phenelzine was found to be efficacious in patients with PD with and without agoraphobia in a study with 35 outpatients during 6 months [33]. The frequency of panic attacks improved 100 % in patients with PD and 94.7 % in patients with agoraphobia. The symptoms anticipatory anxiety and avoidance have not improved significantly in 73.6 % in the group of agoraphobic patients [33].

19.2.1.7 Tranylcypromine

Tranylcypromine (TCP) was effective at doses of 30 and 60 mg/day in a double-blind controlled study of 12 weeks in 36 patients with PD and comorbid social anxiety disorder (SAD) according to DSM-IV [34]. At the end, 13 of 19 patients (68.4 %) in the 30 mg/day group and 12 of 17

patients (70.6 %) in the 60 mg/day group reported to be free of panic attacks. There was no statistically significant difference between both in relation to panic attacks. However, concerning SAD, the group that used 30 mg/day of TCP have not improved [34]. Regarding dropping out from treatment, eight (22.2 %) patients have left the study of which five (13.8 %) because of AE and three (8.3 %) due to insufficient clinical response. The dropouts were 26.3 % (n=5) in the 30 mg/day group and 17.6 % (n=3) in the 60 mg/day and this difference was not statistically significant. All dropouts occurred within the first 4 weeks [34]. AE were generally of low intensity with few cases of moderate intensity and no severe AE. Moreover, the majority of AE regressed after 4 weeks of treatment. The most common AE (>10 %) in both groups were orthostatic hypotension, decreased libido, insomnia, constipation, sleepiness, nausea, drowsiness and dry mouth. In the group of 60 mg/day also occurred: headache, fatigue and delayed ejaculation [34]. The number of studies is quite limited and the samples are small, making it difficult to generalize the results. However, the TCP has shown positive results in severe cases, resistant to previous treatments or with psychiatric comorbidities.

19.2.1.8 Moclobemide

Reversible MAOI like moclobemide do not require dietary restrictions and have fewer drug interactions in relation to irreversible MAOI. Studies that investigated depression suggest that reversible MAOI may be equally effective and better tolerated than TCA [35].

Moclobemide was compared to placebo and cognitive behavioral therapy (CBT) in the treatment of PD with agoraphobia [35]. Fifty-five patients were evaluated for 8 weeks and later there was a 6 months follow-up. Moclobemide was as effective as CBT and more effective than placebo [35]. Also, two multicenter, double-blind, randomized, parallel-group studies compared moclobemide with a TCA or SSRI and both found similar effectiveness [36, 37]. One study compared moclobemide 450 mg/day with clomipramine 150 mg/day for 8 weeks in 135 patients [36]. Response was 49 % in moclobemide group and 53 % in clomipramine group. Whereas improvement in global functioning was 78 % in moclobemide group and 88 % in clomipramine group. AE were more frequent in clomipramine group due to anticholinergic effects [36].

Another study compared moclobemide (n = 182) with a target dose of 450 mg/day to fluoxetine (n = 184) with a target dose of 20 mg/day during 8 weeks [37]. Both showed similar efficacy with a panic free rate of 70 % in fluoxetine group and 63 % in moclobemide group. Patients with much or very much improvement were 74 % in fluoxetine group and 76 % in moclobemide group. Fourteen patients in fluoxetine group and 11 patients in moclobemide group discontinued due to AE. Both groups had similar tolerability. This study had a long-term extension of 1 year in which 74 patients of fluoxetine group and 83 of moclobemide group participated. At endpoint, 65 (87.84 %) patients of fluoxetine group and 61 (73.49 %) patients of moclobemide group completed the trial. Patients of both groups remained much improved after 1 year. There were no severe AE and both medications were well tolerated [37].

19.2.1.9 Tianeptine

Tianeptine is an antidepressant with a mechanism of action that is the opposite of SSRI' mechanism. Tianeptine increases serotonin uptake in the brain [38]. A randomized, double-blind study of 6 weeks compared the vulnerability to 35 % CO_2 in patients with PD before and after treatment with tianeptine or paroxetine [39]. Tianeptine significantly reduced panic reaction to 35 % CO_2 as well as paroxetine. However, tianeptine has not shown significant clinical improvement whereas paroxetine did [39].

19.2.1.10 Nefazodone

A small study of 12 weeks evaluated nefazodone efficacy in 15 patients with PD [40]. At the end, 47 % of the patients that completed the study presented remission. AE were mild and the most common were dizziness and sedation [40]. Another open label study of 12 weeks with ten patients found much improvement in 90 % of patients and 70 % of remission of panic symptoms with nefazodone treatment [41].

A third small open trial evaluated nefazodone in patients with PD and high degree of comorbid major depressive disorder or generalized anxiety disorder (GAD) [42]. After 8 weeks, 10 of 14 (71 %) patients improved from panic symptoms including phobic avoidance, disability and global impression. Five of 8 (62.5 %) patients with comorbid depression and 3 of 5 (60 %) patients with comorbid GAD and 5 of 6 (83.33 %) patients with only PD achieved response. There were no dropouts because of AE [42].

Those few and small open label studies show preliminary evidence of efficacy and tolerability of nefazodone in PD. Although, controlled studies with larger sample are needed to obtain stronger evidence.

19.2.1.11 Milnacipran

A small open trial with 31 patients, evaluated effectiveness of milnacipran in PD with and without agoraphobia for 10 weeks [43]. Daily doses were titrated to 50 mg twice a day. Milnacipran significantly reduced panic symptoms intensity and remission was achieved in 58.1 % of the sample [43]. Another open study evaluated combined sequential therapy of clonazepam and milnacipran in the treatment of PD with comorbid depression. First, patients were treated only with clonazepam that was slowly switched to milnacipran. Patients presented good antidepressant and antipanic response. At the end, milnacipran was effective and well tolerated [44].

19.2.1.12 Agomelatine

Animal studies have shown that agomelatine has anxiolytic action. Agomelatine is a melatonin agonist but the anxiolytic property seems to be related to an antagonistic action at 5-HT2c receptors [45]. Because of this property, agomelatine has being studied for anxiety disorders and there is preliminary evidence of its effectiveness in GAD [46].

In relation to PD, until now, there is only a small open label trial published with agomelatine [47]. This study evaluated 13 patients and 11 of them completed the study. Nine patients had PD with

Table 19.1 Antidepressant doses and guidelines recommendation levels [8, 70]

Antidepressant	Initial dose (mg)	Usual dosage (mg)	Maximum dose (mg)	British guideline (acute treatment) [6]	Canadian guideline [5]	WFSBP [3]
Venlafaxine	37.5	75–225	300	A Long-term +	First-line; Long-term +	1
Duloxetine	30	60–120	120	NC	NC	NC
Trazodone	50	150–400	600	NC	Not-recommended	NC
Tranylcypromine	20	20–60	60	NC	Third-line	NC
Phenelzine	15	45–90	90	D	Third-line	3
Moclobemide	150	300–600	600	B	Third-line; Long-term +	5
Mirtazapine	7.5–15	15–45	45	C	Second-line	4
Reboxetine	4	8–10	10	D	Second-line	5
Milnacipran	12.5	100–200	200	NC	NC	4
Tianeptine	12.5	12.5–37.5	37.5	NC	NC	NC
Agomelatine	25	25–50	50	NC	NC	NC
Nefazodone	200	300–600	600	NC	NC	NC
Bupropion	75–150	150–300	300	Not-recommended	NC	4

Recommendation of: <u>British guideline</u> [6]: A – Directly based on category I evidence (either I [M] or I [PCT]); B – Directly based on category II evidence or an extrapolated recommendation from category I evidence; C – Directly based on category III evidence or an extrapolated recommendation from category I or II evidence; D – Directly based on category IV evidence or an extrapolated recommendation from other categories; S – Standard of clinical care. <u>Canadian guideline</u> [5]: First-line – Level 1 or Level 2 evidence plus clinical support for efficacy and safety; Second-line – Level 3 evidence or higher plus clinical support for efficacy and safety; Third-line – Level 4 evidence or higher plus clinical support for efficacy and safety; Not recommended – Level 1 or Level 2 evidence for lack of efficacy. <u>WFSBP</u>: 1 – Category A evidence and good risk-benefit ratio; 2 – Category A evidence and moderate risk-benefit ratio; 3 – Category B evidence; 4 – Category C evidence; 5 – Category D evidence
NC not classified
Long-term + positive results in long-term study
WFSBP World Federation of Societies of Biological Psychiatry
Evidence levels: <u>British guideline</u> [6]: I [M] Evidence from meta-analysis of randomized double-blind placebo-controlled trials; I [PCT] Evidence from at least one randomized double-blind placebo-controlled trial; II Evidence from at least one randomized double-blind comparator-controlled trial (without placebo); III Evidence from non-experimental descriptive studies; IV Evidence from expert committee reports or opinions and/or clinical experience of respected authorities. <u>Canadian guideline</u> [5]: 1 – Meta-analysis or at least two randomized controlled trials (RCTs) that included a placebo condition; 2 – At least 1 RCT with placebo or active comparison condition; 3 – Uncontrolled trial with at least ten subjects; 4 – Anecdotal reports or expert opinion. <u>WFSBP – evidence levels</u> [3]: A – Full evidence from controlled studies; B – Limited positive evidence from controlled studies; C – Evidence from uncontrolled studies or case reports/expert opinion; C1 – Uncontrolled studies; C2 – Case reports; C3 – Based on the opinion of experts in the field or clinical experience; D – Inconsistent results; E – Negative evidence; F – Lack of evidence

agoraphobia. The reasons of the two patients who did not complete the study were not related to treatment. Regarding psychiatric comorbidity, 9 of the 11 patients had comorbid depression and three had comorbid SAD. The starting dose was 25 mg/day and could be increased to a maximum of 50 mg/day if necessary. Nine patients needed the dose of 50 mg/day. Furthermore, concomitant use of BZD was allowed until a predetermined dose, which occurred with five patients. At endpoint, there were significant improvement of panic symptoms and quality of life compared to baseline [47]. Please see Table 19.1.

19.2.2 Anticonvulsants

There are few studies on the use of anticonvulsants in TP and consequently little evidence to support its use.

19.2.2.1 Gabapentine

Gabapentine has a mechanism of action incompletely understood. Nevertheless, it is known that gabapentine action is related to modulation of GABA synthesis and glutamate [48]. In addition, this antiepileptic and antinociceptive agent also has anxiolytic action [49]. Therefore, the effect of gabapentine is studied for anxiety disorders.

Until now, there is only one study with gabapentin. A randomized, double-blind, placebo-controlled, parallel-group study evaluated the efficacy and safety of gabapentin (600–3600 mg/day) in PD in 103 patients for 8 weeks [50]. At the end, there was no significant difference between patients treated with gabapentin or placebo. However, when assessed only patients severely ill, gabapentin was effective with women showing greater improvement than men. The most common AE were somnolence (31 %), headache (27 %), and dizziness (23 %). No deaths occurred but one patient had severe AE [50]. In conclusion, gabapentin have not been proven effective in the only existing study. Although, since it has superiority in a subgroup of severely ill patients, more studies are needed.

19.2.2.2 Pregabalin

Pregabalin was already studied for the treatment of SAD and GAD with positive results. However, until the moment, there are no studies evaluating its effectiveness in PD. There is only one open label study of 1 year to examine safety and tolerability of pregabalin in a mixed sample of patients with SAD, generalized anxiety disorder or PD [51]. Results were evaluated together and pregabaline was considered well tolerated and safe in long-term treatment [51].

19.2.2.3 Valproic Acid, Valproate and Divalproex Sodium

Until the moment, valproic acid and derivatives are medications that have been effective in open-label studies, but require controlled studies for PD treatment. Two small open-label studies evaluated valproic acid or valproate efficacy in PD [52, 53]. Both of them found significant improvement of panic symptoms and global function

with few AE. One of these studies have done a 6 months follow up and observed maintenance of improvement after this period [53].

Other two open label studies analyzed divalproex or valproate efficacy in PD comorbid to bipolar disorder [54, 55]. They showed improvement of depressive and anxiety symptoms, panic attacks and mood instability. Also with few AE and few dropouts [54, 55].

Another study evaluated valproate in the prevention of lactate-induced panic attacks [56]. Patients underwent lactate before and after treatment. There were significant improvement and valproate blocked panic attacks [56].

19.2.2.4 Tiagabine

A placebo-controlled, double-blind study compared tiagabine (n = 10) and placebo (n = 9) for 4 weeks [57]. Tiagabine showed no superiority to placebo. This study also tested the sensitivity to panic attacks induced by cholecystokinine-tetrapeptide (CCK-4) before and after treatment and there was a reduction in sensitivity [57].

An open-label study evaluated efficacy and safety of tiagabine 2–20 mg/day in moderate to severe PD with or without agoraphobia for 10 weeks [58]. Twenty-three of the 28 patients completed the study. There were no significant clinical improvement in PD and agoraphobia symptoms or global function. In general, tiagabine was well tolerated and only one patient dropped out because of AD [58]. Accordingly, tiagabine has not shown to be a good option in the treatment of PD.

19.2.3 Atypical Antipsychotics

Antipsychotic medications have shown to block conditioned fear behavior. One mechanism of action to decrease anxiety symptoms may be the direct modulation of the dopamine system. Some atypical antipsychotics have serotonin action and because of this may have anxiolytic properties. Therefore, the dual action of antipsychotics suppressing both dopaminergic and serotonergic activity can be interesting to control anxiety [59, 60].

However, antipsychotics are less tolerated than antidepressants and BZD and should be used only after medications with more scientific support have failed.

19.2.3.1   Risperidone

A randomized, single-blind study compared low-dose risperidone (0.25–1 mg/day, average dose = 0.53 mg/day) to paroxetine (30–40 mg/day) in 56 patients with panic attacks for 8 weeks [61]. The sample included patients with PD with or without agoraphobia (76.8%) and with Major Depressive Disorder with Panic Attacks (23.2%). Thirty-three patients received risperidone and 23 received paroxetine. This study evaluated panic attacks, anxiety and depressive symptoms and global functioning. Thirteen (51.8%) patients dropped out in risperidone group and 14 (60.9%) in paroxetine group. This difference was not statistically significant. The main reason for dropouts were lost of follow-up or non-compliance (n = 14, 51.9%) and the other reasons were side effects and lack of response (n = 3, 11.1% for each one) [61].

At the end of the study, there were significant and similar improvement in both groups. In addition, risperidone was as well tolerated as paroxetine. Moreover, they suggest that the risperidone-treated group may have a quicker response than paroxetine [61].

Another study evaluated risperidone augmentation for patients with treatment resistant anxiety disorders for 8 weeks [62]. This open-label trial included 30 treatment resistant patients with PD, GAD and SAD, these patients were treated previously with antidepressants, benzodiazepines or both, with proper dose and duration, and did not achieve remission. Risperidone doses ranged from 0.25–3.00 mg/day (mean dose: 1.12 ± 0.68 mg/day). Seventy percent (n = 21) of patients completed the study and risperidone augmentation showed a significant improvement in anxiety symptoms and global functioning. The main reasons for dropouts were weight gain and sedation [62].

Risperidone showed preliminary evidence suggesting antipanic effect. Larger studies should continue investigating risperidone properties in samples with only PD patients.

19.2.3.2   Olanzapine

Two open-label studies analyzed olanzapine effectiveness in treatment-resistant PD [63, 64]. One as monotherapy and another one as augmentation strategy to SSRI. Olanzapine seems to be beneficial for patients with treatment-resistant PD. In both studies, olanzapine was well tolerated and safe [63, 64].

The first one evaluated ten patients for 8 weeks with an average daily dose of 12.3 mg/day of olanzapine [63]. At the end, 5 of 10 patients (50%) showed remission of panic attacks and 4 of 10 patients (40%) showed significant decrease in the number of panic attacks. Also 6 of 10 patients (60%) were free from anticipatory anxiety. Regarding AE 6 of 10 patients (60%) gained weight but it was not significant during the study [63].

The second study analyzed augmentation with olanzapine 5 mg/day in SSRI-resistant PD with or without agoraphobia [64]. During 12 weeks, 31 patients were followed and 26 completed the trial (dropout rate = 16.1%). Twenty-one patients (81.8%) were responders and all patients improved panic symptoms, anticipatory anxiety, agoraphobia, depressive symptoms and global functioning. Fifteen patients (57.7%) achieved remission. The most common AE were mild to moderate drowsiness and weight gain [64].

The comorbidity of anxiety disorders and bipolar disorder is common [65]. In these cases, bipolar disorder usually has worse disease course. Therefore, because of its pharmacological characteristics and mood stabilization effect, olanzapine is probably a good option for patients with PD comorbid to bipolar disorder or insomnia. However, controlled studies are needed to confirm this hypothesis.

19.2.3.3   Aripiprazole

To date, there is only one open-label study that analyzed efficacy and tolerability of augmentation with aripiprazole in the treatment of GAD (n = 13) or PD (n = 10) in patients that remained symptomatic after a first line treatment [66]. The mean daily dose in this study was 10.5 ± 4.95 mg/day. Patients showed significant improvement of anxiety and depressive symptoms. One PD patient (1%) and 3 with GAD (23%) achieved remission. At endpoint, the reduction of the severity of panic symptoms was not significant. Seventeen patients completed the study. In PD subgroup, there were three dropouts of which two for lack of response and one for chest discomfort [66]. The result of this study is insufficient to evaluate the use of aripiprazole as an adjunctive therapy for treatment-resistant PD. Further studies with larger samples are needed.

19.2.3.4 Quetiapine

Until now, there are no literature support for the use of quetiapine in PD. Some studies have evaluated the effects of quetiapine for GAD and have found a significant effectiveness. However, quetiapine showed more AE and higher dropout rates than antidepressants. Because of these anxiolytic properties, it would be interesting to investigate the effects of quetiapine in PD [67].

19.2.3.5 Sulpiride

Sulpiride is an antipsychotic also used to treat depression [68]. To date, there is only an open label study with 19 resistant PD patients [69]. In this study, patients were considered resistant when there was prior treatment failure with two antidepressants. During 8 weeks, patients have been treated with sulpiride 100, 150 or 200 mg/day in monotherapy. At the end, patients showed significant improvement, especially those with lower doses. Although patients had AE, those were less prevalent than with previous medications [69].

19.3   Conclusion

In recent years, the anxiolytic properties of many medications with different mechanisms of action have been investigated in the treatment of PD. Based on knowledge of PD pathophysiology, medications with serotonergic, noradrenergic and GABAergic actions have been tested.

Besides SSRI, TCA and BZD, venlafaxine is the only medication with a greater number of controlled studies and favorable evidences supporting its use in PD treatment. In addition to the higher number of evidence related to effectiveness, venlafaxine is also a well-tolerated drug.

Owing to these characteristics, this antidepressant is considered as first-line treatment in PD guidelines as well as SSRI.

TCA are considered second line treatments in guidelines because of AE. Other medications with controlled data and positive results are mirtazapine, reboxetine, moclobemide and tranylcypromine. These medications have less evidence and usually more AE than first-line medications. Highlighting that reboxetine seems to be less effective than SSRI and less tolerated and that moclobemide may be less effective than fluoxetine and clomipramine but is as well tolerated as fluoxetine. In addition, it is important to consider that tranylcypromine requires dietary restrictions to prevent AE. Therefore, those antidepressants are the options with more scientific evidence and with better results after SSRI, venlafaxine, benzodiazepines and TCA for PD treatment.

Current studies presents preliminary evidence for the use of the antidepressants duloxetine, nefazodone, milnacipran, phenelzine and agomelatine as well as the anticonvulsant valproic acid and its derivatives and the antipsychotics risperidone, olanzapine and sulpiride for the treatment of PD. These medications seems to have antipanic properties but there are few and uncontrolled studies with small sample sizes. Therefore, there are insufficient scientific evidence until now but these medications are promising for PD treatment. It is also important to evaluate medication tolerability. At this point of view, antipsychotics usually have more side effects and it may be more interesting to use them as an augmentation strategy. Trazodone has controversial evidence and needs further investigation. Medications like tianeptine, gabapentine, tiagabine and aripiprazol have little and negative data. Except by one study that showed improvement of PD patients severely ill with gabapentin but needs to be replicated. Pregabaline and quetiapine have not been studied yet for PD treatment.

Currently, there is no consensus about the definition of concept and management of treatment-resistant PD. In relation to pharmacological treatment, the strategies after the failure of the first treatment choice after an adequate trial are augmentation or switching. However, there is little evidence of the sequence to be followed regarding medication choice because most medications have been little studied. However, in cases of resistant PD and based on available literature, after SSRI, venlafaxine, TCA and benzodiazepines, the following choices should be mirtazapine, reboxetine, moclobemide or tranylcypromine. After trying these antidepressants or in case of augmentation, the options would be valproic acid and derivates, risperidone, olanzapine or sulpiride.

In view of this limited knowledge, more studies are needed to better define the effectiveness and tolerability of medications for PD. Short and long-term studies comparing new and old drugs should be performed. It is also very important to standardize criteria for treatment-resistant PD and study treatment strategies for this disorder. For the moment, it is important to use existing scientific evidences regarding effectiveness, tolerability and safety, combined with clinical experience to design a rational treatment sequence for PD patients.

References

1. Pollack MH, Otto MW, Roy-Byrne PP, Coplan JD, Rothbaum BO, Simon NM, et al. Novel treatment approaches for refractory anxiety disorders. Depress Anxiety. 2008;25(6):467–76.
2. Batelaan NM, de Graaf R, Penninx BW, van Balkom AJ, Vollebergh WA, Beekman AT. The 2-year prognosis of panic episodes in the general population. Psychol Med. 2010;40(1):147–57.
3. Bandelow B, Zohar J, Hollander E, Kasper S, Moller HJ, Allgulander C, et al. World Federation of Societies of Biological Psychiatry (WFSBP) guidelines for the pharmacological treatment of anxiety, obsessive-compulsive and post-traumatic stress disorders—first revision. World J Biol Psychiatry. 2008;9(4):248–312.
4. Stein M, Goin M, Pollack M, Roy-Byrne P, Sareen J, Simon N, et al. Practice guideline for the treatment of patients with panic disorder: study design, sample sizes, subject characteris. 2nd ed. 2015. Available from: http://psychiatryonline.org/pb/assets/raw/sitewide/practice_guidelines/guidelines/panicdisorder.pdf.
5. Katzman MA, Bleau P, Blier P, Chokka P, Kjernisted K, Van Ameringen M, et al. Canadian clinical practice guidelines for the management of anxiety, posttraumatic stress and obsessive-compulsive disorders. BMC Psychiatry. 2014;14 Suppl 1:S1.
6. Baldwin DS, Anderson IM, Nutt DJ, Allgulander C, Bandelow B, den Boer JA, et al. Evidence-based pharmacological treatment of anxiety disorders, post-traumatic stress disorder and obsessive-compulsive disorder: a revision of the 2005 guidelines from the British Association for Psychopharmacology. J Psychopharmacol. 2014;28(5):403–39.
7. Freire RC, Hallak JE, Crippa JA, Nardi AE. New treatment options for panic disorder: clinical trials from 2000 to 2010. Expert Opin Pharmacother. 2011;12(9):1419–28.
8. Freire RC, Machado S, Arias-Carrion O, Nardi AE. Current pharmacological interventions in panic disorder. CNS Neurol Disord Drug Targets. 2014; 13(6):1057–65.
9. Dannon PN, Iancu I, Grunhaus L. The efficacy of reboxetine in the treatment-refractory patients with panic disorder: an open label study. Hum Psychopharmacol. 2002;17(7):329–33.
10. van den Akker OB, Eves FF, Stein GS, Murray RM. Genetic and environmental factors in premenstrual symptom reporting and its relationship to depression and a general neuroticism trait. J Psychosom Res. 1995;39(4):477–87.
11. Nardi AE, Freire RC, Zin WA. Panic disorder and control of breathing. Respir Physiol Neurobiol. 2009;167:133–43.
12. Kjernisted K, McIntosh D. Venlafaxine extended release (XR) in the treatment of panic disorder. Ther Clin Risk Manag. 2007;3(1):59–69.
13. Pollack M, Mangano R, Entsuah R, Tzanis E, Simon NM, Zhang Y. A randomized controlled trial of venlafaxine ER and paroxetine in the treatment of outpatients with panic disorder. Psychopharmacology (Berl). 2007;194(2):233–42.
14. Pollack MH, Lepola U, Koponen H, Simon NM, Worthington JJ, Emilien G, et al. A double-blind study of the efficacy of venlafaxine extended-release, paroxetine, and placebo in the treatment of panic disorder. Depress Anxiety. 2007;24(1):1–14.
15. Pollack MH, Worthington 3rd JJ, Otto MW, Maki KM, Smoller JW, Manfro GG, et al. Venlafaxine for panic disorder: results from a double-blind, placebo-controlled study. Psychopharmacol Bull. 1996;32(4):667–70.
16. Ferguson JM, Khan A, Mangano R, Entsuah R, Tzanis E. Relapse prevention of panic disorder in adult outpatient responders to treatment with venlafaxine extended release. J Clin Psychiatry. 2007;68(1):58–68.
17. Bradwejn J, Ahokas A, Stein DJ, Salinas E, Emilien G, Whitaker T. Venlafaxine extended-release capsules in panic disorder: flexible-dose, double-blind, placebo-controlled study. Br J Psychiatry. 2005;187:352–9.
18. Liebowitz MR. Tranylcypromine treatment of major depression. J Clin Psychiatry. 1987;48(7):303.
19. Papp LA, Sinha SS, Martinez JM, Coplan JD, Amchin J, Gorman JM. Low-dose venlafaxine treatment in panic disorder. Psychopharmacol Bull. 1998;34(2):207–9.

20. Liebowitz MR, Asnis G, Mangano R, Tzanis E. A double-blind, placebo-controlled, parallel-group, flexible-dose study of venlafaxine extended release capsules in adult outpatients with panic disorder. J Clin Psychiatry. 2009;70(4):550–61.

21. Simon NM, Kaufman RE, Hoge EA, Worthington JJ, Herlands NN, Owens ME, et al. Open-label support for duloxetine for the treatment of panic disorder. CNS Neurosci Ther. 2009;15(1):19–23.

22. Charney DS, Woods SW, Goodman WK, Rifkin B, Kinch M, Aiken B, et al. Drug treatment of panic disorder: the comparative efficacy of imipramine, alprazolam, and trazodone. J Clin Psychiatry. 1986; 47(12):580–6.

23. Mavissakalian M, Perel J, Bowler K, Dealy R. Trazodone in the treatment of panic disorder and agoraphobia with panic attacks. Am J Psychiatry. 1987;144(6):785–7.

24. Bertani A, Perna G, Migliarese G, Di Pasquale D, Cucchi M, Caldirola D, et al. Comparison of the treatment with paroxetine and reboxetine in panic disorder: a randomized, single-blind study. Pharmacopsychiatry. 2004;37(5):206–10.

25. Seedat S, van Rheede van Oudtshoorn E, Muller JE, Mohr N, Stein DJ. Reboxetine and citalopram in panic disorder: a single-blind, cross-over, flexible-dose pilot study. Int Clin Psychopharmacol. 2003; 18(5):279–84.

26. Versiani M, Cassano G, Perugi G, Benedetti A, Mastalli L, Nardi A, et al. Reboxetine, a selective norepinephrine reuptake inhibitor, is an effective and well-tolerated treatment for panic disorder. J Clin Psychiatry. 2002;63(1):31–7.

27. Andrisano C, Chiesa A, Serretti A. Newer antidepressants and panic disorder: a meta-analysis. Int Clin Psychopharmacol. 2013;28(1):33–45.

28. Boshuisen ML, Slaap BR, Vester-Blokland ED, den Boer JA. The effect of mirtazapine in panic disorder: an open label pilot study with a single-blind placebo run-in period. Int Clin Psychopharmacol. 2001; 16(6):363–8.

29. Carpenter LL, Leon Z, Yasmin S, Price LH. Clinical experience with mirtazapine in the treatment of panic disorder. Ann Clin Psychiatry. 1999;11(2):81–6.

30. Montanes-Rada F, De Lucas-Taracena MT, Sanchez-Romero S. Mirtazapine versus paroxetine in panic disorder: an open study. Int J Psychiatry Clin Pract. 2005;9(2):87–93.

31. Sarchiapone M, Amore M, De Risio S, Carli V, Faia V, Poterzio F, et al. Mirtazapine in the treatment of panic disorder: an open-label trial. Int Clin Psychopharmacol. 2003;18(1):35–8.

32. Ribeiro L, Busnello JV, Kauer-Sant'Anna M, Madruga M, Quevedo J, Busnello EA, et al. Mirtazapine versus fluoxetine in the treatment of panic disorder. Braz J Med Biol Res. 2001;34(10):1303–7.

33. Buigues J, Vallejo J. Therapeutic response to phenelzine in patients with panic disorder and agoraphobia with panic attacks. J Clin Psychiatry. 1987;48(2): 55–9.

34. Nardi AE, Lopes FL, Valenca AM, Freire RC, Nascimento I, Veras AB, et al. Double-blind comparison of 30 and 60 mg tranylcypromine daily in patients with panic disorder comorbid with social anxiety disorder. Psychiatry Res. 2010;175(3):260.

35. Loerch B, Graf-Morgenstern M, Hautzinger M, Schlegel S, Hain C, Sandmann J, et al. Randomised placebo-controlled trial of moclobemide, cognitive-behavioural therapy and their combination in panic disorder with agoraphobia. Br J Psychiatry. 1999; 174:205–12.

36. Kruger MB, Dahl AA. The efficacy and safety of moclobemide compared to clomipramine in the treatment of panic disorder. Eur Arch Psychiatry Clin Neurosci. 1999;249 Suppl 1:S19–24.

37. Tiller JW, Bouwer C, Behnke K. Moclobemide for anxiety disorders: a focus on moclobemide for panic disorder. Int Clin Psychopharmacol. 1997;12 Suppl 6:S27–30.

38. Wagstaff AJ, Ormrod D, Spencer CM. Tianeptine: a review of its use in depressive disorders. CNS Drugs. 2001;15(3):231–59.

39. Schruers K, Griez E. The effects of tianeptine or paroxetine on 35 % CO_2 provoked panic in panic disorder. J Psychopharmacol. 2004;18(4):553–8.

40. Papp LA, Coplan JD, Martinez JM, de Jesus M, Gorman JM. Efficacy of open-label nefazodone treatment in patients with panic disorder. J Clin Psychopharmacol. 2000;20(5):544–6.

41. Bystritsky A, Rosen R, Suri R, Vapnik T. Pilot open-label study of nefazodone in panic disorder. Depress Anxiety. 1999;10(3):137–9.

42. DeMartinis NA, Schweizer E, Rickels K. An open-label trial of nefazodone in high comorbidity panic disorder. J Clin Psychiatry. 1996;57(6):245–8.

43. Blaya C, Seganfredo AC, Dornelles M, Torres M, Paludo A, Heldt E, et al. The efficacy of milnacipran in panic disorder: an open trial. Int Clin Psychopharmacol. 2007;22(3):153–8.

44. Cia AH, Brizuela JA,Cascardo E, Flichman A, Varela ME. Clonazepam and milnacipran in the treatment of patients with panic disorder and comorbid major depression-Preliminary. Int J Neuropsychopharmacol. 2004;7:S177–S.

45. De Berardis D, Conti CM, Marini S, Ferri F, Iasevoli F, Valchera A, et al. Is there a role for agomelatine in the treatment of anxiety disorders? A review of published data. Int J Immunopathol Pharmacol. 2013; 26:299–304.

46. Levitan MN, Papelbaum M, Nardi AE. Profile of agomelatine and its potential in the treatment of generalized anxiety disorder. Neuropsychiatr Dis Treat. 2015;11:1149–55.

47. Huijbregts KM, Batelaan NM, Schonenberg J, Veen G, van Balkom AJ. Agomelatine as a novel treatment option in panic disorder, results from an 8-week open-label trial. J Clin Psychopharmacol. 2015;35(3): 336–8.

48. Taylor CP. Mechanisms of action of gabapentin. Rev Neurol (Paris). 1997;153 Suppl 1:S39–45.

49. Singh L, Field MJ, Ferris P, Hunter JC, Oles RJ, Williams RG, et al. The antiepileptic agent gabapentin (Neurontin) possesses anxiolytic-like and antinociceptive actions that are reversed by D-serine. Psychopharmacology (Berl). 1996;127(1):1–9.

50. Pande AC, Pollack MH, Crockatt J, Greiner M, Chouinard G, Lydiard RB, et al. Placebo-controlled study of gabapentin treatment of panic disorder. J Clin Psychopharmacol. 2000;20(4):467–71.

51. Montgomery S, Emir B, Haswell H, Prieto R. Long-term treatment of anxiety disorders with pregabalin: a 1 year open-label study of safety and tolerability. Curr Med Res Opin. 2013;29(10):1223–30.

52. Primeau F, Fontaine R, Beauclair L. Valproic acid and panic disorder. Can J Psychiatry. 1990;35(3):248–50.

53. Woodman CL, Noyes Jr R. Panic disorder: treatment with valproate. J Clin Psychiatry. 1994;55(4):134–6.

54. Baetz M, Bowen RC. Efficacy of divalproex sodium in patients with panic disorder and mood instability who have not responded to conventional therapy. Can J Psychiatry. 1998;43(1):73–7.

55. Perugi G, Frare F, Toni C, Tusini G, Vannucchi G, Akiskal HS. Adjunctive valproate in panic disorder patients with comorbid bipolar disorder or otherwise resistant to standard antidepressants: a 3-year "open" follow-up study. Eur Arch Psychiatry Clin Neurosci. 2010;260(7):553–60.

56. Keck Jr PE, Taylor VE, Tugrul KC, McElroy SL, Bennett JA. Valproate treatment of panic disorder and lactate-induced panic attacks. Biol Psychiatry. 1993; 33(7):542–6.

57. Zwanzger P, Eser D, Nothdurfter C, Baghai TC, Moller HJ, Padberg F, et al. Effects of the GABA-reuptake inhibitor tiagabine on panic and anxiety in patients with panic disorder. Pharmacopsychiatry. 2009;42(6):266–9.

58. Sheehan DV, Sheehan KH, Raj BA, Janavs J. An open-label study of tiagabine in panic disorder. Psychopharmacol Bull. 2007;40(3):32–40.

59. Gao K, Muzina D, Gajwani P, Calabrese JR. Efficacy of typical and atypical antipsychotics for primary and comorbid anxiety symptoms or disorders: a review. J Clin Psychiatry. 2006;67(9):1327–40.

60. Inoue T, Tsuchiya K, Koyama T. Effects of typical and atypical antipsychotic drugs on freezing behavior induced by conditioned fear. Pharmacol Biochem Behav. 1996;55(2):195–201.

61. Prosser JM, Yard S, Steele A, Cohen LJ, Galynker II. A comparison of low-dose risperidone to paroxetine in the treatment of panic attacks: a randomized, single-blind study. BMC Psychiatry. 2009;9:25.

62. Simon NM, Hoge EA, Fischmann D, Worthington JJ, Christian KM, Kinrys G, et al. An open-label trial of risperidone augmentation for refractory anxiety disorders. J Clin Psychiatry. 2006;67(3):381–5.

63. Hollifield M, Thompson PM, Ruiz JE, Uhlenhuth EH. Potential effectiveness and safety of olanzapine in refractory panic disorder. Depress Anxiety. 2005; 21(1):33–40.

64. Sepede G, De Berardis D, Gambi F, Campanella D, La Rovere R, D'Amico M, et al. Olanzapine augmentation in treatment-resistant panic disorder: a 12-week, fixed-dose, open-label trial. J Clin Psychopharmacol. 2006;26(1):45–9.

65. Keck Jr PE, Strawn JR, McElroy SL. Pharmacologic treatment considerations in co-occurring bipolar and anxiety disorders. J Clin Psychiatry. 2006;67 Suppl 1:8–15.

66. Hoge EA, Worthington 3rd JJ, Kaufman RE, Delong HR, Pollack MH, Simon NM. Aripiprazole as augmentation treatment of refractory generalized anxiety disorder and panic disorder. CNS Spectr. 2008; 13(6):522–7.

67. Depping AM, Komossa K, Kissling W, Leucht S. Second-generation antipsychotics for anxiety disorders. Cochrane Database Syst Rev. 2010;(12): Cd008120.

68. Ruther E, Degner D, Munzel U, Brunner E, Lenhard G, Biehl J, et al. Antidepressant action of sulpiride. Results of a placebo-controlled double-blind trial. Pharmacopsychiatry. 1999;32(4):127–35.

69. Nunes EA, Freire RC, Dos Reis M, de Oliveira ESAC, Machado S, Crippa JA, et al. Sulpiride and refractory panic disorder. Psychopharmacology (Berl). 2012; 223(2):247–9.

70. Zamorski MA, Albucher RC. What to do when SSRIs fail: eight strategies for optimizing treatment of panic disorder. Am Fam Physician. 2002;66(8): 1477–84.

Index

© Springer International Publishing Switzerland 2016
A.E. Nardi, R.C.R. Freire (eds.), *Panic Disorder*, DOI 10.1007/978-3-319-12538-1

Printed in the United States
By Bookmasters